Geography:
A Modern
Synthesis

Harper & Row Series in Geography
D. W. Meinig, Advisor

Geography: A Modern Synthesis

SECOND EDITION

Peter Haggett

UNIVERSITY OF BRISTOL

HARPER INTERNATIONAL EDITION
Harper & Row, Publishers
New York, Evanston, San Francisco, London

Credits for Opening Photographs

Prologue George W. Gardner. Chapter 1 Helmut Gritscher, DPI. Chapter 2 David Haas. Part One NASA. Chapter 3 Beckwith Studios. Chapter 4 Wide World. Chapter 5 Ron Church, Rapho Guillumette. Part Two Charles Gatewood. Chapter 6 Paolo Koch, Rapho Guillumette. Chapter 7 Paolo Koch, Rapho Guillumette. Chapter 8 Mimi Forsyth, Monkmeyer. Chapter 9 Charles Gatewood. Part Three George W. Gardner. Chapter 10 Ralph Mondol, DPI. Chapter 11 Wide World. Chapter 12 Chris Reeberg, DPI. Part Four Georg Gerster, Rapho Guillumette. Chapter 13 K. W. Gullers, Rapho Guillumette. Chapter 14 Bob West. Chapter 15 Mann, Monkmeyer. Chapter 16 George Hall, Woodfin Camp. Part Five Paolo Koch, Rapho Guillumette. Chapter 17 Beckwith Studios. Chapter 18 United Nations. Chapter 19 Wide World. Chapter 20 NASA. Chapter 21 J. Allan Cash, Rapho Guillumette. Chapter 22 Henry Monroe, DPI. Epilogue Charles Gatewood.

Sponsoring Editor: Ronald K. Taylor
Project Editor: Ralph Cato
Designer: Rita Naughton
Production Supervisor: Stefania J. Taflinska
Photo Researcher: Myra Schachne

GEOGRAPHY: A MODERN SYNTHESIS, Second Edition

Copyright © 1972, 1975 by Peter Haggett

Library of Congress Cataloging in Publication Data

Haggett, Peter.
  Geography: a modern synthesis.

  (Harper & Row series in geography)
  Bibliography: p.
  Includes index.
    1. Geography—Text-books—1945–   I. Title.
G128.H3 1975   910   74-11649
ISBN 0-06-042576-8

Harper International Edition
INT 35–03372 EINT 35–62014 AINT 35–32017
First Printing

Contents

Advisor's Foreword

Geography: A Modern Synthesis is already something of a landmark in the ongoing development of the field. In the three years since it first appeared, it has become widely known as an impressively fresh and creative work from one of the major talents in geography. This revised edition, with much new material and extensive reworking of the original, will extend the reputation of the book and the author.

All through the 1960s the field of geography was in the throes of change. New methods, new concepts, new interdisciplinary relationships led to some major shifts in approaches and emphases, marked by a good deal of contention and divergence over character, structure, and purpose. Increasingly, there was generated a need for new general works which could bring these developments into focus, relate them to the grand themes of geographic inquiry, and give an over-arching coherence to the field once more. In short, there was a need for a new synthesis. It was immediately apparent with the first appearance of *Geography: A Modern Synthesis* in 1972 that Peter Haggett had directly and imaginatively addressed that great task and had produced a highly original and aptly titled work. Here was an attempt to bring to bear the full range of new concepts and techniques upon the full breadth of the field. The spatial, ecological, and regional traditions, the general and the particular, the theoretical and the concrete, were all here, yet cast within a new mold, viewed from new perspectives. So too there was a lively concern for helping the student to see that geography could be an academic field with application to many areas of public concern.

Such a fresh work was exciting. It was also inevitably a challenge and an experiment. It was clearly designed with an eye more on the future than on the past. Since it was not written to fit comfortably into the general run of existing courses, it faced the test of classroom use. This revised edition is the product of that test, reflecting suggestions which have come in from a wide variety of classroom experiences from colleges and universities around the world. The principal changes are: an extensive rearrangement of topics to fit more closely the preferred classroom sequences; clarification of many concepts; the addition of much new material to many topics, especially with reference to cultural geography and urban geography.

These changes have markedly improved the book for the classroom without in any way lessening those qualities which made it such a stimulating work. This has been accomplished because the author has continued to apply the full power of his lively mind to the challenge of making "a modern synthesis." Thus, one can confidently recommend this new edition even more strongly than the original.

D. W. Meinig
Syracuse University

Preface

"If I ONLY knew geography!"
Chico Pacheco kept repeating the phrase
between clenched teeth, lamenting the wasted
days of his youth; He had been a notorious
cutter of classes. And all the time he had lost
during his life, frittering it away on nonsense,
when he could have devoted himself, body
and soul, to the intensive study of geography,
a science whose utility he had only come to
realize!
*   ... "I'll have to send to Bahia for some*
textbooks."

—JORGE AMADO

Of the Drawbacks of Not Knowing Geography,
and the Deplorable Tendency to Bluff at
Poker.

—In OS MARINHEIROS (1963)

GEOGRAPHY: A MODERN SYNTHESIS, Second Edition, is an attempt to present the whole broad spectrum of geography in a modern context and within a single volume. It tries to synthesize at two levels: first, by bringing together the different traditions and themes within the field; second, by stressing the synthesizing role of geography as a whole in relation to neighboring fields. The book is designed to introduce the student with no previous geographic training to a field of rapidly expanding horizons and increasing consequence both as an academic subject and as an applied science. Lying athwart both the physical and social sciences, geography challenges students to abandon familiar and comfortable "straightjackets" and to focus directly on relationships between man and his environment, their spatial consequences, and the resulting regional structures that have emerged on the earth's surface. Geography is uniquely relevant to current concerns both with the environment and ecology and with regional contrasts and imbalances in human welfare.

No single exponent of geography or any other academic field of inquiry can write about the whole of it in detail. Past efforts to do so seem naive in retrospect. The problems that face the beginning student in geography, however, are now so complex that the challenge must be met. It is all too easy to take the view that all one can or should do is to have a student take introductory courses in various easily identifiable subfields—physical geography, cultural geography, and so on—and hope that somehow these will produce an integrated view of geography as a whole. The osmosis, however, by which this is supposed to take place is rarely clearly defined.

To begin with the parts of a field of inquiry and go on to the whole seems to me to be a tactic of convenience, forced on us by the rising tide of research and the continuous fission of new subdisciplines. Surely, to confine larger questions to postgraduate seminars is an inversion of the desirable sequence of scholarship. We owe it to those who are starting in a field, and who we hope will follow us, to look around and ahead just as far as we can. Therefore, in this book I have turned away for a while from my own research patch and forced myself to put the various parts of geography together into what seems to me at this time to be an integrated form.

SECOND EDITION REVISIONS In addition to the "normal" updating and elimination of rough areas, this second edition of *Geography: A Modern Synthesis* contains a number of important changes. First, certain topics have been rearranged into a more coherent pattern. For example, urbanization, the growth of city systems, and the locational theory that stems from their study have been brought together in a new Part Four, "Regional Hierarchies." Advanced material on remote sensing and space perception has been taken out of early chapters and brought together near the end of the book (Chapter 20).

Second, the number of chapters has been increased from 19 to 22. The additional space has been used to give fuller treatment to topics of growing importance in geographic studies: Ecosystems, resources and their conservation, questions of spatial justice, and long-term prospects for man on the earth have all been highlighted. At the same time, some of the original material on formal locational models has been selectively pruned, and some of the more mathematical material has been dropped.

The changes of most personal interest are those designed to bring geography more clearly into the mainstream of ideas in our times. Such matters as zero population growth, economic development, and social equity have an important geographic component which should form part of the wider debate. Study questions, identified as "Reflections," have been added at the end of every chapter to provide a focus for thinking and for reviewing the topics discussed. In formulating these questions, I have tried to strike a balance between simple checks on matters in the text and more demanding questions designed to challenge the reader favored with more time or greater interest. Some questions propose simple projects; others raise issues for class debate.

In addition to these major text changes, there are smaller improvements added here and there that I hope will prove useful. For example, some overly complicated diagrams have been redrawn in a more simplified fashion and marginal material has been extended to include lists of key terms. These not only cover terms used in this book but some that may be met by students who choose to investigate the suggested readings.

To accompany this second edition, an *Instructor's Manual* and *Student Self-Paced Learning Manual* have been prepared by Professor Larry K. Stephenson and Professor John R. Healy of the University of Hawaii at Hilo. These should help instructors using this book as well as students reading independently. Revised suggestions for the use of the book in different teaching contexts are given in a new Appendix D.

ORGANIZATION OF THE SECOND EDITION

Geographers are concerned with the structure and interaction of two major systems: the ecological system that links man and his environment, and the spatial system that links one region with another in a complex interchange of flows.

The five parts of this book explore the internal structure and the linkages between each of the basic components in the two systems. Other geographers might have used different and equally arguable arrangements, but this particular one has two very useful characteristics: enough rigidity to provide a framework on which concepts, methods, and facts can be hung; and enough flexibility to accommodate new discoveries and new perspectives. For the only thing we can predict with confidence is that pressures for expansion and differentiation will continue, albeit perhaps in ways quite unlike those foreseen in this text.

Part One, "The Environmental Challenge," presents a geographer's view of the uncertain planetary environment in which the

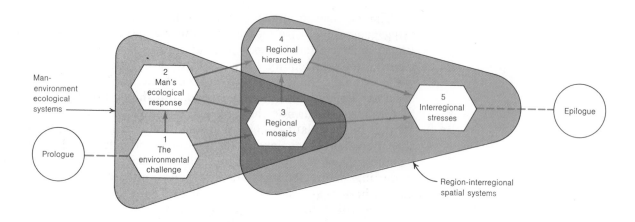

human population has evolved and now lives at ever-increasing densities. Part Two, "Man's Ecological Response," takes up the two-pronged response of man to the environmental challenge: his adaptation of the environment and his own adaptation to it. Part Three, "Regional Mosaics," turns from the ecological balance of man and his environment to the cultural fission and divisiveness within the human population responsible for regional contrasts. Part Four, "Regional Hierarchies," shows how the forces of urbanization work to override some of these regional differences and to organize human settlements into chains of city regions. Part Five, "Interregional Stresses," examines the interactions between the regional structures described in Part Four and the problems to which they give rise. Finally, the Epilogue is concerned with the future in two senses: It explores the increasing concern of geographers for the multiple worlds of the future, and it conjectures upon the future of geography itself. A detailed breakdown of the material covered is given in the introductions at the opening of each part.

The organization adopted in this book allows students to look at geography in an integrated way, abandoning the orthodox classification of geography as either physical or human, regional or systematic. Thus, chapters are not titled in the conventional manner or grouped into familiar patterns. Nonetheless, an initial course based on this book will provide the groundwork for full introductory courses in these areas as given in most university departments.

We can show the links between the chapters of *Geography: A Modern Synthesis*, Second Edition, and these courses in terms of a helix. Each complete circuit of the helix represents one level of approach, from the basic approach presented in this volume to advanced postgraduate courses. Topics introduced on one level may be taken up and expanded on others. Concepts, facts, and techniques introduced on one level may be reinforced and integrated on higher levels.

AREAS OF EMPHASIS An attempt has been made to include the majority of concepts that are likely to be of use to the novice geographer. Although I have tried to cover the field in a catholic and impartial way, space is a prime constraint in any volume of this length. Thus, in the last analysis, there must be a heavy personal bias. An author, although he cannot eliminate this bias, should at least identify his own predilections so instructors can correct and modify the text if they so choose. My own partiality, where it presents itself, is fairly evident. I have

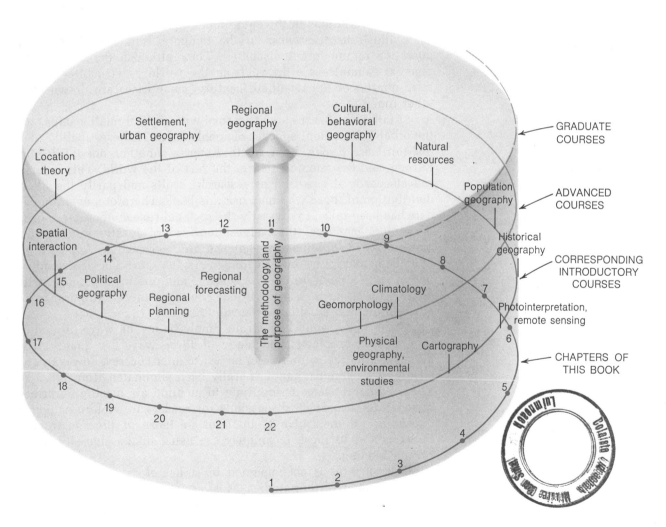

leaned toward systematic theory and hypothesis rather than toward the elaboration of many regional case studies; I have generally adopted contemporary rather than classical statements of such theory; and I have chosen to restrict the discussion of physical geography to those aspects that contribute most directly to the other subfields of geography. Advanced physical geography is taught in most schools in Britain and the Commonwealth, and students there may find this treatment somewhat oversimplified. But many students in the United States and elsewhere may be meeting these topics for the first time. Compromise has therefore been inevitable.

Although the emphasis throughout is on concepts and methods, it would be inconceivable to write an introduction to geography without numerous regional case studies. These range widely to stress the global variations in the environment and its exploitation as well as to illustrate differences in the temporal and spatial scales of operation of the forces discussed. Thus, although the bulk of the regional examples are drawn from the world as structured in the third quarter of the twentieth century, some cases are drawn from other times.

Half of the figures are concerned with general relationships, and the other half present specific regional cases. Of these, slightly over one-third are drawn from North America, another one-third from Europe, and the remainder from the rest of the world. This balance reflects partly the pattern of research results and partly the likely distribution of those who may use this book. Therefore, even though care has been taken to diversify the regional case studies, a minority of readers may feel themselves deprived of locally relevant examples. Ways to adjust this balance are suggested in Appendix C, "Supplementary Regional Reading." As for the scale of examples, one in six is drawn on the world scale, and one in three deals with an area no larger than a single city. Between these extremes, cases are well distributed along a size continuum.

Modern geography lays strong emphasis on the ways of analyzing research problems, and many of these ways are quantitative. Most quantitative methods can be left to later courses, however, and advanced techniques have generally been eliminated from the text. Those which it seems appropriate to mention are usually described in separate "marginal" discussions for readers to explore or ignore depending on their own inclinations or those of their instructor. Fuller guides to these techniques are listed in the suggestions for further reading.

Each chapter is accompanied by a list of references to guide readers in their further studies. Most are standard and widely avail-

able texts found in most college libraries. Journal references, because they change from year to year, are left for more advanced courses.

One's books, like one's children, develop a style and individuality of their own as they grow up. This second edition of *Geography: A Modern Synthesis* reflects two types of maturing influence: first, the many helpful reactions and suggestions of those who have used the book; second, the ongoing trends that continue to work themselves out within geography itself. I hope that readers of this second edition will continue to reinforce both sources through their comments and criticisms.

Peter Haggett

ACKNOWLEDGMENTS

I have a growing list of geographers from around the world who have been kind enough to comment critically on the first edition. Special thanks go to Professor Gerard Rushton of the University of Iowa, Professor C. M. Strack of Henderson State College, Arkansas, and Professor J. Stuart Krebs, Colorado State University, for unsparing reviews, and to Professor Wilbur Zelinsky, Pennsylvania State University, Professor Jonathan Sauer, University of California at Los Angeles, and Professor Richard Hartshorne, University of Wisconsin, for advice on particular sections. Since I've not always been able to make all the changes suggested, none should be blamed for the errors and omissions that may remain.

To Donald Meinig, advisory editor of this series, I owe a special debt for encouraging me to complete this revision and for his foreword. He was one of a series of reviewers, including students, whose critical comments helped to shape the final book.

Students and colleagues at Bristol, Pennsylvania State, Cambridge, and University College, London, where I have taught over the years, have all contributed to the ideas in this volume in ways I hope they will recognize. I am grateful to them and also to the score of universities in North America and Europe that have allowed me, however briefly, to try some of the ideas contained here on their classes. In addition, all general texts draw on such a vast storehouse of published work that to acknowledge the debt in full would mean producing a *Who's Who* of contemporary scholars. The credits given for the figures will, I hope, indicate where my main acknowledgments lie. Books owe more to behind-the-scenes design and production work than authors often concede, and I have been more than fortunate in the talented team at Harper & Row that has worked on this volume.

Prefaces are too public a place to express the personal indebtedness of an author to his wife and family—they will know just how much I have to thank them for. Both my father and mother, who played so decisive a part in my interests and education, died while the first edition was being prepared. I dedicate this book to my second child, Timothy, in the trust that he may fulfill his late grandparents' hopes for him and his generation.

To the Student

To teach I would build a trap such that,
to escape, my students must learn.
—ROBERT M. CHUTE
Environmental Insight (1971)

Starting a course in a new subject at college is like driving into an unfamiliar city. We see the sprawling new suburbs, the bustling freeways, the pockets of decay, but find it hard to get an overall impression of the structure or to know where we are. Geography is a Los Angeles among academic cities in that it sprawls over a very large area and merges with its neighbors. It is also hard to find the central business district.

This book has been written specifically for "newcomers to the city" who have not previously taken courses in geography at college. It attempts to introduce some of the basic concepts geographers use as well as some of the essential environmental facts that form the background to their worth. The emphasis of the book is on ideas, or concepts. But these cannot be applied in a vacuum. Certain technical material has therefore been placed in separate "marginal" discussions. These discussions are set outside the main text, and you may skip or explore them depending on the amount of time at your disposal and your taste.

The approach is essentially nonmathematical, and the book can be understood without any training in mathematics. On the other hand, geographers are using mathematics increasingly in their research, and certain aspects of a topic can be stated more explicitly in mathematical terms. These aspects, too, are presented outside the main text in separate discussions. You may wish to return to this marginal material on a second reading.

For those of you who may be going on to further work in geography, each chapter makes some suggestions for further reading in the section entitled "One step further. . .". These suggestions are largely confined to a handful of books that, in turn, open up other aspects of a subject. The final chapter points out some of the areas in which further training can be obtained and the kinds of courses you may wish to take.

Each of you may have your own favorite method for studying a textbook. Certainly no author can tell you which way is best for you, though many students find it useful to skip through a whole chapter quickly to get the general story. Figures have been designed to be self-contained wherever possible and so have been given somewhat fuller captions than normal. When, after a more lengthy reading of the chapter, you feel confident that you've understood it, you can turn to the concepts listed for review in the "Reflections" section to check yourself.

For those of you whose formal study of geography takes you no further than this book, I hope the brief acquaintance will have been a provoking one. If you take with you even some small part of the concern and fascination geographers experience in their exploration of the earth's environment and man's place in it, then I will feel that my job is done.

P. H.

Prologue

Some Basic Concepts

The Prologue introduces the reader to some of the basic concepts geographers use in their study of the planet Earth and its problems. *On the Beach* (Chapter 1) begins with a familiar scene—people arriving for a few hours of relaxation by the sea—and shows how careful geographic observation mirrors light on man and his ecological and spatial behavior. Analyzing this small slice of the human environment allows us to see how geographers approach their subject and provides a launching pad for work on a more vital global scale. The notion of changing geographic scale is a central one in this book, and the transfer of findings on one scale to applications on another remains a major geographic concern. So in *The World Beyond the Beach* (Chapter 2) we look at some of the ways geographers do this through maps. Maps provide the essential spatial language in which many geographic problems are discussed and conclusions and recommendations are given. Ways in which maps are made and used are illustrated, and the increasingly important role of environmental surveillance from space is introduced. Topics in this Prologue are picked up throughout the book, and we shall review them again in a fuller context at the end of the book, in the Epilogue.

Chapter 1
On the Beach

We shall not cease from exploration
And the end of all our exploring
Will be to arrive where we started
And know the place for the first time.

—T. S. ELIOT
Little Gidding (1942)

n Nevil Shute's compelling novel *On the Beach*, the end of man's occupation of the earth is forecast. If you have read the book or seen the movie, you may recall the small group of survivors in Australia, waiting for radiation clouds to drift over the southern hemisphere—to complete the annihilation already accomplished in the north.

Whatever the merits of Shute's grim forecast, what justifies borrowing his title for the opening chapter of a textbook? Well, there are good reasons for a geographer's selecting this title. Man has historically been a creature of the strandline between water and land. He moves like a crab in the denser bottom layer of gas on the surface of the earth, not occupying the water itself, but never far from it. Prehistoric man used the beaches as a highway; Renaissance man used them as a springboard for colonization and conquest. Even in the latter half of the twentieth century, man's biggest cities are on the strandline: Three-quarters of the world's largest urban centers—those with over 4 million inhabitants—are on the ocean or lake shore. Most of the remainder are on major rivers.

Today's urban man remains in an ecological relationship with the earth's resources which is less intimate, but no less fundamental, than that of prehistoric man. This relationship has always been a finely balanced one, in which quite small swings could bring about discomfort or disaster. But the hazards for early man were essentially local, and new and empty lands could always be found over the horizon. For urban man the hazards are regional or global, and most of the new or empty lands have long since been filled or abandoned.

For over 2000 years geographers have been describing and analyzing the ways in which man has come to terms—or failed to come to terms—with his planetary environment. In this book, we shall try to see what kind of insights into that environment they have achieved.

1-1 | THE CROWDED BEACH

Our opening photograph shows a small segment of a crowded beach. Figure 1-1 shows a more extensive beach scene. These photographs were taken not from ground level, but from a helicopter, producing maplike pictures of the scene below. Because an overhead shot tends to distort the shape of familiar objects (notably ourselves), you may need a moment or two to sort out what is what.

Beach scenes are ordinary enough events, so what makes them of special interest to geographers? Let us answer that question in a roundabout way. If we give three similar rocks to three different people, they may respond in quite different ways: A sculptor may busy himself with shaping his rock into new and more interesting forms, a

Figure 1-1. The crowded beach. A low-level view of Hampton Beach, New Hampshire, taken in late summer. [Photograph by Rotkin, P.F.I.]

mineralogist may start to break his up to examine its chemical structure, and a protester may hurl his through the nearest window. The trains of thought started and the actions taken are determined not by the object at hand (the rock), but by the attitudes of the three individuals to it. People other than Sir Isaac Newton have been hit on the head by apples falling from apple trees; reacting to the blow by pondering the laws of gravity clearly represents only one of many possible responses!

In the same way, a familiar beach may provoke different reactions even among various types of scientists. Geologists may head for the sand particles and the fluid dynamicist for the breaking waves. Sociologists may study the behavior of the groups using the beach and economists the profits of the hot dog stands. How would a geographer react?

Perhaps the first reaction of a geographer to the beach scene in Figure 1-1 would be to try to pin down exactly where events were occurring in space. A photograph taken from a helicopter allows a much more accurate assessment of the location of people on a beach than a photograph taken on the ground. It is for this reason that most of the photographs you will find in this book are aerial photographs. Concern with locations in space is a characteristic of geographers' curiosity; specifying location accurately is one of the prime rules of the geographic game. An inaccurate description of location causes a geographer to wince in the same way as a linguist would over a mispronunciation, or a historian over an inaccurate date.

A second reaction to the beach scene in Figure 1-1 would be to try explaining the spatial variations which are observed. Why are some parts of the beach packed with people while others are deserted? How much does this circumstance relate to differences in the quality of the beach? Questions of this kind lead to a general concern with the relationship of man to his environment.

A third geographic reaction to the beach scene shown here would be to try to sort the various elements in the photograph into some kind of spatial order. One of the simplest ways of doing this is to dissect the area into several zones, each of which has certain characteristics that give it a particular character. For example, we could divide the beach into three zones: a swash zone below the regular high water mark, an upper zone above the regular high water mark, and a belt of sand dunes behind. Figure 1-2 shows a cross section of a typical beach zoned in this way. Geographers use this process of dissection to establish sets of regions. Regions are a shorthand way of describing the variable character of an area in an efficient manner. (See the discussion of regions in Sections 5-2 and 10-5.)

A geographer, then, is concerned with three different but interlocked questions: (1) the question of *location* in which his concern is to establish the precise spatial position of things within a particular area of the earth's surface; (2) the question of *man-environment relations* within the area; (3) the question of *regions* and the identification of the distinctive character of particular spatial subdivisions of the area.

Perhaps at this point we should follow the conventional textbooks and attempt a definition of geography itself based on these three central questions. If you like formal definitions, then please turn to the last chapter and look at those given in Table 22-1. The definitions have been deliberately placed near the end of the book to encourage you to draw your *own* portrait of a geographer as you work through the chapters; this should give you a better likeness of geography than would

Region

A *region* is any tract of the earth's surface with characteristics, either natural or man-made, that make it different from areas that surround it.

Figure 1-2. Variations within a beach environment. This cross section of a representative beach shows its characteristic zoning. The vertical scale has been exaggerated to emphasize the effect of height above sea level.

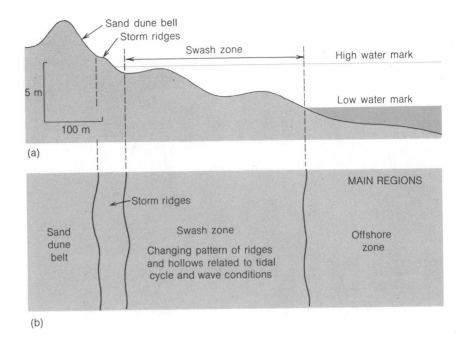

looking up a dictionary entry right away. So let's agree for the moment that "geography is what geographers do" and go on to see them at work on the beach. Later in the book we shall see them working on wider, more important questions and in a global context.

1-2 | SPACE AND TIME ON THE BEACH

If we wish to establish the accurate location of individuals on the beach, we can simply ask the question "Where are they?" However, even this simple question can be answered in many ways. We can answer it, for example, by analyzing the distribution of individuals in terms of their primary location, or in terms of their secondary location. A person's *primary location* is his position in terms of an arbitrary grid system. In Figure 1-3(a) the primary location of individual A is about 9 m east and 6 m north of an arbitrary point of origin at zero. A person's *secondary location* depends on his primary location; individual B is approximately 6 m from A. A grid provides a framework within which the relative position of people can be established.

Ways of mapping
The primary location of things is important in making accurate maps. Figure 1-3 shows various ways of mapping a population. We begin by

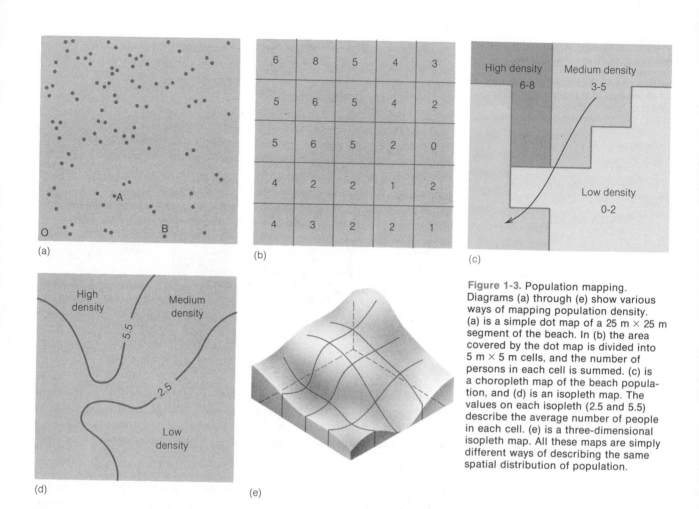

(a)

(b)

(c)

(d)

(e)

Figure 1-3. Population mapping. Diagrams (a) through (e) show various ways of mapping population density. (a) is a simple dot map of a 25 m × 25 m segment of the beach. In (b) the area covered by the dot map is divided into 5 m × 5 m cells, and the number of persons in each cell is summed. (c) is a choropleth map of the beach population, and (d) is an isopleth map. The values on each isopleth (2.5 and 5.5) describe the average number of people in each cell. (e) is a three-dimensional isopleth map. All these maps are simply different ways of describing the same spatial distribution of population.

assigning each individual on the beach to a corresponding location on the map, representing each person by a single dot. We then place a grid over the dots. By counting the number of dots that falls within each square cell of the grid, we can translate the distribution of dots, or people, into an array of numbers that reflect the density of the population [Figure 1-3(b)]. Crowded parts of the beach are represented by cells with high values, sparser parts by cells with low values, and empty stretches by cells with zero values.

Two kinds of maps can be drawn from these cell values. We can make *choropleth maps* (from the Greek *choros*, area, and *plethos*, fullness or quantity) by assigning different shades of color to each of the cells. By linking cells with similar colors, we can create a general picture of the distribution of population on the beach [Figure 1-3(c)]. In doing this, we lose some information. Instead of 9 different values (between 0 and 8), we now have only 3. However, we have conveniently simplified our map, or grid, by reducing the spatial pattern from 25 cells to 4 areas. The second and more common way of mapping population distributions using a grid is to draw lines between all points having the same quality or value [Figure 1-3(d)]. Such lines are known as *isopleths* (from the Greek *isos*, equal), and maps of this type are known as *isopleth maps*. These maps also give an accurate picture of variations in population density. A three-dimensional version of an isopleth map would show areas of denser population as peaks and sparser areas as hollows [Figure 1-3(e)].

Spatial organization

The secondary, or relative, location of the members of a population is also of great interest to geographers. It helps them to understand *why* a population is organized, or distributed, in a particular way. Animal behaviorists like Konrad Lorenz in *On Aggression* and social psychologists like Edward Hall in *The Hidden Dimension* have discussed how man, like other primates, has a strongly developed sense of territoriality. He surrounds himself with visible or invisible "space bubbles" that are sensitive to crowding. Hall distinguishes between four "action zones" based on the distance between people: intimate space, personal space, social space, and public space. *Intimate space* is reserved for physical interactions like loving or fighting, while *personal space* is used for soft talk and friendly interaction. The boundary between these first two zones is about half a meter ($1\frac{1}{2}$ ft). Beyond a distance of $1\frac{1}{2}$ meters (about 4 ft), personal space gives way to the *social space* used for formal business and social contacts. At around 4 meters (about 12 ft), the outer zone of public space, in which the preacher and the ice-cream-cone salesman on the beach may operate, begins. Clearly, these zones may vary from person to person and from culture to culture—the personal conversational space of the French may be seen by the more reserved English as an intrusion into their intimate space!

Because of man's sense of territoriality, it is possible to argue that individuals arrange themselves on a beach to achieve a certain distance from each other. Their object may be to be as near as possible to those they love, as far as possible from those they hate, and at a convenient distance from those about whom they feel less strongly. When the distribution of people in an area is regarded in this way, it is measured in terms of *interpersonal space*, the linear distance separating an in-

Isopleth maps

Isopleth maps are one of the most common ways in which geographers show spatial distributions. Some of the isopleths are given distinctive names.

Isochrone maps show lines of equal time.

Isohyet maps show lines of equal rainfall.

Isoneph maps show lines of equal cloudiness.

Isophene maps show lines of biological events that occur at the same time (e.g., flowering dates of plants).

Isotherm maps show lines of equal temperature.

Isotim maps show lines of equal transport cost.

One of the commonest isopleth maps you will encounter is the *contour* map, which shows lines of equal height of land above sea level. "Contours" are also used by geographers as a general term for any type of isopleth.

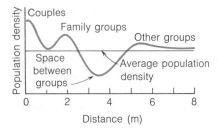

Figure 1-4. Interpersonal distances. This is an idealized profile of the density of population on the beach in terms of the distance between individuals. Note the three characteristic peaks.

dividual from his neighbors. To determine interpersonal space, we ask "How many meters is a given person from his nearest neighbor, his second nearest neighbor, his third nearest neighbor, and so on?" Geographers frequently use this approach in the study of human settlements. (See Section 14-1.)

The result of analyzing the relative location of a beach population is given in Figure 1-4. The three peaks, near zero, at 2 m, and at 5 m, may be interpreted as related to the space between couples less concerned with the beach (let alone geography!) than with each other, to family groups, and to strangers outside the family groups and keeping a respectful distance from them.

Time and spatial diffusion

The scene in Figure 1-1 is unreal insofar as it freezes at one click of the camera's shutter a constantly changing pattern. The photograph is static; the real beach scene is dynamic. A similar photograph taken at another time would show a different picture. Just how different would depend on the timing of the second shot. A picture taken a few seconds later would reveal little change in the population but would catch the movement of the breakers. One taken a few hours later would show an empty beach and a different tide level. One taken some months later in winter might show a change in some of the physical structures. One taken some years later might reveal a substantial change in the form of the beach itself due to erosion.

All geographers work within a specific time context. And as Figure 1-5 makes clear, this context vitally affects the conclusions they draw. In each graph the number of people on a beach is related to the passage of time. Over a period of 100 years (e.g., 1870-1970) the most noticeable change is the increasing use of the beach. On the average, more people were using the beach at the end of the period than at the beginning. This is not surprising when we think of the substantial

Figure 1-5. Changing population densities. These graphs show the impact of the length of the observation period on the trends detected in population density. (a) shows the historical trend, (b) the seasonal cycle, (c) the weekly fluctuation, and (d) the short-term equilibrium.

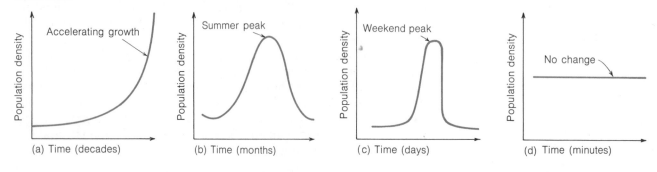

increases in southern New England's population over this period and the changes in social attitudes toward leisure time.

If we reduce the period of observation to a single year, we find a wavelike trend with a peak in late summer and a trough in late winter. If we reduce it to a single week, the waveform narrows to a sharp peak on the weekend. But over a much shorter period (for example, half an hour), the number of people on the beach remains constant and the trend line is horizontal. These general trends—of accelerating growth, wavelike cycles, or stability—are functions of the period over which observations are made.

Consider what we would see if we observed the beach at regular intervals from daybreak. The first arrivals might well occupy what seemed to be the "best" sites—what is best being determined by the requirements of the group: say, near the surf line for the youngsters, and near the parked car for the old folks. As the best sites were taken, new arrivals would have to occupy less attractive areas or crowd into the already occupied ones, reducing the amount of interpersonal space and producing the population pattern shown in the photograph.

Figure 1-6 traces the evolution of this pattern over three hours. This is a simple example of the spatial *diffusion* of a population, in which the location of an individual is related to the time he arrives. We shall look at more complex examples of spatial diffusion in later chapters. Observation of this process over time not only allows us to consider the present spatial pattern in light of its past development; sometimes, it makes it possible to predict spatial diffusion in the future. Given the first two maps in Figure 1-6 (and knowing what has happened on the beach on similar days in the past), could you make a reasonably accurate prediction of the noontime pattern?

Diffusion

In geography, diffusion is the process or spreading out or scattering over an area of the earth's surface. It should not be confused with the physicist's use of the term to describe the slow mixing of gases or liquids with one another by molecular interpenetration. Different kinds of geographical diffusion are discussed in Section 12-1.

1-3 | MAN AND THE BEACH ENVIRONMENT

The concept of relations between man and his environment is basic to geographic thinking. For early man these conditions were largely natural and included such elements as the local climate, terrain, vegetation, and soils. With the rise of civilization, man surrounded himself with artifacts which, because of their sheer scale and longevity, became an integral part of his environment. For today's urban dweller the environment is dominated by the fixed structures of urban life (freeways, city blocks, asphalt surfaces). The natural environment has been either replaced or radically modified.

We can bring the general relationship between man and his environment into focus by going back to the beach. The population density on a beach is partly a function of environmental quality. Good beaches (i.e., beaches with fine sand or good surf) tend to attract more

Environment

By an environment geographers mean the sum total of conditions that surround (literally, environ) man at any one point on the earth's surface.

Figure 1-6. Spatial diffusion. Stages in the diffusion pattern of people arriving on an empty beach at different stages of the morning. Note how the filling-up process is related to the point of access to the beach and to differences in the quality of the beach environment.

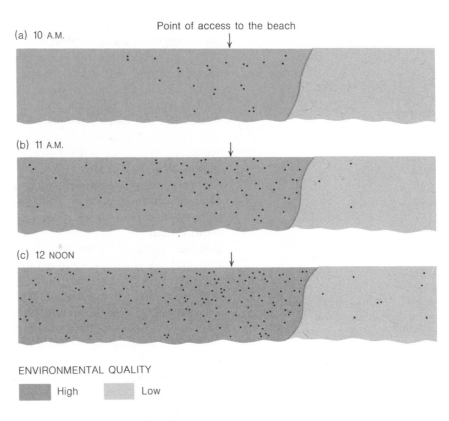

Point of access to the beach

(a) 10 A.M.

(b) 11 A.M.

(c) 12 NOON

ENVIRONMENTAL QUALITY

High   Low

users, while poor beaches (those that are, say, polluted by oil or the local dog population) are shunned. Other things being equal, we can relate the capacity of a beach to attract a population to its environmental quality. Even within the area of our photograph, local variations in environmental quality can lead to variations in population density. Figure 1-7(a) shows the distribution of both the beach population and the beach environment as maps.

The study of two or more geographic distributions varying over the same area is a study of the *spatial covariation,* an idea we shall meet repeatedly in this book. When the two maps look alike and the two distributions "fit" one another closely, we say that the two phenomena are associated by area; that is, high values for population density in one area correspond with high values for environmental quality in the same area, and vice versa. Other hypothetical cases with little covariation are also illustrated in Figure 1-7. Comparing pairs of maps in this manner often tells us a great deal about the spatial covariation of different phenomena. Distributions can also be compared by statistical

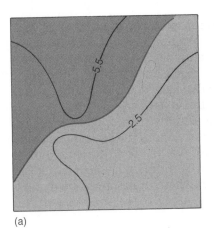
(a)

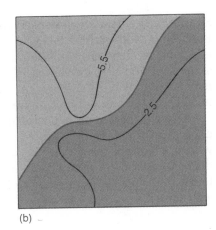
(b)

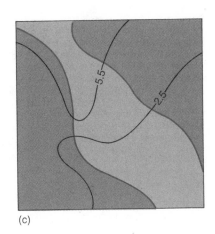
(c)

Environmental quality

High   Low

Figure 1-7. Spatial covariation. Here we have three hypothetical distributions of population density on a beach. In (a) there is a strong positive correspondence between the distribution of population and the quality of the beach environment. In (b) there is a strong negative correspondence, and in (c) there is little correspondence.

methods, but a discussion of these methods lies outside the scope of this book.

Man is affected by his environment, but he also has the capacity to modify that environment. He can, for example, modify the form of a beach by erecting defensive walls and alter its quality by permitting its fouling with oil and debris. We can think of the relationship between man and his environment as a simple two-way *system*. (See Figure 1-8.) As in any such system, there is always the possibility of a feedback effect. For example, overuse of a beach may lead to its pollution, and pollution may cause fewer people to use the beach.

Feedbacks of this kind are complicated by both time and space. They may occur only after a certain amount of time has passed, so that one generation pays for or benefits from the actions of another. In Biblical terms, the sins of the father may be visited on the children.

Figure 1-8. Feedbacks. Man's relationship with his environment is a two-way system. As the diagram on the left shows, this fact introduces the possibility of feedbacks. As the diagram on the right shows, these feedbacks may occur later in time at the same place (*a*) or later in time and at a different place (*b*). In practical terms, sewage discharged off a beach may pollute either that beach or a neighboring one.

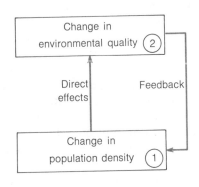

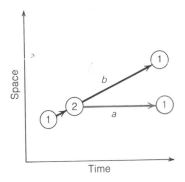

System

A system is a group of things or parts that work together through a regular set of relations. Thus we can regard the beach as a system in which its various parts—shingle, sand, mudbanks, and the like—are each linked together through a set of relations involving the energy of waves, tides, and winds. Geographers are particularly interested in systems which link together man and environment.

Similarly, the benefits of conservation may accrue to the next generation. Then, too, actions taken in one location may produce feedback effect in another location. One area may pay for or benefit from the actions of another. Protection of the beach on one part of the coast may mean erosion at another, sewage discharged at one point in a river may affect the fish population downstream, and so on.

These examples of man's relations with his environment are rather crude. Nevertheless, they provide a clue to one of the basic frameworks within which the geographer looks at his world. For whether he studies distributions in absolute or relative space (primary or secondary locations), his object is not merely to learn where things are but why they are there.

1-4	THE BEACH IN WORLD FOCUS

Beaches are fine and pleasant places, and we might all wish to spend more time there. But it would be misleading to suggest that geographers spend more time at the seashore than anyone else. We have chosen to concentrate on it in this chapter because it represents a microcosm, but *only* a microcosm, of the kinds of phenomena that geographers study.

On the inside of the front and back covers of this book are examples of man's relationship with his environment on two quite different scales. At the front, we see how the human population is distributed throughout the world, and at the back, how the population of the United States is distributed throughout the various regions of the country. The world, and the nations of the world, are certainly more usual arenas for the geographer's work than a beach, which is only a tiny section of the human environment. The modern geographer deals with a continuum of environmental regions of increasing size. These range from the microenvironment of the individual in his local surroundings to the macroenvironment of mankind as a whole.

Throughout this text we discuss geographic studies confined to sizes, or *orders of magnitude,* between these two extremes. Although the variations are considerable, the geographer is concerned with a narrow "window" within the scale of scientific inquiry. In Figure 1-9 the main areas of scientific inquiry are plotted along a centimeter scale. Exponential notation (in which 1000 is written as 10^3, 1 as 10^0, 0.001 as 10^{-3}, and so on) is used in order to present a large range of variation on the same diagram.*

The scale in Figure 1-9 extends from the microworld of the atomic

* See Appendix A for tables of conversion constant from metric to nonmetric measure.

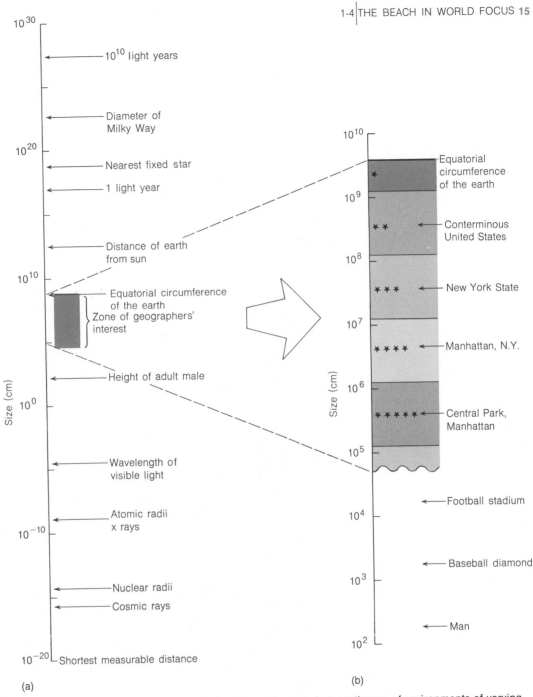

Figure 1-9. Orders of geographic magnitude. Geographers study a continuum of environments of varying size. However, even this continuum is rather limited when one considers the continuum of environments that comprises the universe. This larger continuum, and the zone of interest to geographers, is represented by the scale in Figure 1-9(a). The enlargement of the zone of interest in Figure 1-9(b) enables us to divide it into zones which are roughly parallel to the orders of geographic magnitude described on p. 17.

physicist studying cosmic rays with wavelengths of about 10^{-15} cm to the radio astronomer studying galaxies with diameters of 10^{23} cm and more. By comparison, the geographer's world is confined to a narrow band along the middle range of the scale. The smallest objects studied are about the size of a beach or a city block. They are not less than a few hundred meters across, or approximately 10^4 cm. Conversely, the largest object studied is the earth, with an equatorial circumference of around 40,000 km (24,860 mi), or about 10^9 cm. There is therefore a range of 5 (i.e., 10^9 minus 10^4) in the "diameters" (where diameter is measured by the distance along the longest axis) of objects studied by geographers. To put this another way, the real world shown on the inside front cover is around 100,000 times greater in diameter than the world of the beach.

We use the orders of geographic magnitudes described in the margin of this section to keep in focus the areas we are dealing with. These orders of magnitude serve a purpose similar to that of the astronomer's orders of star brightness and are a happy substitute for the jumble of scales usually used in geography books. They remind us that we are dealing not with the real world but with reduced and simplified models of it.

1-5 | MODELS IN GEOGRAPHY

The real world is much more complex than a beach. In trying to make sense of the structure of a particular region, geographers often attempt to simulate reality by substituting similar but simpler forms for those they are studying. They do this by constructing models. In everyday language the term *model* is used in at least three different ways. As a noun, it signifies a representation; as an adjective, it implies an ideal; as a verb, it means to demonstrate. Thus, we are aware that when we refer to a model railway or a model husband we are using the same term in different senses.

In scientific work the term "model" has, to some extent, all three of the meanings. Scientific model-builders create idealized representations of reality in order to demonstrate certain of its properties. Models are made necessary by the complexity of reality. They are a prop to our understanding and a source of working hypotheses for research. They convey not the whole truth, but a useful and apparently comprehensible part of it.

Model

A model is an idealized representation of the real world built in order to demonstrate certain of its properties.

Maps and other models

We have already seen a simple example of model-building in our study of the crowd on a beach. Our opening aerial photograph, and Figure 1-1, illustrate a first stage of abstraction. They represent the

Orders of geographic magnitude

Geographers have to deal with objects that vary considerably in size. In this book we indicate the size in terms of orders of magnitude:

* First Order of Magnitude. Areas with a range of diameters from that of the surface of the earth itself (with an equatorial circumference of 40,000 km, or 24,860 mi) to 12,500 km (7,770 mi).

** Second Order of Magnitude. Areas with a range of diameters from 12,500 to 1250 km (7,770 to 777 mi). A typical example from the middle of this range is the coterminous United States

*** Third Order of Magnitude. Areas with a range of diameters from 1250 km down to 125 km (777 to 77.7 mi). A typical example from the middle of this range is New York State.

**** Fourth Order of Magnitude. Areas with a range of diameters from 125 km to 12.5 km (77.7 to 7.77 mi). A typical example from the middle of this range is New York City.

***** Fifth Order of Magnitude. Areas with a range of diameters from 12.5 km to 1.25 km (7.77 to 0.777 mi). A typical example from the middle of this range is Central Park in New York City.

It would of course be possible, as Figure 1-9 shows, to continue downward to a sixth order and beyond, but the five classes shown cover the main range of geographers' work. Note that differences between the orders of magnitude are not linear: The contrast between the second and fifth order is not a difference of 3 but one of $10 \times 10 \times 10$ (i.e., 10^3, or 1000).

properties of the people on the beach faithfully, but on a different scale. It is common knowledge that most scientists make things larger in order to study them. The optical microscope, the electron microscope, and the radio telescope were scientific breakthroughs because they permitted increasingly powerful magnifications of reality. Geographers are curious folk in that they follow a reverse process. They bring reality down in size until it can be represented by a map. To shrink the universe to manageable size, they use various standard *linear scales,* which determine the ratio of the length of a line segment on a map to the true length of the line on the earth's surface. (See Table 1-1.) For example, 1 cm on a map may represent 1 km (100,000 cm) on the ground. Usually, the scale of a map is given as a representative fraction, as 1/100,000 or 1:100,000. This ratio applies equally to metric and nonmetric measurements. Thus on a 1:100,000 map, 1 in. is equal to 100,000 in. (about 1.6 mi) on the ground, and 1 cm is equal to 100,000 cm (1 km) on the ground.

Table 1-1 Standard lengths and areas on ten commonly used map scales

Class and scale	Countries using the scale for major map series	Equivalent on the earth's surface of standard measures on the map	
		1 cm on the map	1 in. on the map
Large scales			
1:10,000	European	0.100 km	0.158 mi
1:10,560	British and Commonwealth	0.106 km	0.167 mi
1:24,000	United States	0.240 km	0.379 mi
1:25,000	British and Commonwealth	0.250 km	0.395 mi
Medium scales			
1:50,000	European	0.500 km	0.789 mi
1:62,500	United States	0.625 km	0.986 mi
1:63,360	British and Commonwealth	0.634 km	1.000 mi
1:100,000	European	1.00 km	1.578 mi
Small scales			
1:250,000	International	2.50 km	3.946 mi
1:1,000,000	International	10.0 km	15.783 mi

All the maps in this volume are very selective and partial models of the real world, with all the advantages—and the drawbacks—that simplification brings in its train. Simple scaled-down representations of reality are called *iconic models.* Our opening photograph on page 1 is such a model. It shows the sunbathers' actual bodies and has a linear scale of about 1:400. Figure 1-3(a) constitutes the second stage of abstraction—an *analog model.* Here, real people have become points

Figure 1-10. Maps as models. Three views of the southeastern corner of Oahu, one of the Hawaiian Islands, consisting of the city of Honolulu and surrounding areas. (a) is a satellite photograph of the entire island chain with superfluous ocean and clouds dropped out in order to accent the islands' configurations (Oahu is the third island from the left); (b) is a reproduced section from a U.S. Geological Survey topographic sheet; and (c) is a population density map of the area shown in (b), drawn from 1970 census records. The scale in (b) and (c) is 1:250,000. The scale in (a) approaches 1:6,000,000. [Photograph courtesy of NASA; topographic map courtesy of U.S. Geological Survey; population map from R. Warwick Armstrong, Ed., *Atlas of Hawaii* (University Press of Hawaii, Honolulu, 1973), p. 118.]

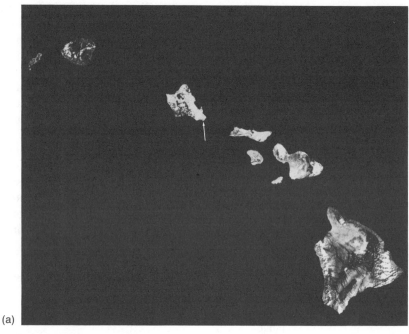

(a)

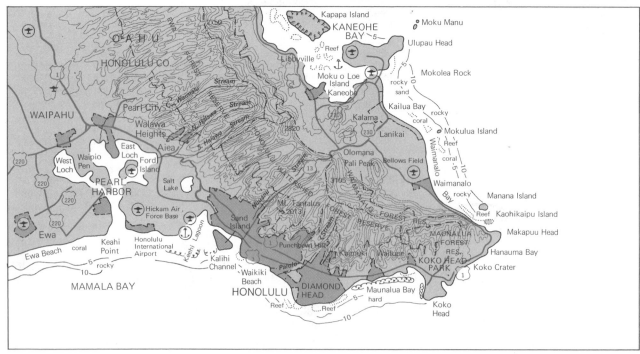

* * * *

(b)

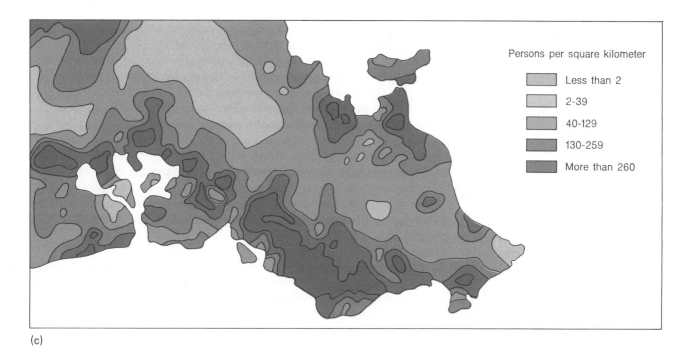

(c)

on a map. Clusters of people on the beach have become point clusters on the map. Abstraction is pushed still further in a third type of model, the *symbolic model,* in which real-world phenomena are represented by abstract mathematical expressions, such as that for the population density of a beach. Figure 1-3(d) is a symbolic model. It takes us a step further from reality than either a photograph or a map for a second example of progressive abstraction. See Figure 1-10. Here the subject is part of the Hawaiian island of Oahu (a) as seen from an earth satellite, (b) as shown on a topographic map on the same scale, and (c) as a population density map.

From models to paradigms

We may conveniently think of model-building as a three-stage process in which each stage represents a higher degree of abstraction than the last. (See Figure 1-11.) At each stage information is lost and the model becomes less realistic but more general. Throughout this book we shall return to the idea of models and look at their actual use. We shall encounter many models that are considerably more intricate

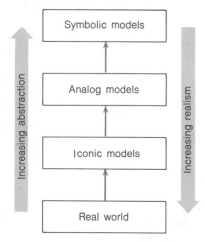

Figure 1-11. Model building. Model building is shown here as a three-stage process of increasing abstraction.

than the three simple ones in the preceding section, but we will postpone discussion of them until we meet specific examples. It is useful at this point, however, to bring forward the idea of a *paradigm*. A paradigm is a kind of supermodel. It provides intuitive or inductive rules about the kinds of phenomena scientists should investigate and the best methods of investigation. This chapter—like the whole of this book—represents a paradigm of geography.

Research in geography, like research in most fields, is based on a shared paradigm; that is, those who pursue this research are committed to probing the same problems, observing the same rules, and maintaining the same standards. Tedious methodological debates, and concern over what constitutes legitimate research or appropriate methods of analysis, are symptoms of transitional periods in the evolution of a science. Once a paradigm is fully established, debate languishes.

In his provoking history of modern science, *The Structure of Scientific Revolutions,* Thomas Kuhn contends that the origin, continuance, and eventual obsolescence of paradigms is the prime factor in the evolution of science. Modern geography has witnessed a massive shift in emphasis from descriptive geography toward more analytic work, in which mathematical models of how regions grow and interact play a dominant role. In this book we try to illustrate both the traditional paradigm and the new one. We regard the current emphasis on mathematical models as the most recent phase in a long history of change, in which ways of looking at man on the beach and in the world have been refined, but the essential questions that geographers ask remain unaltered.

From paradigms to real-world problems

How can we use the idea of spatial organization and man's relations with his environment in ways that are helpful to mankind? The answer depends partly on the paradigm within which geographers work, and there is a noticeable contrast between the traditional and modern views of the field.

The traditional role of the geographer has been the provision of two types of essential information: locational information on the exact position of events and environmental information on the quality of particular areas. In response to demands for this kind of information the great geographic works of the Greeks were written, the exploration societies of the early nineteenth century were formed, and the universal geographies of the Victorian period were assembled.

Today, however, geographers are more concerned with optimization—with finding the "best" location for things and making the "best" use of areas. Where should a new model city be located? What is the best site for a hospital within a city? What is the best dividing line between two hostile communities? What is the best use for the more remote Appalachian areas? They are also interested in forecasting projected trends into the future and monitoring the likely effect of policy decisions in a wide variety of situations.

Geographers work in agencies that range from the World Health Organization or the World Bank on the international level to local city halls and county agencies. They dominate the regional planning sections of several countries (notably Britain) and form a significant element in government agencies from Washington to Moscow to Peking. In the world of business or in private agencies, geographers perform essentially the same role: advising on locational priorities, watching for environmental feedbacks, and providing a geographic view of some part of the world.

How important that geographic view is will depend on the problem being examined. In many instances, the importance of spatial and environmental considerations may be negligible. In such cases the contribution of the engineer, the economist, or the educator will surely outweigh that of the geographer. When large-scale environmental issues are discussed, however, a geographer's viewpoint will certainly be needed.

We do not intend to argue for a purely geocentric view of the world's problems. At a time when the walls between academic subjects are crumbling, the isolationist subject makes about as little sense as an isolationist state. Geography has always been heavily dependent on its academic neighbors such as mathematics, the earth sciences, and the behavioral sciences, and it has everything to gain by remaining so.

Geography is interesting today not because of its solution of past problems, but because of its potential contribution to resolving future difficulties. Geographers as a group have been a little embarrassed to discover that locational and environmental questions, so long a part of their classroom discussions, are now a daily topic in the news media, in Senate committee rooms, and on the campus. For over 2000 years geographers have been studying the world and man's place in it. Suddenly, in the late twentieth century, these seemingly academic preoccupations are being regarded as relevant—to us and to our children. Mankind is taking a new look at itself and its world, and, like T. S. Eliot's explorer, "knowing the place for the first time."

Reflections

1. Note down in a few sentences what you think you are likely to get out of a course in geography. Keep this statement in your file so that you can look back at the end of the term and compare what you expected to gain with what you actually gained.

2. It is sometimes said that disciplines are distinguished from one another by the questions they ask. What special questions do geographers ask? What questions *should* they ask?

3. How much can geographers studying the behavior of man learn from biologists studying the behavior of animals? List (a) some advantages and (b) some disadvantages of such cross-borrowing between disciplines.

4. Consider the graphs of population density over time in Figure 1-5. Try to construct similar graphs for (a) the campus restaurant, (b) the local shopping plaza, and (c) the local airport. Comment on the similarities and differences in these graphs.

5. Use locally available maps with small, medium, and large scales to examine a beach of your choice. (For American students the maps will be at scales of 1:24,000, 1:62,500, and 1:250,000; students in other countries should check with Table 1-1 for guidance.) Familiarize yourself with the conventional symbols used on such maps, and note how the map scales control the richness of the information that can be conveyed.

6. Review your understanding of the following concepts:
   (a) region
   (b) primary location
   (c) secondary location
   (d) interpersonal space
   (e) choropleth maps
   (f) isopleth maps
   (g) feedbacks
   (h) linear scales
   (i) orders of geographic magnitude
   (j) models
   (k) paradigms

One step further . . .

Two paperbacks that provide a lively account of modern geography, with plenty of examples of applied research, are

> Lanegran, D. A., and R. Palm, *An Invitation to Geography* (McGraw-Hill, New York, 1973). See chap. 1.
> Taaffe, E. J., *Geography* (Prentice-Hall, Englewood Cliffs, N.J., 1969). See chap. 1.

You may also like to browse through two excellent sets of collected readings that give some idea of the range and richness of geographic work:

> Dohrs, F. E., L. M. Sommers, and D. R. Petterson, Eds., *Outside Readings in Geography* (Crowell, New York, 1955).
> Detwyler, D. R., Ed., *Man's Impact on Environment* (McGraw-Hill, New York, 1971).

The ideas of Konrad Lorenz and Edward Hall on interpersonal space and territoriality are ably treated in two geographic publications,

> Soja, E. W., *The Political Organization of Space* (Association of American Geographers, Washington, D.C., 1971) and
> Saarinen, T. F., *Perception of Environment* (Association of American Geographers, Washington, D.C., 1969).

The nature of paradigms and the general evolution of academic ideas are expounded in an exciting fashion by

Kuhn, T. S., *The Structure of Scientific Revolutions* (University of Chicago Press, Chicago, 1974).

Beginning students will find it helpful to look through recent numbers of some of the leading geographic serials to see what kinds of work geographers are currently doing. As starters, you might try *Geographical Review* (published quarterly) and *Annals of the Association of American Geographers* (also a quarterly). Longer and more substantial research is published in monograph form. Two representative monograph series are *University of Chicago Department of Geography Research Papers* and *Lund University Studies in Geography* (both occasional publications).

Students who wish to read further on a particular topic in this book or follow up its themes with respect to a particular region might like to consult

Church, M. *et al.*, *A Basic Geographical Library* (Association of American Geographers, Washington, D.C., 1970).

This gives excellent selective lists with some helpful comments on each book recommended.

Finally, it will be useful to have at hand a standard world atlas. More specialized maps and atlases will be suggested at appropriate places in the book.

Chapter 2

The World Beyond the Beach

Map me no maps, sir, my head is a map, a map of the whole world.

—HENRY FIELDING
Rape upon Rape (1745)

19194

The view from the beach is a very limited one. Even if we scramble to the highest point on the sand dunes backing the beach, we can see only a few kilometers out to sea. If, looking inland, the land is also flat, then our total sweep—from horizon to horizon—still covers only a small disk of the earth's vast surface. We are unlikely, even on the clearest of days, to scan more than 0.0008 of 1 percent of the total area of the globe. We could fit over 100,000 disks this size on the earth's surface without them ever touching one another! Even from a high-flying jet the situation is not vastly improved, though we can then see an area half the size of Texas (or about 0.05 of 1 percent of the global area).

It was not until the late 1960s that man first saw the earth from deep space and the kind of hemispheric view shown in our opening photograph was obtained. For most of man's history, his visual picture of the world was limited by his local horizon. What lay beyond was a matter for, first, speculation, then calculation, and, finally, confirmation. (This fascinating trail of conjecture and discovery has gone cold as the pieces of the map have fallen into place.) However, the research methods developed along the trail have proved a very useful legacy for modern geography. To record the characteristics of the earth, and to preserve and exchange this information, a comprehensive *spatial language* was developed. This spatial language describes the absolute and relative locations of places either in terms of conventional maps and in other, maplike ways. It allows us to unequivocally assign events to locations. To put it simply, by providing "a place for everything" it lets us put everything in its place. In this chapter we present some of the basic grammatical rules of this language and show how geographers use it to describe and record the world beyond the beach.

2-1 | FLAT WORLD OR ROUND

It is doubtful if our early ancestors were much concerned with the shape of the earth. As we have seen, their visual horizon, even from the highest peak in the clearest weather, would have been confined to a few hundred kilometers. The facts that observation failed to provide, legend and narrative no doubt found substitutions for. J. R. R. Tolkien's fictional Wilderland (Figure 2-1) very likely recaptures the spirit, if not the substance, of primeval world pictures.

Against this background, the intellectual achievements of the Greek cosmologists are remarkable. Their observations of heavenly bodies led them to deduce a spherical rather than a disklike form for the earth. By 200 B.C. one of the earliest Greek geographers, Eratosthenes of Alexandria, had made rather accurate estimates of the earth's

Figure 2-1. Imagined worlds. J. R. R. Tolkien's "Wilderland," one segment of his fictional Middle Earth, the scene of *The Lord of the Rings*. Early mapmakers of the real world also used imagination and legend to fill in unexplored areas. Maps of Africa, the "dark continent," contained mythical rivers and lakes as late as the middle of the nineteenth century. [From J. R. R. Tolkien, *The Hobbit* (Allen & Unwin, London, 1937), end map. Copyright © 1966 by J. R. R. Tolkien. Reprinted by permission of the publisher, Houghton Mifflin Co., Boston.]

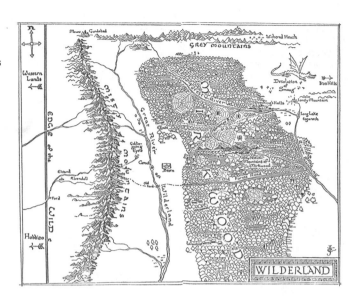

Figure 2-2. The size of the earth. Eratosthenes, about 200 B.C., estimated the circumference of the earth from the angle at which the rays of the noon sun reached Alexandria and Syene. [From A. N. Strahler, *The Earth Sciences,* 2nd ed. (Harper & Row, New York, 1971), p. 150, Fig. 10-1.]

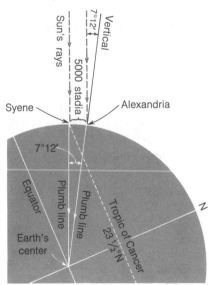

shape. As Figure 2-2 shows, the basic procedures he used in his calculations were simple. Indeed, Eratosthenes' method was (at least until the advent of the satellite) the same in principle as that used to measure the earth in our own time. Assuming that light rays from the distant sun were parallel, as for all practical purposes they are, Eratosthenes calculated the differences in the angle they made with the earth's surface at different points and thereby determined its curvature. Specifically, he compared the angle made by the rays of the noon sun at the summer solstice (June 21) at two places in Egypt: Syene, where the sun is directly overhead and the angle is vertical, and Alexandria, where a shadow of 7° 12′ is cast. Multiplying the known north-south distance between Syene and Alexandria by the difference in the angles gave him an approximate diameter (based on a Greek measure called the *stadium,* which is about one-sixth of a kilometer) for the earth of 46,250 km (28,740 mi). Since current estimates put the diameter at 40,000 km (24,860 mi), Eratosthenes' estimate was extraordinarily accurate.

Little improvement was made on Eratosthenes' measurements for another 1800 years. There was an apparent loss of interest in Greek concepts, and their theories about the earth were temporarily replaced, in medieval Europe, by a theological view of the world. Figure 2-3 shows a typical map of that period, with the Holy City of Jerusalem occupying the center of a disk-shaped world. With the Renaissance and the wave of European overseas voyages, interest in the form of

Figure 2-3. Changing views of the world. Figures (a) through (e) show representative maps from 1770 years of global exploration. On each map the sites of London (A), Jerusalem (B), and Colombo (C) are indicated.

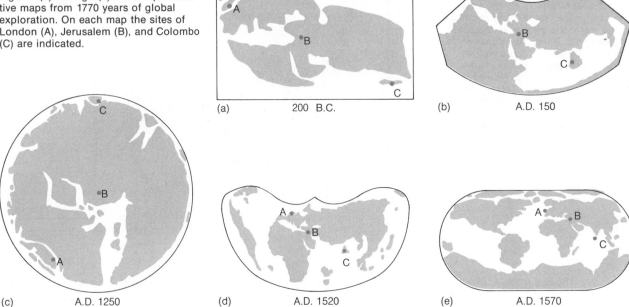

(a)   200 B.C.

(b)   A.D. 150

(c)   A.D. 1250

(d)   A.D. 1520

(e)   A.D. 1570

the planet was rekindled. Further measurements of the earth were made in the early seventeenth century, when astronomer Willebrord Snell's recalculations at Leiden University triggered a succession of increasingly accurate measurements of the distance over ever-longer sections of the earth's curved surface. From the results stemmed a bitter debate; gross variations in the estimated circumference indicated that the earth could not be regarded simply as a regular sphere. Scholarly opinion divided over whether the earth was flattened at the poles or elongated like a football. By the 1730s evidence from geodetic expeditions was overwhelmingly in favor of the former view.

Since the eighteenth century, precise measurements have proved that the earth's polar diameter is shorter than the equatorial diameter, and the polar circumference is shorter than the equatorial circumference. (See Table 2-1.) Moreover, the equator does not mark the fattest part of the earth. The distance around the earth is, in fact, greatest along a circle slightly south of the equator. The fact that the earth is an ellipsoid with a somewhat pear-shaped bulge and slightly flattened poles, rather than a perfect sphere, has complicated mapping in many instances.

Terms used in mapping the earth

Ellipsoids are spheroids (figures like a sphere, but not perfectly spherical) with a regular oval form.

Equator is an imaginary line around the earth which is an equal distance from the north and south poles.

Geodesy is the science that deals with the shape and size of the earth.

Graticules are networks of parallels and meridians drawn on a map.

Grids are arbitrary networks of lines drawn on a map. For example, a network of squares may be used to give a simpler reference system than that of graticules.

Latitude is the distance north or south of the equator measured in degrees.

Longitude is the distance east or west from a prime meridian (usually Greenwich in London) measured in degrees.

Meridians are imaginary half-circles around the earth which pass through a given location and terminate at the North and South poles.

Parallels of latitude are imaginary lines on the earth which are parallel to the equator and pass through all places the same distance to the north or south of it.

Poles are the two ends (North Pole and South Pole) of the axis about which the earth spins or rotates.

Prime meridians are meridians used as a baseline for the measurement of the east-west position of places on the earth's surface in terms of longitude. The most commonly used prime meridian is Greenwich in London.

Table 2-1 The basic dimensions of the earth

Total surface area	510,056,000 km²	(196,934,000 mi²)
Area of land surface	149,137,000 km²	(57,582,000 mi²)
Circumference measured around the poles	40,003 km	(24,857 mi)
Circumference measured around the equator	40,074 km	(24,901 mi)
Highest point on land surface	+8.85 km	(5.45 mi)
Lowest point on ocean floor	−11.03 km	(−6.85 mi)

Estimates of the earth's shape have varied in different parts of the world, and several geodesists have computed what they believe to be the best ellipsoid for specific areas. For mapping purposes, the earth is zoned into several regions, as in Figure 2-4, each with its own ellipsoid of reference. The most recent information about the exact form of the earth has come from earth satellites. Satellites crossing the equatorial bulge slightly increase their velocity in response to the additional gravity. (Gravitational forces are stronger since the earth's mass

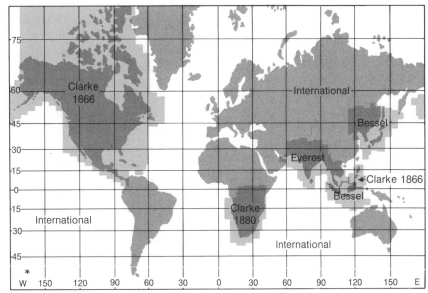

Figure 2-4. Alternative estimates of the earth's shape. Since the earth is *not* a perfect sphere, but an odd-shaped spheroid, cartographers have to use a best estimate of its shape in making their maps. These best estimates are termed "ellipsoids of reference." For mapping purposes the earth is zoned into regions, for which slightly different ellipsoids are used (e.g., Clarke's ellipsoid, calculated in 1886, for the United States, and Everest's, calculated in 1830, for India. [From A. N. Strahler, *Physical Geography,* 3rd ed. (Wiley, New York, 1969).]

is fractionally nearer to the satellite at this point.) By comparing shifts in satellite orbits during successive revolutions, we can make precise calculations that confirm the slightly pear-shaped form of the planet.

2-2 | **LOCATION ON THE FLAT MAP**

Once we have established the shape and size of the earth, we can go on to specifying locations on its surface. We look first at the devices used on simple maps before considering the more complicated problems posed by the globe.

Place names

The simplest way of specifying a location somewhere on the earth's surface is to give its name. In everyday language we almost always talk about places in terms of the names attached to them: Chicago, Kansas, or Tibet. This way of specifying locations is termed *nominal specification*. We used it in Chapter 1 to identify particular beaches, and in the preceding section to describe the cities involved in Eratosthenes' calculations. It can describe places that vary in size (from Mount Vernon to Afro-Asia), it can provide distinctive and memorable place names (like the Donner Pass in northern California or Xochimilco in central Mexico), and it can be made as complex or as simple as necessary. The typical address on an envelope exemplifies a hierarchical method of specifying location in which areas of decreasing size are nested within each other. Our resistance to the replacement of familiar addresses by zip codes indicates the innate attractions of a nominal system for everyday use.

Another advantage of place names is that they may contain a considerable amount of historical or environmental information. Places founded by particular groups tend to have particular types of names. Witness, for example, the many Spanish place names in the American Southwest and Texas. These names give us clues to the previous extent of a population. When other information is lacking, such clues may be vital. In a similar manner, names may indicate the kind of environment settlers found on first moving into an area. In southern Brazil names containing the word "pine" help to depict the original distribution of a type of vegetation now much reduced by deforestation and the clearing of land for agriculture [Figure 2-5(a)]. Work by Zelinsky on the northeastern United States has shown a rich variety in the distribution of stream names. The terms "creek," "brook," "run," and so on each have distinct regional clusters related to the settlement history of the area [Figure 2-5(b)].

If place names are so attractive and useful, why don't geographers use them all the time? Unfortunately, the disadvantages of nominal specification for scientific purposes outweigh the advantages. First, this kind of language is nonunique because several different locations can have the same name. There are scores of São Paulo's in Brazil, and names such as Newport or New Town occur in literally hundreds of forms in the English-speaking world. Second, nominal terms are unstable. The same location can have different names at different times (such as Breuckelen and Brooklyn), and in different languages. (The German Konigsberg is the Russian Kaliningrad.) To correct this problem, many countries have established committees to standardize geographic names throughout the world. In 1890 the United States set up a Board on Geographical Names that published authoritative lists for individual countries. Some idea of the immensity of this task can be gained from the size of standard geographic gazetteers. The Merriam–Webster *Geographic Dictionary* contains approximately 40,000 names, and the *Times Atlas of the World* has 315,000 names. The current London telephone directory, which is actually a list of the "locations" of subscribers within the city, contains about half a million names.

We can only guess at the total number of world place names, which probably lies in the trillions. Such a vast array of names creates formidable problems for those concerned with information storage and retrieval, problems which are complicated by the duplication of place names. Various ways of dealing with these problems are being investigated by experimenters at Oxford University, England, who are developing systems for the coding, computer storage, and rapid retrieval of information from geographic gazetteers.

Another serious disadvantage of nominal specification is its spatial imprecision. The boundaries of an area may change while the name remains the same. Turn to a historical atlas and compare the Poland of 1930 with the Poland of 1970. The imprecision of nominal terms for geographic areas is illustrated in Figure 2-6, which shows a series of changes in the boundaries adopted by geographers in describing "Mitteleuropa" (Middle Europe) over a 40-year period. Areas included by at least one geographer extend well outside the limits set by other geographers, and the only part of Europe not included in Mitteleuropa by any of the definitions is the Iberian Peninsula. The only area not in dispute, in fact, is the small core area containing Austria and Bohemia–Moravia.

Cartesian grids

If we can't use place names to specify location, how do we proceed? The geographer's answer is to use *reference grids* — mathematical devices that specify the location of a point in relation to a system of

Figure 2-5. Place names as historical evidence. (a) Pine (*pinhal*) place names in southern Brazil in relation to the present distribution of Araucaria pine forests. (b) Stream names in the northeastern United States. The approximate areas within which particular terms are dominant are indicated on the map. "Brook," "run," and "branch" are very common terms and strongly associated with particular locations. "Creek" and "stream" occur less frequently but are also strongly associated with distinct localities. "Kill" is a very uncommon term in the Northeast as a whole but very prevalent in areas of early Dutch settlement in New York State. [From W. Zelinsky, *Annals of the Association of American Geographers* **45** (1955), p. 323, Fig. 20.]

• "Pine" place names

Distribution of Araucaria forest

(a)

coordinates. The type of grid used depends on whether the area involved is small or large. We can treat small areas of the earth's surface as if they were flat planes; in the case of larger areas the curvature of the planet must be taken into account. Reference grids for small areas use a *Cartesian coordinate system*. The location of a place is specified in terms of its distance from two reference lines that intersect at right angles. The horizontal reference line is called the abscissa or x axis, and the vertical reference line is called the ordinate or y axis. Inter-

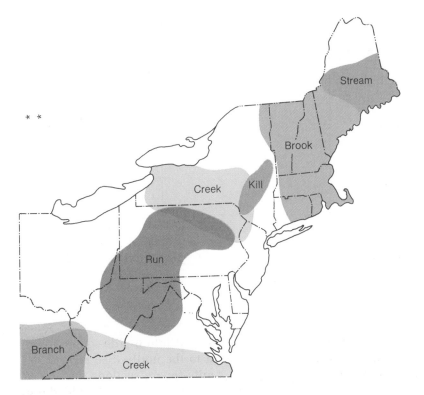

* *

(b)

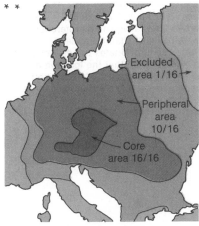

* *

Figure 2-6. Ambiguities in regional names. The degree of coincidence between 16 geographic definitions of the terms "Central Europe," "Mitteleuropa," or "Europe Centrale." Note the small area of agreement and the very wide area of disagreement. [After K. Sinnhuber, *Transactions of the Institute for British Geographers No. 20* (1956), p. 19, Fig. 2.]

section of the two axes is the point of origin of the system. [See Figure 2-7(a)].

Cartesian coordinate systems are commonly adopted by national mapping agencies, and Britain's National Grid system (Figure 2-8) is a typical example. Its origin lies southwest of the Isles of Scilly; its abscissa runs east and its ordinate north. All geographic locations are measured in kilometers, or subdivisions of kilometers, east and north from these reference lines. In practice, references are given first to the 100-km (62-mi) squares into which the grid is divided, second to distances east of the origin, and then to distances north of it. Thus, the precise reference to the United States Embassy in Grosvenor Square, London, is 51 (TQ) 283808; that is, it lies exactly 528.3 km (328.3 mi) east and 180.8 km (112.3 mi) north of the arbitrary origin point.

A modified form of the Cartesian system is the *range, township, and section method*, which originated in a land act conceived by Thomas Jefferson in 1784. This rectilinear grid is the basis of surveys in most of the United States west of the Appalachians and, in a modi-

Figure 2-7. Reference grids. The location of *A* with respect to the origin in two alternative spatial reference systems. Note that in the case of Cartesian coordinates the distance east is given first so that the location of *A* is 20,40 and *not* 40,20. In practice the comma is usually omitted in giving a locational reference so that 20,40 becomes simply 2040.

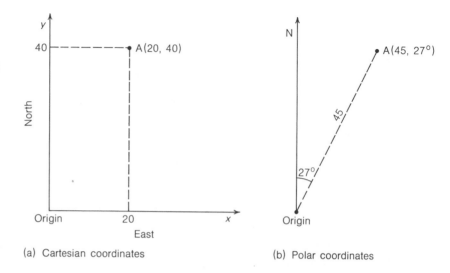

(a) Cartesian coordinates

(b) Polar coordinates

fied form, much of Canada. It divides land into 6-mile-square townships around a Cartesian axis oriented to the north. (See Figure 2-9.) Each township is subdivided into 1-mile-square sections (numbered 1 to 36 in zigzag fashion). Each section can be further subdivided into quarter-sections and then into 40-acre fields. Even today the imprint of the township and range system on the agricultural landscape of much of the United States is evident.

A less common type of locational reference grid is based on a *polar coordinate system* which specifies the location of a place in terms of an angle, or *azimuth*, and its distance from an origin [Figure 2-7(b)]. In mathematical work, angles are specified counterclockwise from a horizontal; in geographic work, angles are measured clockwise from the north, which is designated 0° or 360°. Thus an azimuth of 225° is a line running southwest from the origin. The polar coordinate system is generally used only when the relation of locations to a single origin is important—for instance, in studying distances a population has migrated from a given county.

2-3 | LOCATION ON A ROUND WORLD

So long as the environment under study is a small one, geographers can use very simple devices for recording location. Once we begin to study very large areas or attempt worldwide studies, then a more sophisticated way of specifying location is needed.

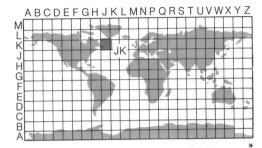

A world geographic reference system

Cartesian coordinate systems are simple to use but are accurate only for small areas of the earth's surface; spherical coordinate systems are accurate for large or small areas but involve tedious directional and angular references. The World Geographic Reference System (commonly known as Georef) combines some of the advantages of both systems.

It divides the global surface into 288 segments created by the intersection of 15° meridional and latitudinal bands, each defined by an index letter. By always giving the letter of the meridional first, geographers can uniquely define each segment by a two-letter code. For example, Newfoundland lies in segment JK. Every segment is, in turn, divided into 225 quadrangles, each 1° × 1°, with a similar lettering system. Thus, the northern part of the island of Miquelon lies in quadrangle DC. Within each quadrangle, location is given by the number of minutes of arc a point lies east and north of the southwest corner. With this system, locational reference can be carried to any desired degree of accuracy. For example, the final Georef value for Cap Miquelon may be given in terms of minutes (JKDC 3908) or tenths of a minute (JKDC 392088), depending on the degree of accuracy required. The horizontal component of location (west–east location) is always given before the vertical component (south–north), and the southwest corner always represents the origin. [See G. C. Dickinson, *Maps and Air Photographs* (Edward Arnold, London, 1969), Chap. 8.]

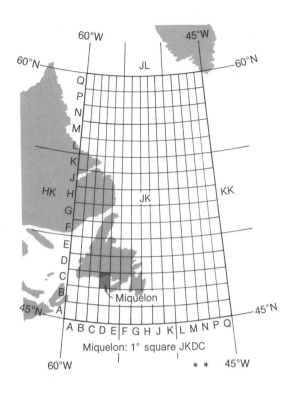

Miquelon: 1° square JKDC

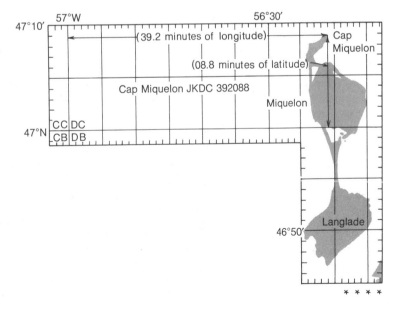

Cap Miquelon JKDC 392088

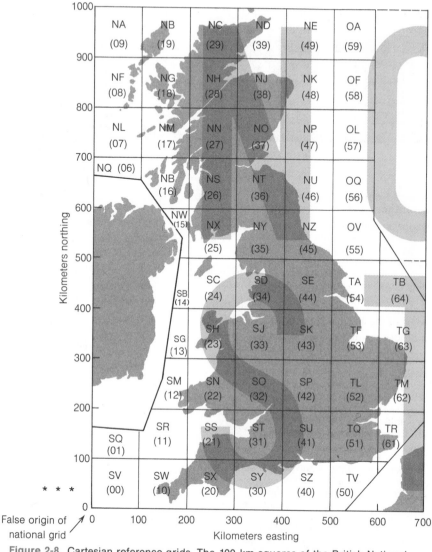

Figure 2-8. Cartesian reference grids. The 100-km squares of the British National Grid. Each grid square has a distinguishing number or pair of letters. Numbers describe the distance east (the "easting") and distance north (the "northing") of the south-west corner of each square when measured in hundreds of kilometers from the false origin (e.g., the south-west corner of grid square 31 is 300 kilometers east and 100 kilometers north of this origin.) The "false origin" is an arbitrary point out at sea chosen for the British map users. It stands in contrast to "true origins" such as the poles or the equator. The letters on the map are part of a wider reference system which ties the British National Grid into other international grids. [Reproduced from *The Projection for Ordinance Survey Maps and Plans and the National Reference System,* with the sanction of the Controller of Her Majesty's Stationery Office. Crown copyright reserved.]

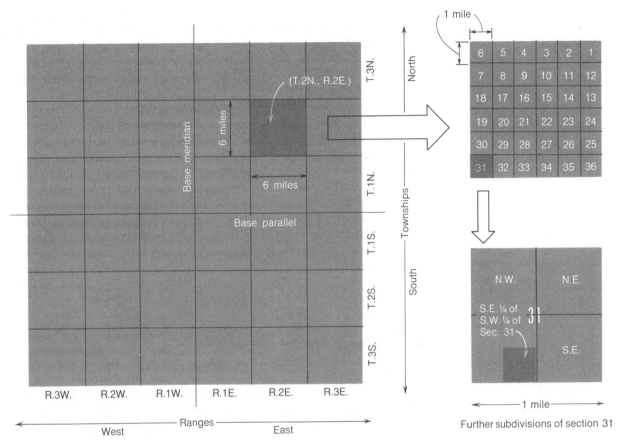

Figure 2-9. A hierarchic reference grid. In much of the United States, the location of parcels of land was specified on a rectilinear grid divisible into ranges and townships. Each square of the range and township grid was further divisible into sections and subsections. The "base meridian" and "base parallel" tie this grid system into the worldwide graticules of latitude and longitude described in Figure 2-10. [From G. C. Dickinson, *Maps and Air Photographs* (Edward Arnold, London, 1969), p. 124, Fig. 46.]

A spherical grid

The curvature of any portion of the earth's surface depends, of course, on its size. As the area being mapped increases, so does the deviation from a true north-south direction of vertical reference lines on a flat surface. Thus the extension of the range and township system used in the United States as people moved west brought about an increasing divergence of range lines from their true direction. To minimize this divergence, new grids had to be established for each state.

To accommodate the spherical form of the earth, the Cartesian system was adapted to create a geographic grid of meridians and parallels. *Meridians* are imaginary half-circles around the earth which pass through a given location and terminate at the North and South Poles. Thus they form true north–south lines joining two fixed points of reference, the geographic poles, where the earth's axis of rotation intersects the planet's spherical surface. Each meridian is actually a half circle (an arc of 180°). *Parallels* are imaginary lines on the earth that are parallel to the equator and pass through all places the same distance to the north or south of it. They form true east–west lines and are full circles (arcs of 360°). Parallels intersect meridians at right angles. (See Figure 2-10.)

Coordinate positions on the meridian–parallel grid are measured in terms of longitude and latitude. The *longitude* of a place is its distance east or west from a prime meridian measured in degrees. It describes the angle between two planes that intersect each other along the earth's axis and intersect the surface of the earth along, respectively, the prime meridian and the meridian of the place to be located. Prime meridians have a value of 0°. Angles are measured east and west, reaching a maximum of 180° at the meridian opposite the prime meridian. The prime meridian in worldwide use is the Greenwich meridian (the former site of the Royal Observatory near London, England), but any meridian can be chosen. Italian topographic series use the Monte Mario meridian near Rome (12°27′E of Greenwich), and some other European countries use the Ferro meridian in the Canary Islands (17°14′W of Greenwich).

The *latitude* of a place is its distance north or south of the equator measured in degrees. It is given by the angle between the plane of the equator and the surface of a cone that has its apex at the earth's center and intersects the surface of the sphere along a given parallel. Unlike the prime meridian, the equator is a uniquely determined and natural reference line, rather than an arbitrary one. It also has a value of 0°, and angles are measured north and south, reaching a maximum of 90° at the North and South Poles.

Figure 2-10. The earth's spherical coordinate system. (a) Parallels of latitude lie in planes oriented at right angles to the earth's axis of rotation. (b) Meridians of longitude lie in planes passing through the earth's axis. (c) Combined parallels and meridians form a spherical geographic grid. [From A. N. Strahler, *The Earth Sciences,* 2nd ed. (Harper & Row, New York, 1971), pp. 13–14, Figs. 1-10, 1-11, 1-13.]

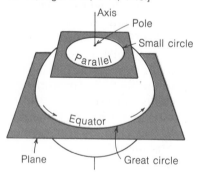

(a) Parallels

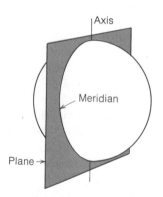

(b) Meridians

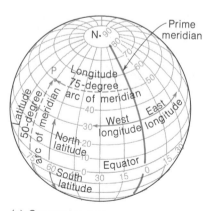

(c) Geographic grid

Geographers give the exact location of a point on the earth's surface by specifying its longitude and latitude. In Figure 2-10, the point *P* has the coordinates 50°N, 75°W.

Degrees can be subdivided into 60 minutes of arc (60'), and minutes into 60 seconds of arc (60"). For simplicity, however, latitude and longitude are generally expressed in decimal parts of a degree. Thus, 77°03'41" is rewritten in decimal terms as 77.0614°. Modern worldwide reference systems are sophisticated enough to successfully combine the advantages of both the Cartesian and the spherical systems of reference.

Determining location on a spherical grid

Developing a spherical reference system for the earth is one thing, and determining where a location actually is within that system is another. Finding latitude involves simply measuring the angular elevation of the sun or a star above the horizon, and it was first done with some accuracy more than 1000 years ago. The determination of longitude is more difficult. The first requirement is a reliable clock, so that the local noon time (when the sun is at its zenith) can be compared to noontime at the prime meridian. The difference between the two gives us our correct east–west position. But the clock must be very accurate. One hour on the clock is recording the time during which the noon-time position of the sun has shifted 1/24th part of its east–west track around the earth. At the equator this would mean that an error in time of only 22 seconds on the clock would cause the navigator to be 10 km (6.2 mi) in error in judging his east–west position. It was not until 1761 that this problem was solved when John Harrison's marine chronometer, made in response to a prize offered by Britain's Board of Longitude, provided the breakthrough that allowed position to be determined accurately.

The job of fixing positions with respect to latitude and longitude has become much less tedious in the present electronic era. Figure 2-11 presents one of the electronic systems currently being developed for this purpose. This *Omega system* depends on a pattern of only six radio stations, the latitude and longitude of which can be given precisely. Each station transmits extremely long radio waves—with wavelengths up to 5 km (3.1 mi)—on exactly the same frequency in a predetermined sequence. Because these waves can be distinguished by the length of their pulse and the sequence in which they are transmitted, ships and aircraft can plot their progress from a known starting position by comparing signals from two stations and using those from others as a check. When it is completed, positional fixes based on the Omega system should be accurate to about 1 km (.6 mi) any-

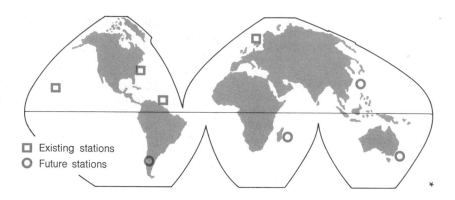

Figure 2-11. Position fixing on the earth's surface. The U.S. Navy's Omega system uses long radio waves transmitted from a few stations to provide position data. Ships using the system compare signals from two stations and use signals from a third station as a check. The system represents the latest in a series of moves to reduce the number of fixed points needed to determine position on the globe.

□ Existing stations
○ Future stations

where on the global surface. Navigational satellites giving more accurate fixes to a few hundred meters may be used in the future.

The determination of positions on the earth's surface has two interrelated phases. The first consists of fixing primary positions with respect to their horizontal location on the geographers spherical grid system (and their vertical position with respect to mean sea level). Once a network of primary positions is established, a swarm of secondary positions can be fixed from the horizontal and vertical coordinates of the primary position. The trend in current survey work has been to reduce the need for primary survey positions by extending the area of secondary fixes that can be calculated from the primary points. The Omega system represents an extreme reduction of the number of primary positions needed for global coverage.

Map projections: A round world on flat paper
Once we have established a system of parallels and meridians and have located places in terms of that system, we have the basic ingredients of a world map. To make one we need only to specify a scale and construct a model globe. If we select a scale of 1:10,000,000, we can represent the real globe by a sphere 127 cm (50 in) in diameter. Making globes like this is not a difficult task; indeed, people have been constructing them for centuries. Martin Behaim made one of the first terrestrial globes at Nuremberg in 1492, and they are still popular for many purposes.

Despite their attractiveness, however, globes are rarely used by geographers in their research. Most globes are less than 1 m (3.3 ft) in diameter and are too small to be of any real value. There are, of course, a few large globes such as the 39-m (128-ft) Langlois globe in France, but larger globes are expensive and unwieldy to work with. The obvious way around the problem is to substitute flat map sheets for the bulky spheres. But how do we get a map off the surface of a globe and onto flat paper without tearing or otherwise distorting it?

Types of projections There is really no easy approach to this subject. The puzzling problem of mapping the round world on flat paper attracted some of the best mathematical minds of sixteenth-to-nineteenth-century Europe, figures like Gerardus Mercator, Hans Mollweide, and Johann Lambert. Their work is a study in the mathematics of compromise. They showed that, though it is impossible to reproduce, faithfully, in two dimensions all the characteristics of the three-dimensional earth, it is possible to reproduce some of them at the expense of others. We have simply to decide which properties are important and which we are prepared to sacrifice.

The dimensional problem is illustrated in Table 2-2, which shows the dimensions of a 1° by 1° area bounded by lines of latitude and longitude at different points on the earth's surface. If we lay out lines of latitude and longitude using a square Cartesian coordinate system, we greatly distort areas away from the equator. (See Figure 2-12.) At 30°N each square on the grid is 1.16 times too large in comparison to the true area on the globe; at 60°N the areal exaggeration is twice as large, and at 89°N it is 59 times as large. But it is not only area that changes when three dimensions are represented on a flat sheet; the angular relations between points also are greatly altered.

In practice, the main decision is whether to select a *conformal projection*, in which the shape of any small area is shown correctly, or an *equal-area projection*, in which a constant areal scale is preserved over the whole map. Although the word "projection" is used here, in a strict sense it should be applied only to maps that are true geometric projections of a sphere. Through usage the word has come to represent any orderly system of parallels and meridians drawn on a map to represent the earth's geographic graticule.

Table 2-2 Dimensions of the global grid at various latitudes

Latitude	Length of 1 degree				Area of 1-degree squares	
	Measured on meridians		Measured on parallels			
	km	mi	km	mi	km²	mi²
Equator	110.6	68.7	111.3	69.2	12,309	4753
30°	110.9	68.9	96.5	60.0	10,696	4130
60°	111.4	69.2	55.8	34.7	6,217	2400
89°	111.7	69.4	1.9	1.2	218	84

SOURCE: *Geographical Conversion Tables* (International Geographical Union, Zurich, 1961), Tables 148, 150, and 153.

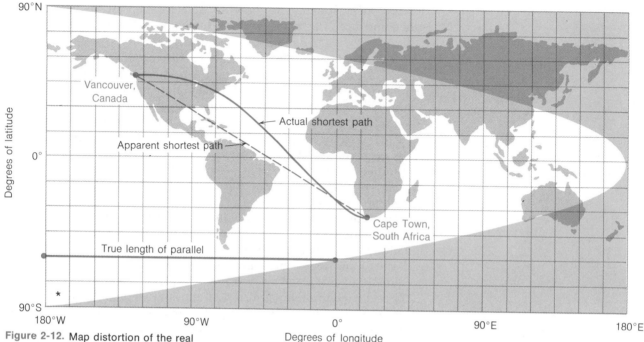

Figure 2-12. Map distortion of the real world. *Cartesian Grid.* A simple cylindrical map of the world on which locations are plotted in terms of degrees of latitude and longitude greatly distorts distances along the parallels. (See Table 2-2.) Only the shortest paths between places on the same meridians or parallels are straight lines. All other shortest paths (like that from Vancouver to Cape Town) are distorted. This map shows just how impossible the map-maker's task is made in wrestling with a spherical world. Any desirable property—like the simple arrangement of meridians and parallels used here—can be obtained only if we are willing to sacrifice all the other desirable properties (e.g., correct areas, correct shape, correct directions).

Table 2-3 summarizes the characteristics of some of the more commonly encountered projections and comments on their uses. Map projection is a highly specialized branch of geography that is likely to attract only those with mathematical minds. It is an evolving field, however. The first projections were invented over 2000 years ago (see the discussion of stereographic projection on page 44), and new projection systems are still being invented today to meet special needs such as satellite tracking.

Selecting a projection The choice of a projection is a complicated one. There are well over 100 projection systems available to geographers, and each reproduces a given section of the earth's surface in a different way. In Figure 2-13, the island of Greenland, which has an actual area of 2,175,000 km² (840,000 mi²), is depicted in ten ways. The first projection (a), an azimuthal equal-area projection based on the North Pole, bears the most similarity to Greenland as shown on a desk globe, and the last projection (j), a Miller cylindrical projection, the least. However, none of the ten outlines is undistorted (since distance, direction, area, and shape cannot *all* be shown accurately at the same time), and each projection is optimal for a selected purpose

Table 2-3 Families of map projections

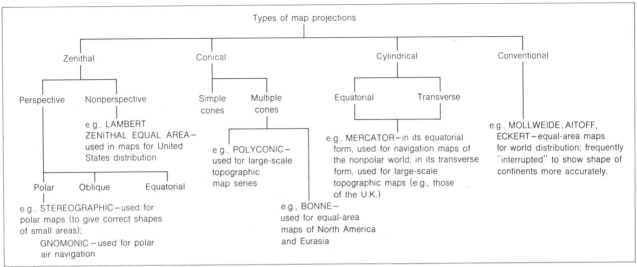

a For a full description of projections, their properties, construction, and use, see the materials suggested in the section "One step further.." at the end of the chapter

(i.e., some of these properties are accurate over part of the globe).

Which projection a geographer chooses will depend on the job the map has to do, and the advantages and drawbacks of the different projections must be carefully balanced.

Mapping agencies for individual countries always adopt projection systems that have the most advantages from the point of view of the maps' potential users. Maps of the whole United States frequently use Albers' projection, while Britain employs a transverse Mercator system for its Ordnance Survey series. By a judicious choice of such factors as local prime meridians or minor modifications of scale, a projection system can be tailored to meet the needs of the expected users of the final map sheets. The choice of a projection is always an exercise in the art—or in this case, the science—of compromise.

2-4 | LOCATION FROM THE AIR

The first manned ascent of a balloon in the United States was from Philadelphia on January 9, 1793. George Washington and a large crowd are said to have watched the event. Today's geographers look back on those early ballooning experiments by the Montgolfier brothers and their friends as a watershed (Figure 2-14). From the late eighteenth century onward, man has been able to view increasing portions of the earth's surface from above and to construct maps from direct aerial observation rather than laboriously piecing them together from measurements made on the surface. As the airplane succeeded the balloon,

Stereographic projection

The problems of depicting the spherical earth on a flat surface can be illustrated by considering one of the oldest map projection systems, the stereographic. This technique was known to the Greek geographer Ptolemy in the second century A.D. To obtain the map of the northern hemisphere shown here, we project the parallels and meridians of the globe from the South Pole onto a plane which is tangent to the North Pole. The easiest way to envisage this is to take a transparent globe and place a sheet of paper just touching the North Pole. If we place a light at the South Pole, then the outline of the continents and the grid of meridians and parallels will be thrown as shadows onto the map. If you follow the dotted lines from the South Pole on the illustration, you can see how this can be done graphically.

Since the meridians and parallels so constructed intersect at right angles, the shape of small areas is correct. The scale of the projection increases rapidly away from the pole, and lower latitude land masses are greatly distorted in area. However, any circle on the sphere remains a circle on the map, a property of use in geophysical investigations.

A stereographic projection can be constructed for a plane tangent to the globe at any point. Thus, in addition to northern and southern *polar* cases there are *oblique* and *equatorial* cases where the grid of parallels and meridians is projected from a point at one end of a diameter onto a plane tangent at the other. Other families of map grids (see Table 2-3) project onto cones (*conical* projections) or cylinders (*cylindrical* projections) that touch or intersect the globe along circles. Nonperspective systems mathematically transform global coordinates into map coordinates to optimize certain properties of shape, area, or direction. [See J. A. Steers, *Introduction to the Study of Map Projections* (University of London Press, London, 1960), Chap. 1.]

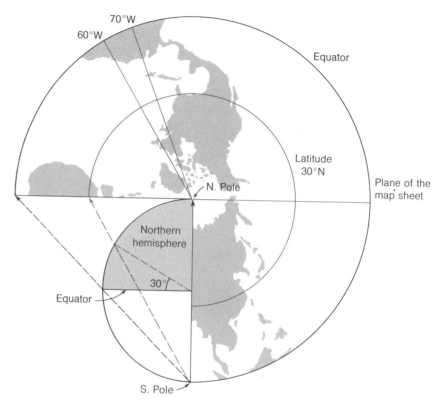

and the space rocket the airplane, the horizon continued to recede until, in the late 1960s, the whole hemisphere of the planet came into view. In this section we shall review this latest phase in environmental reconnaissance and the ways in which it has affected spatial language and the world picture developed by maps.

The single aerial photograph

Photographs of the earth's surface taken from the air present the same problems as those taken from the ground; that is, both give a highly distorted perspective. Each of us, in our first forays with a camera, has come up with prints in which buildings appear to lean drunkenly backwards or portions of the subject's anatomy near the camera are hugely swollen. Identical problems are encountered in photographs taken from the air. Note, for example, the way in which the skyscrapers in midtown Manhattan in Figure 2-15 appear to be distorted. The rules that govern such variations in scale in photographs are fairly simple,

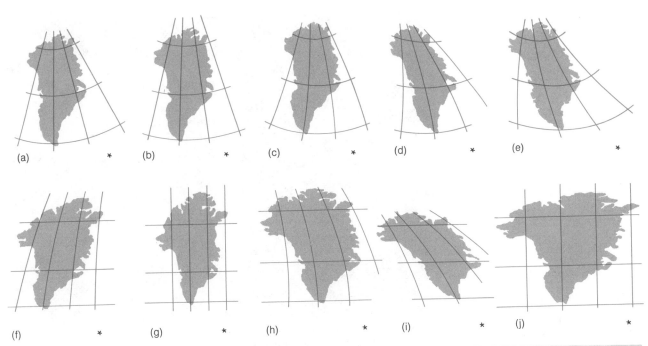

(a)   *   (b)   *   (c)   *   (d)   *   (e)   *

(f)   *   (g)   *   (h)   *   (i)   *   (j)   *

Figure 2-13. Map projections. The impact of the type of map projection used on the shape of an area is evident from the variations in the outline of Greenland in maps constructed using ten different systems. The first projection (a) is the most accurate representation of the island's true shape. The last (j) is the least accurate. [From J. R. Mackay, *Geographical Review* **59** (1969), p. 377, Fig. 2. Reprinted with permission.]

Figure 2-14. The beginnings of air photography. Late-eighteenth-century experiments with ballooning provided the first maplike perspectives of the earth's surface. [From William L. Marsh, *Aeronautical Prints and Drawings* (Halton and Truscott Smith, London, 1924).]

Scale in aerial photographs

The nominal or average scale of a vertical aerial photograph can be simply represented by

$$S = \frac{f}{H}$$

where S = the scale,
f = the focal length of the camera,
H = the height of the camera above the ground surface,

and both f and H are measured in the same units of length. Thus, if the camera has a focal length of 20 cm (about 8 in) and the aircraft is flying at a height of 1000 m (i.e., 100,000 cm, or about 40,000 in), the scale at the ground surface will be 20:100,000, or 1:5,000. Areas of higher ground (H_2) are nearer the camera and hence appear larger; conversely, areas of lower ground (H_1) are farther from the camera and appear smaller. Large distortions in scale can be introduced by tilting the camera so the photograph is not truly vertical. This displaces the perspective center of the photograph from the actual center and makes the estimation of both scale and height considerably more complex. [See D. R. Lueder, *Aerial Photographic Interpretation* (McGraw-Hill, New York, 1959). Chap. 1.]

Figure 2-15. Perspective distortion. A vertical view of Manhattan from a height of 500 m (1640 ft) shows the extreme distortion caused by differences in height. St. Patrick's Cathedral (in the upper center of the photograph abutting Fifth Avenue) appears much smaller than it really is in comparison to neighboring skyscrapers. [Photo by Lockwood, Kessler, and Bartlett, Inc.]

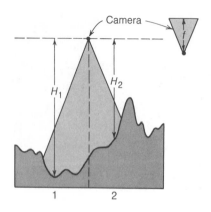

so long as the camera axis is vertical to the ground surface. (See the marginal discussion of scale in aerial photographs.) As the camera is tilted away from the vertical, more difficult problems occur.

Because the distortion in aerial photographs varies with the tilt of the camera, it is useful to place aerial photographs in two main classes: first, *vertical photographs*, in which the camera's axis points directly downward to give a plan of the terrain below, and, second, *oblique photographs*, in which the camera's axis points at a low angle to the ground to produce a perspective view. If the camera is tilted enough to include the horizon, the photograph is termed a *high oblique*; pictures that do not include the horizon are termed *low obliques*. In practice, it is extremely difficult to guarantee an absolutely vertical axis. Thus the first category normally includes photographs whose camera axis is within two or three degrees of the vertical axis. "Verticals" are the most widely used type of photograph because distortions due to scale, tilt, and height can be readily corrected to produce detailed maps. The vertical aerial photograph is not a map, how-

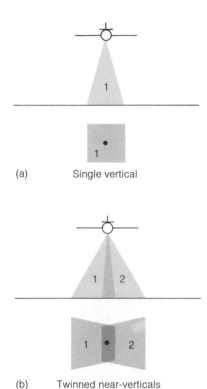

(a)　　Single vertical

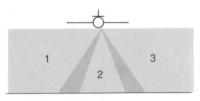

(b)　　Twinned near-verticals

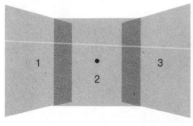

(c)　　Trimetrogon

Figure 2-16. Aerial photograph coverage, showing three typical camera arrangements for aerial photographs of the earth's surface.

ever. Scale can vary not only from one photograph to the next, but also from one part of the same photograph to another as the terrain below varies in height. Oblique photographs cover large areas of ground and present few problems in interpretation because their perspective is more familiar to the viewer. "Obliques" are therefore most popular for illustrative or publicity purposes. They are of limited value for scientific purposes because they contain wide variations in scale, they risk large areas of blocked visibility ("dead ground"), and they possess complex geometric properties from a measurement and mapping (*photogrammetric*) viewpoint.

In geographic studies obliques are frequently combined with the more generally used vertical aerial photographs by "twinning." Figure 2-16 presents two examples of this technique. The first [Figure 2-16(b)] links two slightly overlapping near verticals; the second [Figure 2-16(c)] combines one vertical view with two high-oblique ones. The latter combination is quite useful when detailed work and general reconnaissance must be done in a single flying mission.

Stereoscopic pairs

It is obvious that a flat aerial photograph has only two dimensions, while the terrain it depicts has three dimensions. How is this third dimension, height, shown in a photograph? If we reexamine Figure 2-15, we should find a clue. Not only are the tops of the skyscrapers greatly enlarged, but they also appear to be leaning back from the center of the photograph.

As Figure 2-17 illustrates, differences in the height of objects in a photograph can be determined by observing their horizontal displacement from their true ground position. In a truly vertical photograph this displacement is along lines radial from the center of the photograph. Objects such as hills, which have an elevation above the mean elevation of the photograph, are displaced outward from the center, and low points are displaced inward. The amount of displacement is inversely proportional to the altitude of the camera and directly proportional to the variations in height of the terrain.

Although this horizontal displacement is a nuisance because it prevents the direct use of aerial photographs as maps, it permits direct measurements of relief when photographs are taken in overlapping pairs. Figure 2-18 presents a typical series of photographs in a mapping run. Exposures are timed to allow a 60-percent forward overlap between successive photographs in the same strip. There is also a 25-percent lateral overlap between adjacent strips. Note that the orientation of the photographs is reversed in each successive strip, because of the flight pattern. The location of photographs in such a run is

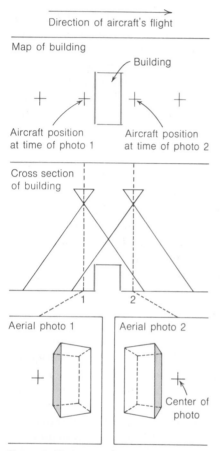

Direction of aircraft's flight

Map of building

Building

Aircraft position at time of photo 1

Aircraft position at time of photo 2

Cross section of building

1 2

Aerial photo 1

Aerial photo 2

Center of photo

Figure 2-17. Stereoscopic effects. Elevation differences in paired aerial photographs. Successive photographs of the same building from two aircraft positions are shown by horizontal displacements of its image on the two photographs.

Figure 2-19. Stereoscopic instruments. (a) A zoom stereoscope for photo interpretation. [Photo courtesy of Bausch & Lomb, Rochester.] (b) An automated stereotrigomat plotter for contour mapping from stereoscopic photographs. Instruments of this kind now allow very detailed maps to be made from aerial photographs at much lower costs than with conventional ground survey methods. [Photo courtesy of Carl Zeiss, Jena.]

facilitated by a *titling strip* automatically included at the top of each print, which gives information on the sortie number, general location, date and time of exposure, focal length of the camera, flying height, and film type.

Pairs of photographs that overlap as in Figure 2-18 are termed *stereoscopic pairs* (from the Greek *stereos*, solid). *Stereoscopes* (Figure 2-19) are devices that allow us to view two overlapping pictures taken from slightly different points of view at the same time. Viewing the two photographs through a pair of binoculars fuses them into a single image having the appearance of solidity or relief. The horizontal displacements are perceived as vertical displacements in a third (vertical) dimension. Advanced stereoscopic equipment permits an observer to convert vertical images into measurements of height and therefore allows the rapid production of contour maps.

Figure 2-18. Survey flight patterns. Aerial photographs used for stereoscopic analysis have both a forward and a lateral overlap. This effect is achieved by a flight pattern similar to the track of a lawnmower.

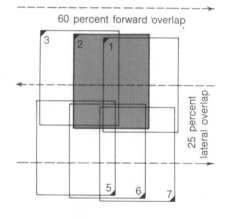

60 percent forward overlap

3 2 1

25 percent lateral overlap

5 6 7

Parallel aircraft flight paths

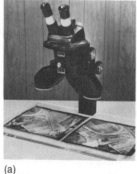

(a)

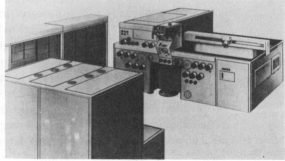

(b)

Cost advantages of aerial surveys

Low cost has been a critical factor in the growing use of aerial surveys for environmental surveillance over the last 30 years. Figures are hard to come by because, although some countries set up special government departments, such as the United States Geological Survey, to perform environmental surveys, others use international agencies like the United Nations Food and Agriculture Organization to conduct surveys, and the rest hire independent contractors. The figures that are available suggest that the cost of mapping an area by stereoscopic photo coverage is only 3 to 10 percent of the cost of conventional ground survey methods.

The actual disparity in costs will depend on a number of factors. One of the most important is the size of the area to be mapped. The United States Forest Service estimates that if the unit cost of mapping an area 25 km² (9.65 mi²) using a scale of 1:20,000 is 100, quadrupling the area surveyed to 100 km² (38.6 mi²) will bring the unit cost down to 37. Unit cost for surveying 5000 km² (1930.5 mi²) drops to about 10. Of course, costs change rapidly with the scale of the map and the fineness of the height detail. Reconnaissance mapping on a scale of 1:250,000 may cost only one-twentieth of what detailed surveys of the same area on a scale of 1:25,000 would cost. The cost of maps without elevation details is only 10 to 15 percent of the cost of contoured topographic maps.

Cost estimates for resource surveys for regional geology, land use, and timber resources studies and the like depend not only on the scale of the maps but on the amount of detailed ground surveying required. Reconnaissance studies may be based primarily on interpretation of aerial photographs, and only to a very limited extent on ground surveys by scientists working in the field. On scales of about 1:250,000, such studies may be remarkably inexpensive. For example, integrated surveys involving photography, mapping, and preliminary resource appraisals (of land capability and use, surface geology, soils, and vegetation cover) for the 1960s ranged from about $20 to $50 per km². These figures are for essentially underdeveloped areas where the samples include such outback places as the Guayas Basin, Ecuador, and lowland New Guinea. However, even large-scale studies of urban areas are increasingly being carried out by aerial survey.

In this chapter, we have moved from a view of early man gazing at an unknown world to modern air and satellite surveys. In the next three chapters we shall see what kind of world the centuries of exploration that lay between the two revealed.

Reflections

1. Consult a topographic map of your own locality, and list the names given to streams and water courses. Then do the same for a different part of the country. Are there any variations between the two lists? What do you think might have caused these variations?

2. Have the members of your class check off on a list of "possibles" (a) the countries they consider part of the "Middle East" and (b) the states they consider part of the United States' "Midwest." Can you identify a firm core area on which everyone agrees? Are there any "difficult" borderline cases?

3. Review the section on the earth's spherical coordinate system. How is the location of the earth's (a) equator and (b) poles determined?

4. What effects has the "township and range" system used for land surveys had on (a) the rural and (b) the urban landscapes of the central and western parts of the United States?

5. Consult two atlases in your college's map library which show your own country's outline on different map projections. (The title of the projection is usually given in the lower margin of the atlas page.) Compare these map outlines with the "true" shape of your country on a globe. Which projections appear to produce (a) the most and (b) the least distortion? Does this depend at all on where your country is on the map? Why?

6. Review your understanding of the following terms:
   (a) meridians
   (b) parallels
   (c) place names
   (d) reference grids
   (e) spherical coordinates
   (f) latitude and longitude
   (g) map projections
   (h) horizontal displacement
   (i) stereoscopic pairs
   (j) stereoscopes

One step further . . .

For a basic but brief account of the variety and range of maps used by geographers, see

Tyner, J., *The World of Maps and Mapping* (McGraw-Hill, New York, 1973)

and browse through one of the two standard works introducing the principles and practice of mapmaking:

Robinson, A. H. and R. D. Sale, *Principles of Cartography* (McGraw-Hill, New York, 3rd ed., 1969) and

Thrower, N. J. W., *Maps and Man: An Examination of Cartography in Relation to Culture and Civilization* (Prentice-Hall, Englewood Cliffs, N.J., 1972).

The history of man's efforts to pin down the exact distribution of the world's geographic features is told at length in

Bagrow, L., in R. A. Skelton, Ed., *History of Cartography* (Harvard University Press, Cambridge, 1964)

and summarized more briefly in

Abler, R., J. S. Adams, and P. Gould, *Spatial Organization: The Geographer's View of the World* (Prentice-Hall, Englewood Cliffs, N.J., 1971), Chap. 3.

The basic principles of photointerpretation and examples of the use of aerial photographs in research are given in

Colwell, R. N., Ed., *Manual of Photographic Interpretation* (American Society of Photogrammetry, Washington, D.C., 1960) and

Leuder, D. R., *Aerial Photographic Interpretation* (McGraw-Hill, New York, 1959).

Its geographic applications are discussed in

James, P. E. and C. F. Jones, Eds., *American Geography: Inventory and Prospect* (Syracuse University Press, Syracuse, N.Y., 1954), Chaps. 25 and 26, and

St. Joseph, J., Ed., *The Uses of Air Photography* (John Day, New York, 1966).

Excellent examples of earth photographs from spacecraft are available in a number of NASA publications. For the most recent developments in mapping, see the *International Yearbook of Cartography* (published annually) and the *Journal of Cartography* and *Surveying and Mapping* (both quarterlys). See Appendix C for a list of national atlases illustrating map production in a variety of countries.

Part One

The
Environmental
Challenge

A geographer's view of the uncertain planetary environment in which the human population evolves and lives at ever-increasing densities is presented in Part One. *The Fertile Planet?* (Chapter 3) estimates the inherent quality of the earth as a permanent home for man. It shows that the fertility of the planet varies enormously from one place to another, and it considers the major factors that underlie this spatial variation at different geographic scales. *Environmental Risks and Uncertainties* (Chapter 4) looks at how the environment is changing over time. Slow, long-term swings of climate and sea level are intermixed with short-term, season-to-season, and year-to-year variations. Superimposed on both are the sudden and catastrophic effects of extreme geophysical events—earthquakes, volcanic eruptions, floods, cyclones—that interrupt regular cycles of change. *Ecosystems and Environmental Regions* (Chapter 5) provides a context for viewing the earth's environment in biological terms relevant for man. It notes the ecological dependence of man on a series of fundamental cycles—such as the carbon cycle or the hydrologic cycle—and shows our own species as one of many interdependent organisms, each relying on worldwide exchanges of material and energy. Geographers find it convenient to look at this interdependence in terms of a series of environmental regions, each forming a module within which a particular set of ecological relationships is dominant. This first part provides an essential "stage" on which the drama of human activity described in the following parts will be played out.

Chapter 3

The Fertile Planet?

*Behold, a sower went forth to sow; and when
he sowed, some seeds fell by the way side . . .
some fell among stony places, where they had
not much earth . . . but others fell into good
ground, and brought forth fruit, some an
hundredfold, some sixtyfold, some thirtyfold.*

*The Gospel According to
St. Matthew, XIII. 3-8*

As we move outward into deep space, the intimate world of the beach disappears. Viewed from this remote distance, the earth is simply a pale blue planet, much of it wreathed in clouds, with the outlines of tawny continents dimly visible. But even though the beach is invisible, relations between man and his environment are still observable. The scale of observable relations is vastly different, though; our range of vision now includes the whole of man's 3.6 billions on the earth's 510 million square kilometers of surface. In place of individual sunbathers on the beach, we now have great clusters and nebulae of population set against a continental background. In place of wet sand and dry soil, we have an environmental range that runs from the Amazonian forest to Arctic ice caps.

In this chapter we shall look at the earth from a global perspective and assess some of its qualities as an environment for man. We shall follow the mapmakers of the past in their search to fill in the missing details of the global picture. In the millions of years in which man has been present on the earth, he has moved far beyond the beach, spreading far and wide over continent and island. But, like the sower's seed in Christ's parable, in some areas he has found the living hard and the land unfriendly and his stay has been brief; in others there has been a steady growth of population to massive levels. How fertile is the planet earth? How do its regions vary, like the sower's field, from good to bad as a home for man—and why do these variations occur? What part do they play in shaping the distribution of human populations?

| **WORLD PRODUCTIVITY PATTERNS**

The visual evidence of strong environmental contrasts is so compelling that we should begin by reminding ourselves that, compared to other wanderers in the solar system, the earth is a uniform planet. It is almost spherical, with a radius of 6365 km (3955 mi) that varies by less than 0.02 percent. In height, no two places on its surface vary by more than 20 km (about 12 mi), or the length of Manhattan island.

Despite this uniformity compared to harsher planets the earth presents considerable environmental variety in terms of man's specialized needs. In this section we look at two critical questions: How shall we measure the environment, and what kind of pattern do we find in it?

Measures of environmental contrast

The geographer's approach to measuring environmental diversity will be somewhat different from that of a "pure" earth scientist such as a geophysicist. Like our questions about the beach in Chapter 1, our questions here relate not to the abstract physical properties of the environment, but to its interpretation in human terms. It is a straightforward matter to show, on a map, the differences between the tropical equatorial zone and the polar zones. (See Figure 3-1.) We can, for example, show the constancy of warmth in one, and the constancy of cold in the other, in terms of the amount of *permafrost,* subsoil that remains permanently frozen all year round. Turning the pages in the

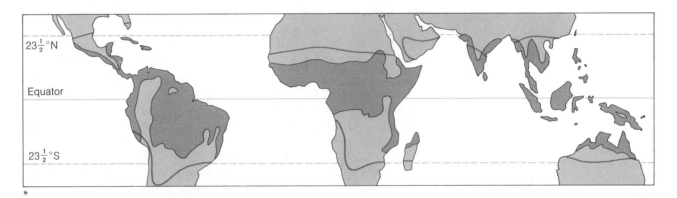

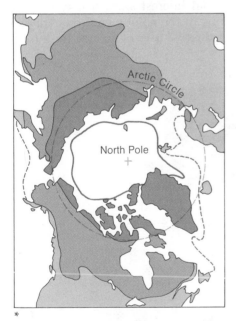

■ Temperature above 21°C (70° F) all year
—— Average daily range of temperature greater than average annual range

Figure 3-1. Global contrasts in warmth. The *tropics* are formally defined as the areas of the earth between the Tropic of Cancer (latitude 23½°N) and the Tropic of Capricorn (latitude 23½°S). Thus the tropics lie in those latitudes where the sun is vertically overhead at some time of the year. These limits correspond very roughly with the two measures of warmth shown. At the other extreme the *polar* zones lie north of the Arctic Circle (66½°N) and south of the Antarctic Circle (66½°S). Two thermal measures of the Arctic polar zone, *permafrost* (subsoil that remains frozen all the year round) and *pack ice* (large areas of ice covering the entire sea surface) are shown. Note that these extend south of the Arctic Circle in Siberia but lie well to the north over the sea areas north of Scandinavia. [Based on R. Common, in G. H. Dury, Ed., *Essays in Geomorphology* (Heinemann, London, 1966), p. 68, Fig. 7.]

■ Continuous permafrost
■ Sporadic permafrost
—— Limits of year-round pack ice
---- Average spring maximum of pack ice

environmental section of an atlas will show you dozens of ways of indicating evident contrasts in heat and cold, wetness and dryness, constant and fluctuating climate conditions, and the like.

Any yardstick we propose for measuring the environment must place man in his proper context as one of the living organisms on the earth's surface. Figure 3-2 presents a simple, schematic representation of the links between the living and nonliving, the *biotic* and *abiotic*, parts of the environment. We can further subdivide the abiotic environment into: (1) the solid earth, or *lithosphere*; (2) the liquid waters, or *hydrosphere*; and (3) the gaseous air, or *atmosphere*. Each of these subdivisions of the environment has specific properties and rates of change (see Chapter 4), but it is the interactions between them that determines the special character of a local environment.

Many geographers have wrestled with the problem of combining biotic and abiotic elements into a single measure of environmental differences. Some have concentrated on the lithosphere and developed terrain-related schemes; more commonly, climate or climate-plus-vegetation has been used as an indicator of environmental variations. Vegetation has attracted special interest since it may be seen as (a) a biotic response to variations in the abiotic environment and (b) an indicator of the potential productive uses of an area for human

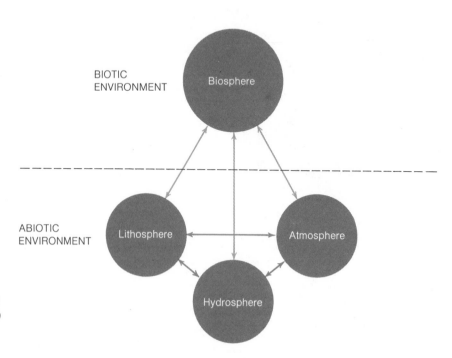

Figure 3-2. Major divisions of the environment. Geographers divide the environment into two broad divisions of the living world (the *biotic* environment) and the nonliving world (the *abiotic* environment).

settlement. Figure 3-3 shows how changes in the climate, the soils, and the terrain of the Great Plains of North America are closely reflected in changes in the natural vegetation. (Note that we use the term "natural" vegetation. Man's activities have greatly modified the actual vegetation in large areas of the world.) During the settlement of the Great Plains by both North American Indians and Europeans, these environmental differences were reflected in the choice of livestock and crops.

One crude measure of variations in vegetation is the sheer mass of plants that grow in a particular area. Suppose we could bulldoze

Figure 3-3. Interlocking environmental factors in the Great Plains. This idealized cross profile of the Great Plains of North America shows related changes in climatic, vegetational, and soil conditions. Note that as the moisture situation changes from humid to dry, notable changes occur in soils and vegetation. Soils change from black in the humid East to dark brown and brown on the dry western margins of the Plains region. Very critical changes are caused by a layer of salts in the subsoil at the lowest limit to which seasonal rainfall penetrates. This layer is as deep as 1.2 m (4 ft) in the East but rises to only 0.2 m (8 in.) in the West. An important dividing line occurs between the 98th and 100th meridians of west longitude, where the layer of salts rises above the 0.8 m (2.6 ft) mark. Along this line black soils give way to brown, and tall prairie grass gives way to short bunch grass. As a result of grazing and wheat farming, much of the original vegetation in the West has disappeared. But the dividing line created by the changing depth of the salt layer remains, separating the productive ex-prairies from the western areas where cultivation is still risky without specialized irrigation or dry-farming techniques. [After W. R. Meade and E. H. Brown, *The United States and Canada* (Hutchinson, London, 1962), p. 216, Fig. 41.]

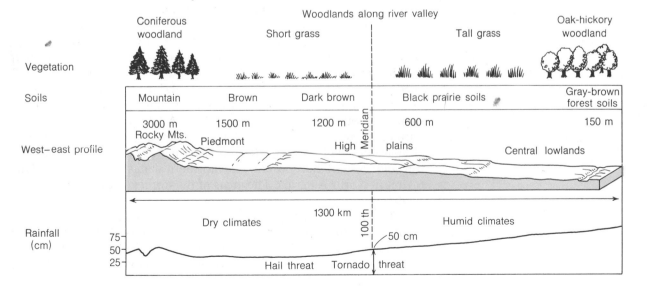

Paterson's index of plant productivity

The index of plant productivity on which the map in Figure 3-4 is based is calculated using the following equation, the right-hand side of which is a combination of basic climatic elements:

$$I = \frac{T_m PGS}{120(T_r)}$$

where I = an index of plant productivity,
T_m = the average temperature of the warmest month in degrees centigrade,
T_r = the annual range in average temperature between the coldest and warmest months in degrees centigrade,
P = the precipitation in centimeters,
G = the growing season in months, and
S = the amount of solar radiation expressed as a proportion of the radiation at the poles.

The growing season is calculated by counting the number of months in which the average monthly temperature reaches or exceeds the threshold needed for plants to grow (assumed to be at 3°C). Thus, Portland, Maine, with an average temperature in the warmest month of 19.7°C, a temperature range of 24.9°C, a rainfall of 106 cm, a growing season of 8 months, and a radiation value of 0.56, has an index of 3.13. For a full discussion see S. S. Paterson, *The Forest Area of the World and Its Potential Productivity* (Geography Department of the Royal University of Göteborg, Göteborg, Sweden, 1956). Note that this index is only one of many ways of estimating the potential plant growth in an area from climatic data. All give a similar picture of global productivity patterns but differ in various details.

all the vegetation growing in sample kilometer-square patches throughout the globe and weigh the resulting piles of trunks, stems, and leaves. The pile that was once a tropical rain forest would weigh many thousands of tons; the pile gathered in the open woodland of the savannah would weigh between one-tenth and one-hundredth as much. The muskeg swamps of northern Canada would provide still smaller piles, and the Arctic ice caps and the most arid parts of the deserts would provide nothing.

Geographers have found various relationships between plant productivity and climatic statistics. For example, the Swedish geographer Sten Paterson has determined that productivity increases with the length of the growing season, the average temperature of the warmest month, the annual precipitation, and the amount of solar radiation and decreases with the range of temperatures in an area. By combining values for these factors into a single index, he was able to give numbers to different places on the surface of the earth indicating their potential for plant growth. (For a more detailed description of Paterson's work, see the marginal discussion). Consider an example: Portland, Maine, has an index of 3.1. By contrast, Miami, Florida, has an index of 22.3; and Belem, at the mouth of the Amazon River in Brazil, has one of 118.0. Many measures similar to Paterson's are available, and differences between the world maps based on these various indexes are relatively minor.

A six-zone world map

Figure 3-4 presents a world map whose contours are based on Paterson's index of potential productivity. It has six zones, and we label them, like examination grades, from A to F. F is a nonproductive zone with index values up to 0.25. Areas in this zone are readily identifiable as the cold and dry areas of the world: ice caps, the fringes of northern tundras, midlatitude deserts. Zone E, a band of very low productivity with index values between 0.26 and 1.00, borders on Zone F. It is a relatively narrow band around the deserts but covers significant sections of North America and Soviet Asia. A low-productivity band (D) with index values of 1.01 to 3.00 is confined mainly to cool, temperate climates having the most rain in the summer. This band includes central and eastern Europe and the northern states of the Great Plains region in North America. The areas of the tropics in this band, too, are relatively small.

The fourth band (C), with index values of 3.01 to 10.00, is one of medium productivity and covers some of the most densely populated areas in the world. (Cf. the map on the inside front cover.) It includes the eastern United States, western Europe, and much of India and

South China, as well as the savannah areas of Africa. A higher-productivity band (B), with index values of 10.01 to 50.00, is restricted to the tropics; most of this area is in South America. Finally, there is a zone (A) of very high productivity (index values greater than 50.00) in the equatorial belt: The Amazon basin in South America, the Congo basin in Africa, and the Indonesian archipelago in South Asia lie in this zone.

The photographs in Figure 3-5 show how dramatic zonal differences in vegetation can be. However, it is important to remember that we are talking about zones of *potential* productivity; Paterson's index measures the plant life the climate of a particular environment *could* support. The actual plant cover of an area will reflect other factors such as its vegetational history or the degree of human interference with the ecosystem. Men may reduce plant growth (e.g., by pollution) or increase it (e.g., by irrigation). Furthermore, broad zonal maps such as Figure 3-4 do not reveal local areas whose productivity is affected by factors other than the ones included in an index.

3-2 GLOBAL KEYS TO THE PATTERN

What factors determine the productivity patterns in Paterson's maps? Like the population patterns on the beach, the *productivity contours* of Figure 3-4 can be unraveled by locational analysis. Geographers have found that the natural fertility of any location on the earth's surface is governed by four main factors: (1) its solar climate (corresponding to its latitude), (2) its location relative to the general atmospheric circulation of the earth, (3) its location relative to the continents, the oceans, or other major terrain features, and (4) local environmental factors. We shall begin by looking at the first two of these factors, which operate at the global level, before turning to the part played by the other, smaller-scale elements in the puzzle.

Latitudinal factors

Geographers in ancient Greece regarded the climatic environment of any part of the earth's surface as largely a result of its latitude. The earth was thought to slope (the Greek word *klima*, used in climatology, means slope) *away* from the sun north of the latitude of Greece and the Mediterranean Sea, making the northerly climate increasingly colder. The part of the planet to the south was thought to slope *toward* the sun, eventually becoming a torrid zone too hot for human life.

Figure 3-6(a) is a simplified picture of the Greek world view. The earth lies in a beam of solar radiation whose angle of incidence is directly related to latitude. As we move from the equator toward the poles, equivalent amounts of energy are spread over ever wider areas of

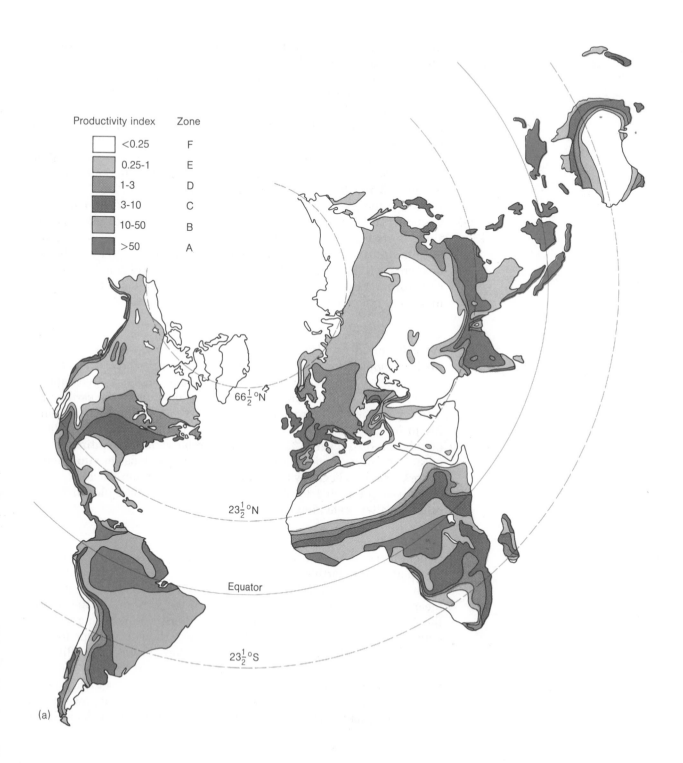

Productivity index   Zone

	Productivity index	Zone
	<0.25	F
	0.25-1	E
	1-3	D
	3-10	C
	10-50	B
	>50	A

$66\frac{1}{2}°N$

$23\frac{1}{2}°N$

Equator

$23\frac{1}{2}°S$

(a)

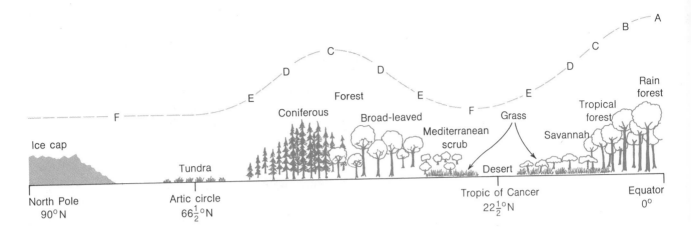

(b)

Figure 3-4. World productivity. (a) The earth can be divided into zones of varying productivity, based on the potential for plant growth. Here, that potential is estimated from climatic elements included in Paterson's index of productivity. Note that the "A" zone is the most productive and the "F" zone least productive. (b) Idealized cross section from the North Pole to the equator. This shows the six zones (A–F) and the corresponding types of plant growth. For a detailed description of the vegetation, see Table 5-1. [After S. S. Paterson, *The Forest Area of the World and Its Potential Productivity* (Royal University of Göteborg Geography Department, Göteborg, 1956), p. 144, Fig. 33.]

the globe, producing an equatorial *torrid* zone, a midlatitude *temperate* zone, and a high-latitude *frigid* zone.

To understand this low-latitude to high-latitude gradient we must look at the energy relations between the earth and the sun. Each day the earth intercepts a massive beam of solar energy (estimated at 17×10^{13} kilowatts). This energy is emitted on different wavelengths; the peak intensity is the visible part of the spectrum (daylight), but the beam also includes important shortwave (ultraviolet) and longer wave (infrared) emissions. If the earth were like the moon and had no atmosphere, the impact of these latitudinal variations in energy would be catastrophic. Daytime temperatures near the equator would rise to some hundreds of degrees centigrade, while in the winter night at the poles temperatures would fall almost to the absolute zero of outer space. Actually, as Table 3-1 shows, the most extreme shade temperatures ever recorded at the earth's surface (in 1933 and 1960) differ

Table 3-1 Global extremes of temperature

Thermal characteristic	Temperature		Location
	(°C)	(°F)	
Individual readings			
Highest	58.0	136.4	San Luis Potosí, Mexico (1933)
Lowest	−88.3	−126.9	Vostok, Antarctica (1960)
Range of readings			
Widest	88.9	160.0	Verkhoyansk, Siberia
Narrowest	13.4	24.1	Fernando de Noronha, South Atlantic
Annual averages			
Hottest	31.1	88.0	Lugh Ganane, Somalia
Coldest	−57.8	−72.0	78°S, 96°E, Antarctica

by less than 150°C, and the average temperatures of the hottest and coldest locations on the earth vary by less than 90°C. Moreover, these are extremes. The average contrasts between any pair of locations are much more subdued.

Clearly, latitudinal variations in temperature are dampened and modified in some way. To explain this softening of thermal contrasts on the earth's surface, we must bring a thin coating of air into our model of the earth's relationship with the sun.

Earth's atmospheric filter

The planet earth is surrounded by a thin but critically important shell of gases—its *atmosphere.* This shell is held to the earth's surface by gravitational attraction and, as we should expect, is densest at the

(a)

(b)

(c)

Figure 3-5. Contrasts in environmental
production. Major global extremes in
Paterson's productivity regions are
represented by (b) the tropical rain
forest (Zone A) and (c) the arid dune
fields (Zone F). Within major zones
important variations in quality may be
related to secondary factors (such as
elevation) or to tertiary factors (such as
geology, soil, and drainage). Both (a) an
Alpine ice field and (d) alluvial swamp-
lands illustrate the mosaic of local
variations within the area labeled Zone
C on the world map in Figure 3-4.
[Photographs (a) courtesy of the Swiss
National Tourist Office, (b) courtesy of
the American Museum of Natural
History, (c) courtesy of the U.S.
Department of the Interior, National
Park Service Photo, George A. Grant,
and (d) by Rotkin, P. F. I.]

(d)

bottom, thinning rapidly as we move upward. The zone of greatest importance to man is the *troposphere* [see Figure 3-6(b)], which gets its name from the Greek phrase for a "turbulent realm." The upper boundary of the troposphere, the *tropopause*, varies seasonally in height but on the average is 9 km ($5\frac{1}{2}$ mi) above the poles and 17 km ($10\frac{1}{2}$ mi) above the equator. Above the tropopause lies the stratified layers of the *stratosphere*.

The troposphere is so thin, relative to the size of the earth, that it would be barely detectable on the regular classroom globe. Why then is it so important in any analysis of our planet as a home for man? There are four overriding reasons: (a) It contains the invisible and odorless gas, oxygen (about 20 percent of its volume), which is essential to human life. Climbers above about 6 km (about 20,000 ft) find that they need additional oxygen as the concentration of the gas in the air diminishes. High-flying aircraft (and of course spacecraft) must create their own oxygenated atmospheres. (b) Carbon dioxide, which is critical to plant life, is also present in the atmosphere, although in minute quantities (less than half of 1 percent). The role of carbon dioxide in plant growth and the food chains dependent on it is discussed in Chapter 5, Section 1. (c) Water vapor is drawn up from ocean and sea surfaces into the troposphere to be circulated and redistributed over the earth's surface as precipitation. Like oxygen and carbon dioxide, this water vapor is essential to the life and growth of organisms. (d) The gases in the troposphere and the layers above it act as a filter and a blanket. Harmful short-wave radiation from the sun is absorbed and reflected, while long-wave radiation from the earth itself is retained.

A detailed account of what happens to solar energy when it strikes the earth's atmospheric shell is given in Figure 3-7. A small amount (6 percent) is reflected by the atmosphere and a larger amount (27

Figure 3-6. General circulation of the earth's atmosphere. Here we see three stages in the historical evolution of concepts of atmospheric circulation and climatic zonation. The Halley model (c) names the three main wind belts of the world—the *polar easterlies,* the *westerlies* of the middle latitudes, and the *trades* of the tropical zone. (The word "trades" comes from a Latin word indicating that the winds blow in a constant direction, and has nothing to do with trade in the sense of commerce.) Two important lines of air convergence are also shown. The *polar front* marks the junction of the polar easterlies and the westerlies in both the northern and southern hemispheres. Storms marked by low pressure move west along this front, playing a major part in the weather of the United States and Western Europe. The *inter-tropical convergence* is a zone of low atmospheric pressure separating the two trade wind systems. It lies more or less along the equator but moves slightly north and south with the seasons. Areas of air divergence with high pressures occur at both poles (the *polar highs*) and again at about 30°N and 30°S (the *subtropical highs*).

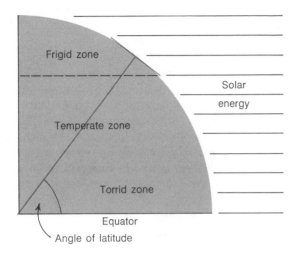

(a) GREEK MODEL

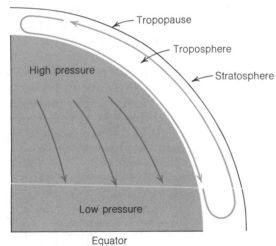

(b) ATMOSPHERIC MODEL WITH NO ROTATION

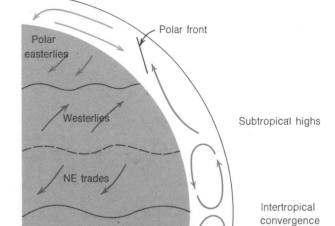

(c) HADLEY MODEL

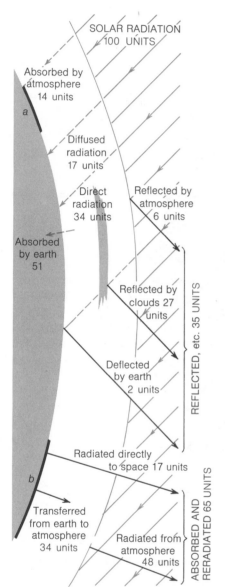

SOLAR RADIATION
100 UNITS

Absorbed by
atmosphere
14 units

a

Diffused
radiation
17 units

Direct
radiation
34 units

Reflected by
atmosphere
6 units

Absorbed
by earth
51

Reflected by
clouds 27
units

REFLECTED, etc. 35 UNITS

Deflected
by earth
2 units

Radiated directly
to space 17 units

b

Transferred
from earth to
atmosphere
34 units

Radiated from
atmosphere
48 units

ABSORBED AND
RERADIATED 65 UNITS

Figure 3-7. The earth's solar radiation budget. The diagram shows how 100 units of solar radiation are affected by the thin layer of atmosphere surrounding the planet. (For demonstration purposes, the height of the atmosphere in relation to the earth's curvature is exaggerated.)

percent) by the world's cloud cover. The remaining two-thirds is absorbed by the earth and its atmosphere, and reradiated. If you follow the arrows around the diagram you will find that the amount of energy received (100 units) is exactly balanced by the amount of energy reflected and reradiated. This *global energy balance* is the engine that powers not only the circulation of air and water (ocean currents), but the food chains of which man himself is a part.

The variation between the torrid, temperate, and frigid zones recognized by the Greeks (and shown in Figure 3-6) is also related to this cycle of energy. Latitudinal variations in temperature indicate the average angle of incidence of the sun's rays, which is at a maximum in the tropics but decreases toward the North and South Poles. The energy beam from the sun must pass obliquely through the earth's atmosphere and falls on a wider surface area. Compare areas *a* and *b* in Figure 3-7 to see the effect of these combined factors on the amount of energy received in the second area.

The general circulation of the atmosphere

Like any gas or fluid, the atmosphere responds to temperature changes —increasing in density when it is cooled, decreasing in density when it is warmed. As we might expect, therefore, the warmer (and lighter) air near the equator rises and flows poleward, to be replaced by cold (and dense) air near the poles moving equatorward along the surface. These two flows together make up a *convective circuit*, illustrated in Figure 3-6(b). The unequal heating of the air by the sun sets up compensating currents that act like escalators, redistributing heat across various latitudes, and tempering the simple Greek pattern of hot and cold zones.

As early as 1686, the English astronomer Edmund Halley had outlined a circulational model like Figure 3-6(c) to explain why the lower latitudes, despite receiving more heat, did not become progressively hotter. By incorporating the effects of the earth's rotation into the model, it was possible to give a reasonable explanation of the trade winds, those steady air currents blowing toward the equator from the northeast in the northern hemisphere and from the southeast in the southern hemisphere.

Halley's model was far from satisfactory, however. If we look at the motions of the earth's atmosphere as revealed by telltale cloud patterns in satellite photos (Figure 3-8), we can see two important features he left out. First, there are extensive cloudless areas of very dry descending air at latitudes of about 30° N and 30° S (above the Sahara Desert, for example). Second, there are belts of strong westerly winds in the middle latitudes of both hemispheres. The westerly drift

Figure 3-8. Earth's cloud cover. Major climatic zones are suggested in this picture of the entire earth's cloud cover on a single day (February 13, 1965). It is made up from 450 photographs received from the weather satellite Tiros 9 on its pole-to-pole orbit of the earth. Note the relative absence of cloud cover from the desert areas of North Africa, central Australia, and the middle section of the west coast of South America. [NASA photograph.]

of clouds across the North Atlantic illustrates this circulatory force.

The missing elements in Halley's explanation of atmospheric movements were provided by another English scientist, George Hadley, in 1735. His model [Figure 3-6(c)] replaced the single pole-to-equator convective circuit by a series of *three* convective circuits in each hemisphere. Hadley's model—although much modified in detail—still forms the basis of modern concepts of atmospheric circulation. It not only accounts for the puzzling feature of the Westerlies (now seen as caused by the westward rotational deflection of poleward-moving air), but throws light on global patterns of precipitation. Each of the main belts of winds are defined in the caption of Figure 3-6.

Though a number of factors affect precipitation, its general cause is the cooling of air that contains water vapor. Since water vapor is present throughout the troposphere, it is the location of cooling and drying processes at the global level which explain the wet and dry areas of the earth. Such processes are indicated in Hadley's model by areas of upward and downward motion in the general atmosphere. As Figure 3-9 shows, there are three main zones of upward motion: (a) an *equatorial convergence* zone, where the two trade-wind systems meet, and (b) two *polar front* zones at around 60°N and 60°S, where warm tropical westerly air overrides the colder air flowing outward from the two poles. A map of world precipitation (Figure 3-10) confirms that these are precisely the locations where high values are observed.

Using the reverse argument, we should expect zones of low pre-

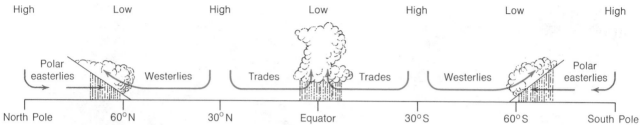

High   Low   High   Low   High   Low   High

Polar easterlies   Westerlies   Trades   Trades   Westerlies   Polar easterlies

North Pole   60°N   30°N   Equator   30°S   60°S   South Pole

Figure 3-9. Circulation and precipitation patterns. A simplified cross section of the earth's atmosphere shows the general pattern of circulation. Note the association of the three zones of upward air motion (low pressures) with belts of rain. The four high-pressure areas are marked by descending air motion and very low precipitation. The sharp discontinuities at latitudes of 60°N and S mark the polar fronts.

cipitation in areas of downward-moving air. As Figure 3-9 shows, such zones coincide with the two poles and with two zones of divergence at 30°N and 30°S; again, we find that these are zones of low precipitation on the world map. Note on Figure 3-10 the location of the world's main desert zones in relation to Hadley's simple model.

3-3 | CONTINENTAL KEYS TO THE PATTERN

We began our locational analysis of global productivity patterns in Figure 3-4 by assuming a simple, airless planet. We have seen how adding a thin but critical layer of atmosphere to the harsh Greek model of torrid, temperate, and frigid zones produced a more realistic model that went a long way toward explaining the overall pattern of warmth and wetness. We now add another thin but critical layer—the oceans—to our world picture.

Oceans and the world water cycle

Water vapor was noted earlier as one of the critical components of the earth's atmosphere. Water as a gas or small droplets in the air forms only 1 part in 100,000 of the planet's overall water resources. Over 97 percent is concentrated in the great water sheets of the world's oceans. Not only do the oceans cover over 70 percent of the earth's surface, but their lowest point (Marianas Trench in the western Pacific, 11.03 km, or 6.85 mi deep) greatly exceeds the highest point on the land surface (Mount Everest in the Asian Himalayas, 8.85 km, or 5.5 mi high). Perhaps the simplest way to remember the oceans' size is to recall that if the earth had a smooth surface it would be covered everywhere, to a depth of about 3 km (1.86 mi), by water.

Water vapor is carried and distributed over the earth's sea and land surface by the global wind belts described in Hadley's model of the atmosphere. Each year an estimated 33,600 km³ (8,100 mi³) of water evaporates from the ocean surface. Figure 3-11 shows what happens to that water. About 89 percent of it is returned directly to the ocean by precipitation. The remaining 11 percent moves over the earth's

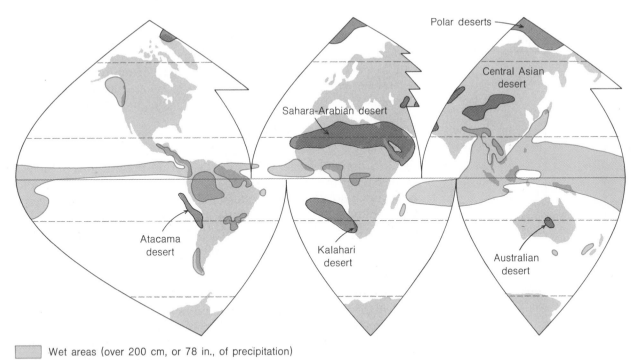

Wet areas (over 200 cm, or 78 in., of precipitation)

Dry areas (below 10 cm, or 3.9 in., of precipitation)

Figure 3-10. The world's wettest and driest areas. The world's wettest areas straddle the equator and the windward margins of continents. The location of the driest areas is related to tropical high-pressure cells. [After *The Times Atlas* (The Times, London, 1958), Vol. 1, Plate 3.]

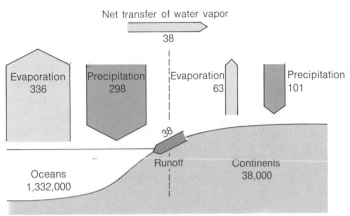

Figure 3-11. The world water balance. The diagram shows major water flows between the oceans and the continents. The figures indicate flows in 100 km³ of water. [After A. N. Strahler, *The Earth Sciences,* 2nd ed. (Harper & Row, New York, 1971), p. 586, Fig. 33-2.]

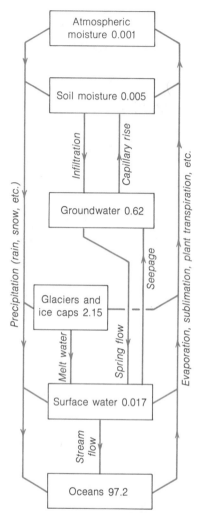

Figure 3-12. Earth's hydrologic cycle. The diagram shows the main components in the global hydrologic cycle. The figures refer to the percentage of all terrestrial water in each of the six main stores. [After R. J. More, in R. J. Chorley and P. Haggett, Eds., *Models in Geography* (Methuen, London, 1967), p. 146, Fig. 5-1.]

land surfaces before precipitating out. Precipitation falling on land may be either returned directly to the atmosphere by evaporation and transpiration, or temporarily stored (in lakes, ice caps, in the upper soil layers, or more deeply as groundwater), or returned to the oceans by flowing streams and melting glaciers. Hence, the balance between the moisture leaving the oceans as water vapor and returning as a liquid is maintained.

This global circulation of water is termed the *global hydrologic cycle*. Figure 3-12 shows the earth's main stores of water, as well as the direction of transfers. Outside the world's oceans most of the water is locked up in glaciers and ice caps. Only 0.001 percent of the total is in the atmosphere. To understand the varying pattern of precipitation around the world, we must understand something of how that small atmospheric fraction is shunted from one location to another by the general circulation of the atmosphere.

Continents and islands

Temperatures vary more on land than they do at sea. But only 29.4 percent of the earth's surface is land, and this land is distributed in an asymetric way in relation to latitude. Most of it is in the northern hemisphere. The climatic implications of the division of the planet into areas of land and water are dramatic. The thermal conductivity of the surface layers of land and sea is quite different. Land heats and cools much more quickly than the sea; bodies of water act as heat stores whose temperatures fluctuate much less than those of adjacent bodies of land.

Thus there are considerable global variations in the seasonal range of temperatures. The *average* annual range is lowest at the equator and increases with latitude, from about 3°C to 60°C at the South Pole. The smallest ranges are in oceanic islands near the equator; the largest ranges are in midcontinental locations in high latitudes. On Saipan in the Mariana Islands of the western Pacific, the highest and lowest temperatures ever recorded, the *extreme* range, differ by only 12°C. By contrast, Olekminsk in central Siberia has an extreme range of 105°C, from −60°C to 45°C. The conventional distinction between continents and smaller land masses is an arbitrary one. All can be regarded as "islands" of different sizes, and within each an increased range is detectable, though it weakens as the area becomes smaller.

A broad distinction can be drawn between the climates of continental interiors and areas near the sea. *Continental* climates are characterized by great ranges of temperature (both between day and night and between winter and summer), low humidities, and very variable precipitation. *Maritime* climates have the reverse characteristics:

smaller temperature ranges, higher humidities, and more uniform precipitation. These contrasts are not symmetrical on a continent, but are related to the latitudinal location of the land area with respect to the circulation pattern of both the atmosphere and the oceans. We can combine the effects of latitude and continentality and superimpose them on an idealized continent of low and uniform elevation. The resulting distribution of precipitation on the continent [Figure 3-13(a) should be interpreted in terms of the airflows shown in Figure 3-9. The dry areas coincide with the subtropical high-pressure areas, and the wet areas with the storms in the tropical zones. The westerly circulation of air brings a sequence of midlatitude storms over the western margins of the continent. If you compare the distribution of wet and dry zones with the direction of air movements worldwide, you will see why precipitation decreases away from the continent's oceanic boundary.

Because of the great changes in temperature and air pressure over continents, we must add a further seasonal effect to our model of precipitation patterns [Figure 3-13(b)]. The differential heating of the continental air masses causes maritime air to be drawn inward in summer to replace the warm, light, and rising air over the continents. In winter, the colder, heavier, and descending air above the world's land masses flows outward toward the seas. The impact of these flows on the world's largest land mass, where the most striking seasonal reversals occur, is discussed in Chapter 4. (See Section 4-3 on monsoon India.)

Figure 3-13. Continental precipitation patterns. The graphs show the distribution of (a) total precipitation and (b) seasonal deficits in precipitation on a hypothetical continent of low, uniform elevation. [After C. W. Thornthwaite, in R. J. Chorley, Ed., *Water, Earth and Man* (Methuen, London, 1969), p. 124, Fig. 3-1(i), 5.]

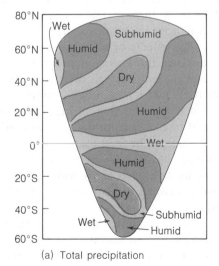

(a) Total precipitation

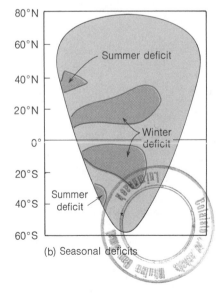

(b) Seasonal deficits

3-4 | **SOURCES OF LOCAL VARIATION**

Within the overall climatic framework formed by the effects of latitude and continentality we can detect a hierarchy of smaller scale factors. The geographic scale of their impact is minor, but they may play a decisive part in determining the use that can be made of an area.

The impact of elevation

We have seen that the earth is almost a perfect sphere and that variations in its surface elevation are equal to only a small fraction of its radius. On a classroom globe, these variations would appear as almost imperceptible bumps and hollows. Table 3-2 shows the proportion of surface area in each altitude zone on the earth. We shall ignore for the present the variations in areas below sea level (although we shall return to them in Section 18-1 when we consider territorial claims on the shallow waters of the continental shelves). Land above sea level ranges up to 8.9 km (5.53 mi) in height, but over two-thirds of it is at elevations below 1 km (.62 mi), and less than one-tenth is above 2 km (1.24 mi). The city of Denver, the highest major city in the United States, is at an elevation of 1.6 km, or almost 1 mile.

The direct effect of these relatively small differences in elevation on the characteristics of lowland and highland environments is striking. At an altitude of 8 km (roughly 5 mi), the density of the atmosphere is less than one-half its density at sea level. High elevations have a thinner shell of atmosphere above them and receive considerably more direct solar radiation than sea-level locations, but they lose much more heat by radiation from the ground surface. Within the lower layers of the atmosphere, temperature decreases with elevation at an average rate of about 6.4°C per kilometer. This rate of loss in temperature with height is termed the *thermal lapse rate*. It is generally the same throughout the troposphere.

The effects of elevation on temperature are twofold: As elevation increases, the average temperature of an area decreases, and the daily range in temperature increases. Both effects are due to clearer, more rarified air, which allows more solar radiation to reach the ground surface (raising the midday temperature) and also permits a more rapid heat loss from the ground at night. The net effect is to reproduce the changes in temperature we may encounter as we move from one latitude to another over a short vertical distance. If we live on the equator and wish to see some snow, we can either travel 8000 km (5000 mi) poleward to find snowbanks at sea level or climb 4.5 km (15,000 ft) up!

The symmetry between changes in climate due to latitude and changes due to elevation is shown in Figure 3-14. Because snowlines change with the seasons, the upper limits of plant growth have been

Table 3-2 Distribution of the earth's surface area by height

Continents		
Over +4 km	(+2.5 mi)	0.5 percent
+2 to +4 km	(+1.2 to +2.5 mi)	3.3 percent
0 to +2 km	(0 to +1.2 mi)	25.3 percent
Oceans		
−2 to 0 km	(−1.2 to 0 mi)	11.5 percent
−4 to −2 km	(−2.5 to −1.2 mi)	18.7 percent
−6 to −4 km	(−3.7 to 2.5 mi)	39.7 percent
Below −6 km	(−3.7 mi)	1.0 percent

used as a substitute measure of climatic differences. Timberlines reach their greatest sea-level extent at a latitude of about 72° in the northern hemisphere and 56° in the southern hemisphere. A mean temperature of 10°C for the warmest part of the year appears to be a prominent factor limiting forest growth, and this limit is met at varying elevations in different parts of the world.

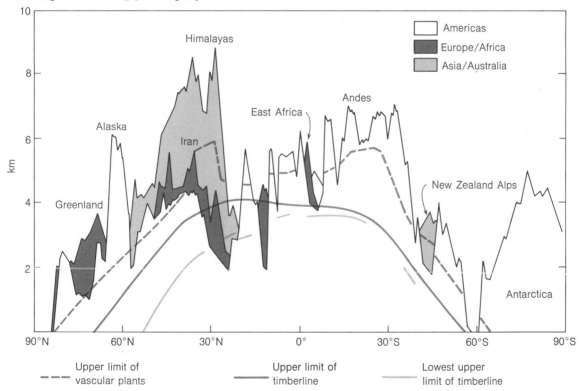

Figure 3-14. World timberline variations. The altitude at which plants will grow in mountain areas changes at different latitudes, as also does the timberline. [After L. W. Swan, in W. H. Osburn and H. E. Wright, Jr., Eds., *Arctic and Alpine Environments* (Indiana University Press, Bloomington, Ind., 1968), p. 32, Fig. 1.]

We can spot the effect of altitude on plant cover within small areas as well as on the world scale. Figure 3-15 shows the variations in zones of vegetation on a small tropical island. Note that the effect of elevation is complicated by the direction in which moisture-bearing air is traveling and by variations in cloud cover with elevation. The simple relationship between height and variations in vegetation is disturbed on the windward side of the island by a characteristic "cloud forest" belt. The leeward side of the island is much drier, and the forest belts start at different elevations. Similar rain-shadow effects, related to the prevailing moisture-laden winds from the Pacific, are visible in the forest levels in the southwestern United States (e.g., in the Sierra Nevada or White Mountains).

Minor contrasts: Regional examples

Latitude, continentality, and altitude combine to account for the largest share of systematic environmental variation on the global and continental levels. On local levels, broad patterns are broken up by the effects of another interlocking set of variables: terrain, slope, drainage, surface geology, and soils. Instead of isolating the effect of each of these variables we shall illustrate their combined impact by two regional examples. A little fieldwork on local scale should produce other examples of the effect of these factors in your own locality.

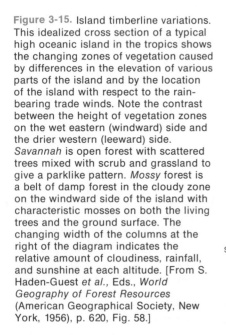

Figure 3-15. Island timberline variations. This idealized cross section of a typical high oceanic island in the tropics shows the changing zones of vegetation caused by differences in the elevation of various parts of the island and by the location of the island with respect to the rain-bearing trade winds. Note the contrast between the height of vegetation zones on the wet eastern (windward) side and the drier western (leeward) side. *Savannah* is open forest with scattered trees mixed with scrub and grassland to give a parklike pattern. *Mossy* forest is a belt of damp forest in the cloudy zone on the windward side of the island with characteristic mosses on both the living trees and the ground surface. The changing width of the columns at the right of the diagram indicates the relative amount of cloudiness, rainfall, and sunshine at each altitude. [From S. Haden-Guest *et al.*, Eds., *World Geography of Forest Resources* (American Geographical Society, New York, 1956), p. 620, Fig. 58.]

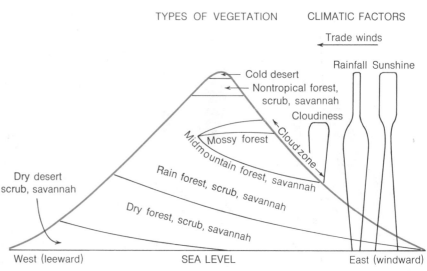

The Great Smoky Mountains Our first example is taken from the Great Smoky Mountains, a national park in the Blue Ridge Mountain area of Tennessee and North Carolina. The area has a high annual rainfall (over 135 cm, or about 53 in), and temperatures range from a January average of daily lows of 0°C to a July average of daily highs of 32°C. The area lies entirely within zone C in Figure 3-4. As Figure 3-16(a) shows, it consists basically of a heavily forested set of ridges. The main contrasts in vegetation are related not simply to elevation but to a combination of elevation and terrain. Figure 3-16(b) shows how the main forest types may be located on an elevation-terrain diagram. The different types of landscape are ordered according to the dryness or wetness of the immediate environment. The dry end of the gradient is marked by ridges and peaks, whereas the wetter end consists of deep ravines with flowing streams. Between these two extremes lie other types of terrain. Even within these broad classes, wetness varies with the direction of a slope; an open slope is much drier if it faces southwest (where it gets sun and is in the lee of rain-bearing wind) than if it faces northeast. By adding an elevation axis, we can sort the major types of vegetation into distinctive groups with specific elevation—terrain preferences. For example, untimbered grassy areas occur in dry sites at elevations over 1300 m (4265 ft), whereas hemlock forests form a distinct, diagonal band.

The north Somerset plain The north Somerset plain [Figure 3-17(a)] covers a 400-km² (154-mi²) area in southwest England. The average rainfall over the area is around 60 cm (24 in). The temperature range is about one-half that of the Great Smoky Mountains. Differences in elevation are modest (from sea level to 400 m or 1300 ft). Despite its uniformity with respect to climate and elevation, the ways that the land has been used by people who settled and farmed it over the last 4000 years reflect strong but highly localized environmental contrasts.

Soil conditions vary from place to place in four respects. First, soils differ substantially in *depth*. The vital topsoil layer, with its high organic content, is rarely more than a ploughshare's depth over most of the area. But it tends to be thinner on hilltops and ridges (where bare rock outcrops also occur), and at its deepest on the lower slopes and in the valley bottoms. Second, the soils vary greatly in *texture*. Small hilltop areas consist of coarse pebbles and gravels, and the dune areas along the coast are made up of coarse wind-blown sands. At the other extreme are lowland areas whose soil is largely composed of very fine clay-sized particles. The coarse soils do not hold enough water for plants to root, and the clay soils absorb too much water,

(a)

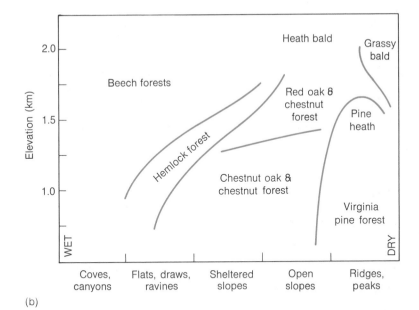

Figure 3-16. Local variations in environmental quality. (a) The apparently uniform timber cover of the Great Smoky Mountains in the southeastern United States is actually broken up into a complex mosaic of vegetation related to local differences in the height of ridges and in their exposure. (See Figure 3-17.) (b) Boundaries between vegetation types are related to two environmental factors: elevation and exposure. In the diagram, different types of terrain, from protected and wet sites to open and dry sites, are arranged in sequence along an exposure axis. [Photograph (a) by Rotkin, P. F. I. (b) after R. H. Whittaker, *Ecology Monographs* **26** (1956), p. 11, Fig. 5. Copyright © 1956 by Duke University Press.]

(b)

becoming like butter after heavy rains and drying out in hot, dry weather to a bricklike consistency. Between these two extremes lie the best soils, made up of loamy brown earths, mixtures of sand, silt, and clay.

Soils in the lowlands also differ in their *chemical make-up*. Those on the limestone rocks have too much calcium for most crops; conversely, some of the lowland clay soils are too acid and must be treated

Terms used in the study of soils

A-horizons are the upper layers of a soil, often rich in organic material and subject to leaching as moisture seeps downwards.

B-horizons are the next layers of a soil below the A-horizon where some of the chemical elements (notably iron) leached from the top soil accumulate.

Brown earths are rich brown soils formed in the middle latitudes where the prevailing natural vegetation is deciduous woodland.

C-horizons are the lowest layer immediately below the B horizon. They are made up of decomposed rock from which soil has not yet begun to form.

Catenas are sequences of soils which vary with relief and drainage though normally derived from the same parent materials.

Chernozems are rich, black soils formed in middle-latitudes where the prevailing natural vegetation is grassland.

Gley soils are waterlogged soils.

Horizons are the main layers or strata within a soil.

Laterites are red-colored soils formed in tropical regions. They are heavily leached and consist largely of aluminum and iron oxides.

Leaching is the removal of soluble chemical compounds from the upper layers of a soil by moisture seeping downwards.

Pedalfers is a general term for soils formed in humid regions where some compounds (notably calciums) have been removed by leaching, leaving aluminum and iron as the main constituents.

Pedocals is a general term for soils formed in dry areas where there is little leaching and the soils remain rich in calcium carbonates.

Pedology is the scientific study of soils, including their characteristics, their origins, and their use.

Podzols are the soils formed under cool, moist conditions where the natural vegetation is coniferous forest or heath. They are poor and very heavily leached where the soils are sandy.

Figure 3-17. Local variations in environmental quality: the North Somerset Plain. An aerial view (a) of this part of southwest England shows a sharp contrast between the flat, waterlogged valleys like those of the rivers Brue and Axe and the steep-sloping hills of the Mendip uplands. These differences in terrain (b) are also reflected in the detailed pattern of soils (c). Village settlements were established here by the late Saxon period (about A.D. 700), being located on the lower slopes of the hills so as to gain the advantage of both types of soils. Large areas of the lowlands were reclaimed from their original marshland state about four centuries ago, for, with good drainage, their potential agricultural productivity is high. [Photo by Aerofilms. Maps from D. C. Finlay, *Soils of the Mendip District of Somerset* (Soil Survey of England and Wales, Harpenden, 1965), p. 4, Fig. 3; p. 38, Fig. 8.]

(a)

* * * *

(b) TERRAIN

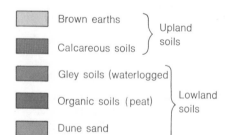

Direction of photographic view in (a).

Upland areas with slopes indicated by hatched lines.

Lowland areas which are flat are subject to flooding.

(c) SOILS

Brown earths ⎱
Calcareous soils ⎰ Upland soils

Gley soils (waterlogged ⎫
Organic soils (peat) ⎬ Lowland soils
Dune sand ⎭

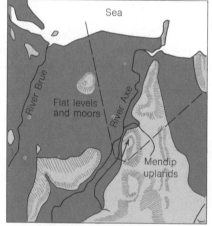

(b) TERRAIN

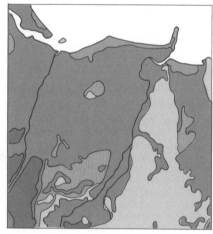

(c) SOILS

with lime. Small areas are lacking in a single trace element like boron, which may lead to deficiencies in pasture land. There are also differences in soils related to *drainage*. The productivity of fertile peat soils in low-lying river basins is considerably reduced by a water table too high for crops to grow.

Figure 3-17(c) shows the six principal soil types found in the lowlands. The overall pattern reflects the large landform division evident in Figure 3-17(b). Local variations in soils may also be related to different drainage conditions on the upper and lower slopes of a valley side. These local variations, illustrated in Figure 3-18, are usually referred to as *catenary* sequences (from the Latin *catena*, or chain). Water percolating downward through the soil removes soluble elements (an action termed *leaching*) and produces leached upper levels or horizons in the top few centimeters of the soil; these soluble elements may accumulate in the lower levels of the soil as hard layers. Waterlogged soils on the lower slopes are called *gley* soils. Organic soils develop by peat and bog formation on both the level uplands and the lowlands. Soils have a complex terminology and you may find it useful to check through the list given in the margin before exploring the literature suggested in "One step further . . .".

In this chapter we have taken a measure of our planet's productivity and used it to look at variations in the geographic environment. We have illustrated how the intricate and complex spatial pattern of our environment can be broken down into a series of patterns of different sizes. Within each fragment a different set of local environmental factors comes into play.

In raising the fundamental questions about the geography of the earth that we have in this chapter, we are entering the field of *physical geography*. This branch of geography analyzes the physical structure of our planetary environment—its land forms, climate, vegetation, soils, and so on. Partly because physical geography is closely linked

Figure 3-18. Soil catenas. This idealized cross profile of soils in the higher areas of Figure 3-18 shows changes associated with differences in elevation. The vertical depth of the soils is greatly exaggerated for clarity. Note that peat can form under waterlogged conditions in the river valleys (basin peat) and also on the higher parts of the uplands where rainfall is higher and where low slopes may impede drainage (hill peat). [After J. A. Taylor, *Geography* **45** (1960), p. 55, Fig. 1.]

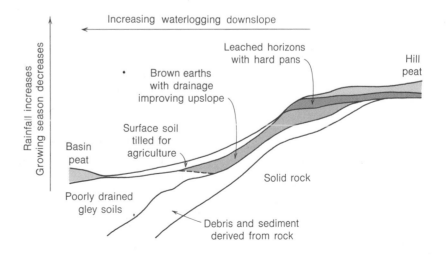

Increasing waterlogging downslope

Rainfall increases
Growing season decreases

Leached horizons
with hard pans

Hill
peat

Brown earths
with drainage
improving upslope

Basin
peat

Surface soil
tilled for
agriculture

Solid rock

Poorly drained
gley soils

Debris and sediment
derived from rock

to other natural sciences (geophysics, geology, meteorology, botany, etc.) and partly because of its longer history, it is currently one of the strongest and most developed branches of geography. (Cf. Section 22-2, especially Figure 22-9.) It has the most highly developed theoretical models, and its predictive capability has already reached levels that are unlikely to be matched in the rest of geography for some decades, if at all. Anyone attempting, therefore, to summarize its concepts within a few chapters must do so with a broad brush. Students who have already taken introductory earth-science courses or an elementary course in physical geography will wish to move quickly through the next two chapters or concentrate on the readings suggested in the section "One step further . . .". If the concepts discussed there are new to you, remember that they are the first steps in a field of geography you will be able to study in much greater depth in advanced courses.

Reflections

1. Consider Paterson's map of potential productivity (Figure 3-4). List (a) the advantages and (b) the disadvantages of measuring potential fertility in this way. What additional information would you like to have before deciding on the productivity of a particular environment?

2. What are the main wet and dry areas of the continent in which your country is located? Why are these areas located where they are?

3. What kind of climate would the United States have if it were shifted several thousand miles south, so that the equator passed through Washington, D.C.? Would the country be more or less productive, assuming that productivity is measured adequately by Paterson's index?

4. Look at some atlas maps of January and June temperatures for North America. Why do the thermal contours bend equatorward in winter and poleward in summer?

5. Use Figure 3-12 to trace what happens to rain falling on the earth. How long do you think it takes rain falling in your own locality to get back to the ocean? How might you figure this out?

6. How are the timberline limits on high mountains like those one sees as one moves poleward? List the ways in which the climates of poleward and high-mountain regions are (a) similar and (b) dissimilar.

7. Draw a cross section like that in Figure 3-3 for your own local region. Does the diagram show evidence of associated changes in (a) climate, (b) terrain, and (c) vegetation or land use?

8. Review your understanding of the following terms:
   (a) permafrost
   (b) atmosphere
   (c) lithosphere
   (d) hydrosphere

(e) troposphere
(f) tropopause
(g) westerlies
(h) trade winds
(i) intertropical convergence

(j) Hadley's model
(k) solar-radiation budget
(l) hydrologic cycle
(m) catenary sequences

One step further . . .

Descriptions of the globe's physical environment and the basic findings of the earth sciences are given in numerous sources. Among the best introductions to this subject are

> Strahler, A. N., *The Earth Sciences* (Harper & Row, New York, 1971), 2nd ed., and Earth Science Curriculum Project, *Investigating the Earth* (Houghton Mifflin, Boston, 1967).

Major zonal variations around the world and the climatic factors that lie behind them are discussed in an interesting way in

> Trewartha, G. T., *Earth's Problem Climates* (University of Wisconsin Press, Madison, Wis., 1961).

There is a long list of descriptive works about each of the major zones. (See Appendix C under "Tropic, Arid, and Polar Zones.") The interlinking of the major physical systems at the surface of the earth is described in

> Miller, D. H., *The Energy and Mass Budget at the Surface of the Earth* (Association of American Geographers, Washington, D.C., 1968).
> Manners, I. R. and M. W. Mikesell, Eds., *Perspectives on Environment,* (Association of American Geographers, Washington, D.C., 1974).

Chapter 4

Environmental Risks and Uncertainties

The world's a scene of changes, and to be
Constant, in Nature were inconstancy.
—ABRAHAM COWLEY
Inconstancy (1647)

Once every few months the newspaper headlines stop concentrating on the usual round of politics and diplomacy to report a natural disaster. It may be a famine in Bihar, India, a cyclone in Bangladesh, an earthquake in Turkey, a flood in Missouri. Readers are jolted by the reminder that someone else's environment is not so house-trained as we assume ours to be, but the disaster rarely stays in the headlines for more than two days. It slips back into small type on inside pages, and finally drops out of the news altogether.

Such cataclysmic reminders of the instability of the environment are reinforced by smaller scale changes. The slow variations in the levels of land and sea, the silting of a reservoir, the cycle of the seasons —each underlines the constancy of change in the world around us. In occupying the earth, man was forced to learn to cope with the uncertainty of nature, with a multitude of possible risks and changes rather than a consistent opponent. In this chapter we view some examples of environmental instabilities and try to piece them into a broad pattern of change. We distinguish between the forces that lead to long-term environmental swings over millions of years and the sources of short-term rhythms repeated each year. Since not all change falls into these simple classes, we also include a short section on more mysterious middle-term changes.

4-1 LONG-TERM ENVIRONMENTAL SWINGS

The environmental patterns described in Chapter 3 are but a frame in a feature-length film. The earth is probably over 4.5 billion years old, and in that time all the major environmental boundaries—of land and sea, mountain and lowland, tropic and pole—have never ceased to change. Many of these changes occurred so far back in time that they are of interest only to geologists. But others, which have been taking place during the last million or so years, are of direct interest to us all. They continue, slowly but perceptibly, to transform our environment.

Some evidence of change

How do we know what environmental changes have been taking place since man's emergence on this planet? Dim memories of change have been passed down through the ages in oral traditions and written records. The account of the flood in *Genesis* may not be literally true, but there is no doubt from related evidence that major floods did occur in appropriate parts of Asia Minor in Biblical times. However, very precise evidence of climatic change, in the form of written records, is available only for a very short period. By including early occasional records of heavy storms or extended cold periods, we can construct long runs of meteorological observations for a few parts of Europe.

Figure 4-1. Climatic change. This graph of winter mean temperatures in central England from 1680 to 1960 serves as a historical record of climatic change. Temperatures have been averaged over ten-year periods to produce a smoother curve. (See the marginal discussion on page 104 for an explanation of averages.) [From G. Manley, *Archiv for Meteorologie, Geophysik und Bioklimatologie* **9** (1959).]

Figure 4-1 presents a remarkable 280-year record of winter temperatures for central England compiled by climatologist Gordon Manley. But even with a record of this length, finding a pattern is difficult. It is not unlike trying to make sense of changes in Wall Street share prices.

In practice, therefore, geographers find that they have to build up a picture of environmental change from a wide variety of indirect sources. Let us examine some of these sources. Early researchers were aware of significant climatic changes primarily from macroscopic organic remains. All sorts of evidence, from the excavation of elephant and rhinoceros skeletons on the edge of the tundra to the discovery of warm-water shells in cold-water streams, pointed to a substantial climatic change in the recent past. In the early nineteenth century, Victorian scientists reported the growing size of the Alpine glaciers in central Europe. They also described small streams meandering through great valleys that, judging by their cross sections, must once have carried much larger flows of water. Similarly, old beach levels were mapped many meters above existing lake levels.

Disputes occurred not over the degree of change but the order and time of changes. One of the most important sources of evidence on the order of climatic changes is provided by *pollen analysis*. As all hay fever sufferers know, plants that depend on the wind for pollination produce many thousands of microscopic pollen grains. (See Figure 4-2.) We can detect a changing climatic pattern from statistical analysis of the relative abundance of different types of grains preserved

Figure 4-2. Pollen types. The analysis of different types of pollen provides vegetational evidence of recent climatic shifts. The statistical frequency of pollen grains (magnified about 200 times in the diagram) preserved in peat deposits allows us to reconstruct the probable plant cover during different time periods. [From H. Vedal and J. Lange, *Traer og Buske* (Politikens Forlag, Copenhagen, 1958), p. 208.]

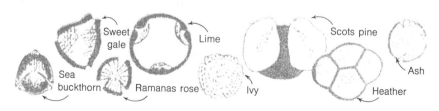

in lakes, peats, and muds. Table 4-1 summarizes the main sequence of climatic and vegetational conditions in western Europe, indicated by pollen analysis, since the end of the last major expansion of the ice caps. The present cool, rainy climate, which began about 400 B.C., is the ninth phase in a series of postglacial oscillations. About 5000 B.C., this part of Europe had a rather warm continental climate conducive to extensive pine and hazel forests.

Table 4-1 Main Postglacial Ecological Changes[a]

Period	Time (B.C.)	Climate	Dominant Cover (Main Species)
Sub-Atlantic	Since 400	Cool coastal	Woodland: beech and hornbeam Cultivation and clearings
Subboreal	2500–400	Continental (cold winters, warm summers)	Woodland: oak and ash Cultivation and clearings
Atlantic	5500–2500	Warm coastal	Woodland: oak and elm
Boreal	8000–5500	Warm continental	Woodland: pine and hazel
Later Dryas	9000–8000	Arctic	Tundra
Allerod	10,000–9000	Cool subarctic	Scrub: birch and aspen
Early Dryas	15,000–10,000	Arctic	Tundra: large barren areas

[a] General sequence typical of lowland areas of northwest Europe.

Pollen evidence of recent shifts in belts of vegetation has been studied mainly since the 1920s. It helps us to sort out the order of environmental changes, but it leaves unsolved the problem of exactly when they occurred. That we can give actual dates to within a few decades is due to the remarkable advances made in the early 1950s by physicist F. Willard Libby at the Institute for Nuclear Studies in the University of Chicago. By 1947, carbon-14, a radioactive form of carbon that loses half its radioactivity in the first 5750 years of its existence and half the remainder in each 5750 years that follow, had been discovered in nature. Its constant rate of decay enabled Libby to devise a method of dating organic material. This technique, called *carbon dating*, lets scientists correlate organic evidence of how the world looked in the past with other biological, geologic, or archaeological evidence. Carbon dating has proved to be extremely accurate for a period of 1000 years, but evidence from tree-ring counts from the very old Bristlecone pine suggests that corrections are needed for periods beyond 2000 years.

Carbon dating has now been supplemented by other radioactive

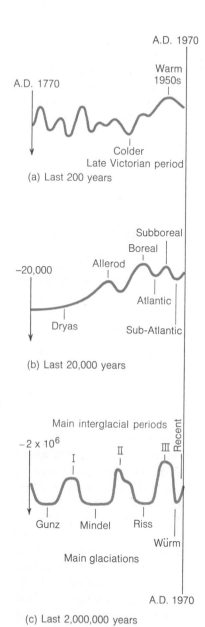

Figure 4-3. Continuities in climatic change. The graphs show the general pattern of temperature changes over three time periods: (a) the last 200 years, (b) the Recent period, and (c) the Pleistocene epoch. Note that each period is 100 times longer than the period immediately above it.

dating methods. Most new dating methods have been used to study the top few centimeters of deep-sea sediments because they hold the greatest promise as an archive of environmental change. Microscopic analysis of these sediments confirms the postglacial warming indicated by pollen analysis: a rise of about 8°C in mean ocean temperatures in the North Atlantic during the last 15,000 years, and an even greater warming (12°C) over a similar period in the Mediterranean. Most recent work is on the magnetic orientation of old sediments, which shows the position of the earth's magnetic field at the time they were laid down and allows us to date material as much as 20,000 years old.

Patterns of change: The Pleistocene epoch

What kind of environmental changes have these techniques revealed? In discussing this, it is helpful to use geologists' terms and confine ourselves to the Quaternary Period, the fourth and last of the four major geologic divisions of earth history. This period is conventionally divided into two epochs: the longer Pleistocene epoch of some 3.5 million years and the Recent epoch, covering the last 25,000 years. Man himself probably developed from his primate forebears sometime in the second half of the Pleistocene epoch. The human species emerged in a period that, from the standpoint of previous geologic periods, was one of intense environmental contrasts and rapid changes. Global differences in elevation, climate, and vegetation, for example, were sharper than they had been for the last 250 million years.

The earth's climate, which had been cooling slowly for the last 65 million years, grew much colder about 2 million years ago. The effect of this cooling was to lock up more of the world's water in the form of ice. In North America ice caps formed over central Canada and Labrador and spread as far south as Missouri and southern Illinois. In Europe ice caps formed over Scandinavia and advanced south into England and east almost to Moscow. In the southern hemisphere and in the tropical highlands the evidence of glaciers is less clear, but a large expansion of the ice fields is indicated. This expansion did not occur in a single surge, but entailed several slow advances and retreats, separated by mild and sometimes long *interglacial periods.* [See Figure 4-3(c).] In Europe there were four main periods of glacial expansion: the Günz, Mundel, Riss, and Würm glaciations. These correspond timewise to the Nebraskan, Kansan, Illinoian, and Wisconsinan glaciations in North America.

The impact of the ice sheets on the planet was threefold. The first effect we can predict from our knowledge of the earth's hydrologic cycle. (Refer to Figure 3-12 for an outline of this cycle.) As more water was stored as ice, the return flow to the oceans lessened and sea levels fell. At their largest, the ice caps lowered the ocean levels by approximately 100–125 m (328–410 ft). Although the vertical drop in water levels was relatively small, the horizontal effects were striking. The shallow continental shelf fringing the main land masses was exposed. For example, the shoreline of the northeast United States was extended by 100–200 km (62–124 mi). As a result, new routes between continents and islands were opened. Although the archaeological evidence on this point is conflicting, it seems probable that man's entry into the New World by way of a land bridge with East Asia (now the Bering Straits) was in this period.

A second environmental change produced by the ice sheets was the compression of broad climatic and vegetational belts toward the equator. The productivity zones mapped in Chapter 3 were sharpened and realigned. For example, the Sahara Desert in North Africa (zone F) may have moved as far south as latitude 10°–15°N, compressing the savannah and equatorial zones into a narrow area. Figure 4-4 shows

Figure 4-4. Changing desert margins. Ancient longitudinal dunes, now under cultivation in the savannah belt of northern Nigeria, were formed under desert conditions and provide evidence of major climatic shifts in the Pleistocene epoch. Note that the dunes form the parallel narrow strips; the large square patterns are simply the individual air photographs in the mosaic. [From A. T. Grove and A. Warren, *Geographical Journal* **134** (1968), p. 244, Plate 1. Crown copyright photos by Directorate of Overseas Surveys.]

the sand dunes of this Pleistocene desert, now covered with vegetation and lying well south of the present desert margin.

Expanding ice also bulldozed and realigned the river systems of much of North America and Europe. The Great Lakes system emerged and was shaped as a frieze of vast seas on the edge of the retreating ice field. Northern Canada and Finland have a landscape fretted with millions of small lakes and pools left among the jumbled debris at the base of the ice sheets. Debris, ranging in size from a few centimeters to over 50 m (164 ft), was scraped and gouged from one region to be smoothed and plastered on another. It reshaped the terrain of the northern halves of both Europe and North America.

Patterns of change: The Recent epoch

Almost 10,000 years ago (about 8000 B.C.), the latest of the poleward shifts of climatic zones began. The continental ice caps contracted, and the glacial tundra climates of lowland North America and Europe gave way slowly to the present middle-latitude climates. The general warming continued until about 6000 B.C., when the *Atlantic* climatic stage, characterized by temperatures 2.5°C higher than the temperatures we experience today, began. This stage lasted until about 3000 B.C. [See Figure 4-3(b).] From this climatic optimum there has been a general but irregular deterioration. The *Subboreal* stage (2500 B.C. to 400 B.C.) was colder than now, and sea levels were generally what they are today. The evidence of climatic changes in the last 2000 years is more detailed, and we can detect continuing swings of temperature. A low point in the record was reached in the northern hemisphere in the middle of the eighteenth century, and temperatures remained low into the nineteenth century. Whether the Recent epoch is a separate and warmer phase of earth history is a matter of dispute. The present warmer conditions may be merely a prolonged interglacial stage.

These later stages of retreating ice caps have encompassed two opposing phenomena: landward and seaward movements of shorelines. As the ice caps have melted, there has been a general worldwide rise in ocean levels of about 30 cm (1 ft) a century. This has resulted in a considerable net loss of land, particularly because the coastal plains that emerged during the glacial maxima (the most intense periods of glaciation when the greatest volume of water was backed up in ice) provided some of the most attractive sites for early human settlement and communication. The rate of loss is trivial, however. For example, human populations in the Gulf of Mexico have been driven inland only 15 km (9.3 mi) every 1000 years.

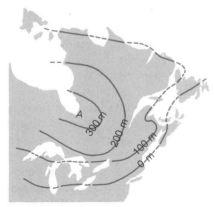

(a) Eastern North America

(b) Scandinavia

Figure 4-5. Changing beach levels. Contour lines show the height of old beach lines above the *present* sea level. They indicate how the melting of two huge ice caps (centered over points A and B) was followed by an upward warping of the land surface after the weight of ice had been released. Since the greatest weight was near the centers of the caps, it is here that the upward adjustment of the land surface has been greatest. [From B. W. Sparks, *Geomorphology* (Longmans, London, 1960), pp. 329, 332, Figs. 190, 192.]

Around centers of former ice caps, shorelines have moved seaward and land has risen. At their maximum, the centers of the Labrador and Scandinavian ice caps may have been up to 3 km (1.87 mi) thick. This enormous weight caused a compensating downward displacement of the earth's crust. As the ice has melted, so the crust has recovered, but slowly and haltingly. Today the land around the former ice cap areas (e.g., the Hudson Bay and Scandinavia) is still slowly rising, and the sea is still retreating, leaving lines of old marine beaches inland from the present coast. (See Figure 4-5.)

Why change?
One cannot be aware of the massive long-term changes in our global environment without wishing to know their cause. However, to answer all our questions on this subject would take us well outside the bounds of geography into the realm of sunspots, mountain-building cycles, and other geophysical events. (See "One step further . . ." at the end of the chapter.) We can, however, give two brief illustrations of the ways in which change continues to occur today. Both involve the notion of *cycles* and *succession*.

Cycles of erosion We noted in the last chapter how millions of km³ of water evaporates from the oceans, moves over the continents as water vapor, precipitates out, and returns to the seas as rivers flow and ice melts. (Refer to Figure 3-11 for the details of this transfer.) But the rivers do not return empty to the sea. On the average, the Mississippi River brings back 1 ton of sediment in every 1200 tons of water; in floods this figure may rise to 1 ton in 400 tons of water. We can see one dramatic result of this process in the delta, at the mouth of the Mississippi, which looks like a growing bird's foot. Much less dramatic, but equally inexorable, is the general wearing away of the land surface within the area drained by the river (i.e., within its *catchment*, or *watershed*). In one man's lifetime the effect is miniscule, perhaps 3.6 mm is worn away in a 70-year span. But over a million years this adds up to lowering of the land of over 51 m (167 ft).

We should be careful not to extend our calculations too far. The slow reduction of the continental surfaces by this return flow of the hydrologic cycle does not necessarily mean that the elevation of the continents is decreasing. There are two reasons for this. To understand the first, we must look more carefully at the earth's crust. This crust is made up of two layers. The first is an upper layer of granitelike rocks called *sial;* below this is a heavier layer of rocks, called *sima,* which extends under the continents and the ocean basins. Thus, the lighter continents appear to float on the heavier rocks of the earth's mantle. Slow reductions in the continental mass by erosion are compensated for by upward movements of the land surface. But the balance is not achieved instantaneously, and the stresses caused by these adjustments are among the many reasons for earth tremors and earthquakes.

A second reason the elevation of the continents tends to remain the same is that the erosional forces set in motion by the hydrologic cycle are self-limiting. High regions are more easily eroded than lower ones, so as the elevation of the land is reduced the rate of lowering also declines. The interaction among the processes of erosion, sedimentation, and compensating uplift can be seen as part of yet another basic environmental cycle (Figure 4-6). Physical geographers have been particularly interested in the erosion phases of the cycle. The Davisian school, named after the American geomorphologist W. M. Davis, stressed the role of declining rates of erosion over time and developed a series of *geomorphic cycles,* each related to the shaping of land masses under different climatic conditions. Most erosion was found to be due to the action of water rather than glacial action or wind. (See Table 4-2.) Although these processes are, in the main, very slow, they can be punctuated by rapid changes that severely affect man. The shifting of the Chinese Hwang Ho River from an outlet north of Shantung in the years 1192 and 1938, and its reversal in 1852, resulted not only in many deaths but in long-term changes in patterns of settlement and land use. Similarly, compensating movements of the earth's crust may cause dramatic surface effects such as earthquakes and volcanic activity. (See Section 4-4.)

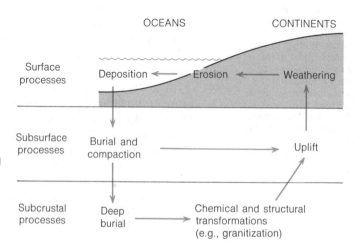

Figure 4-6. Erosion cycles. This idealized cross section shows major elements in the cycle of erosion, sedimentation, and uplift. Geographers are concerned with the surface phases of the cycle; geologists and geophysicists are mainly interested in the subsurface phases.

Table 4-2 Estimated transfers of mineral matter from continents to oceans

Transfers	Millions of Metric Tons per Year
Eroded from Continents	
By streams	9.3
By wind	0.06 to 0.36
By glaciers	0.1
Total	9.46 to 9.76
Deposited in Oceans	
Shallow waters (less than 3 km)	5 to 10
Deep waters (more than 3 km)	1.2
Total	6.2 to 11.2

SOURCE: Data from S. Judson, *American Scientist* **56**, No. 4 (1968), p. 371.

Vegetational successions A second type of slow change that is directly observable occurs among plants. When a field is first abandoned, it is bare. But it does not remain that way very long. Pioneer plants rapidly take root and flourish, establishing a simple community of weeds. Gradually, more brushy plants invade the community; a few quick-growing trees emerge, and some of the earlier plants are forced out. Over the decades a mature woodland may develop, but slow changes in the plant community continue. In the very long run (which may be hundreds or thousands of years), an equilibrium state may be achieved in which no further change occurs unless it is externally induced by a climatic change. This vegetational equilibrium is termed a *plant climax*. The stages by which plants colonize an area and replace one another is termed a *plant succession*.

Frequently, the very long-term changes of the erosional cycle and the shorter term changes in vegetation may be intertwined. This is very clearly shown in the history of many lakes. Lakes may begin as very deep bodies of clear water with few plant nutrients. As sediments are carried into the lake by inflowing rivers, the water becomes chemically enriched and the depth of the lake lessens. Plant productivity may build up in the lake itself, and vegetation may begin to encroach on the lake margin, with rooted plants near the shore acting as a sediment trap. Mosses and sedges build up, and great floating rafts of vegetation extend into the lake itself. (See Figure 4-7.) In the final stages of the lake's history, it may slowly fill with vegetation and sediment until eventually a forest is established and land plants take over from aquatic ones. Figure 4-8 shows just such a generalized succession of vegetation.

Although many environmental changes are very slow, we can clearly see them in operation in the contemporary world if we look

Figure 4-7. Plant succession in a lake environment. Lakes are often good examples of the intertwining of long-term erosional changes and short-term vegetational changes. [Beckwith Studios.]

carefully enough. We now turn to a different set of phenomena, to the rapid, short-term changes that demand our attention and of which we can never be unaware.

4-2 | **SHORT-TERM ROUNDABOUTS**

Cambridge University philosopher Ludwig Wittgenstein liked to illustrate the motions of the earth by spinning himself around, while at the same time circling around one of his friends. In the meantime, the friend was supposed to walk, following a leisurely, curving path, across the lawn. His biographers don't record how long this giddy game was kept up, but it would be as relevant today.

The earth has three motions. First, it moves with the sun as it orbits the center of the Milky Way once every 200 million years. Second, it travels around the sun once every 365.26 days. Third, it spins like a top around its own axis once every 23.94 hours. The planet's motion along the solar orbit is largely of astronomical interest, but its second and third motions are of direct and vital significance to man.

STAGE I

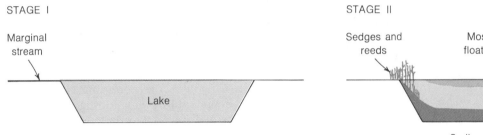

Marginal stream

Lake

STAGE II

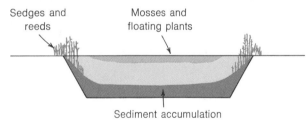

Sedges and reeds

Mosses and floating plants

Sediment accumulation

STAGE III

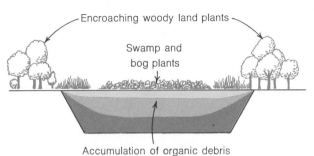

Encroaching woody land plants

Swamp and bog plants

Accumulation of organic debris

STAGE IV

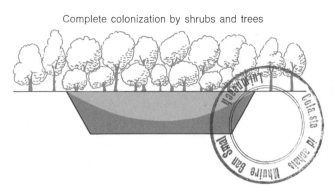

Complete colonization by shrubs and trees

Figure 4-8. Plant succession as a cause of environmental change. Pictured are four stages in an idealized sequence of plant succession in a lake environment. These stages are not of equal length: Commonly, the later phases of a succession are slower than the earlier ones. In a geological sense, all lakes are ephemeral and have a finite life span.

The daily round

The regular sequence of darkness and light that accompanies the daily rotation of the earth is so familiar that we ignore it. Yet man has evolved biologically in phase with this regular cycle, and his heartbeats, his blood pressure, his urine flow, even his sexual awareness, all have a distinct daily rhythm. In Chapter 13 we go on to look at the basic rhythms of human communal activity and find that all our settlements are adapted to this same 24-hour beat.

From a strictly environmental viewpoint, the main effect of the earth's turning away from the sun is to cut off darkened areas from massive inputs of solar energy. Thus the nighttime is a period of energy loss by radiation from the land surface and falling temperatures. From dawn onwards, the average amount of incoming solar energy increases; it reaches a noontime peak, then declines again as evening approaches. Average air temperatures follow the same pattern, but peak in the early afternoon.

On some warm summer days we can observe this daily buildup and decline of temperatures by watching cycles of cloud formation. Figure 4-9, a Gemini V satellite photograph, shows the view looking south along Florida's Atlantic shoreline. Note especially (a) that the clouds consist of thousands of isolated cells, giving the whole expanse a mottled appearance, and (b) that the clouds stop at the ocean's edge and are apparently absent from the main lake areas within the peninsula. This cloud pattern illustrates the noontime stage in the development of *tower* clouds, deep, rapidly developing clouds with small bases but considerable vertical depth. Such clouds are formed from the cooling of vertical columns of moist air that develop as the land surface heats up rapidly on a summer day. As some of the vertical columns grow stronger, larger and taller clouds predominate, some giving heavy rain showers. As the land cools toward evening, the tower clouds flatten and decay, so that the sky is mostly clear by nightfall. At dawn a new cycle of cloud formation will begin. Clear areas between the clouds over land are related to countervailing downward movements of air that compensate for the rising cloud columns. Clear areas over the sea and lakes are related to the different rates of warming of the land and sea.

Seasonal rhythms in temperature

Night and day are related to the rotation of the earth; winter and summer are related to the earth's revolution around the sun. Figure 4-10 shows that the planet's path around the sun lies on an imaginary flat surface (termed the *orbital plane*) that cuts through the sun. The earth's axis, around which it spins (shown as a line connecting the two poles), does not stick up vertically into the orbital plane but is tilted at an angle of 23½°. It is this tilt, together with the earth's motion around the sun, that produces our seasons. In late December the northern hemisphere is tilting away from the sun, so on each revolution it receives solar radiation for less than half a day. On December 21 and 22 the sun is vertically overhead at a latitude of 23½°S (the Tropic of Capricorn), but locations north of the Arctic Circle (latitude 66½°N, i.e., 90° − 23½°) receive no direct sunlight. If you follow the diagram around, you can work out the seasonal alteration through the northern spring, summer, and autumn, and trace the reverse patterns in the southern hemisphere. Thus the traditional spring months of March, April, and May in the northern hemisphere herald colder weather in the southern hemisphere and form its autumn.

Outside the tropics the essential characteristic of the seasonal cycle is a swing in temperatures. Thus for crops winter is the dormant period, spring that of sowing and germination, summer that of growth and maturity, and autumn that of harvest. As Figure 4-11 shows, this familiar cycle is related to changes in solar radiation. How much radia-

Figure 4-9. Diurnal cycles of environmental change. This view looking south over Florida's Atlantic coast with Cape Kennedy in mid-photo was taken from Gemini V. The mottling over the land occurs as tower clouds form in response to rising warm-air currents. When the temperature falls and these thermal currents die away, the clouds will dissipate. Note that clouds are absent both from the sea and from the main lake areas. [Photograph courtesy of NASA.]

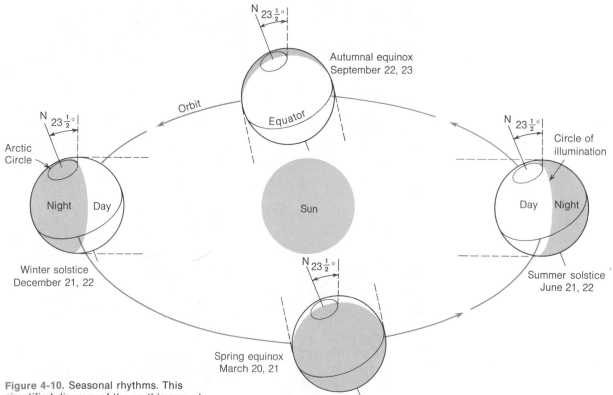

Figure 4-10. Seasonal rhythms. This simplified diagram of the earth's annual orbit around the sun shows how the fixed orientation of the earth's axis in relation to its orbital plane gives rise to the familiar sequence of seasons. [From A. N. Strahler, *The Earth Sciences,* 2nd ed. (Harper & Row, New York, 1971), p. 50, Fig. 4-1.]

tion is received at any point on the earth's atmospheric surface is related to its location in terms of latitude. Note the shifting position of the latitude where the sun is vertically overhead in the diagram. The amount of radiation received is not greatest at this latitude because, in summer, areas in higher latitudes have a longer day (i.e., more hours when they fall within the illuminated area shown in Figure 4-10). The annual flux of temperatures over the earth's surface lags behind that of solar radiation. As Figure 4-11(b) shows, this lag is caused by air temperatures that continue to rise so long as the incoming energy from solar radiation exceeds the energy reradiated from the earth. Thus, New York has its highest average air temperatures not late in June but in the middle of August. Offshore, the lag in sea temperatures is still longer.

Seasonal water deficits

The regular swings of peak solar radiation cause a continuous north-to-south, south-to-north shift of the atmospheric circulatory system de-

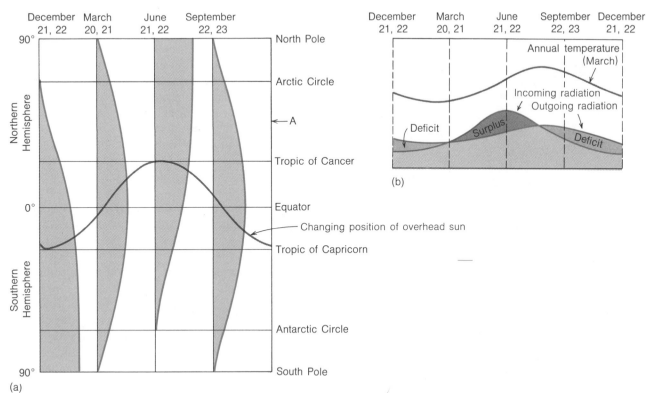

(a)

(b)

Figure 4-11. Seasonal variations in solar radiation. (a) Variations with latitude. Note the changing position of the sun and the variations in solar radiation received at different latitudes in the northern and southern hemispheres at different times of the year. (b) Seasonal variations for a single middle-latitude location in the northern hemisphere, A. Note the lagged pattern of temperatures — the late January minimum and the late July maximum.

scribed in Figure 3-9. In late June, the northern summer, the whole zonal sequence is shifted north by up to 20° of latitude. This shift brings subtropical high-pressure areas with dry, warm, descending air over areas like California and the Mediterranean, but sends the trade wind belt, with damp, unstable air, into northern Nigeria and Venezuela. By late December the system has been shifted 40° of latitude southward, bringing winter rainfall to California and a winter drought to northern Nigeria. We must therefore modify our general continental pattern of precipitation to include areas of seasonal deficits arranged symmetrically in the northern and southern hemispheres.

Because of its significance for vegetation and crop production, geographers are interested in measuring the seasonal variation in these environments. Let us consider the seasonal balance in Berkeley, California (Figure 4-12), as typical of the summer-deficit areas. Berkeley receives about 63 cm (25 in) of precipitation in an average year, over half of it, as Figure 4-12(a) shows, in the three winter months. If we compute the area's potential water loss from evaporation over the year, we find that it is only slightly more, 70 cm (about 28 in); hence, the moisture deficit appears to be 7 cm (close to 3 in). However, most of the evaporation loss [Figure 4-12(b)] comes in the hot, summer months when rainfall is at its lowest. Some of the precipitation that falls in winter can be stored in the soil as soil moisture, and this can be drawn on by a growing crop to compensate for a lack of rainfall. [The actual evaporation and transpiration curve for Berkeley is given in Figure 4-12(c).] Still, there is not enough water from this source for plant growth to reach its full potential. From April on growth is inhibited, until by August the monthly deficit reaches 5 cm (2 in). These monthly deficits add to a yearly sum of 18 cm (7 in). This large moisture deficit is caused not by a lack of rainfall, but by its seasonal concentration.

Information on the moisture deficit of a particular environment is useful in assessing its irrigation needs. If we extend our calculations, we can show other significant deficit areas over the western part of the United States.

In tropical latitudes more complex patterns of seasonal moisture deficits are encountered. Annual swings of temperature are less significant, and the daily range in values is often greater than the seasonal range. Important seasonal changes are related more closely to periods of rain and drought than to variations in temperature. Rainy seasons are directly related to the weather at the zone of convergence between the trade winds of the two hemispheres as it swings

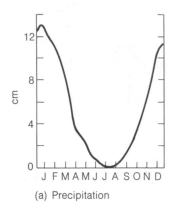

(a) Precipitation

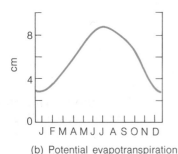

(b) Potential evapotranspiration

Figure 4-12. Seasonal moisture balances. This series of graphs shows monthly changes in the relationships between precipitation, potential evapotranspiration, and moisture stored in the soil for Berkeley, California. "Evapotranspiration" is the term used to describe the return of water to the atmosphere from the soil surface and from transpiration from plants. [After C. W. Thornthwaite and J. R. Mather, *The Moisture Balance* (Climatology Laboratory, Centerton, N.J., 1955).]

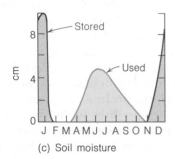

(c) Soil moisture

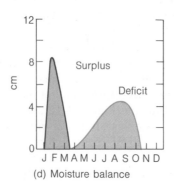

(d) Moisture balance

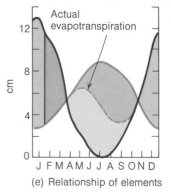

(e) Relationship of elements

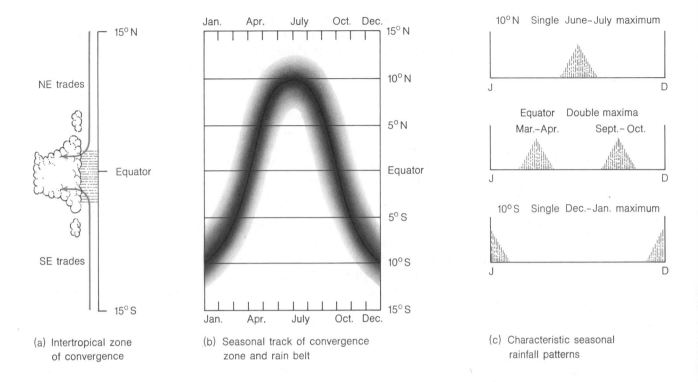

(a) Intertropical zone of convergence

(b) Seasonal track of convergence zone and rain belt

(c) Characteristic seasonal rainfall patterns

Figure 4-13. Seasonal patterns of precipitation in the tropics. Tropical rainy seasons (c) are associated with the rhythmic northward and southward oscillation (b) of the high-rainfall belt associated with the intertropical zone of convergence where the trade winds from the two hemispheres meet (a). However, the uneven distribution of land and sea in the tropical zone, as well as monsoon effects, tend to blur the simple seasonal patterns shown here.

first northward and then southward in its annual cycle. Figure 4-13 shows an ideal cycle of shifts with distinctive two-season peaks of precipitation at the equator, the first in March-April and a second in October-November. North and south of the equator the two peaks merge into a single rainy season.

In interpreting Figure 4-13, we should recall that it shows an ideal situation. In reality, irregularities of air flow may blur the picture and convert a regular seasonal cycle into a tragically uncertain pattern of precipitation.

4-3 | MID-TERM MYSTERIES

Long environmental swings like postglacial cooling are too remote to worry us; short daily and seasonal rhythms are so repetitive that man has learned to adapt to them. Even year-to-year fluctuations can be coped with if the surplus of one year can be carried over to the next. Irregular and uncertain fluctuations are what hit man hardest. But they are also one of the most difficult things for geographers to interpret.

The records are too short for statistical patterns to appear, and the link to physical theory too tenuous to provide reliable guides for forecasting. Here we shall illustrate the difficulties caused by irregular geographic phenomena by two regional examples.

The great plains: Midlatitude uncertainties

In the middle latitudes the boundaries between the major wind systems are continually shifting. Thus the polar front we described in Section 3-2 (see especially Figure 3-9) may wander considerably about its average location at about 60° N and 60° S. Fluctuations in the westerly circulation of the atmosphere in the middle latitudes are associated with waveforms that follow a four- to six-week cycle. The cycle begins with a zonal latitudinal flow in which waves of increasing amplitude build up to produce poleward and equatorward movements of air. Circulation then breaks these air movements into cellular patterns before the zonal flow is slowly reestablished (Figure 4-14). During the wave maximum [Figure 4-14(c)], strong incursions of freezing air from the north and warm, tropical air from the south may greatly distort "normal" climatic conditions. Past records of the world's climates reveal that these cycles are part of much larger swings which can last for several years. These cycles, in turn, are part of longer-term climatic shifts in the Recent period. Thus, the drought typical of arid zones may extend well outside the normal desert boundaries in one year, and precipitation characteristic of wetter regions may make incursions into an arid zone in the next year. A map of the world climate based on the meteorological records for 1975 would not be the same as one based on similar records for 1974, even if exactly the same classification criteria were used.

These shifts in climate become critical in areas where agriculture is carried on at the margins of humidity. For example, we saw in Chap-

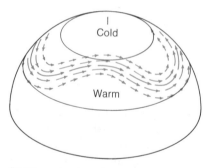

(a) The air stream begins to undulate

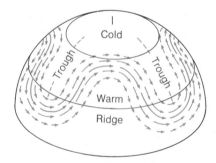

(b) Waves begin to form

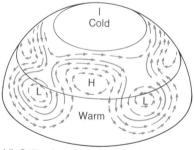

(c) Waves are strongly developed

(d) Cells of cold and warm air are formed

Figure 4-14. Changes in the westerlies. This six-week cycle of wave formation and dissipation in westerly air flow (Rossby waves) is one of the more regular of the short- and long-term fluctuations that lead to the characteristic instability of midlatitude climates. The general pattern of the westerlies was shown in Figure 3-6. [After J. Namias, from A. N. Strahler, *The Earth Sciences*, 2nd ed. (Harper & Row, New York, 1971), p. 247, Fig. 15-26.]

Figure 4-15. Climatic variations in the Great Plains. The maps show the stark contrasts between (a) the "normal" climatic pattern with that experienced in (b) the drought year of 1934. The second map shows the kind of conditions that were described so vividly in the opening chapters of John Steinbeck's *Grapes of Wrath*. [From C. W. Thornthwaite, in the U.S. Department of Agriculture's *Climate and Man* (Government Printing Office, Washington, D.C., 1941), Fig. 2, p. 182.]

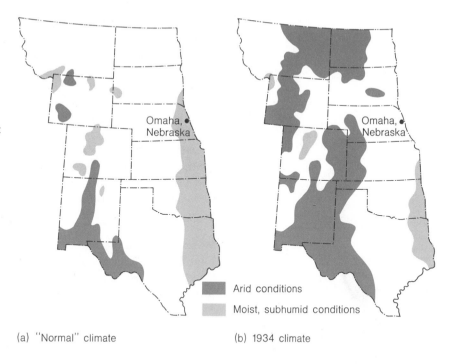

Arid conditions

Moist, subhumid conditions

(a) "Normal" climate

(b) 1934 climate

ter 3 (Figure 3-3) that rainfall in the Great Plains of North America decreases from around 125 cm (49 in) in the humid east to 25 cm (10 in) in the dry west. (See the map of the region's "normal" climate in Figure 4-15.) But rainfall may fluctuate not only from year to year but from decade to decade. The 1930s saw a disastrous run of dry years in the plains; conversely, the 1940s were generally wetter there than average. In the 1950s the regional pattern varied still further, with little rainfall in the southern plains but average conditions in the north.

What do we mean when we talk of "average" conditions? One way of interpreting records is to filter out small variations and leave only the principal swings. Figure 4-16 displays a 70-year record of precipitation for a Great Plains location, Omaha, Nebraska. Graphing the original yearly values produces an irregular pattern, but by selecting 5-, 10-, or 20-year means we can smooth it out to emphasize the general decline in rainfall over the period. Smoothing out data can be useful in checking trends; but, as investors in stocks often find, the curves that result are of limited value in predicting what will happen next. Even wholly random data can have deceptively plausible rhythms and trends. (See the marginal discussion of averages and trends for more comments on this subject.)

The problem of variability in rainfall in the Great Plains has its counterpart in the other midlatitude grasslands of the world—the South American Pampas, the South African Veld, the Australian Murray-Darling Plains, and so on. In humid regions the annual range of precipitation is small and poses few problems for agriculturalists; in the desert the drought is expected and plans are made accordingly. It is in semiarid areas like the Great Plains that settlers have frequently been fooled because these locations are sometimes desert, sometimes humid, and sometimes a hybrid of the two. Good years draw optimistic individuals into very marginal areas where a run of bad years has caused failure and tragedy in the past. A knowledge of the climatic environment's intrinsic variability may prevent future misfortunes.

Monsoon India: Tropical uncertainties

Figure 4-17(a) is a conventional rainfall map of the Indian subcontinent. It shows the average amount of rain that may be expected to fall in any one year and distinguishes between very wet areas (southwest

Averages and trends

We have already seen in Figure 4-15 how misleading maps of *average* environmental conditions can be. Their unreliability is directly related to the way averages are calculated. When we wish to determine the average of a distribution of values, we usually find the *arithmetic mean*. If we have a string of five rainfall values of 57 cm, 69 cm, 85 cm, 96 cm, and 116 cm, we obtain the average value by summing all the values and dividing the result by the number of observations. In this case the sum of the values is 423 cm and the number of observations is 5, so the arithmetic mean, or average, is 84.6 cm. This is a reasonably satisfactory result, as it is very close to the *median* of the distribution (the middle value in the series when the observations are ranked from lowest to highest) of 85 cm.

In studying Great Plains or monsoon rainfall we characteristically get less well-behaved sets of observations (i.e., sets in which there are some exceptionally high values due to exceptionally wet years). The same thing happens when we measure flood levels in rivers or, for that matter, the income of individuals. What happens when we try and use averages to describe these "skewed" distributions? We can illustrate the problem by going back to our rainfall records and making the last value much bigger—say, 196 cm

rather than 96 cm. If we now recalculate our mean, we find that it has risen to 104.6 cm (i.e., the new sum of 523 cm divided by 5). In this case we are less happy with the mean. Four of the five observed rainfall values lie *below* this average; 104.6 cm seems unrepresentative either of the four "normal" years or of the one "abnormal." Note, however, that the median of the distribution is unaffected by the change in the final value; it remains at 85 cm. Clearly, the arithmetic mean is poorly representative of the many natural events that tend to have a few very high values and a large number of low ones. In these conditions the median is probably a preferable proxy.

Another type of average, one we met in Figure 4-1, is the *moving average*. Moving averages are used in the study of environmental trends and may be calculated from either arithmetic means or medians, depending on the skewness of the observations.

Moving averages

Moving averages are a simple means of smoothing time series by adding the values at regular intervals over a period and dividing the result by the number of observations. If we have a set of yearly values for rainfalls (y), the 5-year moving average for a particular year (t) is

$$\frac{y_{t-2} + y_{t-1} + y_t + y_{t+1} + y_{t+2}}{5}$$

Thus, if our rainfall values in cm for the first seven years are 57, 69, 85, 96, 116, 141, and 124, the corresponding 5-year moving averages are −, −, 84.6, 101.4, 112.4, −, and −. Note that moving averages cannot be computed for the end values of the series. They can, however, be calculated for any length of time, depending on the length of the series available and the degree of smoothing required. As Figure 4-16 shows, the longer the period of the moving average, the greater the amount of smoothing. It is preferable to use an odd number of years in calculating this kind of average, so that the midpoint of the period to which the average refers will be an actual year. Moving averages can also be extended to two dimensions to smooth map series.

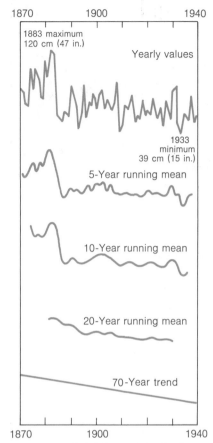

1870   1900   1940

1883 maximum
120 cm (47 in.)

Yearly values

1933
minimum
39 cm (15 in.)

5-Year running mean

10-Year running mean

20-Year running mean

70-Year trend

1870   1900   1940

Figure 4-16. Rainfall trends in the Great Plains. The graph shows the smoothing effect of running means on the record of annual precipitation in Omaha, Nebraska, from 1871 to 1940. [From E. E. Foster, *Rainfall and Runoff* (Macmillan, New York, 1949).]

India, the eastern Himalayas and Assam, and the Burma coast) and very dry ones such as the Thar Desert. Like all maps based on averages, it should be viewed with suspicion until we know how much variability is concealed by the averages. We can illustrate the difficulty of working with averages by looking at maps for the two months of January and July [Figures 4-17(b) and (c)]. The January map shows the situation during the *winter monsoon* period, when India is dominated by dry, colder air moving from the cold high-pressure cell over central Asia. (See the discussion of "continentality" in Section 3-3.) The June map shows the contrasting period at the height of the *summer monsoon*, with moist, warm tropical air moving from the Indian Ocean as southwesterly winds in response to the low-pressure cell developing over central Asia.

This regular seasonal reversal of wind directions and precipitation patterns lies at the heart of the Indian agricultural system. The rainy summer season from June to September provides some 90 percent of the annual water supply of the subcontinent and is especially critical for crops like rice, which depend on waterlogged conditions. The end of the dry season and the sudden "burst" of the summer monsoon is awaited with anxiety. Figure 4-17(d) shows the average dates for the onset of the monsoon rains. Ceylon in the south begins its wet season some two months later than the Indus Valley in the northwest.

The special anxiety over the monsoon relates to (a) its timing and (b) its character. Delays in its onset affect planting conditions, jeopardize irrigation regimes, and may—if followed by poor rains—lead to famine and millions of deaths through starvation. On the other hand, exceptionally heavy rainfalls may lead to flooding, wash seeds from the soil, cause landslips, and so on.

The intense variability of the Asian monsoon environment is illustrated by Figure 4-18, which shows 40 years of July rainfall records for Anuradhapura in the dry zone of Ceylon. (The location of Anuradhapura is in north-central Ceylon.) Note how misleading is the average July rainfall of 3 cm (1.9 in). In 15 years no rainfall at all was recorded in this month, while in one year nearly 20 cm (7.9 in) fell. The high values for a few years distort the average upward, so that the midpoint of the distribution, the median, is a better indicator of the probable rainfall in future Julys.

Both the examples we have chosen, the Great Plains and the Indian subcontinent, illustrate the puzzling nature of middle-term environmental changes. In both cases the causes of change are complex and are only now beginning to be unraveled. But the human reactions and consequences are clear. Both examples also underline the care needed in the interpretation of maps and help to explain why modern geog-

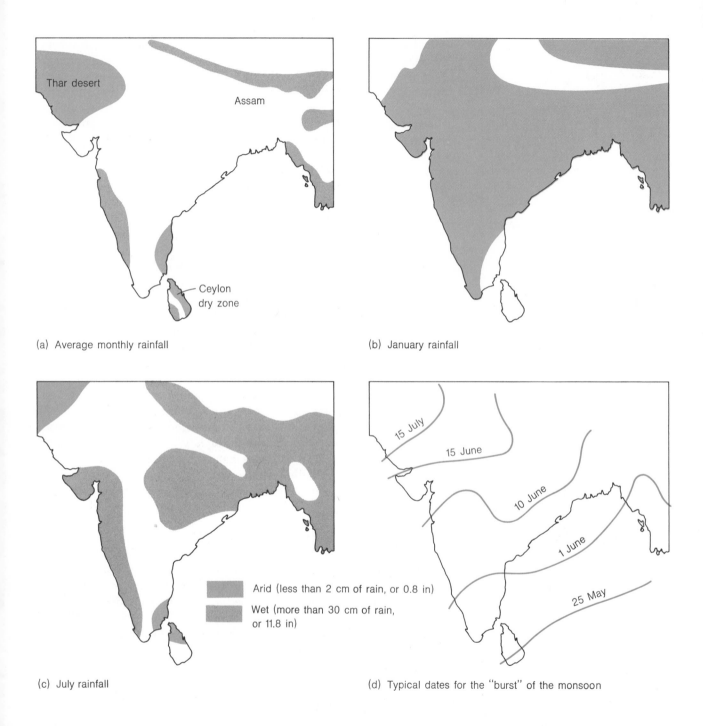

(a) Average monthly rainfall

(b) January rainfall

(c) July rainfall

Arid (less than 2 cm of rain, or 0.8 in)

Wet (more than 30 cm of rain, or 11.8 in)

(d) Typical dates for the "burst" of the monsoon

raphers are so interested in probability theory. In the kind of environments we have just described, a knowledge of the odds is the first step to wisdom.

4-4 | EXTREMES, HAZARDS, AND DISASTERS

Table 4-3 summarizes some of the losses incurred in the United States in this century from extreme geophysical events of one kind and another. In this final section of Chapter 4, we shall look at some of the hazards of sudden and extreme environmental change.

Table 4-3 Losses from natural hazards[a]

Hazard	Average No. of Lives Lost Per Year	Period for Which Data Are Given	Annual Property Damage (in billions)
Extremes of heat, cold, insulation	551	1955–1964	—
Hurricanes, tornadoes, etc.	304	1915–1964	$0.68–$1.05
Lightning strikes & fire	160	1953–1963	$0.10
Floods	70	1955–1964	$1.00
Tidal waves	18	1945–1964	$0.009
Earthquakes	3	1945–1964	$0.015

[a] Estimates of losses for selected geophysical hazards for the United States. Annual property damage is for the most recent year for which data are available; figures for hurricanes, tornadoes, etc., refer to insured losses only. Data from I. Burton, R. W. Kates, and G. F. White, *Natural Hazard Research*, Working Paper No. 1 (1968), Table 1.

Figure 4-17. The Indian monsoon. The map shows contrasts between (a) average monthly conditions, (b) dry-season conditions, and (c) wet-season conditions. The dates in the fourth map (d) are averages; major delays in the "burst" occur in some years.

Hurricanes

As far as hurricanes and tornadoes are concerned, the United States' losses are dwarfed by those of Asian countries. For example, on November 13, 1970, the greatest natural disaster of the century struck the low-lying delta areas of the Ganges and Brahmaputra rivers at the head of the Bay of Bengal. A tropical cyclone with a storm surge and winds of over 160 km (100 mi) per hour destroyed 235,000 houses and 265,000 head of cattle, and led to over 500,000 human deaths. Violent vortices of this type are termed *tropical cyclones,* or *hurricanes.* They form in moist tropical air between 5° and 15° from the equator and move poleward along characteristically sickle-shaped paths. (See Figure 4-19.) For example, in the North Atlantic region hurricanes form between Africa and South America and move west and north into the Caribbean, the Gulf of Mexico, and the offshore areas of the eastern United States before curving back toward the northeast. The most com-

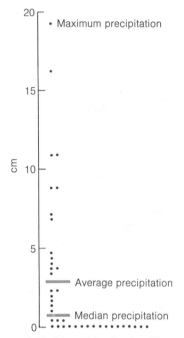

Figure 4-18. Monsoon climates. The graph shows year-to-year variations in July rainfall recorded at Anuradhapura in the dry zone of Ceylon between 1906 and 1945. Note that the average value seriously overestimates the rainfall likely to occur in a typical year. Each point indicates the July rainfall for one year. [After B. H. Farmer, in R. W. Steel and C. A. Fisher, Eds., *Geographical Essays on British Tropical Lands* (George Philip, London, 1958), p. 238, Fig. 4.]

monly affected areas in this region are the Caribbean islands, but Florida may also be hit. Very occasionally, a hurricane will not cross the land until as far north as New England. Considerable research is being conducted on ways of controlling hurricanes by seeding them at an early stage of development to trigger precipitation before they reach critical land areas. At present, however, we can only reduce the damage by more accurate tracking and forecasting of approaching storms.

Earthquakes

However difficult hurricanes are to forecast, they appear relatively predictable compared with the abrupt and cataclysmic environmental changes that occur with earthquakes and volcanoes. Although only one or two earthquakes causing much damage occur each year, about 150,000 earth tremors are detected every year. If we measure tremors by the area over which their effects were felt, then the largest in recent times was probably the Assam earthquake of 1897, which affected an area of 4.2 million km² (over 1.6 million mi², half the continental United States!). The largest earthquake in the United States was the San Francisco earthquake of 1906. In addition to the loss of life, damage by fire, and destruction of buildings caused by such events, there are also longer-term environmental changes resulting from the displacement of sediments and the redirection of river courses.

Although the specific timing of earthquakes is unpredictable and no effective countermeasures (other than appropriate building restrictions) are possible, the location of earthquake-prone areas is well established. Figure 4-20 shows that earthquakes occur principally in two elongated belts. The first passes around the Pacific Ocean and includes the Aleutian Islands, southern Alaska, and the Pacific coast of Canada and the United States. This circum-Pacific belt is estimated to account for some 80 percent of all the earthquake energy released on the planet. A second major belt runs from Portugal through the Mediterranean, the Middle East, and the Himalayas and meets the circum-Pacific belt in the Indonesian islands.

As more oceanographic exploration is completed, the presence of other earthquake belts associated with midoceanic ridges is being established. These areas form the boundaries between a series of great structural plates rather like the sutures of a human skull. Movements and adjustments in pressure appear to take place in these critical tension zones within the earth's crust.

Volcanic eruptions

More lasting environmental changes than those produced by earthquakes probably stem from volcanic activity. The largest such ex-

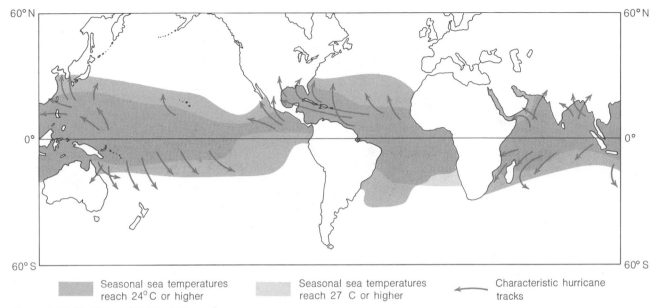

Seasonal sea temperatures reach 24°C or higher

Seasonal sea temperatures reach 27 C or higher

Characteristic hurricane tracks

Figure 4-19. Hurricane danger zones. Tropical cyclones (hurricanes) form in areas where the sea surface temperatures are high. Most originate in latitudes between 5° and 15° where the sea temperature in the hurricane season is 27°C (81°F) or higher. Characteristic tracks of hurricanes are shown on the map by arrows.

plosion in historic times was probably the Krakatau explosion of 1883, which blew away two-thirds of an island and triggered a tidal wave estimated at 45 meters in height (almost 150 feet) that broke upon the adjacent coast of Java with great destructive force. However, volcanic activity can also have beneficial effects. Slow accumulations of volcanic lava and dust may lead to the creation of entirely new land areas. The Hawaiian Islands, for example, were formed in this way. And while some ejected materials remain rocklike and sterile for centuries, others decompose into exceptionally fertile soils. Parts of Java, the Japanese island of Kyushu, and south India typify volcanic areas with fertile soil structures that support high populations.

The location of volcanic activity broadly follows that of earthquakes. The specific locations of areas of active volcanoes are superimposed on the earthquake zones in Figure 4-20.

Floods

Flood hazards occur in two distinct zones: coastal areas and areas bisected by rivers. *Coastal flooding* follows above-average sea levels caused by (a) unusual atmospheric conditions (e.g., the high seas created by onshore hurricane or tornado winds) or (b) earth tremors or volcanic eruptions that set up huge tidal surges. *River flooding*, a more frequent hazard, is related to heavy precipitation, rapidly melting snow, and—very rarely—the collapse of natural or manmade dams and the release of impounded waters.

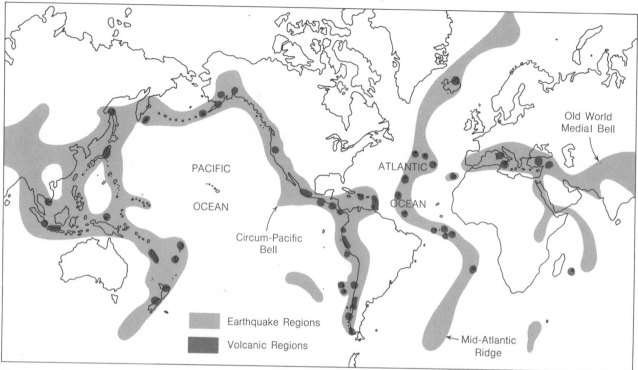

Figure 4-20. Earthquake-volcanic hazard zones. Schematic map of main areas of earthquakes and volcanic activity during the most recent geological period (i.e., that most relevant to man). Note the concentration into three main zones. The most important is the circum-Pacific zone or girdle which accounts for about 80 percent of all earthquake activity. The Old World medial belt running from the Mediterranean to Indonesia accounts for most of the remaining activity with the mid-Atlantic ridge forming a third and less active region.

In both coastal and river flooding the hazards are made greater by the attractiveness of such locations as places for human settlement. Some 12 percent of the United States' population elects to live in areas where there are periodic floods. The fact that flood losses have topped $1 billion in recent years (See Table 4-3) must be weighed against the advantages—fertility, flatness, etc.—that make areas near the water so attractive. The *floodplain* of a river is created by water spilling over the normal channel limits (often banks or levees) and depositing sediment over the surrounding plain. Under natural conditions, this floodplain, or "spillplain," will be covered by water for a small but rather regular number of days each year. Where there are human settlements, this natural overspill and sedimentation process is interrupted by building artificially high levees and protective dykes. This has the desired effect of keeping a stream within its main channel, but it means that additional debris is deposited on the stream bed, raising the height of the channel and creating a need for still higher artificial levees. Many of the world's major streams now run across their lower floodplains in artificially constrained channels some meters above the level of the surrounding densely settled land. When floods occur under these conditions, the depth of the flooding and its impact on the human population are immense. The effects are particularly severe in densely used floodplains.

The frequency of extremes

How reliable are the earth's environmental conditions? An agriculturalist may pose this question in terms of averages: How likely is a crop failure in any 10-year period? But to the settler on the coast or floodplain, the strength and frequency of extremes may be more important: What will be the highest flood or the strongest wind in any given period?

One way to answer both these questions is to try to calculate a *return*, or *recurrence, period*, the average interval within which one event of a specified size can be expected to occur. To do this, we first rank all observations of a particular phenomenon according to their magnitude, from the largest (1) to the smallest (n). The return period is then equal to $(n + 1)/r$, where r is the rank of a particular observation. Suppose we have a 49-year flood record which gives us the highest flood level each year. Then the average return period of the tenth-largest flood will be (49 + 1) divided by 10; a flood as large or larger should recur on average once every 5 years. Because the timing of floods is irregular, there may be several large floods in our 49-year record. One way around this problem is to plot the magnitude of an event against its return period on a graph. Thus we can average the recorded observations by drawing a straight line through them, as in Figure 4-21. This method allows us to estimate the most likely return ratios on the basis of all the records available. Figure 4-21 shows that we can expect

Figure 4-21. Predicting extreme values for natural hazards. The graph shows the frequency with which the greatest amount of rainfall in a 24-hour period reached particular levels in any one year during the period 1950-1956 in Nantucket, Massachusetts. Each year's record is shown by a point on the graph with the lowest value at lower left and the highest part at upper right. Note that the numbers on the horizontal axis of the graph are not regularly spaced. By drawing the axis in this way the sequence of years is found to fall along a line (labeled the "apparent line"). This apparent line can be approximated by a straight line (the "theoretical line"). Extending the straight line allows an estimate to be made of the greatest rainfall to be expected once in a century —about 14 cm (5.51 in.). Further extensions beyond the 100-year mark on the graph are possible but are likely to be of uncertain accuracy. [From D. M. Hershfield and M. A. Kohler, *Journal of Geophysical Research* **75** (1960), p. 1728, Fig. 1. Published by the American Geophysical Union.]

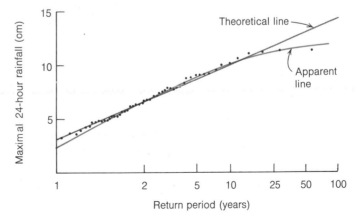

Nantucket to have a heavy rainfall of 10 cm/day once in a decade and a 14 cm/day rainfall once every century. Of course, these estimates are only averages. The rainfall that comes once every 1000 years may still fall next week!

This type of frequency analysis depends on rather simple assumptions. It provides a useful first approximation of the size of the risks a given environment is likely to pose. We assume, for example, that the pattern of floods or heavy rainfalls is not undergoing cyclic changes. If floods of a certain river are getting steadily worse, possibly because of deforestation, then this method will underestimate the size of a 100-year flood. Like rainfall, floods may tend to occur in clusters. The great floods on the Ohio and Mississippi rivers in 1936 and 1937 were paralleled three decades later by the disastrous floods of 1964 and 1965.

We began this chapter by looking at some of the massive swings in the environment that have occurred since man first began his tenure on this planet. We end by considering the immediate hazards and risks that may dominate tomorrow's headlines. Throughout, the emphasis has been on change, both the predictable kind that man can anticipate and turn to his advantage and the unpredictable kind against which he can only insure. To paraphrase Cowley's lines with which we began this chapter, "in nature, change is the only constant."

Reflections

1. Set up a debate on whether very slow and long-term environmental changes are (a) too remote to concern modern man or (b) give us an essential perspective on man's tenure on the planet Earth. Which side would you support? What arguments would you use?

2. Using newspapers and recent magazines, make a list of any extreme geophysical events (floods, earthquakes, eruptions, etc.) reported. Then look at the spatial distribution of these events. Is there any pattern to them? How would you explain such patterns?

3. Do you consider that you live in a "very safe" or "very unsafe" environment so far as natural hazards go? Explain your reasoning. Can you suggest any measures that individuals, corporations, or government might take to make conditions safer? Would you support these measures?

4. Consider what is happening in Figure 4-10. If the earth's axis were tilted further from the orbital plane (say $33\frac{1}{2}°$ rather than $23\frac{1}{2}°$), what effect would this have on the planet's climate? Where would the Tropics of Cancer and Capricorn and the two polar circles be? What changes would you expect in the climate of your own college town?

5. Collect some figures for monthly rainfall in your own locality and plot them

on a dispersion graph, using Figure 4-18 as a guide. Comment on the distribution of values. Is the mean a good description of the usual monthly rainfall?

6. Use an atlas map to look at the seasonal pattern of precipitation in the tropics between latitudes 10° N and 10° S. To what extent do the maps resemble the idealized model in Figure 4-13? How would you explain the discrepancies?

7. Review your understanding of the following terms:
   (a) pollen analysis   (f) plant succession
   (b) carbon dating   (g) seasonal moisture balance
   (c) Pleistocene epoch   (h) monsoon climate
   (d) Recent epoch   (i) hurricanes
   (e) erosion cycles   (j) earthquakes

One step further . . .

The long-term environmental swings in the later Pleistocene period are summarized in

    Strahler, A. N., *The Earth Sciences*, 2nd ed. (Harper & Row, New York, 1971), Chap. 41,

and treated in detail in

    Sawyer, J. S., Ed., *World Climate from 8000 to 0* B.C. (Royal Meteorological Society, London, 1966).

    Shapley, H., Ed., *Climatic Change: Evidence, Causes, and Effects* (Harvard University Press, Cambridge, Mass., 1954) and

How environmental variability is measured, its global patterns, and its effect on man's use of the earth are brought together in

    Chorley, R. J., Ed., *Water, Earth and Man* (Methuen, London, 1969), Chaps. 3, 5, 9, 10, 11, and

    Maunder, W. J., *The Value of Weather* (Methuen, London, 1970).

For a closeup of man's attempts to cope with a changeable geographic environment begin with a true classic,

    Webb, W. P., *The Great Plains* (Ginn College, Waltham, Mass., 1959),

and supplement this by a modern account of similar problems in an Australian context:

    Meinig, D. W., *On the Margins of the Good Earth* (Rand McNally, Chicago, 1963).

For a study of human reactions to environmental risks (a topic expanded in Chapter 20), see

    White, G. F., *Choice of Adjustment to Floods* (University of Chicago, Department of Geography, Research Paper No. 93, Chicago, 1964).

Research in the areas treated in this paper is summarized in the regular geographic journals. Research on climatic topics is summarized in *Weatherwise* and *Weather*, both monthly publications.

Chapter 5

Ecosystems and Environmental Regions

We will now discuss in a little more detail the
struggle for existence.

—CHARLES DARWIN
The Origin of Species (1859)

Ugly, sixteen-armed starfish up to two feet wide were first observed in large numbers at the resort of Green Island on the edge of Australia's Great Barrier Reef in 1963. Within a decade sightings of this *crown of thorns* starfish had been reported from Ceylon to Hawaii. On the small island of Guam the number of these creatures shot up from an estimated few hundred to more than 20,000 in three years. Since an adult starfish of this species lives on coral and can devour an area its own size in one day, the impact on the coral reefs was devastating. For Guam it meant the destruction of 24 miles of living reef. With the corals dead, the complex and dependent web of marine life broke down. The unprotected reefs were pounded and broken by constant wave attack.

Gloomy and pessimistic forecasts came from marine biologists as an ever-widening area of Indian and Pacific Ocean reefs were affected. However, by the early 1970s the outlook was more hopeful. Starfish populations were falling abruptly, and it now looked as if the crisis had been caused by one of the unexplained cyclic eruptions of animal numbers that affect many species. Thus in the summer of 1970 southern New England suffered a plague of leaf-eating caterpillars that stripped its maple trees, and in early fall of that same year unprecedented millions of Monarch butterflies could be seen pouring south from Canada en route to the Gulf states and Mexico.

It is unusual and headline-catching incidents of this nature that draw attention to the delicate balance between one animal species and another, and between each species and its environment. For the most part, we accept the web of life as a constant, reliable, and well-balanced system. In this chapter we look first at the insights into the links between animal populations and local environments provided by the "ecosystem" approach. Second, we extend the notion of the ecosystem to include a set of world *ecological regions*. Third, and finally, we draw together some of the threads of Chapters 3, 4, and 5 and look at the environment in terms of man's special needs and characteristics as an animal—and as a member of our global ecosystem.

5-1 | ECOSYSTEM CONCEPTS

The earth we have studied in the last two chapters is an *ecosystem*. That is, it contains an intricate and delicate network of cycles and feedbacks that have both nonliving elements (the atmosphere, hydrosphere, and lithosphere of Figure 3-2) and living elements. The term ecosystem is simply a contraction of the phrase "ecological system." It is used to describe a system that has much in common with the systems of energy and water circulation we have already encountered. The ecosystem differs from these systems in its deliberate inclusion of

Key terms in the study of ecosystems

Biomes are the major environmental zones of the earth marked by a distinctive plant cover (e.g., the subarctic tundra biome).

Carrying capacity is the largest number of a population that the environment of a particular area can carry or support.

Climax is the state of equilibrium reached by the vegetation of an area when it is left undisturbed for a long period of time.

Communities are groups of animals and plants that live in the same environment and depend on each other in some way.

Ecological efficiency measures the ability of organisms in a food chain to convert the energy received into living matter.

Ecology is the study of plants and animals in relation to their environment.

Ecosystems are ecological systems in which plants and animals are linked to their environment through a series of feedback loops.

Food chains describe the series of stages that energy goes through in the form of food within an ecosystem.

Food webs are complex networks of food chains.

Predator-prey relations describe the links between the population of one set of animals (the prey) that are hunted for food by another set (the predator).

Seres are the transitional stages in plant succession.

Succession describes the orderly sequence of change in the vegetation of an area over time as it passes through transition stages (seres) toward an equilibrium (or climax).

Trophic levels are the main stages in the food chain, green plants occupying the first level, plant-eaters the second, animal eaters the third.

living matter—including man himself. The word *ecology* dates back to 1868, when the German biologist Ernst Haeckel used it in discussing his studies of plants in relation to their environment. It stems from the simple Greek word *oikos*, meaning "a house" or "a place to live in" and serves as a direct link to the geographer's concern with the earth as the home of man. You will note close links between geography and ecology at many points throughout this book.

Ecosystems: A small-scale example

The easiest way to unravel the structure of an ecosystem is to take a small-scale, familiar example. We could stay on the beach and look at the reactions of plant and animal life to the twice-daily changes of environment in the tidal zone. A still better example can be found by moving inland to a small lake like that in Figure 5-1.

A lake is a body of standing fresh water. What physical inputs and outputs does it receive? As a participant in the hydrological cycle, it derives inputs of fresh water from stream inlets and from rainfall and loses water through stream outlets and evaporation. Its most important input is, however, the solar energy it receives from direct sunlight. This will warm the upper layers of the lake very strongly in summer and set up important vertical differences in water temperatures over the seasons.

In addition to the physical processes by which water flows, sediment is deposited, temperatures change, and so on, there are also far more complex biological processes going on in and about the lake. Sunlight furnishes energy used by microscopic green plants in the lake (the *phytoplankton*) to convert inert chemicals in the water into food. These organisms provide food for small larvae and crustaceans (the *zooplankton*), which are eaten by small fish. (See Figure 5-2.) These small fish are eaten by larger fish, which may eventually provide food for animals and for man himself. Plants and animals die and decay, releasing chemicals back into the lake waters. We show these links in the lake ecosystem in Figure 5-3(a). Of course, this is a highly simplified picture of a process that may involve hundreds of living species and very complex chemical chains. Figure 5-3(b) shows some of the inhabitants of a typical lake.

Not all ecosystems are as clearly defined as lakes. Many have boundaries which are hard to establish and which conceal very important internal variations. We shall look at some of the major world-wide ecosystems in Section 5-2, on "Environmental Regions." The simple lake ecosystem does, however, illustrate three elements which are critical in all ecosystems—from the smallest to the largest. These are (a) the cycling of chemicals (especially carbon) through biological

Figure 5-1. Lakes as environmental features. Outdubs Tarn (center) in the Lake District of northwest England is one example of the lake ecosystems discussed in this chapter. This small lake (about 50 m across) is gradually being filled in by sediment from the stream flowing into it (left) and by encroaching vegetation. Mats of reed and mosses are shown as a light ring around the dark waters of the lake itself; scrub timber (darker gray in tone) forms the outer ring, encroaching in turn on the reed-moss layer. The final infilling of this small lake is a slow process in human terms, perhaps a few centuries, but rapid in terms of geological time. Note also the outflowing stream draining to the much larger and deeper lake, Esthwaite Water (right). [Aerofilms photo.]

populations, (b) the linking of these biological populations into food chains, and (c) the impact of the populations' size of positive and negative feedbacks. We now look at each element in turn.

The carbon cycle

One essential element in the lake ecosystem is the conversion of solar energy into living matter. How is this piece of alchemy achieved? Let us use the *carbon cycle* to illustrate one of the most important aspects of this conversion process.

We have already seen that carbon is available in the earth's lower atmosphere as carbon dioxide (CO_2). This gas forms a small but vital 0.033 percent of the total volume of air. It is important climatically as a heat-absorbing blanket, helping to regulate air temperatures near the earth's surface. Biologically, carbon dioxide is essential to plant growth: Green plants with the pigment chlorophyll combine carbon dioxide and water through *photosynthesis* (from the Greek, "putting together with light") to produce all the food materials necessary for life. (See the marginal discussion of photosynthesis in the carbon cycle.) Photosynthesis is actually a cluster of interrelated chemical reactions activated by solar radiation at the wavelength of visible light. Green plants may be regarded as the basic *producers* in the carbon cycle

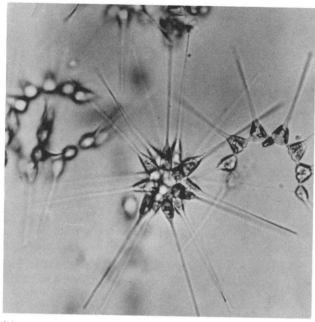

(a)

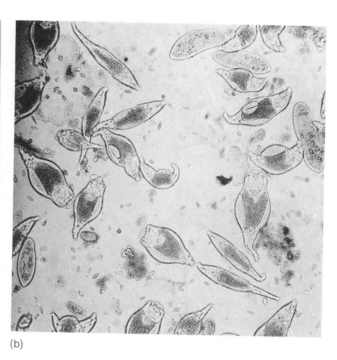

(b)

Figure 5-2. Ecosystem elements. The critical lower links in aquatic food chains are provided by (a) microscopic plant life (phytoplankton) and (b) animal life (zooplankton), magnified here to many times their actual size. [Photo (a) by Walter Dawn and (b) by Hugh Spencer, National Audubon Society.]

because they manufacture consumable energy (food in the form of carbohydrates) from atmospheric carbon and solar energy.

The carbon cycle is completed and carbon dioxide returned to the atmosphere by the processes summarized in Figure 5-4. Consider the food produced by land plants. These are eaten by animals (here termed *consumers*), and the energy stored as food sustains activity at high rates. Some of the carbon from carbohydrates is stored in the body, and the rest is excreted by respiration as carbon dioxide. Consumers can be divided into herbivores, carnivores, and omnivores, depending on whether they eat only plants, only animals, or a mixture of the two (as man does). The final role in the carbon cycle is played by *decomposers*. These are bacteria and fungi that break down the carbon stored in the tissues of dead plants and animals. In the decomposition process, carbon is again returned to the atmosphere or to soil water.

Not all producers and consumers are decomposed as soon as they die. Organic matter is stored and concentrated geologically for millions or billions of years as peat, lignite, coal, petroleum, and natural gas. (See Section 8-2 on fossil fuels.) Plants are also burned as fuel by man. Like eating, burning separates the elements in carbohydrates and returns carbon to the atmosphere as either carbon monoxide or carbon dioxide.

Photosynthesis in the carbon cycle

We can describe the overall process of photosynthesis in the carbon cycle by a simple chemical equation,

$$\text{Light} + n\,CO_2 + n\,H_2O \xrightarrow[\text{of chlorophyll}]{\text{in the presence}} (CH_2O)_n + n\,O_2$$

In other words, green plants extract carbon dioxide (CO_2) and water (H_2O) from their environment, return the oxygen (O_2) to the environment, and incorporate the remaining substances into carbohydrates (represented here by CH_2O). These carbohydrates are decomposed to provide energy or passed on to other parts of the food chain. (See Figure 5-4.) Rates of photosynthesis are critically related to the intensity of light. At low light intensities, the rate of photosynthesis is slower than the rate of plant respiration; respiration involves the oxidation of the carbohydrates and the release of carbon dioxide and water. At a slightly higher light intensity, the two rates are equal. Above this point, the rate of photosynthesis surpasses the plant respiration rate and carbohydrate products accumulate. The greatest, or saturation, rate of photosynthesis is reached in full sunligt. In addition to light, photosynthesis requires adequate moisture and proceeds most rapidly at temperatures between 10°C and 50°C (50°F and 122°F).

In any event, carbon from the atmosphere is circulated through a chain of living organisms to return eventually to the atmosphere. At each stage in this process it combines with different elements in various chemical forms, and each of these combinations is accompanied by energy transfers. Rearrangements of molecules and energy transfers (by photosynthesis in plants, and by metabolic synthesis in animals) are the essential processes that allow human life on the earth to continue. We have selected the carbon cycle, as an illustration of how energy transfers are accomplished, but we would need to sup-

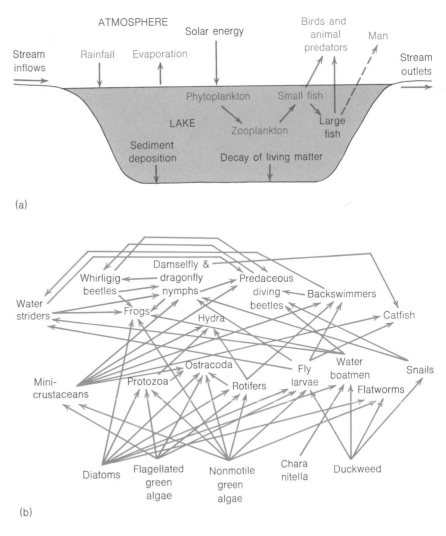

(a)

(b)

Figure 5-3. Lake ecosystems. (a) In this simplified flow diagram of the main links in a lake ecosystem, different colors are used to indicate physical and biological processes. (b) A more detailed diagram of a food web shows the many organisms that take part in just one segment of the system. Ecologists find that the more complex the food web is, the more stable the ecosystem is. [From W. B. Chapham, Jr., *Natural Ecosystems* (Macmillan, New York, 1973), p. 113, Fig. 4-7.]

Figure 5-4. Carbon cycles and the world carbon balance. The figures indicate the estimated stores (boxes) and annual flows (arrows) of carbon in units of 10^9 tons. *Carbonification* is the conversion of dead plant and animal remains into coal, oil, and similar fossil fuels. *Diffusion* refers here to the interchange of carbon dioxide gas between the atmosphere and the oceans by molecular mixing. (Note the different use of the term *diffusion* in Chapter 12.) The carbon cycle shown is only one of the major cycles of important chemical elements through the environment. Similar flows occur in the *nitrogen cycle*, the *phosphorous cycle,* and the *potassium cycle.* [After J. McHale, *The Ecological Context* (George Braziller, New York, and Studio Vista, London, 1971), p. 52, Fig. 21. Copyright © 1971 by J. McHale. Reprinted with permission.]

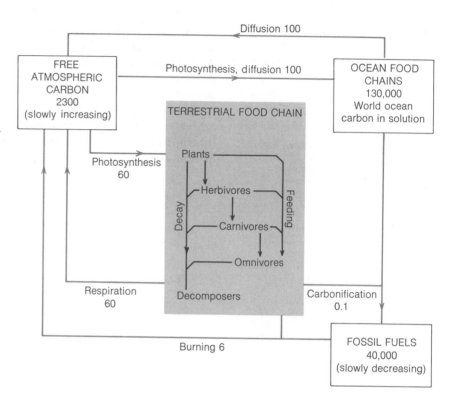

plement our description of it by a description of other cycles, like the nitrogen cycle, to fully explain the exchange processes involved. Each cycle represents an essential link in the ecosystem, for it includes biological elements (producers, consumers, and decomposers) as well as inorganic elements (e.g., carbon dioxide gas in the atmosphere and the carbons stored as fossil fuels).

Food chains

All animals get their food from plants, either directly or indirectly by feeding on other animals that feed on plants. Thus the process of photosynthesis and mineral cycles like the carbon cycle provide the basis of lengthy *food chains.* We have already encountered a simple example of a food chain, stretching from the millions of microscopic plants [the phytoplankton of Figure 5-2(a)] on a lake surface to human fishermen.

In the world's oceans, fish like tuna that are caught and consumed by man are directly dependent on a three- or four-link chain. Phytoplankton are consumed by larvae and shrimps, which are in turn eaten by squids and small fish, which form part of the food consumed by tuna. However, in each case it takes from 5 to 10 food units (calories)

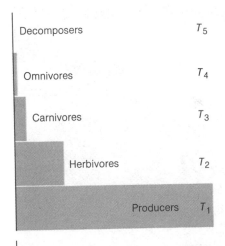

Mass of living materials per unit of area

Figure 5-5. Trophic levels. The pyramid shows the relative dry weights of living materials typical on the five main trophic levels of an ecosystem.

of the prey to produce one unit of the predator; this difference is termed a *food-conversion ratio*. One unit of tuna consumed by man represents an estimated 5000 units of phytoplankton.

It is useful to represent the levels in a food chain as a series of food pyramids, as in Figure 5-5. Each step of the pyramid is termed a *trophic* level (from the Greek *trophe*, food). The first level (T_1) at the base of the pyramid is composed of green vegetation with energy contained in the plant tissues. The second level (T_2) consists of herbivorous animals that feed on the plants; the third level (T_3), of carnivorous animals that feed on herbivorous animals; the fourth level (T_4), of carnivorous animals like man that feed on other carnivorous animals and all the lower tiers. The fifth and final level (T_5) is made up of decomposers that break down the dead tissues of organisms at all the other levels of the food chain.

Biologists have shown us the exact structure of trophic levels for individual communities. For example, they have analyzed the food chains and conversion ratios of over 200 species of fish in coral reefs in the Marshall Islands in the Pacific Ocean. By estimating the dry mass of organisms, from plankton and algae to sharks, they showed that the base of the pyramid (T_1) consisted of producers with a weight of 703 grams (g) per square meter. Above these organisms were herbivores (132 g) and finally carnivores (11 g). Other workers have tried to estimate the actual energy flows between the different species in a community.

Positive and negative feedbacks

In studying ecosystems, we have described relationships in terms of a series of links (e.g., links within the carbon cycle or links between one animal species and another in a food chain). Some of these links may build into a series of loops that feed back into themselves. We can illustrate these *feedback* loops by considering coral reefs (Figure 5-6).

Corals are minute marine animals that live in huge colonies in shallow tropical seas where their combined limy skeletons form reefs and atolls. They have attracted great interest from geographers since Charles Darwin's Pacific voyages in the *Beagle* in the 1830s, and they are still one of the most fascinating of marine ecosystems. Many of the organisms that secrete calcium carbonate to make coral reefs are sensitive to the depth of water. As the depth decreases, sunlight becomes more abundant and the rate of growth of the reefs increases. This accelerated growth further decreases the depth of the water, increases the light, accelerates the growth of algae, and so on, in a *positive feedback* relationship. In positive feedbacks the links in the loop amplify a change. However, the inability of the organisms that make up

Figure 5-6. The coral reef puzzle. Apollo 7 photo of the Tuamotu archipelago in the southern Pacific Ocean. *Atolls* are elliptically shaped reefs of coral enclosing a lagoon—those shown on this photo vary from between 15 and 40 km (10–25 mi) across their longest axis. The reefs are composed of coral limestone, the accumulated skeletons of corals and associated organisms. Exactly why corals form an atoll shape has intrigued scientists over the last century. Charles Darwin suggested that atolls began as reefs fringing an island which sank over a long geological time. Others thought that coral reefs could build up on their own debris and would form atolls by growing most actively on the outside of the reef. The balance between the building activity of the corals and the destructive energy of the surf breaking on the reefs is shown in Figure 5-7. [Photograph courtesy of NASA.]

Figure 5-7. Feedbacks in the coral reef system. The role of the sea level in regulating the growth of algae on a hypothetical coral reef is shown via a flow diagram. [From J. W. Harbaugh and G. Bonham-Carter, *Computer Simulation in Geology* (Wiley-Interscience, New York, 1970), p. 268, Fig. 7-4.]

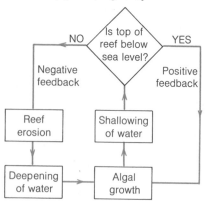

the reefs to grow above sea level and the breakup of the reefs by pounding waves introduces an effective *negative feedback* that limits growth. With negative feedbacks, the links in the loop dampen a change. Figure 5-7 presents a flow chart in which the elevation of the reef with respect to the sea level acts to regulate its growth, initiating either positive or negative feedback loops.

We can also illustrate the concept of feedbacks by going back to our initial example of an ecosystem—a lake. If we study a lake over several years and note the size of its different animal populations, we will find evidence of fluctuations. In some cases, like that of the number of nesting herons on the lake borders, the fluctuations may be quite small; in others, there may be wild fluctuations. Some types of algae may increase hugely in numbers, forming great colored "blooms" in the water which may last for some months. As we have already noted, biologists working on coral reefs notice similar eruptions in animal numbers.

What controls these fluctuations in animal numbers? Figure 5-8 shows an example of two kinds of feedback loops that may help to explain the phenomena. In both cases, we start with an initial breeding

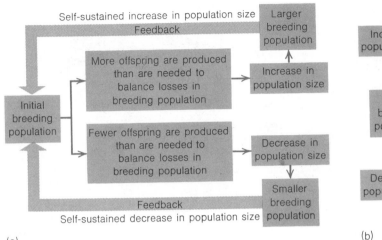

(a)

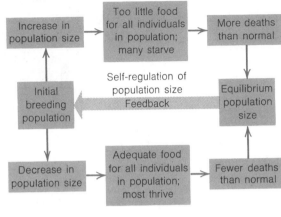

(b)

Figure 5-8. Feedbacks and population control. The diagrams show how feedback mechanisms regulate population size. In (a), positive feedbacks lead to changes that are self-sustaining. In (b), negative feedbacks lead to self-regulation. [After W. B. Clapham, Jr., *Natural Ecosystems* (Macmillan, New York, 1973), Fig. 1-4, p. 12.]

population. If you follow the arrows around the loops, you will see that the positive feedback loop leads to unstable outcomes. (The animal population either declines or explodes.) Here the critical factor is whether more breeders die each winter than are replaced each year. In the second diagram the situation is stable, and the food supply controls the number of deaths.

The ecosystems biologists have worked out in tracing the relations of plant and animal populations to their environment have some implications for human populations as well. In the next chapter, we shall see how man too fits into an ecosystem.

5-2 | ENVIRONMENTAL REGIONS

Geographers have long been intrigued with the possibility of reducing the infinite variety of the globe to a single, comprehensive scheme of regions. We look with some envy at the relative order the labeling of species brought to the previously jumbled world of botany and the periodic table brought to chemistry. Can any similar way be found to classify the mosaic of different environments and ecosystems geographers study?

Nine basic zones

In an earlier chapter we discussed the varying productivity of the global environment in terms of six zones. These ranged, like grades, from A to F. Areas in zone A were highly productive; areas in F had almost no potential. (See Figure 3-4.) We can now take this idea somewhat further by distinguishing three main types of biological environ-

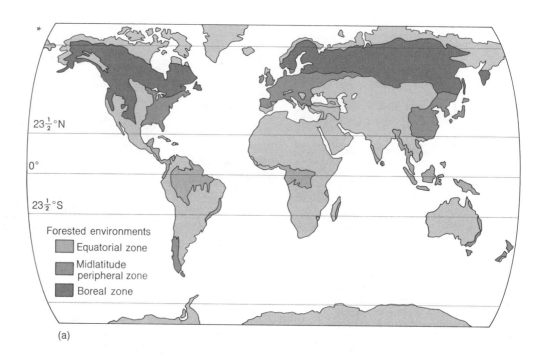

Forested environments
- Equatorial zone
- Midlatitude peripheral zone
- Boreal zone

(a)

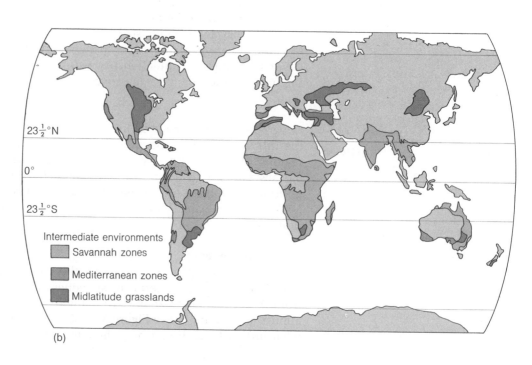

Intermediate environments
- Savannah zones
- Mediterranean zones
- Midlatitude grasslands

(b)

ments—forested, intermediate, and barren—and recognizing distinctive types within each of these broad classes.

The three maps in Figure 5-9 show the land areas of the globe divided into nine basic environmental zones, or *biomes*, ranging from the *polar zone* at high latitudes to the *equatorial zone* at low latitudes. Rather than review each zone's characteristics at length, we have summarized them in Table 5-1. You may find it useful to match up these zones with the productivity ratings given in the extreme right-hand column of the table and already encountered in Chapter 3.

The table indicates each zone's share of the total land area, but this is not necessarily an indication of its importance to man. For example, the *Mediterranean zone* (only 1 percent of the earth's land surface) has played a part in human development out of all proportion to its small size; whereas the largest zone, the *savannah zone* (24 percent of the land surface), has played a minor role.

Each of the nine regions is designed to relate in one simple scheme as many different physical conditions as possible. Thus, the savannah zone has the environment it has because of a mixture of climatic, vegetational, hydrologic, and soil factors. It lies within about 30° of the equator. Its largest single area is a horseshoe-shaped belt in Africa, about one-half of that continent, but it also covers much of South Asia.

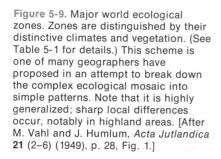

Figure 5-9. Major world ecological zones. Zones are distinguished by their distinctive climates and vegetation. (See Table 5-1 for details.) This scheme is one of many geographers have proposed in an attempt to break down the complex ecological mosaic into simple patterns. Note that it is highly generalized; sharp local differences occur, notably in highland areas. [After M. Vahl and J. Humlum, *Acta Jutlandica* **21** (2–6) (1949), p. 28, Fig. 1.]

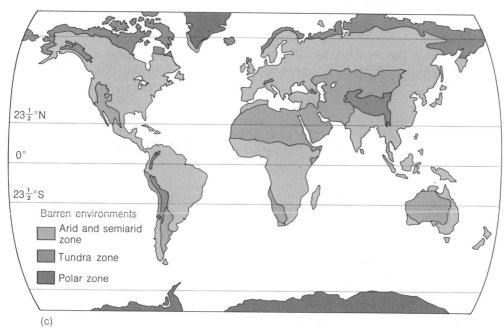

23½°N

0°

23½°S

Barren environments

Arid and semiarid zone

Tundra zone

Polar zone

(c)

Table 5-1 Major environmental zones

Type	Zone name	Percentage of land surfaces[a]	Main areas[b]	Dominant vegetation cover
Forested	Equatorial	8	Amazon Basin, S. America (51) Indonesia, S. Asian peninsulas (21) Congo Basin, Africa (19)	Natural broadleaved evergreen forest; wide variety of species; swamp forests on floodplains and coast
	Midlatitude peripheral	7	Europe (38) E. China (30) E. United States (26)	Broadleaved, deciduous, and mixed forests, merging with warm, temperate, evergreen forests on eastern periphery
	Boreal	14	Russia and Scandinavia (62) Canada, Alaska, N.W. United States (37)	Needleleaf forests; relatively uniform stands with small number of species (e.g., spruce, fir, pine, larch)
Intermediate	Savannah	24	African Tropics (48) S. America (21) S.E. Asia (20)	Ranges from open, tall-grass savannah to deciduous monsoon forest; gallery forest along stream systems
	Mediterranean	1	Lands around Mediterranean Sea (49) S. Australia (31)	Evergreen drought-resistant hardwoods and shrubs
	Midlatitude grasslands	9	C. Asia and E. Europe (42) C. North America (23) E. Australia (15)	Grasslands varying from tall-grass prairie to short-grass steppe with decreasing humidity
Barren	Arid and semiarid	21	C. Asia (42) Sahara and S.W. Asia (30) C. and W. Australia (10)	Widely dispersed, drought-resistant shrubs; salt flats, plantless sand, and rock deserts
	Tundra	5	N. Canada and Alaska (53) Russia and N. Scandinavia (42)	Low herbaceous plants, mosses, and lichens
	Polar	11	Antarctica (87) Arctic (13)	Ice caps; no plant life

[a] Computed on the basis of boundaries shown in Figure 5-9. They do not correspond exactly with the proportions given in Table 5-2, which were computed from boundaries drawn on a slightly different basis. Note that characteristics of all the zones are highly general; for examples of internal variations within zones, see the text.
[b] The figures in parentheses give the percentage of the total land surface of each zone in each area.

Main man-made changes in vegetation	Main precipitation characteristics	Main thermal characteristics	Productivity rating on Paterson's Map[c]
Very variable: ranges from extensive clearing and cultivation (e.g., in Java) to little impact (e.g., in Amazonia); low to very high population densities	High rainfall (over 100 cm, or 39 in) throughout year; heaviest at equinoxes	Uniformly high temperatures; little seasonal variations	A to B
Very extensive clearing and cultivation; medium-to-high population densities throughout	Moderate precipitation (75–100 cm, or 30–39 in) all year; heaviest in winter or autumn on western periphery, in summer maximum in eastern warm temperate area	Cool to warm temperate; seasonal range increases with continentality	C to D
Limited clearing on equatorward fringes; very low population densities	Light precipitation (25–50 cm, or 10–20 in) mainly in summer	Short, cool summers; very large annual temperature range	E
Burning, grazing, variable clearing, and cultivation; high population densities on flood plains in monsoon Asia, otherwise low	Variable rainfall (25–200 cm, or 10–79 in) with pronounced spring or summer maximum	Warm; small seasonal variations	B to E
Extensive clearing and cultivation, especially in lands around Mediterranean Sea; variable population density	Low to moderate rainfall (50–75 cm, or 20–30 in) with pronounced summer drought	Warm temperate; moderate annual range	D to E
Hunting and grazing; main settlement and cultivation in last 150 years; low population density	Low to moderate rainfall (30–60 cm, or 12–24 in), mainly in spring and summer; considerable year-to-year variation	Very strong seasonal variations; cold winters dominated by invasions of polar air	D to E
Little impact outside small irrigated areas	Very low rainfall (0–25 cm, or 10 in) with considerable year-to-year variation	Very high summer temperatures; seasonal variations range from moderate in tropics to very high in midlatitudes	F
Little impact	Low annual totals (10–40 cm, or 4–16 in) with late summer or autumn maximum; light winter snowfall	Severe cold; short, cool summers	F
No impact	Low annual totals; little detailed precipitation data	Extreme cold; no months with above-freezing average temperatures	F

[c] See Figure 3-4.

A look at a sample location (Timbo, in West Africa) reveals that while temperatures are high around the year, there are striking contrasts in precipitation. (See Figure 5-10.) There is a pronounced summer rainfall and a dry winter season. Annual rainfall totals range from 25 cm (10 in) to 200 cm (79 in) and vary moderately from year to year. In southern Asia, West Africa, and northern Australia the rainfall is associated with disturbances in the monsoon flow south of the equatorial low-pressure trough. Late summer hurricanes add significantly to rainfall totals in various parts of this zone.

The vegetation of the savannah zone is highly differentiated. It has heavy forests near its boundary with the equatorial zone and sparse shrubs and grasses near its arid border. Savannah vegetation consists of an open network of drought-resistant shrubs and trees with expanses of tall, coarse grasses. Variations on this theme include the thorn forest of East Africa and the dense, semideciduous jungle of Thailand and western Burma. Variations in the length and intensity of the rainy season relate both to the variety of vegetation and to soil and hydrologic conditions. Alternating precipitation regions lead to seasonal variations in soil characteristics and intense differences in the flow levels of rivers draining the savannah and monsoon areas.

Boundaries between zones

The zoning system outlined in Table 5-1 and mapped in Figure 5-9 is a useful capsule guide to the world environmental mosaic. Terms like "equatorial," "savannah," and "Mediterranean" serve as a way of summarizing the dominant characteristics of regional climates and soil-vegetation complexes. Unfortunately, there is not complete harmony on how many zones there should be or what criteria should be used to separate one zone from another.

We can see the reasons for this lack of harmony by looking more closely at another of the major zones, the *boreal* zone. Generalizing, we can say that the boreal zone is a climatically determined ecological unit covered (except in settled areas) by forests dominated by coniferous growth. The boundary of this zone is not clearly marked by any abrupt change in the environment. Coniferous forests are widely dispersed in regions outside the boreal zone, in the Mediterranean area and parts of Central America, for example, so a definition of the zone based on vegetation alone is difficult. The poleward limit is conventionally marked by the Arctic tree line. In practice, however, this "line" is really a belt where trees grow only in the most favorable sites; *muskeg* (waterlogged depressions filled with sphagnum moss) occupies the hollows while *tundra* (dwarf shrubs, herbs, lichens and mosses) occupy the more exposed ridges.

In the early part of this century, geographers Alexander Supan and

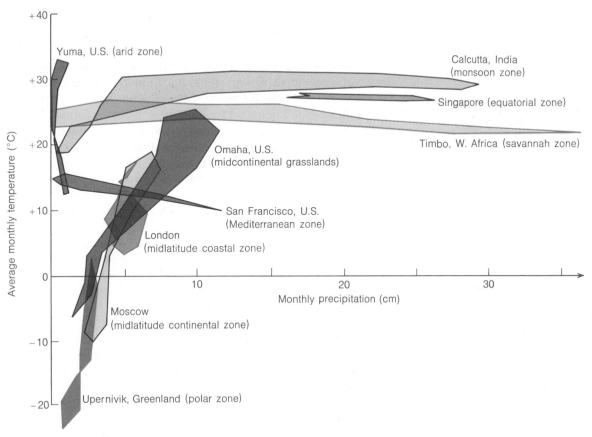

Figure 5-10. Climatic differences in ecological zones. Temperature and precipitation rhythms for representative locations in sample zones are shown by twelve-sided polygons, formed by joining the average temperature and rainfall points for each month of the year.

Vladimir Koppen equated this transitional belt with an isotherm of 10°C or 50°F for the mean daily temperature for the warmest month. Later geographers confirmed that the northern forest boundary follows essentially a thermal limit, though they found a growth threshold of 6°C or 42°F. Similar constraints apply to the equatorward limits of the zone, at least in the humid sections. For example, the southern boundary roughly approximates the line along which there is a mean daily temperature of 6°C in six months of the year. It is here that the broadleaved forests of the midlatitude zone begin. However, the boundary between the boreal zone and the midlatitude grasslands in Siberia and the Canadian prairie areas is controlled by the dryness of the environment rather than by its temperature.

Local contrasts within zones
Even if geographers were in complete agreement on how to draw zone boundaries, another difficulty would still remain. The zones we have

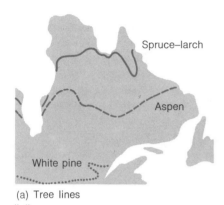

(a) Tree lines

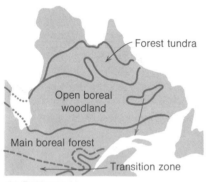

(b) Regions

Figure 5-11. Variations within major ecological zones. Tree lines correspond roughly with subdivisions of the boreal zone in eastern Canada. [After F. K. Hare, *Geographical Review* **40** (1950), p. 617, Fig. 4. Reprinted with permission.]

described give only a broad-brush picture of an immensely detailed mosaic of environments. Environmental variations occur on many scales, and these variations tend to break up and differentiate the major, subcontinental zones. For instance, within the boreal zone itself we can distinguish sharp contrasts. Canadian geographer Kenneth Hare recognizes three subzones within the boreal zone in the northern hemisphere. (See Figure 5-11.) The first subzone is formed by the close main boreal forest, where the crowns of the trees touch. This *closed-crown forest* occupies at least half the area of the zone, except in the driest sites. In the drier areas such as the Mackenzie Basin of Canada the species of trees found changes, grassy openings occur, and soil tends to be alkaline rather than acid (Section 3-4). The second subzone is the *woodland* subzone, where lichen breaks up the closed-crown forest. The open, almost savannahlike woodlands are sometimes called *taiga*, though this word is used by Russian authorities to designate the whole boreal zone. The third subzone is the *forest-tundra* subzone, a mixture of tundra on the drier ridges and woodlands in the valleys. This area is a classic example of an *ecotone*, or transitional belt, where two major environmental zones—the tundra and the boreal—interpenetrate and blend into each other.

The division of the global land surface into the nine major zones listed in Table 5-1 has three limitations. First, strong internal variations (related to elevation, geology, or groundwater) may occur within zones and boundaries between zones may be fuzzy. Second, the boundaries of zones are slowly but continuously changing because of long-term alterations in the climate of the present postglacial period. Third, the zonal vegetation described in Table 5-1 describes only the undisturbed plant cover that is either known to exist or assumed to be able to grow should human intervention in the ecosystem cease. In some zones, like the midlatitude woodland zone, little or no natural plant cover remains; in others, like the savannah zone, the exact role of man is difficult to assess.

Watersheds as alternative regional units

Because environmental variation occurs on many geographic scales, geographers have developed systems of regions that can be modified to fit any scale. One of the most versatile regional units created so far has been the *watershed*, or *catchment area*, of a stream. (See Figure 5-12.) Stream watersheds form a convenient unit because they can be simply and unambiguously defined from a topographic map. They are independent of scale in that large river basins like the Amazon can be broken into a hierarchic system of smaller basins; like a toy Polish doll within larger dolls, each smaller basin fits exactly within the

Figure 5-12. Watershed hierarchies. Pictured are three examples of the hierarchical breakdown of large watersheds into smaller units. Figure (a) shows the Arroyo de los Frijoles basin near Santa Fe, New Mexico. Figure (b) shows a small segment of this watershed, the Arroyo Caliente basin, in greater detail. Figure (c), a further enlargement of a single catchment, illustrates *stream ordering*. A hierarchy of stream segments can be ordered in several ways. One of the commonest ordering systems (Strahler ordering) designates the fingertip tributaries as order 1 channels. Order 2 channels are formed by the junction of two first-order channels; order 3 channels by the junction of two second-order channels; and so on. The watershed in (c) is therefore a third-order unit. [From L. B. Leopold *et al., Fluvial Processes in Geomorphology* (Freeman, San Francisco, 1964), p. 139, Fig. 5-4. Copyright © 1964.]

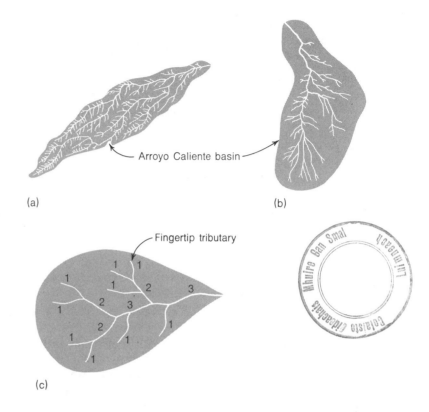

(a)

Arroyo Caliente basin

(b)

Fingertip tributary

(c)

next larger one. Each subdivided basin can be identified and numbered in a way which provides an alternative measure of its size. (See the marginal discussion of the nature of systems.)

Watersheds also have other advantages as a basis for regional divisions. Soil-related changes in vegetation reflect location within the watershed, since the physical features of a basin directly affect the hydrologic characteristics of the streams draining it. A rainstorm falling on a long, narrow basin is likely to produce a lower peak in the water level of the stream draining it than a similar storm falling on a rather broad, almost circular basin. Watersheds therefore form an appropriate unit for agencies interested in flood control, navigation, hydroelectric power production, or soil conservation.

The founding of the Tennessee Valley Authority in 1933 established a trend in using river basins as planning units that has spread around the world. Schemes for the São Francisco River in Brazil, the Snowy River in Southeast Australia, or the lower Mekong Delta all involved a combination of planning measures keyed to the use of water resources. Their adoption allowed the competing demands for water—

The nature of systems

Geographers analyze four types of systems:

1. Morphologic systems, in which relations between individual components are built up by statistical association to produce positive or negative bonds. Changes in the level of one component cause associated changes in other components. Such systems vary in the number of components they have, the strength of the links between them, and the arrangement of the links into positive or negative feedback loops.

2. Cascading systems, in which relations between individual components involve transfers of mass or energy. The *output* from one component becomes

the *input* for another. Inputs and outputs can be controlled by *regulators*. Feedbacks between components occur from the sequence of inputs and outputs, which may be lagged in time.

3. Process-response systems, a hybrid of the first two types of systems in which both statistical associations and input-output transfers form links. Such systems vary in their capacity for *self-regulation* and in the length of time they need to adjust to changes (*relaxation time*).

4. Control systems. Process-response systems can be modified by human intervention. This intervention may take the form of restricting the

levels of individual components or governing the flows of inputs and outputs.

Watersheds are examples of systems analyzed on all four levels. Morphologic relations between aspects of channels and slopes can be linked to the input-output relations of rainfall and streamflows to form a combined process-response system, which in turn is controlled to reduce flood hazards. [For a wealth of examples, see R. J. Chorley and B. A. Kennedy, *Physical Geography: A Systems Approach* (Prentice-Hall, Englewood Cliffs, N.J., 1971), Chap. 1.]

for irrigation, flood control, hydroelectric power production, and navigation—to be dealt with through a single controlling authority. Watershed units are employed in heavily urban areas as water pollution control problems worsen. For all these purposes, the watershed provides a convenient and natural spatial unit. Watershed research has also helped identify *terrain regions*. Geographers divide the physical landscape of an area into particular features and then recombine these features to create a *terrain type* typical of a given locality. For instance, Figure 5-13 shows the landscape around Yuma in southern Arizona. Using topographic maps, we can measure many terrain characteristics such as slope, the amount of local height contrast, the shape of an area's cross profile, and the occurrence of steep slopes. We can combine these features to define a particular type of terrain whose extent can be mapped. By comparing topographic maps of Yuma with maps of areas outside the United States, we can find Yuma types in other parts of the world and paint a global picture of variability in terrain.

5-3 | **ENVIRONMENTAL APPRAISALS**

The geographer's approach to describing and categorizing the earth's physical environment is somewhat different from that of a "pure" earth scientist such as a geophysicist. Along with the technical difficulties of measuring and mapping environmental characteristics like precipitation or soil types are the socioeconomic ones of estimating which properties are relevant or irrelevant to humans. Different human groups with different technologies view the *same* environment in different ways. Can we identify any common elements that will produce the same reactions among all groups?

GROSS LANDSCAPE

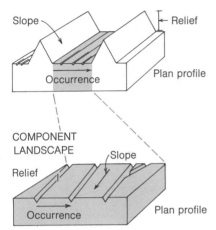

Figure 5-13. Terrain analogs. The breakdown of gross landscapes in the Yuma area of Arizona into a series of comparable units (component landscapes) is based on similarities in four factors: slope, relief, frequency of occurrence, and general shape (plan profile). Such units allow worldwide mapping of the occurrence of Yuma-type terrain. [From J. R. van Lopik and C. R. Kolb, *U.S. Army Water Exp. Stn., Vicksburg, Miss., Tech. Repts.*, No. 3-506 (1959).]

Environments and physiological constraints

The most fundamental question man asks about an environment is, "Will it support human life?" The long evolution of *Homo sapiens* as an animal species has given him some highly specific environmental requirements. Like an atmospheric fish, he swims through an oxygen-rich gas found only near the surface of one of the minor planets. In view of the multiplicity of physical and chemical conditions in the known universe, a human is a remarkably specialized creation, with only a tenuous toehold on survival. If we reduce the oxygen level of his surroundings, he starts to pant; if we increase the proportion of hydrocarbons, he starts to cough; if we immerse him in water, he drowns within a few seconds; if we deprive him of water, he atrophies and dies within a few days. But as a denizen of the planet Earth he appears to be much more robust. Given an unpolluted atmosphere, man's tolerance of climatic conditions (precipitation, wind, and solar radiation) actually found on earth is reasonably good. The climatic element to which he shows the greatest sensitivity is probably temperature. Man is a warm-blooded mammal with an average body temperature around 37°C (98.6°F). Prolonged exposure to conditions that raise or lower this normal body temperature more than a few degrees lead to permanent tissue damage and death.

How can we measure whether an environment is tolerable for human life? Many attempts have been made to assess climatic environments using simple combinations of temperature, humidity, radiation, and wind-speed measurements. One simple index used by the U.S. National Weather Service is the temperature–humidity index (THI). This index is based on temperature readings on two Fahrenheit thermometers, the bulb of one of which is kept permanently damp. Evaporation will make the temperature recorded on the wet-bulb thermometer lower than that recorded by the normal dry-bulb thermometer. When the air is humid, there is little evaporation and the readings on the two thermometers are similar. The THI is the sum of the two readings multiplied by a constant (0.40) and added to another constant (15). If the two thermometers record temperatures of 70°F and 65°F, the THI will be 0.40 × (70 + 65) + 15, or 69. At a THI of 75 in still air, about half the people in an office feel discomfort; at 80 few remain comfortable; and at 86 regulations suggest that all workers (at least in federal buildings) be sent home.

These comfort levels pertain to still air. As Figure 5-14 shows, winds reduce the effect of high temperatures and humidity and make conditions more bearable. On the other hand, when temperatures are low, strong winds considerably increase the level of discomfort. Thus any index of human comfort must clearly include a chilling factor related to air speeds.

Figure 5-14. Human tolerance of climatic ranges. The graph shows comfort, discomfort, and danger zones for inhabitants of temperate climatic zones. Dry-bulb temperatures in °F are shown to the right of the diagram. [After V. Olgay, from R. G. Barry and *R. J. Chorley, Atmosphere, Weather, and Climate* (Methuen, London, 1968), p. 251, Fig. 7-1.]

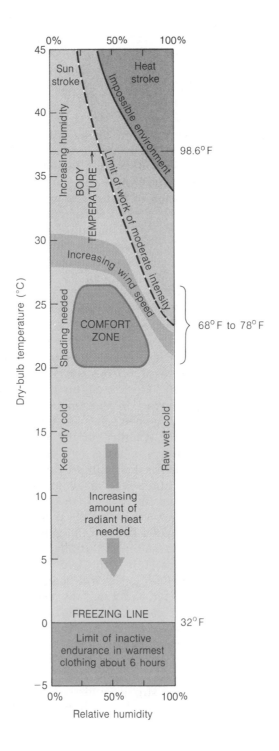

However sophisticated an index is, it can only describe average reactions. Individuals vary considerably in their ability to withstand stress. Our sex, body characteristics, genetic heritage, degree of acclimatization, and cultural background all effect our environmental tolerance. Most indexes have been tested on urbanized North Americans, and we would expect the reactions of Nepalese, Kikuya, or Eskimo groups to be somewhat different.

Environments and trophic levels

Environments of direct interest to man can be directly incorporated into some of the ecological frameworks we met in Section 5-1. For instance, we can estimate the production of food crops in terms of millions of calories per year per square kilometer. The caloric levels of tropical crops such as cassava (280) and sugar cane (254) tend to be high compared with those of midlatitude crops like sugar beets (60) or wheat (8). Food values of less than one calorie per hectare per year are typical of grazing areas, a fact which underlines the poor food-conversion ratios achieved by animals. However, the much higher caloric figures for food crops in the tropics do not necessarily mean that more food is available for human consumption. For an edible crop is also the beginning of a natural food chain, in which predatory and parasitic organisms compete for energy on every trophic level and divert calories away from human food supplies.

We can combine our knowledge of the carbon cycle and trophic levels with the information in Table 5-2, which provides a guide to environmental productivity on the lowest level (T_1), to obtain a rough indication of the food available on higher levels, up to that of man himself (T_4). More organic matter is created by photosynthesis in the tropical rain forests (Paterson's zone A) than in any other environment on earth. Forest regions as a whole account for 40 percent of the total plant productivity of the globe, and the oceans for another 20 percent.

However, the figures in the table are only estimates, and they refer only to the production of organic matter on the first trophic level. Organic matter is not the same as edible food. Most of the organic products of forests are forms of wood, and, with our present technology, very little of these products can be converted to an edible form. As for the organic largesse of the oceans, it is of little benefit to man because we lack the technology to harvest it properly. The long food chains in the sea, the poor food-conversion ratios at each point, and the present wasteful fishing methods mean that we actually derive little food from the oceans. In practice, about 70 percent of the food we eat comes from cultivated land, and the world outlook for food production in the immediate future is going to depend essentially on improving the yield of land already under cultivation. The likely contribution of new virgin lands is actually quite marginal. In the longer term (A.D. 2000 and beyond), the vital areas for increasing man's food supply are the tropical forests and ocean shelves. We return to the issue of food production in the discussion of population in Chapter 6.

Table 5-2 The productivity of global environments[a]

Environment	Area	Dry organic matter	
		Maximal production per year	Maximal *edible* production per year
Forest			
Tropical rain forest	3.9	30.2	6.1
Temperate deciduous forest	1.0	2.4	1.0
Temperate coniferous forest	2.9	12.7	3.1
Taiga	0.8	1.5	0.2
Total	8.6	46.8	10.4
Grassland			
Humid grasslands	2.9	11.2	10.2
Arid grasslands	4.3	7.5	5.1
Total	7.2	18.7	15.3
Cultivated land	2.0	7.8	71.1
Other land			
Wetlands and swamp	0.8	3.0	—
Tundra	1.8	0.9	—
Hot desert	4.3	0.6	—
Cold desert	3.5	—	—
Total	10.4	4.5	
Ocean and lakes			
Deep sea	65.5	19.9	
Shelf, lagoon	5.1	2.7	2.5
Fresh water	0.8	0.3	—
Total	71.4	22.9	2.5

[a] Theoretical estimates of maximal production by photosynthesis in different types of environments. All figures in the table are percentages of the world total. Because of rounding, the total for each column may not equal 100%.

SOURCE: Data from R. U. Ayres, *Science Journal* **3,** No. 10 (1967), p. 102.

Environments and production potentials

During World War II, analog models of terrain enabled troops to be trained in the United States to cope with the problems they would meet in moving vehicles over different types of terrain in foreign areas. Like many wartime ideas, such models have valuable peaceful applications. They now form part of an international effort to classify land in terms of its potential for agriculture and settlement. In this classification process terrain types are combined with information on vegetation and soils from aerial photographs to provide a series of *ecological types.* Field studies of sites in each ecological type are then done to assess the type's agricultural and settlement value. By extrapolating these sample results, geographers can estimate the likely productivity of such "outback" areas as Papua, the Northern Territory of Australia, the Goias Plateau of Brazil, and parts of Nigeria. Assessments made from maps and aerial photographs are, of course, only provisional and may be revised as more detailed field studies are done.

Interregional analogs, which use environmental information from one region in predicting the potential use of another, can also be used to compare climatic conditions. For example, the American Institute of Crop Ecology has established crop analogs to pinpoint climatically similar areas in the United States and the Soviet Union. These climatic analogs have two functions. First, the climatic requirements of a particular U.S. crop variety can be established, and foreign areas can be screened to determine zones where it might grow. Second, the climatic needs of foreign crops can be identified and used to determine suitable growing ranges in the United States. Either way, the analog method identifies areas where new crops can be planted with some hope of success.

As an example, consider Figure 5-15, which depicts areas within the United States whose climate is similar to other areas that produce cork oak. The natural range of this oak species is the lands around the Mediterranean Sea. By using climatic data about the trees' existing range, we can construct a model of the area's thermal and precipitation characteristics and identify analogous areas elsewhere. The map in Figure 5-15 indicates analogous areas within the United States. As more detailed climatic information is added to the model, the number of potential areas for the tree is reduced. Only a small area in California meets all the main climatic requirements. Figure 5-15(c) shows how analogous environments can be screened out by overlapping sieve maps. Of course, the fact that areas are climatically analogous does not imply that they are the same in other ways. The soil in one area may be unsuitable for crops that would grow in the other, or the land may be more valuable for other purposes. Like terrain analogs, interregional analogs do provide a useful first approximation of potential crop areas, however.

In this first section, "The Environmental Challenge," we began with man's concern with what lay over the horizon. We now come back to focus on man himself, for the environmental contrasts we saw in Chapter 3 and the swings and uncertainties we met in Chapter 4 make sense only in human terms. Man does not lie outside the global ecosystem, but is an intrinsic part of it. How he answers a particular environmental challenge depends largely on his own animal physiology, though his response is modified to some degree by his culture and his economic system. Hence we cannot directly equate the utility

(a)

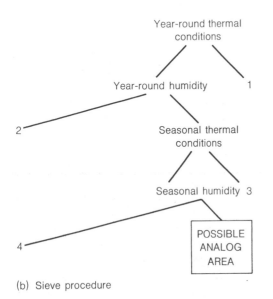

(b) Sieve procedure

(c) Sieve map of the southwest United States   * *

Figure 5-15. Climatic analogs. Geographers use climatic analogs to "match" growing conditions in different parts of the globe. The photograph (a) shows the cork oak (*Quercus suber*) in its native area in the western Mediterranean basin of the Old World. A study of the tree's natural range allows its specific climatic characteristics to be identified. These can then be used to "sieve out" areas where the tree will not do well. Thus area (1) is unsuitable because of year-round climatic conditions, while area (2) passes on this count but lacks the right year-round humidity. Possible analogs must survive all four tests. Application of this sieving process to the southwestern United States produces the "sieve map" in (c). Utah is eliminated in round 1, but much of eastern Texas survives until round 4. Only limited parts of California appear promising, and cork-oak groves have been successfully established there. Nonenvironmental factors (high land prices and labor costs) make the extension of the crop unlikely. [From P. Haggett, *Ecology* **45** (1964), p. 624, Fig. 4. Copyright © 1964, Duke University Press.]

of an area with its innate environmental qualities. Desert areas with low natural productivity (at the bottom of Paterson's scale, in zone F) may sometimes be made highly productive if their soil structure permits irrigation. In the next part of the book, we introduce man into the global ecosystem. We see how he has reacted to it, and how he has changed some parts of it. In moving on, however, we shall still need to look back, occasionally, at this critical early section of the book to check how the pattern of natural environments on the globe underlies and affects its current resources, its hazards, and its future possibilities.

Reflections

1. What do you understand by the term *ecosystem*? Using the diagram in Figure 5-3(a) as a guide, trace the main links in one other typical ecosystem.

2. Do you think that the term ecosystem should be used only for natural plant and animal communities? List the (a) advantages and (b) dangers of viewing human communities as ecosystems. What would be the closest parallel to phytoplankton in a collegiate ecosystem?

3. Debate the view that the major ecological regions in Table 5-1 are too general. Take any one of these regions and list its internal subdivisions.

4. Consider the reactions of man to variations in temperature indicated in Figure 5-14. What do you think would be the limits of human settlement on your own continent if man were unable to construct artificially heated shelters? For how many months of the year would your own college town be habitable?

5. Review your understanding of the following concepts:
   (a) ecosystems   (g) major ecological zones
   (b) food chains   (h) food-conversion ratios
   (c) trophic levels   (i) watershed hierarchies
   (d) carbon cycles   (j) terrain analogs
   (e) positive feedbacks   (k) climatic analogs
   (f) negative feedbacks   (l) sieve maps

One step further . . .

The recent widespread interest in ecology has led to the publication of a large number of excellent brief introductions to the field. See, for example,

Clapham, W. B., Jr., *Natural Ecosystems* (Macmillan, New York, 1973), Chaps. 1 and 2, and

Chute, R. M., Ed., *Environmental Insight* (Harper & Row, New York, 1971), Part 2.

Geographic perspectives on ecological systems are provided by

Stoddart, D. R., in R. J. Chorley and P. Haggett, Eds., *Models in Geography* (Methuen, London, 1967), Chap. 13, while a more advanced approach to ecology, stressing the quantitative aspect and its direct relevance to man, is given in

Watt, K. E. F., *Ecology and Resource Management: A Quantitative Approach* (McGraw-Hill, New York, 1968), Chaps. 4 and 5.

More detailed climatic and environmental analyses of all the main ecological zones are available in most standard physical geographies. A good reference is

Trewartha, G. T., *An Introduction to Climate* (McGraw-Hill, New York, 4th ed., 1968).

Current research is regularly reported in the leading geographic journals. You might also like to look through some of the increasing number of serials devoted to ecological topics, such as *Ecology* (quarterly) or the more popular *Your Environment* (a monthly).

Part Two

Man's Ecological Response

The human side of the man-environment system is considered in Part Two by viewing man in an ecological context. Man's response to his environment in terms of population growth is analyzed in *The Human Population* (Chapter 6). We examine there alternative models of population growth in the face of different environmental constraints, and we note the explosive consumption of material resources that accompanies accelerating population growth. *Man in the Ecosystem* (Chapter 7) looks at humanity's place in the natural world and the effects of man's increasing numbers on the natural environment, both direct and indirect. These effects range from minor intervention through population densities to the firing of grasslands by hunting societies to the far-reaching pollution caused by our urban, industrial society. Special consideration is given to the geographer's view of the many-sided pollution problem. *Resources and Conservation* (Chapter 8) defines natural resources and shows how we estimate the size and probable duration of reserves. Both optimistic and pessimistic views of the energy crisis are explored, and the role of conservation in planning the future use of resources is explored. Finally, in *Man's Role in Changing the Face of the Earth* (Chapter 9) we look at the total effect of man's ecological dominance in terms of the changing landscape. As our numbers have increased, so the environment itself has been modified. But finding out how much it has been changed, or in what ways, is a more complex problem than it appears at first sight. We look at the ways geographers measure and estimate changes in man's use of the land; and, since change is a continuing phenomenon, we close Part Two by looking at ongoing trends in land use.

Chapter 6

The Human Population

A finite world can support only a finite population; therefore population growth must eventually equal zero.

— GARRETT HARDIN
The Tragedy of the Commons (1968)

The optimist proclaims that we live in the best of all possible worlds; and the pessimist fears this is true.

— JAMES B. CABELL
The Silver Stallion (1926)

J ust when man first appeared on the earth is a matter of speculation. We know that several manlike primates emerged during the last 3.5 million years of the earth's most recent geological period — the Quaternary period — but much of our theorizing about man's origins depends on the interpretation of a small number of critical skeletal remains. *Homo sapiens*, with his larger brain (the dividing line is conventionally drawn at 1000 cc), can be traced back to an interglacial period about one and a half million years ago.

Though archaeologists may argue over the exact date of evolutionary advances, two broad ecological generalizations can be made. *First*, man is a very recent arrival on the biological scene. The earth itself is about 4.5 billion years old. The first living forms, algae and bacteria, originated about 2.2 billion years ago, and the first primitive mammals about 0.22 billion years ago. By these standards, the species of mammal we call *Homo sapiens* is a Johnny-come-lately! To use a familiar analogy, his arrival occurred in the last second of an hour-long global history.

Second, the human population of the earth has grown in numbers since its emergence to a staggering level of 3.6 billion, and — what is more important — half that growth has come in the last 35 years. It is not difficult to see that the present explosion of the human population can be only a short-lived ecological phenomenon. Stanford biologist Paul Ehrlich has shown that if the world's present population continued to grow at its present rate (i.e., doubling every 35 years), by A.D. 3000 there would be 2000 people piled on every square meter of the earth's surface — land, sea, and ice included. If we extended this nightmare projection far enough, the universe would eventually consist of a ball of closely packed people expanding outward at the speed of light!

In this chapter we look at the critical facts behind the immense growth of the human species on the earth. First, we review the process by which growth occurs. How is it affected by the relationships between births and deaths? How do we measure population change? Second, we look at the checks on population growth. Here the questions center on how other species control their numbers, and whether man is in any sense a special or an exceptional case. Third, we look at the facts of world population growth. What happened in the past? What is happening now, and what are its long-term implications? Is zero population growth possible — or even desirable?

This chapter is a necessary forerunner to the remaining three in Part Two, which look at the impact of man's increasing numbers on the world's ecosystems (Chapter 7), on resources (Chapter 8), and on the landscape (Chapter 9). An understanding of population and its growth are so important to our understanding of man's place on the earth that we shall be returning to this theme constantly in the remainder of the book.

6-1 DYNAMICS OF POPULATION GROWTH

While the facts of birth and death at an individual level are clear, their effect on the growth and decline of a *population* (that is, a collection of individuals) is more opaque. Here we look at the processes which shape population growth and the kind of yardsticks we use to measure it. In this chapter we shall be concerned mainly with the human population. Much of the reasoning we shall use could, however, be applied to animal populations as well.

Births, deaths, and growth
The total population of any area of the earth's surface represents a balance between two forces. One is *natural change*, caused by the difference between the number of births and deaths. If births are more numerous than deaths in any period, the total population will increase. If they are less numerous, it will decrease. This simple relationship is distorted by a second force, *migration*. When immigrants are more numerous than emigrants, there will be a population increase. When emigrants are more numerous, there will be a population decline.

As Figure 6-1 shows, net changes in population totals are caused by the interaction of four elements: Births and immigrants tend to push the total up; deaths and emigrants tend to bring the total down. Although migration may be the most important factor in small areas (for example, in a small village or a city block), it is less significant on the national level. For the world as a whole, migration is irrelevant because all movements take place within the limits of the recording area. We shall therefore concern ourselves largely with the natural change component in population growth.

We can illustrate the effects of natural change on a small scale by looking at a single island population. Figure 6-2 shows the effects of births and deaths on the total population of the small island of Mauritius in the Indian Ocean. In 1900 the island had a total population of approximately 0.3 million, which increased rather slowly to around 0.4 million by 1950; since then it has increased sharply to nearly 0.7 million. The diagram shows the changes in births and deaths over the period, expressed as birth and death rates per thousand inhabitants per year. Births and deaths were almost in balance until about 1920, when general medical improvements began to lower the death rate. Individual peaks in the two rates are associated with both natural disasters, such as hurricanes and epidemics, and economic fluctua-

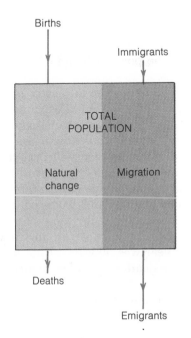

Figure 6-1. Population change. The total size of the population of any part of the earth's surface may be thought of as like the water level in a bath. It is the result of inflows (shown at the top of the diagram) and outflows (shown at the bottom).

Figure 6-2. Natural components in population change. Changing patterns of fertility and mortality are shown for the Indian Ocean Island of Mauritius. Related changes in social and economic conditions and natural hazards are indicated. [From Population Reference Bureau, *Population Bulletin* **18**, 5 (1962), Fig. 1.]

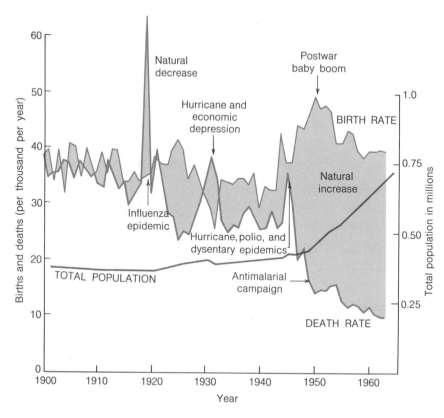

tions, such as the 1929 depression and the postwar boom of the late 1940s.

In describing change on the island we have been talking in terms of "rates." How are these rates measured, and how should we interpret them? There are several ways of measuring the vital rates of a population, of which the crude rates of birth, death, and growth are the simplest. The *crude birth rate* is defined as the number of births over a unit of time divided by the average population. The *crude death rate* describes the number of deaths per unit of time; *crude growth rate* describes the difference between the number of births and deaths per unit of time, each divided by the average population in the time interval. Thus if there were 25 births and 18 deaths in a year on an island whose average population during that year was 500, the crude birth rate would be 50 per thousand, the crude death rate 36 per thousand, and the crude growth rate 14 per thousand.

These rates are described as *crude* because they fail to take into account such factors as the age and sex of the members of the popula-

Exponential population growth

Exponential models of population growth describe a simplified situation in which growth (or a decline) is unchecked and the rate of change is constant. We express this simply as

$$\frac{dN}{dt} = rN$$

where N = the number of people,
r = the rate of natural increase (a constant), and
$\frac{d}{dt}$ = the rate of change per unit of time.

The expression states that the amount of growth is related to the size of the population; the larger a population is, the faster it grows.

To simplify the computation, we can rewrite this as

$$N_t = N_0 e^{rt}$$

where N_t = the number of people at time t,
N_0 = the number of people at time 0, and
e = 2.71828

The constant e is the base of Napierian, or natural, logarithms and is the sum of the infinite series

$$1 + \frac{1}{1} + \frac{1}{2 \times 1} + \frac{1}{3 \times 2 \times 1} + \frac{1}{4 \times 3 \times 2 \times 1} + \cdots$$

If we start with 1000 people (N_0) and assume a growth rate of 1 percent per annum ($r = 0.01$), then we can show, by substituting these values in the equation, that after 70 years ($t = 70$) the original population will have doubled ($N_t = 2000$). In another 70 years the population will have doubled again. Exponential models show the critical importance of small changes in the rate of natural increase. If we halve this rate ($r = 0.005$), the population doubles only every 140 years. Despite their simple structure, exponential models are a useful way of describing recent phases of human population growth. [See A. S. Boughey, *Ecology of Populations* (Macmillan, New York, 1968), Chap. 2.]

tion, or migration. We should certainly expect an island with a large number of young adults to have a higher birth rate and a lower death rate than an island inhabited only by octogenarians! Hence demographers have defined and developed much more sophisticated measures of change called *net* rates, which take into account the structure of the population. They are somewhat complex to go into here, but interested students will find references to discussions of these more refined measures in "One step further . . ." (p. 165). One of the most useful of these measures is the *net replacement ratio*. This relates the number of female births to the number of potential mothers in the existing population (i.e., the number of women between 15 and 45 years of age) and is weighted to allow for the chances of female babies surviving to the reproductive years. If the net replacement ratio is less than one, the number of potential mothers in the next generation is being reduced.

Growth rates and doubling times

The uncertainties involved in estimating future survival and marriage rates, together with the scarcity of data for much of the world, often force us back to a simpler view of population growth. For example, between October 1, 1967, and September 30, 1968, there were 3,453,000 live births and 1,906,000 deaths in the United States. If we take the estimated midyear population of the country on March 31, 1968, as 198,400,000, simple arithmetic indicates that for every 1000 Americans there were 17.4 births and 9.6 deaths. The excess of births over deaths was 7.8 per thousand, and the annual rate of natural increase was less than 0.8 percent.

If we were simply to add eight new individuals each year to the 1000 still living, it would take 125 years for the population to double ($125 \times 8 = 1000$). But this is not what happens. The people added to the population also increase at a rate of 8 per thousand. Population grows exponentially, just like money earning compound interest in a bank. (See the marginal discussion of exponential population growth on p. 147.) The doubling time is, therefore, shortened from 125 years to only 87 years. As the rate of natural increase rises, the doubling time decreases sharply. When it is 2 percent (the current world rate), the doubling time is 35 years. For some of the populations of tropical Latin American countries with rates over 3.25 percent, the doubling time is only a little over 20 years!

Survivorship curves and age pyramids

In order to understand the rate of growth of a population, we must know something about its age and sex structure. By studying the ages at which the members of a population die, we can establish *survivorship curves*. As Figure 6-3 shows, these tell us the number of survivors of an original group (say all those born in a given year) according to their age at death. If this were a perfect world from which all accidents and infections had been eliminated, so that we all lived into our 80th year, the curve would have an abrupt right angle [as in Figure 6-3(a)]. If all members of a given population have exactly the same capacity for survival, their survivorship curve has this shape. In practice, the curves for real populations have complex forms, but there is a tendency for the populations of advanced countries to have curves closer to the hypothetical right-angled one than primitive populations or those of underdeveloped countries. Figures 6-3(b) and (c) go on to illustrate the estimated survival curves for three populations with low survival rates (those of the Stone and Bronze Ages and of China in 1930) and others with medium or high survival rates (like New Zealand and the Netherlands in the 1950s).

Another useful and often-used method of portraying the structure of a population is the *population pyramid*. This is a vertical bar graph showing the proportion of individuals in various age ranges. Figure 6-4 shows some idealized age pyramids for various human populations. The number of males is measured left of the axis and the number of females right of the axis. We shall come back to population pyramids later in this chapter and also in Chapter 19, where we look at the economic implications of highly skewed population-age distributions for developing countries.

6-2 | ECOLOGICAL CHECKS ON GROWTH

The compound course of population growth we have described is really representative only of the *biotic potential* of a population, that is, of its theoretical rate of growth when it is allowed to develop in an optimal environment of unlimited size. Under natural conditions, we would expect this potential for growth to be checked either by some natural limits or, as seems likely in the case of both animal and human populations, by cultural constraints. Here we look at some of the models used to understand these checks and how they operate.

Ecological feedbacks and the Malthusian hypothesis

In an earlier chapter we came across an example of an animal population that increased very greatly in numbers, the crown of thorns starfish. If detailed figures are available, we can chart these kinds of eruptions in animal numbers. Figure 6-5 shows one example of two animals whose numbers follow a cyclic pattern.

We already have feedback models that show the kind of forces

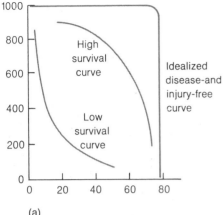

(a)

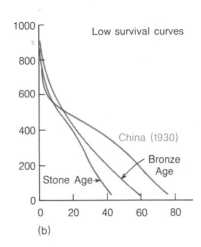

(b)

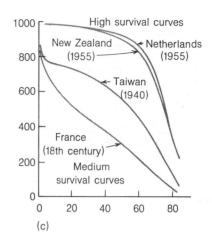

(c)

Figure 6-3. Survivorship curves. Here, idealized survivorship curves (a) are contrasted with actual examples of low survival curves (b) and medium and high survival curves (c) from different cultures and from different time periods. The vertical axis measures the number of surviving members of a population of 1000 births in relation to their age. Age is measured on the horizontal axis in years. [From C. Clark, *Population Growth and Land Use* (Macmillan, London, and St. Martin's Press, New York, 1967), pp. 39-40, Figs. 11A, 11B.]

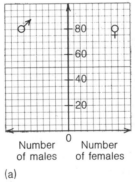

(a)

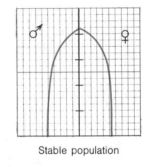

Stable population

(b)

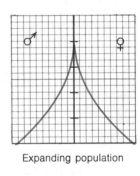

Expanding population

(c)

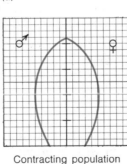

Contracting population

(d)

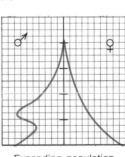

Expanding population
with strong immigration
of young adult males

(e)

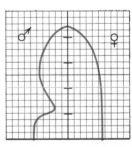

Stable population
after major war losses

(f)

Figure 6-4. Population pyramids. Characteristic pyramids for different populations at different stages of growth.

Figure 6-5. Fluctuations in animal populations. The graph shows changes in the abundance of lynx (predators) and snowshoe hares (prey) over a 90-year period, determined from the number of pelts received by the Hudson's Bay Company. [Data from MacLulich; reprinted, with permission, from E. P. Odum, *Fundamentals of Ecology,* 3rd ed. Copyright © 1971 by the W. B. Saunders Company, Philadelphia.]

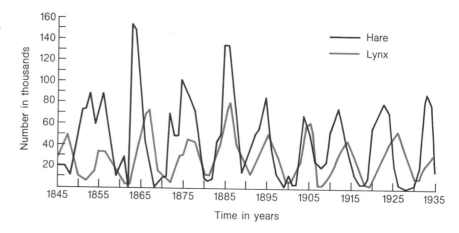

Figure 6-6. The malthusian equation. Nearly 200 years ago, the English demographer Thomas Malthus proposed that a population would always outrun its food supply in the long run since population grows geometrically while food supplies grow arithmetically. In the diagram above, the food supply is originally at a level of 10 units and increases by 3 units in each time period; the population is originally at a level of 0.1 but doubles in each time period. Whatever figures are chosen, in the long run the exponential curve will eventually intersect the arithmetic curve.

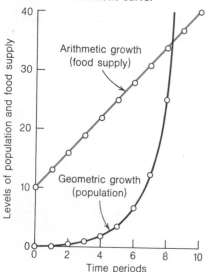

which bring numbers back into line with the capacity of the local environment to support them. As Figure 5-8(b) (p. 123) showed, an increase in population size may mean less food per individual, more deaths than normal, and a subsequent reduction in numbers. Conversely, a decrease in population may set in motion a chain of events that raises the number in a group to the original level. Clearly the model in Figure 5-8(b) is a very rough one, but it broadly fits the observed facts for many animal species.

But what about man? The world population has been uniformly increasing over the last 500 years at least. Does this mean simply that the wavelength of the human population cycle is a very, very long one—that it will hit a peak at some point in the next few centuries and then stabilize or decline? Such questions are impossible to answer with confidence without going into questions of human culture, human economics, and human politics—the substance of Parts Three, Four, and Five of this book. We should, however, at this point, note the analogies between human and other animal populations.

These analogies troubled Thomas Robert Malthus, the English demographer, in writing his now-famous *Principles of Population,* published in 1798. Malthus saw dire ecological consequences in the continuing growth of the human population. He claimed that population has a tendency to increase geometrically (by increasing amounts, as does the series 1, 2, 4, 8, 16, . . .) while the food sources for that population, even with improving agricultural methods, increase arithmetically (by a constant amount, like the series 10, 20, 30, 40, 50, . . .). As Figure 6-6 shows, given these assumptions he was able to demonstrate that any rate of population increase (however small) would eventually exceed any conceivable food supply. When growth reached that point, it could be kept in check, according to Malthus, only by "war, vice, and misery." (Figure 6-7.) Yet the basis for the arithmetic growth of agriculture was never made clear. Moreover, in the 1817 edition of his book Malthus paid considerably more attention to the curtailment of population increases through birth control than to the gloomy devices of war, vice, and increasing human misery.

The carrying capacity of environments

We can explain a simple Malthusian check on population growth by imagining a fixed point above which numbers cannot expand. At this saturation level, the population exactly equals the *carrying capacity* of the local environment (i.e., the number of members of a given species it has the biological capacity to provide food for). This is represented in Figure 6-8 by a population ceiling.

What will happen as population growth approaches this ceiling? Three situations are conceivable. First, the rate of increase may be unchanged until the ceiling is reached, and then abruptly drop to zero. Second, the rate of increase may decline as it approaches the ceiling, eventually falling to zero. Third, the population may overshoot the ceiling periodically, only to be reduced by food shortages, and oscillate above and below the carrying capacity (as in Figure 6.5).

The instantaneous adjustment implied in the first solution seems rather improbable, not least because the mechanism by which such a sudden change might be achieved is unclear. It is unsupported either

Figure 6-7. Malthusian checks on population. Increasing crowding of population would put such pressures on the food supply, according to the Malthusian hypothesis, that only hunger, disease, and war could bring numbers back to supportable levels. [Photograph by Werner Bishof, Magnum.]

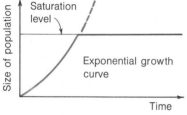

(a) Instantaneous adjustment

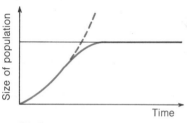

(b) Progressive adjustment

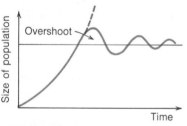

(c) Progressive approximation

Figure 6-8. Environmental constraints on growth. These graphs illustrate three hypothetical relations between a population growing exponentially and an environment with a limited carrying capacity (saturation level).

by empirical evidence on human numbers or by the growing number of studies of other animal populations. The second solution, in which the rate of increase tapers off as numbers approach the critical level, is more plausible. (See the discussion of logistic population growth in the margin.) Such a solution does, however, imply more knowledge about environmental limits and more social control over births than we presently have.

The third possibility as population approaches the critical level is illustrated in Figure 6-8(c). Here, the relationship between the population and the carrying capacity of the environment is reflected in changes in both birth and death rates. Too many people (i.e., a population above the carrying capacity) leads to deaths through starvation and fewer births; this brings the population down. This overshooting and undershooting of the saturation level, as population fluctuates above and below the carrying capacity of the environment, is commonly encountered in animal populations. Periods when a species is abundant follow periods when it is scarce, in a rather regular rhythm.

Because accurate census data for human populations became available only in the late eighteenth century, it is difficult to determine which, if any, of these simple models are appropriate to the human situation. Historical trends in world population reveal that the current exponential pattern of growth is relatively recent. The early period of man's tenure on the earth was one in which the Malthusian constraint of hunger played a key role, and the first two graphs in Figure 6-8 may be more relevant to that time.

Malthusian checks on population: Famine

Here we look at how the food supply acts as a check on the human population in two contexts: during specific local famines and in the longer run on a global scale.

Famines and local food shortages We can find some rough indications of how environmental checks operate by piecing together historical records of local breakdowns of food–population balances. Famines may be closely related to environmental events (as in the case of a drought) or largely unrelated (as was the extensive famine in the refugee population of central and eastern Europe at the close of World War II). We can argue, however, that a poor region with limited food stockpiles whose population has a food intake only slightly above the starvation line and whose climate varies greatly from year to year is more likely to experience famine than other, more fortunate regions.

Although records of famines are difficult to assemble, early historical accounts support the view that famines were once much more pervasive and extensive. The combination of high population densities

Table 6-1 Population losses from major famines

Location	Dates	Estimated deaths[a] (millions)
India	1837	0.8
Ireland	1845	0.75
India	1863	1.0
India	1876–1878	5.0
East China	1877–1879	9.0
China	1902	1.0
China	1928–1929	3.0
USSR	1932–1934	4.0

[a] Famines before the nineteenth century are poorly documented, and estimates of the number of deaths vary widely.

in rural areas and a low caloric intake, in addition to occasional failures of monsoon rains to arrive on schedule, make south and east Asia the world's main disaster areas for famines. The world's worst recorded famine occurred between 1877 and 1879, when an estimated 9 million people died in China. (See Table 6-1.) In Europe, the worst disaster of this type was the Irish potato famine of 1845. Large areas with high populations were supported at subsistence levels by a single plant species, the potato. Blight followed by crop failures in 1845 and again in 1846 brought famines of Malthusian dimensions; deaths, massive emigration (800,000 people moved out of Ireland in the next 5 years), plus a sharply reduced birth rate lowered the population of 8 million (based on the 1841 census) to around half that figure by the end of the century.

When the forces that trigger famine are environmental, as when the monsoon rains fail, we can regard them as fluctuations in the carrying capacity of the environment itself. Thus, we can discard the idea of a fixed limit on population growth (used in Figure 6-8) and replace it with a variable limit. Figure 6-9 shows a series of changes over time in the carrying capacity of an area. Environmental changes are of three kinds. First, there are *nonrecurrent changes* that may be (1) abrupt [Figure 6-9(a)], such as those following the overrunning of fertile fields by a flow of lava, or (2) more gradual, such as those caused by a deteriorating climate or eroding topsoil. Second, there are *periodic regular changes* [Figures 6-9(b) and (c)], including annual variations in productivity connected with seasonal variations in growing conditions. Changes of this type are caused by, for example, low winter temperatures in the boreal zone or summer droughts in the Mediterranean zone. Third, there are *periodic but irregular changes* [Figure 6-9(d)]. That is, environments may have irregular periods of low productivity induced by irregular natural events like river plain flooding. We have already encountered examples of all three types of environmental instability in Chapter 4.

Logistic population growth

When a population is allowed to develop in an optimal environment of unlimited size, its growth follows an *exponential* curve. If we now introduce a fixed *carrying capacity*, or *saturation level* (K), the potential for biological growth, or the biotic potential, will be modified by environmental pressures.

We can introduce this environmental pressure into the exponential growth model

$$\frac{dN}{dt} = rN$$

described earlier by subtracting

$$\frac{K-N}{K}$$

from rN. We then have

$$\frac{dN}{dt} = rN - \frac{K-N}{K}$$

where N = the number of individuals in the population.

K = the maximal number of individuals allowed by the carrying capacity,
r = the rate of growth per individual, and the rate of $\frac{d}{dt}$ = change per unit of time.

Both N and K also can be expressed as population densities. This modified growth curve is termed a *logistic* growth curve and has a characteristic S-shape. Its calculation is described in Section 12-2. [See also A. S. Boughey, *Ecology of Populations* (Macmillan, New York, 1968), Chap. 2.]

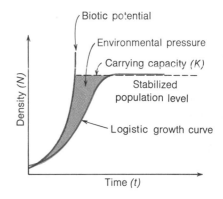

Figure 6-9. Environmental change and population size. The graphs show hypothetical responses of population numbers to changes in the carrying capacity of a given region. In (b) the population copes with periodic regular changes by storing food during good years or seasons. In (c) inward and outward migration keep the population in line with the food supply. In (d) both strategies are used to cope with irregular fluctuations in the food supply.

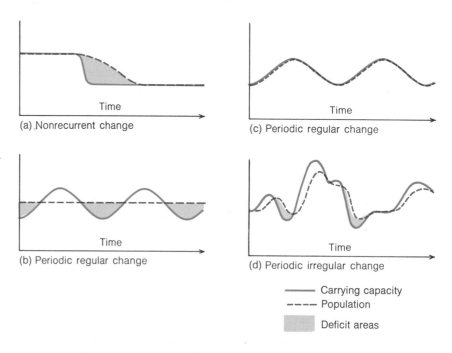

(a) Nonrecurrent change

(c) Periodic regular change

(b) Periodic regular change

(d) Periodic irregular change

——— Carrying capacity
- - - - Population
▨ Deficit areas

Famine and migration Local human populations respond to changes in the carrying capacity of their environment in different ways. Regular seasonal changes may be coped with either by storing food for use during the season of low productivity or by regular migrations to other areas. (Livestock, for example, are moved from low to high pastures in the European Alps as the seasons change.) Periodic but irregular changes pose more severe problems. If the change is for a relatively short period (as in the case of river flooding), temporary abandonment of the area may solve the problem. More serious climatic changes may be both too lasting and too widespread for evacuation to provide a solution. Here the classic famine symptoms set in. Often, the pattern is reinforced by the consumption of the next season's seed corn for food and the resulting loss of productive capacity in the following season. Longer-term declines usually lead to steady emigration and a cumulative fall in population.

All these responses to environmental change involve some spatial (migratory) movements. There may be outward movements (seasonal, periodic, or permanent) of population from areas where there is a food deficit or inward movements of food from areas of surplus. These strategies obviously apply only in the case of local famines. Such spatial reshuffling of population and resources would not help in the event of a global famine.

Global food shortages Setting aside for a moment critical local shortages, how far is the earth as a whole able to support the increasing demands for food that a growing population makes? We can gain some rough idea of the present situation by taking current demands and comparing them with estimates of the earth's ultimate food-producing capacity. The United Nations World Health Organization (WHO) has estimated that the world's 10^9 people today consume around 10^7 tons of food per year. Forecasters at Resources for the Future (RFF) predict that, given the present levels of solar energy and the present distribution of world climate, the maximum amount of organic matter that could be produced by photosynthesis is about 10^{11} tons per year. A comparison of these two estimates suggests that only a trivial portion (about one-hundredth of 1 percent) of the earth's ultimate food-production capability is being used.

But how relevant are these estimates really? Let us see how the RFF estimate was produced. The forecasters reasoned as follows. The prime source of energy on our planet is the sun, which radiates electromagnetic energy waves and high-speed particles into space. Since this constant emission represents almost all the energy available to the earth (except for a small proportion from the decay of radioactive minerals), it can be used to estimate the total amount of energy available to man. Green plants store solar energy through photosynthesis, and thus we can estimate the theoretical totals of dry organic matter (i.e., plants minus moisture) that the earth could produce. But dry organic matter is not always edible. Even in the case of croplands, well below half of the gross product may be edible. As for grazing lands, an energy-conversion factor of 12 to 1 must be used to convert the energy consumed by stock to the food value of the stock to man. Thus, the original figures must be revised downward to a maximum of around 10^9 tons per year. If we also raise the WHO estimate of food consumption to allow for such things as preharvest losses (30 percent), postharvest losses (30 percent), edibility, conversion factors, and noncropland, then the totals converge rapidly. It would be more realistic to say that the farming industry is operating at less than 15 percent of its maximal productive potential.

But the load on the earth's food-producing area is not evenly spread. Most food is produced from a small part of the global surface. The intensive pressure on the cultivated land of the world was indicated in Table 5-2. Cultivated land constitutes only 2 percent of the earth's total area but it is capable of producing nearly three-quarters of the world's potential output of edible matter. The grasslands are second in the productivity of edible material. The greatest gap between total productivity from photosynthesis and productivity of edible material

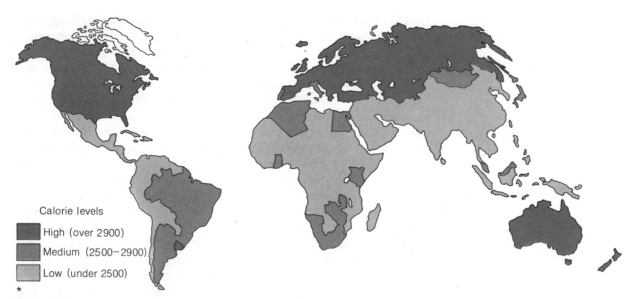

Calorie levels

■ High (over 2900)

■ Medium (2500–2900)

□ Low (under 2500)

*

Figure 6-10. The geography of hunger. This map of world caloric intakes shows the tropical areas, despite their fertility, to be considerably worse off than other areas in terms of food consumption. However, the map, based on United Nations estimates for the mid-1960s, uses national data that differ widely in accuracy. Variations within countries are not shown.

is in forest areas. The edible fraction from the oceans and inland waters is quite small; there is little evidence that the world's oceans (despite their enormous area) will make anything but a marginal contribution to world food production in the next generation or two.

Using the most conservative estimates, we can say that the food-producing capacity of this planet is vast, even with our existing technology. Given improved standards of production, its capacity to feed a population much larger than 3.6 billion is not in doubt. Unfortunately, calculations of global demand conceal immense local differences in food consumption. (See Figure 6-10.) Broadly speaking, the ratio of the well-fed to the undernourished is about 1:6. About 20 percent of the people in the underdeveloped countries are undernourished (i.e., receive less than the minimal number of calories they need per day), and some 60 percent lack one or more of the essential nutrients, commonly protein.

Malthusian checks on population: Crowding and conflict

In our models so far, we have envisaged a simple situation in which a single, homogeneous world population has access to all the planet's (into "haves" and "have nots," for instance) and allow different groups to compete for resources? We can, of course, follow Malthus and assume that competition will lead to conflict, conflict to wars, and wars to a reduction in population. But despite the immensity of war losses in the last 250 years, there is no evidence that they have checked the exponential increase in population. To be sure, individual countries, like France after World War I, experienced severe checks on growth, but these lasted little more than a single generation. Wholesale checks on population growth from international and other conflicts would demand far more deaths than the biggest conflicts have yet produced. Note, however, that Table 6-2 includes only estimated deaths and does not allow for low birth rates due to a reduction in the male population.

Table 6-2 Population losses from major conflicts

Conflict	Dates	Estimated deaths (millions)
World War II	1939–1945	7.3
World War I	1914–1918	7.2
Taiping Rebellion	1851–1864	6.3
Spanish Civil War	1936–1939	6.3
1st Chinese Communist War	1927–1936	6.1
La Plata War	1865–1870	6.0
Indian Communal Riots	1946–1948	5.9
U.S. Civil War	1861–1865	5.8
Russian Revolution	1918–1920	5.7
Crimean War	1853–1856	5.4
Franco-Prussian War	1870–1871	5.4
Mexican Revolution	1910–1920	5.4

SOURCE: Data from L. F. Richardson, *Statistics of Deadly Quarrels* (Boxwood Press, Pittsburgh, Pa., 1960).

Because the historical record is unhelpful in assessing the losses from any future nuclear conflict, we can only turn to more speculative evidence. Ecologist L. B. Slobodkin has considered theoretical patterns of competition between two populations living in the same area and having different rates of growth and saturation levels. Each population's logistic growth curve flattens as its density reaches the carrying capacity of the environment. Figure 6-11 shows representative density

Figure 6-11. Competition and population growth. The graphs show two possible outcomes when exponentially growing populations (N_1 and N_2) with fixed saturation levels (S_1 and S_2) must depend on the same resources to grow.

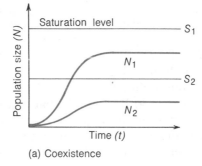

(a) Coexistence

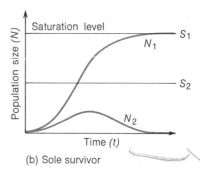

(b) Sole survivor

curves over time for the two populations. An equilibrium is established when both cease to grow. But both populations compete for the same resource, so the growth of one is dependent on the growth of the other. Slobodkin's aim was to specify, given these conditions, whether the two populations could coexist in a state of equilibrium or whether one would progressively dominate the available resources to the exclusion of the other. With his theoretical model, two outcomes are possible. Either both populations coexist, but their numbers remain below the saturation levels that would otherwise prevail, or only one population survives and reaches its relevant saturation level. In other, and more complex cases, both populations may fluctuate cyclically. Note the example of the predator (lynx) and prey (snowshoe hare) populations in Figure 6-5.

Although the Slobodkin model refers only to two populations in highly simplified conditions, its ecological implication for the competitive relations between different populations and between subgroups within the same populations is important. It illustrates how we can simulate future conditions of conflict and thus try to avoid them. Furthermore, it underscores the fact that it is not so much the level of resources as our ability to share and distribute them that lies at the heart of the population–resources dilemma.

6-3 | THE HISTORY OF WORLD POPULATION GROWTH

How far do actual patterns of population growth follow the abstract models discussed above? We shall continue to consider the situation at the world level because this is the only level at which we can legitimately ignore migratory transfers of population. In this sense, then, despite its size, the world is one of the simplest population systems.

Past patterns of growth

Any conclusions about past population growth must begin with Oxford University demographer Colin Clark's axiom that most of the historical (still more the prehistorical) evidence on population is not very accurate and that, as we go back in time, its accuracy generally diminishes further. It is therefore extremely difficult to estimate the size of early populations. Archaeologists suggest that, at the beginning of agriculture, the population of the world was not more than 10 million. For the beginning of the Christian Era the estimate is 250 million, and for A.D. 1650 it is double that figure, 500 million. As we noted earlier, for periods from the late eighteenth century on, the increasing number of national censuses makes the estimating process somewhat easier, and for the early 1970s the United Nations estimate of world population

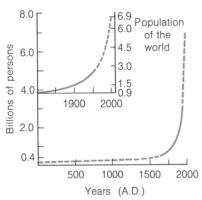

Figure 6-12. World population. The graphs show the general pattern of increase in the estimated world population over the last 2000 years and (inset) the last 200 years. The solid line indicates periods for which reasonably good census material is available; the dashed line indicates estimates. [From H. F. Dorn, in P. M. Hauser, Ed., *The Population Dilemma* (Prentice-Hall, Englewood Cliffs, N.J., 1963), p. 10, Fig. 1. Copyright © 1963 by The American Assembly, Columbia University. Reproduced by permission of Prentice-Hall, Inc.]

is around 3.6 billion. Figure 6-12 presents a general picture of the increase from the beginning of the Christian Era projected to A.D. 2000.

Viewed in a historical context, the present expansion of the world population by about 2 percent per year must be extremely rapid. If the expansion is projected backward in time, the population reduces to a single human couple by only 500 B.C. In fact, we know that human populations were inhabiting the earth in 500,000 B.C. Thus, the average rate of population increase in this early period must have been extremely slow. Rates of increase as low as 0.01 percent per year have been proposed, but any model assuming continuous growth is probably unrealistic and unhelpful. By comparing man's population with other mammalian populations, we can conjecture that primitive man's population underwent major fluctuations, and a fluctuating model on the lines of Figure 6-9(d) may be more appropriate.

The demographic transition

In the last two hundred years industrialization and urbanization appear to have been bringing about a transition in the ways in which population grows. This *demographic transition* can be represented as a sequence of changes over time in vital rates. (See Figure 6-13.) We can recognize four connected phases in this sequence. In the first, or *high-stationary*, phase, both the birth and death rates are high. Although both rates vary, we can assume that the greatest variation is caused by deaths stemming from famines, wars, and diseases. Because the gains in population during a period when death rates are high are canceled by the losses when death rates are low, the population remains at a low but fluctuating level.

The second, or *early-expanding*, phase is characterized by a con-

Figure 6-13. The demographic transition. The graph shows four stages in a demographic sequence in which industrialization and urbanization are important factors.

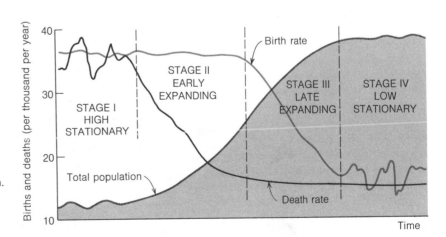

tinuing high birth rate but a fall in death rates. As a result, the life expectancy increases and the population begins to expand. The fall in the death rate is ushered in by improvements in nutrition, in sanitation, in the stability of the government (which means fewer wars), in medical technology, and so on. The third, or *late-expanding*, phase is characterized by a stabilization of the death rate at a low level and a reduction in the birth rate. As a result, the rate of expansion slows down. The fall in the birth rate is associated with the growth of an urban–industrial society in which the economic burden of rearing and educating children tempers the desire for large families, and birth control techniques make family planning easier.

The fourth, or *low-stationary*, phase is a period when birth and death rates have stabilized at a low level; consequently, the population is stationary. This period is unlike the high-stationary phase in that the death rate is more stable than the birth rate.

To what extent do countries today fit this pattern? The high-stationary phase characterizes countries with an uncertain and low level of food production; most of the population in these countries engages in agriculture. This phase was universal among the human population during most of its early history, but now it is restricted to the more isolated and primitive groups. Most of the populations of the Third World of Latin America, Africa, and South Asia fall into the early-expanding phase. Population is expanding rapidly, as environmental and medical technology have brought substantial improvements in life expectancy. Some of the countries in that group have already experienced substantial drops in the birth rate. This decrease is related to socioeconomic changes where people work and in their occupations and reinforced by family planning techniques; thus, the populations of some of these countries appear to have moved into the late-expanding phase. Western Europe, the United States, Canada, and Australia are all examples of countries whose populations appear to be moving into the fourth stage. In Section 18-2 we review the global variation in the stages now reached by different countries and relate these to levels of economic development.

Current trends

The accelerated increase in population in the twentieth century suggests that the exponential model of population growth discussed earlier in this chapter may not be wholly inappropriate for the modern period. It is extremely difficult, however, to proceed from a general recognition of the type of expansion that is occurring to a precise forecast. For example, Figure 6-14 presents six estimates of world population changes before A.D. 2000 that give end-of-the-century populations

Figure 6-14. World populations projected to A.D. 2000. These six estimates of the total number of people likely to be around at the end of the century are based on evidence available in the 1960s. The highest estimate assumes that fertility will remain constant as mortality declines. The U.N. provides three separate estimates (high, medium, and low), based on varying assumptions. Most recent information available to the U.N. is now causing demographers to lower their estimates slightly. [From N. Keyfitz, in *Resources and Man: A Study and Recommendations* by the Committee on Resources and Man of the Division of Earth Sciences, National Academy of Science–National Research Council, with the cooperation of the Division of Biology and Agriculture. (W. H. Freeman, San Francisco. Copyright © 1969).]

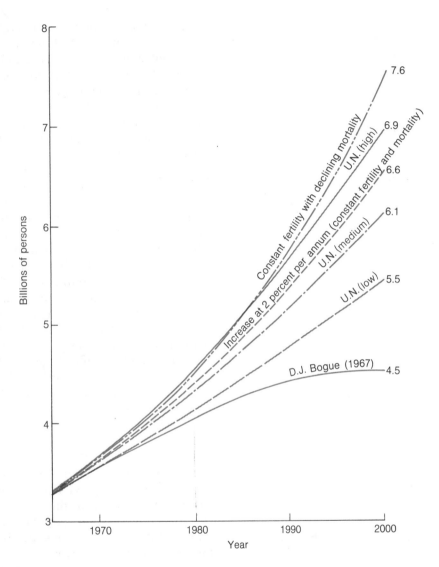

varying from 4.5 to 7.6 billion. The considerable range of these estimates underlines the immense variation in logical projections of the long-term trend, especially when we allow for possible innovations in birth control techniques and attitudes to family size over the remaining years of the century. Each projection tends to reflect the trends existing when the forecast was made. There is evidence that improvements in sanitation and medicine, which allowed a decline in death rates, are now leveling off. Birth rates are likely to be more volatile in the future as both fertility drugs and contraceptive devices permit greater control over family size.

Current demographic projections are concerned largely with the next 25 years, the period up to A.D. 2000. This range is justifiable for two reasons. First, it approximates the extreme limits to which current trends extrapolated. Beyond that date, the assumptions, let alone the calculations, become so fraught with error that longer term forecasts become somewhat absurd. Second, 25–30 years represents one biological generation for a slow-breeding species like man. We hope that our children can improve the range of their own forecasts.

But the world should not end in A.D. 2000, and we can speculate, although in a negative way, on some characteristics of the period beyond it. It seems certain that, in a historical context, the period from 1700 to 2000 is one of exceptionally high increases in population (and high rates of natural resource extraction). To extrapolate present rates of increase into the future is dangerous, but it is instructive. For the United States, a continuation of the rates of population increase prevailing in the mid-twentieth century would result in a density of one person per acre of total land surface—cropland, desert, land of all kinds—by A.D. 2100. The density would rise to around one person per square meter by A.D. 2600. Of course, such projections represent only statistical mumbo jumbo because of the extremely high probability of an intervening decrease in population growth. Although the precise mechanisms that will be operating are not clear, there are some grounds for thinking that the next century will be one of considerable limitations on population growth.

Zero population growth?

As more and more people realize the serious consequences of continued exponential growth, public opinion is shifting rapidly toward a concern for putting on the brakes. ZPG (zero population growth) is becoming an increasingly popular goal, particularly among the younger folk of the developed nations. But is zero growth possible? And if so, how soon can it be achieved?

In the strictly mathematical sense, the achievement of a stable, nongrowing population is a straightforward matter. If all the breeding couples in a population together produce just as many children as are needed to replace the present generation, we have what the demographers term *replacement reproduction*. In current terms, this means that the average number of children per married couple should be 2.3. Two children per couple would presumably not be enough to replace the current population because not everyone gets married, not everyone has children, and not all children live to reproductive age.

Even if replacement reproduction rates were obtainable immediately, it would take many years for population growth to slow down. The present shape of the population pyramid for the whole world shows very large numbers of children below the reproductive age who will eventually move into the critical childbearing age-band (15–45 years). If all were to adopt a family size of 2.3 as a target, the population would still continue to rise to a peak 1.6 times its present level. If the present programs for birth control through family planning and socioeconomic change were successful in achieving a slow reduction to the replacement rate by A.D. 2000—a *very* optimistic assumption— then the world population would rise to a peak at about 2.5 times its present size. Thus there is a momentum about population growth which is very hard to change very quickly.

Does this mean that an immediate reduction in population growth is impossible? Research by demographer Tomas Frejka has shown that, for the United States, this is only possible if the target family size is first set well *below* the replacement rate—as low as 1.2 children per family. If this rate were to prevail for a couple of decades, the total population would decline and the average family size would then need to be raised to above the replacement level. Figure 6-15 shows, on the left, the actual variations in the United States' birth rate over the last sixty years and, on the right, Frejka's calculation of the seesaw birth rates necessary to keep the population at the 1970 level over a 400-year period. According to Frejka, this long period is necessary for the violent cycles of population increases and declines to be smoothed out.

In the first half of this period, the remedy is as unwelcome as the

Figure 6-15. Zero population growth. On the left are the *actual* birth rates in the United States over the 60 years since 1910. On the right are Frejka's estimates of the fluctuating birth rates needed to maintain the U.S. population at its *present* level (i.e., for ZPG) for the next 400 years. Note that the time scales for the curves on the left and the right are different. [Data from Population Reference Bureau and from A. Frejka, *Population Studies* **22** (1968), p. 383, Fig. 1.]

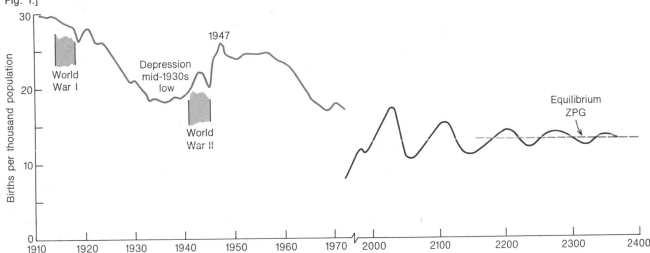

cure. The effects of attempting to stabilize the population at present levels are scarcely less awful than those of unbridled growth. If Frejka's calculations are correct, the United States would go through alternating phases in which its age pyramids fluctuated violently from a dominance of old folk to a dominance of the young. As Figure 6-15 suggests, these phases would have a period of about 80 years and would become less pronounced as the centuries passed. Enormous economic and social problems would be created in the meantime.

It would appear, then, that despite an increasing awareness that man cannot go on reproducing at his present rates for many more years, any substantial reduction in the rate of population growth is likely to be a slow process if the present cultural barriers to birth control are maintained. Changes in birth rates to a replacement level will not produce zero population growth immediately; moreover, an immediate reduction of the growth rate could generate a series of alternating increases and decreases which would last for the next ten or twenty breeding generations. The most likely outcome, from the viewpoint of the 1970s, is that the world population will continue to grow for the next few generations of *Homo sapiens*, but there will be a slowdown in the rate of growth.

Reflections

1. Using the latest national census figures for your own country or state, map the pattern of birth rates and death rates for each province, state, or county. Attempt to explain the spatial pattern which results.

2. Construct hypothetical population pyramids for populations which are (a) expanding rapidly, (b) declining sharply, (c) static, or (d) recovering from major wars. Can you match these hypothetical populations with the populations of real countries today?

3. How would you define the term "overpopulation"? Do you think your own country or state is overpopulated? Why? Suggest and defend an optimum population for your area. How does it compare with the optimum populations suggested by others in your class?

4. Set up a debate in the class on the motion that: "The ideas of Malthus on the balance of population and food supply are no longer relevant." On which side would you prefer to speak? Why?

5. Do you think the United States will achieve zero population growth in your lifetime? Which other countries seem able to achieve this goal?

6. Look closely at Figure 6-13. Which phase in the demographic sequence best describes the situation in your own country today? Which phase describes the situation 200 years ago? What factor might account for a change in the pattern of population growth?

7. Review your understanding of the following concepts:
   (a) crude birth rates
   (b) crude death rates
   (c) crude growth rate
   (d) survivorship curves
   (e) population pyramids
   (f) the Malthusian hypothesis
   (g) carrying capacities
   (h) saturation levels
   (i) the demographic transition
   (j) zero population growth

One step further . . .

Two basic texts by geographers that outline the main concepts of population geography in a systematic manner are

Zelinsky, W., *Prologue to Population Geography* (Prentice-Hall, Englewood Cliffs, N.J., 1966) and

Clarke, J. I., *Population Geography* (Pergamon, Elmsford, N.Y., 1966).

The dynamics of population growth, ways of describing population statistics, and ways of making demographic projections are described in

Hauser, P. M. and O. D. Duncan, Eds., *The Study of Population* (University of Chicago Press, Chicago, 1959) and

Bogue, D. J., *Principles of Demography* (Wiley, New York, 1969).

For a classic survey of population trends, now somewhat outdated but still a basic reference text for historical trends, see

Carr-Saunders, A. M., *World Population: Past Growth and Present Trends* (Barnes & Noble, New York, 1965), first published in 1936.

The relationship of population to food supply is ably, if controversially, argued in

Ehrlich, P. R. and A. H., *Population Resources and Environment: Issues in Human Ecology* (Freeman, San Francisco, 1970).

Problems of population control and the issue of zero population growth are well treated in

Westoff, L. A. and C. F. Westoff, *From Now to Zero: Fertility, Contraception, and Abortion in America* (Little Brown, Boston, 1971).

In addition to the regular geographic journals, look at *Demography* (published semiannually) and *Population Studies* (a quarterly) for substantive reports on current research. The Population Reference Bureau (Washington, D.C.) publishes very useful bulletins and annual data sheets. The United Nations Statistical Office publishes an annual *Demographic Yearbook*, an indispensable guide to world data on population.

Chapter 7

Man in the Ecosystem

Fair-swooping elbow'd earth—
rich apple-blossom'd earth!
Smile, for your lover comes.

— WALT WHITMAN
When Lilacs Last in the Dooryard Bloom'd (1860)

inimata before 1953 was a tiny, unimportant fishing village off the coast of Japan. In that year, it began to gain worldwide notoriety as a fearful symbol of the consequences of human intervention in natural ecosystems. In that year, many residents of the village went down with a mysterious and deforming disease of the nervous system. Termed the *Minimata disease*, it was later traced to concentrations of a deadly mercury compound — methyl mercury — in human body tissues. Altogether, 900 people in the neighborhood of the village were affected by mercury poisoning; of these, 52 died and nearly twice that number were crippled beyond recovery.

Once the cause of the disease had been identified, it was not hard to trace the source of the mercury to the wastes discharged into Minimata Bay from a giant chemical plant. But Minimata was not an isolated case. The same problem of mercury poisoning was later to force a 1967 ban on fishing in 40 Swedish rivers and lakes. In 1970, a scare developed in North America when a graduate student at the University of Western Ontario turned up dangerously high mercury levels in his research on the fish populations of Lake St. Clair. (Lake St. Clair lies on the Canadian–U.S. border northeast of Detroit). Elevated mercury levels were subsequently found in 30 other states in the United States.

Minimata and mercury provide only one example of the problems that can arise because of man's intervention in the tangled web of ecosystems, on which the survival of human populations on the earth ultimately depends. In this chapter we take a longer view and try to place recent problems in a historical context. We shall look first at the scale and pattern of human intervention and then at the increasing degree of intervention as human population densities were built up. Finally, we shall look at the most recent pollution problems to examine why the Minimata disease was at first so puzzling and to speculate on what further environmental backlashes may lie ahead of us.

7-1 INTERVENTION: BENEVOLENT OR MALIGNANT?

Before considering the ways in which man intervenes in natural ecosystems, readers may find it useful to have a recap of some of the points made in Chapter 5. (See Section 5-1, "Food Chains".) There we noted (a) that ecosystems were structured webs connecting the material environment and its biological population, (b) that the main linkages in the ecosystem were food chains running from simple phytoplankton to higher animals, and (c) that the size of biological populations appeared to be controlled by complex feedback mechanisms related to the food supply.

As one of the higher animals, *Homo sapiens* comes at the end of both terrestrial and marine food chains. Both a herbivore and a carnivore, he is a consuming predator of both plant and animal products. Though once a prey to a few other higher animals, he is now almost entirely removed from this role if we discount the few deaths due to sharks, tigers, and the like. He remains vulnerable, however, to a host of microorganisms—most notably the disease-carrying virus and bacillus populations. This position of the human species in the ecosystem has been reinforced by two further critical factors: first, the dramatic exponential increase in numbers of the species, and second, the growing power of man to modify food chains through his technology. To put it simply, man's "natural" position in the ecosystem gave him a potential for dominance; his subsequent technological development and increase in numbers enabled him to capitalize on this potential.

Designs for an improved ecosystem

Most human changes in ecological systems have had a benevolent purpose. For in most cases, intervention has been directed at improving the productivity or the habitability of a given environment.

Consider the example of the lake ecosystem discussed in Chapter 5. (See Figure 5-3.) How could man modify the lake so as to "improve" it for his own purposes? If we assume the objective of the improvement to be the basic one of increased food production, then we can easily draw up a list of means. At the beginning of the list will come simple schemes like that of the selective killing of fish or animals that prey on species with a food value to man. Thus we might try to eliminate species like pike in order to increase the numbers of trout or carp. At the end of the list will come major schemes for environmental reorganization like draining the lake and using the fertile lake-bed soils for direct crop production. The first type of intervention demands only primitive resources (plus some basic understanding of the ecosystem); the last type demands an advanced technology and a high input of resources, for it entails sweeping a whole ecosystem away to replace it with another.

All such processes result in one or more of three main categories of change. First, there is the impact on other animal populations. Thus the expansion in the number of *Homo sapiens* has been accompanied by a great expansion of some animal species (usually domesticated ones such as the horse, the cow, and the chicken) which are of direct use to man. At the same time, some species have been severely reduced in number or wholly eliminated. Figure 7-1 shows the systematic destruction of the North American bison herds on the midcon-

Figure 7-1. Man's impact on animal populations. The maps show the diminishing range of the North American bison over a 75-year period of European settlement, from (a) before 1800 to (d) 1875. A conservative estimate of the numbers of bison when the first European settlers arrived in North America is 60 million, probably the largest aggregation of large animals known to man. The bison (or "plains buffalo") provided the mainstay of the economy of Plains Indian tribes but was slaughtered at an increasing rate during the nineteenth century as European agricultural settlements spread westward across the continent. By 1900 a low point had been reached and the species was on the verge of extinction. Since then bison have been protected on government reserves and the number in managed herds now runs to several thousands of animals. The dangerous decrease in range and number of the bison has parallels with the current situation for other large animals (notably whales and rhinocerus) in other environments today. [Data from J. A. Allen, in R. H. Brown, Ed., *Historical Geography of the United States* (Harcourt Brace Jovanovich, New York, 1948), p. 379, Fig. 9.]

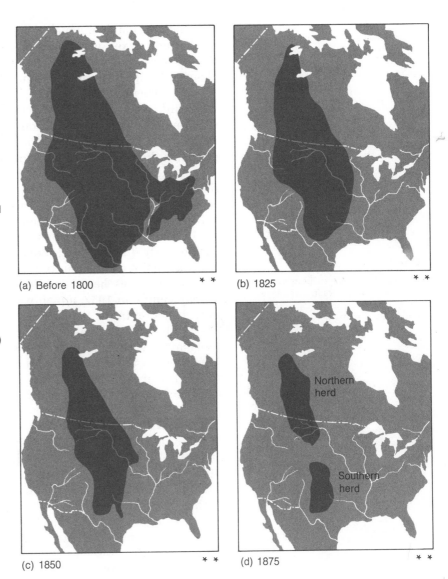

(a) Before 1800   (b) 1825   (c) 1850   (d) 1875

tinental plains during the nineteenth century. At the same time, cattle were being introduced into the same environment and today exist in far greater numbers than the species they replaced.

Second, there is the impact on plant populations. A similar process of selective destruction and expansion of plant populations has led to reorganizations in the balance of plant life. These reorganizations range from rather complete zonal changes, such as the replacement of the midlatitude mixed woodlands of Western Europe by an intensely cultivated mosaic of cropland, grassland, and woodland, to strictly local changes in species composition. These changes in land use are reviewed at length in Chapter 9.

Third, man affects ecosystems by directly altering the inorganic environment. This has traditionally been achieved by interventions in the hydrologic cycle. Irrigation schemes ranging all the way from the primitive diversion of streams to massive basinwide projects are one aspect of such intervention (i.e., bringing water into dry areas to artificially boost plant production). The other side of the coin is the removal of surplus water from marshy or waterlogged areas by drainage and reclamation. Perhaps the most dramatic examples of reclamation are in coastal areas. In Holland, the history of land reclamation goes back to early dike-building projects in the eighth and ninth centuries. Reclamation of a large part of the Zuider Zee by the creation of extensive polders was begun in the early 1920s and is still continuing. More recently, in 1957, a plan was adopted for a twenty-year program of reclamation in the Schelde and Rhine estuaries of south Holland.

As his technological reach increases, the capacity of man to alter inorganic environments is accelerating. Projects for the major remodeling of terrain with massive earth-moving equipment and by the controlled use of explosives and attempts at small-scale climatic modifications are examples of this trend.

Accidental side effects

To the changes that follow man's directed and purposeful efforts at change we must add indirect impacts that have occurred without his being aware of them. These accidents range from the spectacular, such as the much-publicized fouling of Lake Erie, to the insidious, such as the slow rise in DDT levels in some living species. Books with titles like *Rape of the Earth*, *The Population Bomb*, and *Silent Spring* have attracted public attention to these growing byproducts of human occupation. But it seems likely that many second- and third-order effects of man on his environment remain undetected and surface only as links in complex ecological systems.

Any list of these indirect impacts would probably be incomplete. We can certainly include the following: (1) accelerated erosion and sedimentation following changes in the vegetational cover of watersheds; (2) physical, chemical, and biochemical modifications in soils following cultivation or grazing; (3) changes in the quantity and quality of groundwater, surface water, and inland waters; (4) minor modifications of rural microclimates and major modifications of urban micro-

Polder

Polder was originally a Dutch word for an area of land at or below sea-level which is reclaimed by dike or levee building and drained by pumping.

climates; (5) alterations in the composition of animal and plant populations, including both the elimination of species and the creation of new hybrids.

The extension and expansion of all five categories is possible. For example, the third category might be expanded to include not only proven effects on groundwater levels such as the lowered level of water in the Texan aquifers (water-bearing rocks), but also the uncertain impact of toxic chemicals on lake waters. In a similar manner, human modifications of microclimates might conceivably be extended to higher regional levels if Soviet plans for intervention in central Asian water balances are ever carried out. As an extension of the fifth category, the long-term effects of fissionable materials on the genes of animals and plants can only be guessed.

Because a complete roster of impacts, both planned and accidental, is impractical, we shall examine some case studies to highlight the character of the interventions. We shall organize these cases around the idea of population densities; for, in general, the density of man's numbers is an approximate indicator of the degree of environmental alteration. Other things being equal, the most crowded parts of the globe are those that have experienced the greatest environmental change. Of course, there are difficulties in using this organizing principle. Average population density figures may conceal critical local variations. For instance, the average population density figure for Egypt is around 20 people/km², not unlike the average for the continental United States. But almost all the Egyptian population is concentrated in the Nile Valley, about 3 percent of the total land area; in other areas of Egypt there are average densities of over 600 people/km², among the highest rural densities in the world. If the figures for urban areas are calculated separately, the range of densities in a country may run through several orders of magnitude. Within the United States, population densities per square kilometer vary from magnitudes of 10^5 for Manhattan to 10^0 for Nevada.

7-2 HUMAN INTERVENTION AT LOW DENSITIES

The earth's land surface has an area of around 140 million square kilometers (54 million square miles). Two thousand years ago, at the beginning of the Christian Era, the world's total population was still probably only around 250 million, that is an average density of less than 2 people/km². At that density Manhattan would have a population of around 50 people. Of course, we know that population was not distributed evenly, and that the Arctic and Antarctic ice caps, the high mountains, the most arid parts of the desert, and the most isolated islands were empty. These figures underline the fact that for most

Figure 7-2. Shifting cultivation in the humid tropics. This low-oblique photo shows secondary forest regrowth at various stages in New Guinea. Two existing patches of shifting cultivation are shown in the upper part of the photo, while the irregular outline of now-abandoned patches is shown below. [From J. W. B. Sisam, *Use of Aerial Survey in Forestry and Agriculture* (Imperial Forestry Bureau, Oxford, 1947), Fig. 59.]

of man's history on the globe he has existed, by present standards, at very low population densities.

If we turn to the inside front cover of this book, we can see that there are still extensive areas of the world with low population densities. If we were to remove the scientists from the ice caps, the astronomers from the mountain tops, the oil men from the deserts, and the military men from the isolated islands — none of these being settlers in the true sense — then we would return to a persistent pattern of empty areas. In these areas, locally induced environmental change has been minimal, though these regions have not been exempt from globally induced changes like increased atmospheric strontium levels.

The role of fire

More typical of the permanent occupation of the earth at low densities are the tropical grasslands and forests. In both of these areas one of the most important ways of modifying the environment has been by fire. Even before the advent of man, occasional fires were started by lightning bolts or volcanic action in most vegetated areas except tropical rain forests. The long-term impact of such periodic fires is difficult to judge, but there are some indications that species native to the chaparral of the summer-dry Mediterranean zones and the savannah of the winter-dry subtropical zones evolved in association with fire.

Primitive man undoubtedly caused fires himself. Apart from accidental campsite fires, there were two basic types of purposeful burnings. The firing of native grassland in the dry season provided fresh growth for grazing herds; forest trees in the humid tropics were burned to permit cultivation. Vegetation in the humid tropics would not burn naturally, but by girdling or felling trees they could be dried sufficiently in the dry season to burn. The newly opened areas of a forest allowed crops to grow for a few years before production dropped and the plot was abandoned. (See Figure 7-2.)

This *swidden*, or *slash-and-burn*, type of cultivation releases large quantities of nutrients, but the soil and the vegetation can take 15 to 20 years to recover before the process can be repeated. (See Figure 7-3.) If the cultivation cycle becomes too short, as it is likely to do when the population rises and crowding increases, then there may not be enough time for the lands to regain their original levels of fertility. Burning a 40-year-old tropical rain forest provides massive doses of nutrients that can be used by planted crops. The main elements, in proportion, are calcium (100 units), potassium (32 units), magnesium (13 units), and phosphate (5 units). Burning savannah woodland releases less than one-tenth as many nutrients, but a relatively greater share of potassium.

Figure 7-3. Land rotation and population density. The graphs show the relationship of soil fertility levels to cycles of slash-and-burn agriculture. In (a), fertility levels are maintained under the long cycles characteristic of low-density populations. In (b), fertility levels are declining under the shorter cycles characteristic of increasing population density.

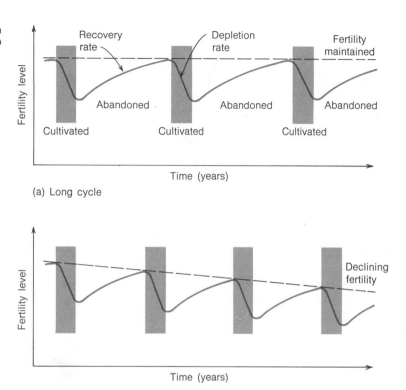

Fire continues to play an important part in crop strategies in many parts of the world. It eliminates unpalatable species and restricts the growth of woody and herbaceous plants. The exact role of fire in the creation and maintenance of grassland areas remains one of the puzzles of modern biogeography. The current trend in research has been toward concentration on the ecological role of man in the creation of such areas, and we discuss this role in the next section.

The creation of new biotic communities

There are two general effects of low-density human occupation on the composition of biotic communities, whether forest or grassland. First, humans tend to eliminate the more conservative, or aristocratic, elements in the biotic population—that is, those species with a low tolerance of fluctuations in moisture levels, high nutrient requirements, or little ability to withstand disturbance. Second, humans usually expand the numbers of the less conservative plants that have higher tolerances of drier, lighter, and more variable conditions. Where man has been active for a long time, plant communities tend to be composed of a small number of extremely vigorous and highly specialized weeds; the secondary forest (jungle) typical of much of the tropics typifies this kind of biotic community. Many of the weeds are widely distributed and originated outside the areas where they now grow. Indeed, their distribution is itself a function of human intervention and the spread of *Homo sapiens*.

The creation of new types of plants, either domesticated ones or weeds, was probably a slow and continuing process in man's post-Pleistocene occupation of the earth's surface. Figure 7-4 shows the results of St. Louis botanist Edgar Anderson's reconstruction of stages in the spatial evolution and hybridization of one of these weeds, the

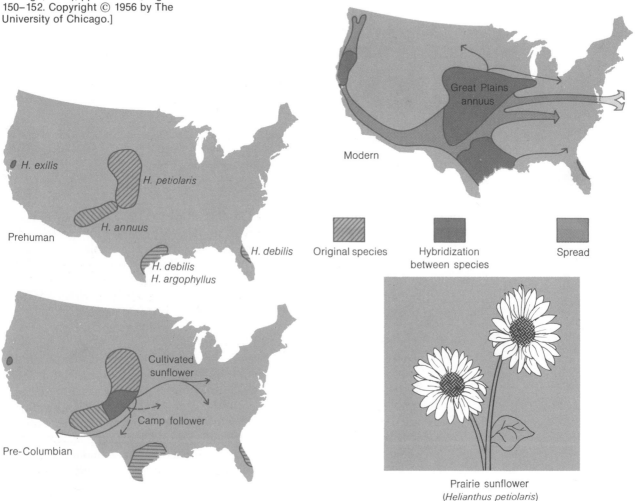

Figure 7-4. Man's role in creating new biotic communities. The maps show the spatial extension and hybridization of the original distinct types of sunflowers in the United States. The longest arrow in the third map represents the overseas spread of sunflower hybrids, widely introduced into European gardens in the nineteenth century and now forming an important agricultural crop. [After Edgar Anderson, in W. L. Thomas, Jr., Ed., *Man's Role in Changing the Face of the Earth* (University of Chicago Press, Chicago, 1956), pp. 768–769, Figs. 150–152. Copyright © 1956 by The University of Chicago.]

Prairie sunflower
(*Helianthus petiolaris*)

various species of sunflower (*Helianthus*), in the United States. In this figure we see the mixing of two of the original five species in pre-Columbian times being followed by intercrossings between four of the five species in the present period. The evolution of the common weed sunflowers into what Anderson terms "superweeds" is continuing today. Mongrelization has increased their ability to colonize new areas like the Great Valley of California and the sandy lands of the Gulf Coast of Texas.

Human intervention helped in forming the species by the creation of disturbed environments and by providing, either deliberately or accidentally, the possibility of hybridization between previously isolated species. Such intervention has important implications not only for the plant world but for the spread of microorganisms. New microorganisms, some of them disease carriers, may also evolve and hybridize in much the same way as Anderson's sunflowers.

The overall effects of man's intervention at low densities appear to have been highly significant at the local level, but trivial globally. There certainly were changes; but, insofar as we can judge, they were largely beneficial and did not affect the long-term productivity of the areas occupied. Lest we should look back on this period as an ecological Eden, it is worth recalling that the technological achievements of these low-density populations also appear to have been somewhat sparse; civilization as we conventionally think of it was associated with a ruder disturbance of the natural environment.

7-3 | HUMAN INTERVENTION AT MEDIUM DENSITIES

Between the extremes of the empty lands and the crowded, heavily urbanized areas lie zones with a medium population density. Medium densities encompass all the ranges of population density that support permanent agriculture. These densities may, in fact, vary by a factor of 100. For example, the swidden agriculture of the northern Congo supports a density of about 8 people/km²; by contrast, the intensive paddy lands of the Mekong Delta support about 800 people/km².

Agriculture: Short-term rotation cycles

As population densities increase, the type of agriculture practiced tends to change accordingly. H. Boserup has proposed a simple five-stage progression in which each step represents a significant increase in both the intensity of the cultivation system and the number of families it can support. Stage 1, *forest-fallow cultivation*, consists of 20–25 years of letting fields lie fallow after 1 or 2 years of cultivation. Stage 2, *bush-fallow cultivation*, involves cultivation for 2 to as many as 8 years followed by 6–10 years of letting lands lie fallow. In

(a)

(b)

(d)

(c)

(e)

Figure 7-5. Changes in land use in the humid tropics. European settlement of the humid tropics for plantation cropping initiated an intricate cycle of changes in land use. (See the diagram in Figure 7–6.) For the Paraíba Valley of southeast Brazil, the main plantation crop was coffee (a). With the aging of the coffee bushes (b) and the progressive abandonment of the hillside plantations (c), the area has become largely grassland. The grasslands are maintained by intensive grazing (d) and regular burning (e). [Photos by the author.]

stage 3, *short-fallow cultivation*, there are 1–2 years when the land is fallow and only wild grasses invade the recently cultivated fields. In stage 4, *annual cropping*, the land is left fallow for several months between the harvesting of one crop and the planting of the next. This stage includes systems of annual rotation in which one or more of the successive crops sown is a grass or other fodder crop. Stage 5, *multicropping*, is the most intensive system of agriculture. Here the same plot bears several crops a year and there is little or no fallow period.

We can find examples of such systems by taking cross sections through time or space. Thus, in Western Europe we can trace the change from the forest-fallow system (stage 1) of the neolithic farmers to the short-fallow cultivation of the medieval three-field system (stage 3), in which one third of the land area was left uncultivated each year. Present intensive cropping involving multicropping and supplemental irrigation represents a midpoint between stages 4 and 5. In the humid tropics, a cross section through space reveals all five stages operating today.

Agriculture: Longer term land-use cycles

Not all agricultural systems fall neatly into simple rotational patterns. In some parts of the world man's intervention has been abrupt and episodic, with periods of intensive use being followed by periods of abandonment.

One example of such *episodic cycles* is provided by the history of some plantation crops in the humid tropics. For example, the growing of coffee (*Coffee arabica*) was introduced into southeast Brazil in the late eighteenth century. By the beginning of the nineteenth century, coffee growing was still confined to the coastal lowland around Brazil's capital city of Rio de Janeiro. With escalating world demand, the area under cultivation increased rapidly; and by 1850 the coffee plantations had crossed the coastal mountain belt of the Sierra do Mar and had become well established in the foothills flanking the Paraíba River. (See Figure 7-5.) Geographers have mapped the spread of coffee plantations in the ensuing decades. Forests were felled and burned as the coffee frontier advanced for some 300 km (186 mi) along the Paraíba River to within a short distance of the city of São Paulo itself. Within another generation the coffee frontier had moved northwest to Campinas and Ribeirao Preto, and with astonishing rapidity the plantation tract along the Paraíba collapsed. The coffee groves were abandoned to weeds and cattle, and the plantation houses and slave quarters to cattle ranches or to decay and the encroaching forest.

The environmental changes that followed the introduction and abandonment of coffee plantations are shown in Figure 7-5. The

Figure 7-6. Land-use cycles. This flow diagram reconstructs the sequence of land use in the Paraíba Valley of southeast Brazil since 1800. Only representative plant species are shown. The numbers indicate how the environment was altered: (1) by abandonment, (2) by clearing and planting, (3) by burning, (4) by heavy grazing, or (5) by cutting trees for charcoal, timber, and so forth. [From P. Haggett, *Geographical Journal* **127** (1961), p. 52, Table 1.]

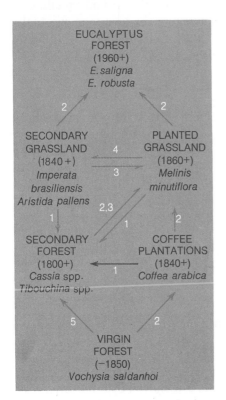

original forest cover which existed in the early 1800s was either cleared and planted with coffee or used as a source of charcoal and construction timber. Figure 7-6 identifies five ways in which the environment could be altered, which lead in turn to six main types of land use. Abandonment of areas once planted would, in the short run, bring about secondary forest areas dominated by rapidly growing species and, in the long run, some reestablishment of the slow-growing tropical rain forest appropriate to the area's prevailing climate and soil structure.

This example gives only a small indication of the rich variety of adjustments between agricultural systems and local environments. Ecologist Clifford Geertz, in his study of agricultural practices in Indonesia, has demonstrated clearly that the systems of swidden cultivation there actually simulate the exchanges of elements (among the atmosphere, vegetation, and soils) that occur in the tropical rain forest under natural conditions. Conversely, the terraced rice paddies in the same area represent an artificial system in which elaborate control of water, fertilizers, and weeds are necessary. Thus, although the swidden systems seem intuitively to be an unstable method of cultivation, and rice paddies a very stable one, the reverse may actually be the case. Replacement of the natural environment by a different ecology greatly increases agricultural production, but continuous work is required to maintain it.

7-4 | HIGH-DENSITY IMPACTS

Despite the very small percentage of the earth's surface occupied by cities, they have a profound effect on the environment. Within and around urban areas, the replacement of rural land by city blocks proceeds faster as the cities grow larger. United States cities with populations of 10,000 have average densities of around 1000 people/km^2, those with populations of 100,000 have densities around 2500 people/km^2, and the larger cities of one million have densities as high as 3500 people/km^2. At these higher densities the proportion of open space in the downtown area dwindles; concrete and asphalt are almost everywhere. In the central parts of large cities, up to 40 percent of the area may be covered with highways alone. (See Table 7-1.)

Cities and climatic modification

The construction of large cities represents man's most profound influence on the climate of specific localities—an influence no less dramatic because it represents a massive side effect rather than an intentional change. For cities destroy the existing microclimates of an environment and create new ones. This is achieved by three processes:

Table 7-1 Land use and town size

Cities	Percentage of land devoted to roads at various distances from downtown			
	Downtown	1 km	2 km	3 km
Large U.S. Cities				
Detroit	47	42	37	34
Chicago	36	34	32	30
Middle-Sized U.K. Cities				
Nottingham	25	16	8	5
Luton	10	7	4	3

SOURCE: D. Owens, *Road Research Laboratory Report* LR 154 (1968), p. 9, Fig. 3.

the production of heat, the alteration of the land surface, and the modification of the atmosphere.

The generation of heat within a city generally results directly from the combustion of fuels and indirectly from the gradual release of heat stored during the day in the city's material fabric (brickwork, concrete, etc.). Temperature studies reveal urban *heat islands,* caused by the fact that city temperatures are generally higher than those of surrounding rural areas. For example, central London has a mean annual temperature of 11°C (58.8°F), the surrounding suburbs have a mean annual temperature of 10.3°C (50.5°F), and the rural areas have a mean annual temperature of 9.6°C (49.3°F). (See Figure 7-7.)

These differences are at a maximum with calm conditions or low wind speeds; wind speeds above 25 km or 15 mi per hour tend to blot out the heat island effect. Contrasts for London are at their sharpest during the summer and early autumn, thus indicating that thermal contrasts depend more on heat loss from buildings by radiation rather than on combustion. There are, however, marked variations between cities in different microclimates and in different topographic situations. In Japan the continued expansion of cities is paralleled by increases in their mean annual temperature (e.g., Osaka's temperature has risen by 2.5°C [4.5°F] over the last century), but it is difficult to isolate the effect of suburbanization from other influences on temperatures.

Cities also affect microclimates through their rugged artificial terrain of alternating high and low buildings and streets. Even though we are conscious of gusting winds being channeled along the canyon-like streets between high buildings, the terrain of the city lowers average wind speeds compared with those in surrounding rural areas. Average wind speeds for a site in central London (7.5 km or 4.7 mi per hour) are substantially lower than for a suburban site (London Airport, 10.2 km or 6.3 mi per hour), although there are considerable variations

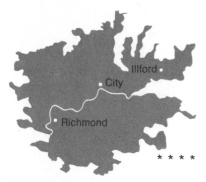

(a) Built-up area

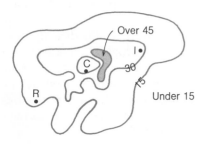

(b) Smoke concentration (mg/100 m³)

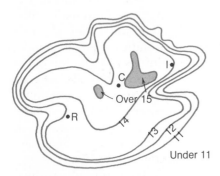

(c) Minimum temperature (°C)

Figure 7-7. Urban climates. The map of Figure (a) shows the impact of London's built-up area on its atmospheric environment. Figure (b) shows the average smoke concentration from October 1957 to March 1958. Figure (c) shows the minimum temperature on June 4, 1959. [From T. J. Chandler, *Geographical Journal* **128** (1962), pp. 282, 295, Figs. 2, 13.]

according to the season and time of day. The effect of urbanization on rainfall is uncertain, but there are strong indications that, under certain conditions, cities in middle latitudes can cause sufficient local turbulence to trigger rainstorms. Significant variations among cities in different climatic zones are probable, and more research on comparative urban climatology is required. It would be dangerous to judge all the world's cities on the evidence provided just by London or Los Angeles.

Atmospheric pollution

The impact of cities on the atmosphere is particularly evident in the context of pollution. City atmospheres are polluted mainly by the emission of smoke, dust, and gases (notably sulfur dioxide). Pollution has three primary effects: It reduces the amount of sunlight that reaches the surface, it adds numerous small particles to the air that serve as nuclei for condensation and hence promote fogs, and it alters the thermal properties of the atmosphere. These three effects often combine to reinforce one another. Smogs, for example, reinforce the reduction of sunlight. The seriousness of this effect is indicated by the fact that British cities are estimated to lose between 25 and 55 percent of the incoming solar radiation from November to March. Although certain common features cause a high concentration of pollution (low wind speeds, temperature inversions, high relative humidities), there is considerable global variation in pollution conditions. For example, although Los Angeles smog is at its worst in summer and autumn, London smog is a winter phenomenon. (See Figure 7-8.)

Although air pollution is most obvious when it soils buildings and reduces sunlight, its most important effects are on human health. The toxic effects of many pollutants are well known, although they reach dangerous concentrations only under unusual meteorological conditions. The most serious accumulation of pollutants over a large metropolis occurred during the London smog of December 5 to 9, 1952. During a strong temperature inversion with little air movement, concentrations rose to six times their normal level, and visibility was reduced to a few yards over large areas of the city. (See the discussion of smog formation in the margin.) During the 5-day period, 4000 deaths were attributed to cardiac and bronchial ailments caused by the smog. The size of this environmental disaster led to legislation (the Clean Air Act of 1956) directly controlling the emission of pollutants through a series of "smokeless zones" where the burning of certain types of fuels was banned.

Cities and accelerating demands for water

Outside the immediate limits of a city, the demands of its urban population lead to a series of impacts that are no less acute for being

(a)

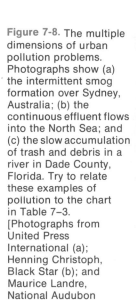

Figure 7-8. The multiple
dimensions of urban
pollution problems.
Photographs show (a)
the intermittent smog
formation over Sydney,
Australia; (b) the
continuous effluent flows
into the North Sea; and
(c) the slow accumulation
of trash and debris in a
river in Dade County,
Florida. Try to relate
these examples of
pollution to the chart
in Table 7–3.
[Photographs from
United Press
International (a);
Henning Christoph,
Black Star (b); and
Maurice Landre,
National Audubon
Society (c).]

(b)

(c)

Smog formation

When winds are strong there is seldom noticeable air pollution. Smoke, dust, and gases are rapidly mixed with a large volume of air and dispersed over a large region so that concentrations at any one point remain low. But the calm conditions and extremely light winds typical of high-pressure (*anticyclonic*) conditions favor the buildup of large and dangerous concentrations of pollutants. Normally, air temperature decreases with height (the average lapse rate is about 6.4°C or 43.5°F per km), so that the warm, polluted air over cities tends to rise and mix vertically [Figure (a)].

During anticyclonic conditions two types of *inversion layers* can interrupt this normal vertical dispersal of pollutants. First, a *high-level inversion* at 1000 or more meters (about 3300 feet) can be formed when calm, upper air falls to lower elevations where it is compressed and its temperature rises. Strong and persistent high-level inversions are typical of the eastern end of the Pacific anticyclone that extends over the Los Angeles Basin, particularly in summer. Second, a *low-level inversion* can be formed overnight by the rapid cooling of the ground. These shallow diurnal inversions affect only the lowest 100 m or so (about 330 feet) of the atmosphere.

Inversion layers, from whatever cause, prevent the vertical dispersal of pollutants and raise the concentration levels [Figure (b)]. Topographic conditions such as narrow river valleys further constrain any dispersal and accentuate concentrations, as in the Donora Valley disaster in western Pennsylvania (October 26 to 31, 1948) and the Meuse Valley disaster in Belgium (December 1930). The map [Figure (c)] shows the average number of days per year when there are inversions and light winds (i.e., *potential air pollution conditions*) in various parts of the United States. [See R. A. Bryson and J. E. Kutzbach, *Air Pollution* (American Association of Geographers, Commission on College Geography, Resource Paper 2, Washington, D.C., 1968).]

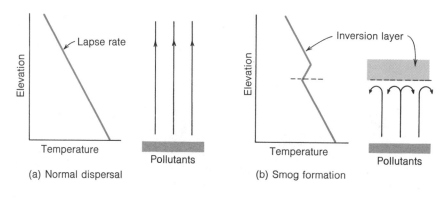

(a) Normal dispersal   (b) Smog formation

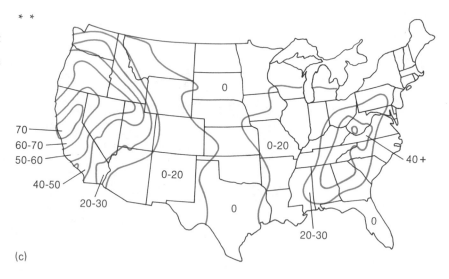

(c)

spatially distant. The insatiable needs of an urban economy for water, food, building materials, and minerals lead to long-range exploitation of the environment. For example, water may be supplied by flooding a remote valley. The creation of artificial bodies of water by damming for a variety of purposes is rapidly increasing. Although the areas used for this purpose are small in relation to the total land area, this type of land use is important locally. For the United States as a whole, an area of not less than 40,000 km² (over 15,000 mi², an area larger than Belgium) is now covered with artificially impounded water.

The average per capita amount of water used in the world's Western cities is now around 600 liters a day. But the demands of the public for water are overshadowed by the needs of industry. Each ton of steel requires 100,000 liters, and each ton of synthetic rubber needs over 2,000,000 liters. Overall, the use of water is growing: It has trebled in the last 30 years and is expected to treble again in the next 30 years.

The most acute problems now being faced are not so much in the provision of water as in the disposal of polluted water. The average city of half a million people now produces over 1800 tons of solid waste each *day* and a further 190 million liters of sewage. The disposal of organic sewage poses less difficulty than the disposal of the inorganic metallic wastes of industry, and it is to this problem that we turn in Section 7-5.

7-5 | POLLUTION AND ECOSYSTEMS

In studying the effect of the very dense concentrations of people characteristic of our urban-industrial civilization on the environment, we are returning to the concern with pollution with which we began this chapter. Pollution *does* occur at lower population densities, but at such low rates and in forms so easily broken down that if fails to form the kind of ecological threat posed by the by-products of urban man.

The pollution syndrome

Any historian looking for a word to summarize the 1970s is likely to find the term "pollutant" (from the Latin *pollutus*, defiled) on his list. But despite the constancy with which it comes up, the term remains difficult to pin down. What do rising mercury levels in the ocean, rising noise around airports, higher temperatures in streams, and higher carbon dioxide levels in the atmosphere have in common? Perhaps the simplest answer is that each represents a substance which, in terms of man's environment, is in the wrong place, at the wrong time, in the wrong amounts, and in the wrong physical or chemical form.

One of the simplest illustrations of this definition is provided by thermal pollution in streams. Heat added to water is in no obvious sense a pollutant, and yet it changes the characteristics of that water as an environment just as surely as chemical contaminants or atomic radiation. Where does the heat come from, and just how does it affect an ecosystem?

Nearly all the waste heat entering our streams comes from industrial processes, and over three-quarters of this comes from electric-power generation. A single 1000-megawatt nuclear-generating plant may need around a million gallons of water each minute for cooling. (See the discussion of nuclear-power generation in Chapter 9.) The water emerging from the outlet pipes may be 11°C (almost 20°F) warmer than that entering the intake pipes.

Important terms used in the study of pollution

Biodegradable is the adjective applied to pollutants that can be decomposed by biological organisms.

Biological concentration is the process by which organisms concentrate certain chemical substances to levels above those found in their natural environment.

Biological magnification is the repeated concentration of chemical substances in food by each organism of a food chain.

DDT is the popular name for *dichloro-diphenyl-trichlorethane*, a powerful insecticide developed in Switzerland in the 1930s.

Dioxin is a very powerful poison used in some weed-killers and found to cause certain deformities in fetuses.

Eutrophication is the excessive growth of algae in nutrient-rich waters leading to reduced oxygen levels and the death of many organisms.

Fallout is radioactive material that settles over the earth after an atomic explosion. Radiation from intense fallout can cause sickness and death and may affect inheritance in living organisms.

Greenhouse effect is the excessive accumulation of heat and water vapor in the earth's atmosphere due to the increased retention of solar energy by polluted air.

Methyl mercury is a highly toxic compound of mercury widely used as a pesticide.

Minimata disease is the name given to symptoms of mercury poisoning first recognized in Japan.

Particulates are tiny particles of solid or liquid matter released into the atmosphere through air pollution.

Photochemical smog is air pollution caused by the reactions between sunlight and particulates that produce toxic and irritating compounds.

Recycling is the reprocessing of waste products for reuse.

Teratogenic pollution is pollution causing birth defects.

Thermal pollution is the discharge of heat into waterways causing reduced oxygen levels and disrupting natural biological cycles.

The critical problem with warm water is what happens to its chemical structure. Warm water holds less dissolved oxygen than cold water but speeds up the metabolic rates of decay organisms in the water, which *increases* their demand for oxygen. These twin effects cause a marked reduction in the water's oxygen level and a fouling of the stream environment. The warmth of water is also critically related to the metabolic rates of fish populations. Spawning and egg development among salmon and most trout occurs at around 13°C (55°F). Even small increases in heat may change the timing of a fish hatch and bring young fish prematurely into an environment in which their natural food sources have yet to arrive.

It would be misleading, however, to regard thermal pollution of streams as only harmful. Many fish species (e.g., catfish, shad, and bass) spawn and develop eggs at higher water temperatures (from 24°C [75°F] to 26°C [79°F] and grow very rapidly in temperatures up to 35°C (95°F). Broad changes in a stream's ecology following thermal pollution may actually increase its overall productivity if we count the increased growth of algae. Future research might well be directed toward making such increases in productivity available to man, thereby changing thermal pollution to thermal enrichment.

Chemicals in food chains

In the discussion above, we used thermal pollution to illustrate the sometimes equivocal nature of a "pollutant." There are other pollutants, however, whose damaging effect is direct and one-sidedly harmful. For example, of the 103 elements in the chemical table, about one in eight plays a prominent part in environmental pollution. Table 7-2 lists the so-called sordid sixteen.

We can divide these elements into three main groups. In the first group are elements like carbon, oxygen, phosphorus, and nitrogen, which are vital to all forms of biological life but which *can* form harmful compounds. In the second group come elements like strontium or uranium, which are important in radioactive pollution. In the third group are toxic chemicals like chlorine and arsenic used in insecticides and toxic heavy metals like mercury and lead.

It is elements in this last group that have proved the most dangerous to the structure of ecosystems in that their effects are insidious. We noted in opening this chapter the mystery that surrounded the Minimata disease. Although the chemical plant producing plastics on the edge of Minimata Bay was known to be discharging dangerous mercury compounds into the sea, the concentrations were so low— about 2 to 4 parts per billion—as to be harmless if drunk in fresh water. Mercury levels in the affected fishermen were about four thousand times greater than this.

Table 7-2 The sixteen most common elements in pollution[a]

Element	Symbol	Main link to pollution
Hydrogen	H	Constituent in pesticides
Carbon	C	Constituent in atmospheric pollution (carbon monoxide) and pesticides
Nitrogen	N	Constituent in photochemical smog
Oxygen	O	Constituent in atmospheric pollution (carbon monoxide and sulfur dioxide)
Phosphorus	P	Causes water pollution by excessive algal growth
Sulfur	S	Constituent in atmospheric pollution from coal-burning power plants
Chlorine	Cl	Constituent in persistent pesticides
Arsenic	As	Constituent in pesticides
Strontium	Sr	Radioactive isotope
Cadmium	Cd	Heavy metal; water pollutant from zinc-smelter wastes
Iodine	I	Radioactive isotope
Cesium	Cs	Radioactive isotope
Mercury	Hg	Heavy metal; toxic water pollutant from the manufacture of some plastics; pesticide
Lead	Pb	Heavy metal; toxic by-product of burning gasoline
Uranium	U	Radioactive element
Plutonium	Pu	Radioactive element

[a] Arranged in order of increasing weight.

The process by which the concentration of mercury built up from harmless levels in seawater to crippling levels in human tissue is termed *biological concentration*. Each creature in a food chain collects in its tissues the mercury present in its own food, and it eats many times its own body weight in nutrients. Since the element is not excreted or broken down, it is passed up the chain in increasing concentrations. Predators like man, at the end of the food chain, are especially vulnerable to mercury poisoning since they consume large amounts of "mercury-enriched" food in which the levels of the metal are already high. In the case of Minimata, the situation was made worse by the local diet of fish, and particularly shellfish. Oysters have been shown to contain concentrations of insecticides as much as 70,000 times the concentration in seawater.

The widely publicized cases of mercury poisoning could be paralleled by the results of the buildup of persistent pesticides in food chains. Figure 7-9 shows the buildup of DDT in part of the food chain in the Long Island estuary in the United States. Note the numbers indicating the DDT levels. These range from as low as 0.04 parts per million for plankton, at the bottom of the food web, to levels up to a thousand times greater in bird populations toward the top of the food web.

The effects of insecticides on ecosystems are often complex and indirect. Birds at the top of the predator tree (i.e., eagles and hawks) suffer the most. Yet the diminishing range of such birds is due less to direct poisoning than to the thinning of eggshells, which fail to hatch successfully. Both the bald eagle and the peregrine falcon have been eliminated from the northeastern United States.

Figure 7-9. Pollution in food chains. In this diagram of a food web in the Long Island estuary the green arrows show the first links and the brown arrows the subsequent links in chains of feeding. The DDT levels found in each organism is given in parts per million. Note the *biological magnification* of these DDT levels by the repeated concentration of this chemical substance by each organism in the food web. [After G. M. Woodwell, *Scientific American* **216**, No. 3 (1967), p. 25, Fig. 1. Copyright © 1967 by Scientific American, Inc. All rights reserved.]

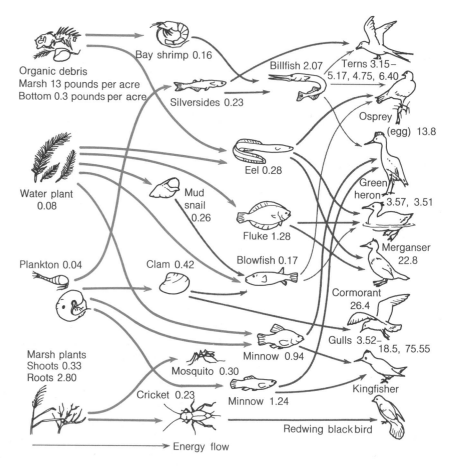

Dimensions of pollution in time and space

Studies of the lead content of the ice layers in the Greenland ice cap (Figure 7-10) show the sharp upturn in worldwide atmospheric lead associated with the beginning of the Industrial Revolution and with the widespread use of gasoline for automobile fuel. Lead has been one of the most useful heavy metals for some 5000 years. It has been widely used in pottery, household plumbing, paints, and insecticides. Unfortunately it is also extremely toxic and the ingestion of excessive amounts can severely affect the human kidneys and liver, as well as the reproductive and nervous systems.

Although most lead is heavily concentrated near its source—so that downtown city dwellers, in areas where the concentration of automobiles is high, have levels of lead in their blood twice that of their suburban neighbors—it also gets caught up in the general atmospheric

Figure 7-10. Long-term lead pollution. The chart shows the lead content of the Greenland ice cap due to atmospheric fallout of the mineral on the snow surface. A dramatic upturn in worldwide atmospheric levels of lead occurred at the beginning of the Industrial Revolution of the nineteenth century and again after the more recent spread of the automobile. [From M. Murozumi *et al., Geochim. Cosmochim. Acta* **33** (1969), p. 1247.]

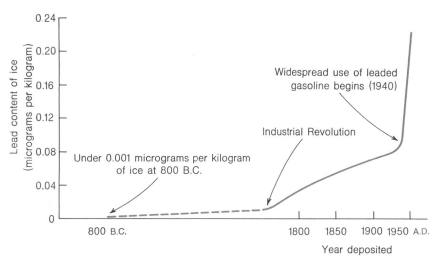

circulation of the earth. Fallout from the atmosphere affects all parts of the globe, and the fallout on Greenland shown in Figure 7-10 represents a gross underestimate of lead levels in the urban areas of the globe.

Pollution problems need to be evaluated in terms of four factors: (1) the nature and properties of the pollutant, (2) the space-time context of the emission, (3) the specific environments affected by it, and (4) the impacts on ecosystems. It is important in studying the lead pollution problem to know that lead is a highly toxic heavy metal which has been continuously released into the environment for some thousands of years and that, with the advent of leaded automobile fuels, the amount of lead in the atmosphere has been increasing at an accelerating rate. It is also crucial to our understanding of the lead pollution problem to know that atmospheric lead is absorbed by animals through their lungs, and that its impact is long term and insidious rather than short term and dramatic.

Table 7-3 summarizes this multidimensional approach to pollution problems. It uses a single example, the oil pollution resulting from the wreck of the supertanker *Torrey Canyon* off the southwest shore of England, to illustrate the dimensional complexity of such problems. Other instances of pollution, such as noise pollution around a major airport or DDT buildups in marine food chains can also be analyzed by using the approach in the table.

One difficulty we face in analyzing pollution problems is that the term "pollution" is now so overworked in the press and TV that it is difficult to establish a balanced view of the degree of environmental threat it represents. The amount of accurate information available on

Table 7-3 Multiple dimensions of a single pollution problem[a]

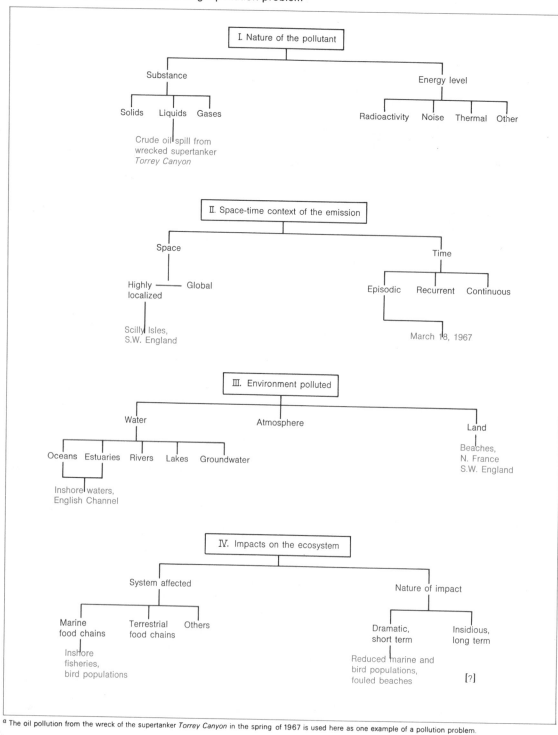

[a] The oil pollution from the wreck of the supertanker *Torrey Canyon* in the spring of 1967 is used here as one example of a pollution problem.

such subjects is less than the strong positions adopted by many people would lead us to suppose. More monitoring of the environment using various sensing techniques is clearly needed. Books like Rachel Carson's *Silent Spring* (published in 1962) certainly played a critical role in awakening both the scientific community and the political lobbyists to the actual and potential hazards that surround us. At the same time, we must note that much pollution is very short-lived and that many ecosystems have remarkable powers of recuperation. Pollution control may have to be carefully priced and the undoubted gains from a cleaner environment set against other desirable goals of the human population.

Reflections

1. Los Angeles has a worldwide reputation for atmospheric pollution. List (a) the manmade causes of this pollution and (b) the natural environmental factors that add to the problem. Do you think that a solution can be found? Why or why not?

2. Debate the proposition that pollution is a luxury—a problem that exists only in the developed one-third of the world.

3. Is shifting cultivation a specifically *tropical* form of agriculture? If so, why? Do you consider it (a) wasteful and destructive or (b) a logical reaction to environmental conditions?

4. Using Table 7-3 as a guide, analyze the multiple dimensions of any single pollution problem you encounter in your own local area.

5. Review your understanding of the following concepts:
   (a) shifting cultivation   (e) atmospheric pollution
   (b) rotation cycles   (f) thermal pollution
   (c) land-use cycles   (g) biological concentration
   (d) heat islands   (h) fallout

One step further . . .

A number of the books we encountered on ecosystems in Chapter 5 also include some consideration of man's role. In particular, look at the introductions provided in

   Chute, R. M., Ed., *Environmental Insight* (Harper & Row, New York, 1971), esp. Part 3, and

   Clapham, W. B., Jr., *Natural Ecosystems* (Macmillan, New York, 1973), Chap. 7.

A good review of man's polluting impact on the main environmental zones is given in

   Boughley, A. S., *Man and the Environment* (Macmillan, New York, 1971), esp. Chaps. 10–12, while more detailed chemical relationships are presented in readable but rigorous form in

> Giddings, J. C., *Chemistry, Man and Environmental Change* (Canfield, San Francisco, 1973).

In addition to the regular geographical journals, keep a weather eye on *Scientific American* (a monthly), which gives good coverage to ecological issues. Some of the major papers from past issues have been drawn together in

> Ehrlich, P. R., *et al., Man and the Ecosphere: Readings from Scientific American* (Freeman, San Francisco, 1971).

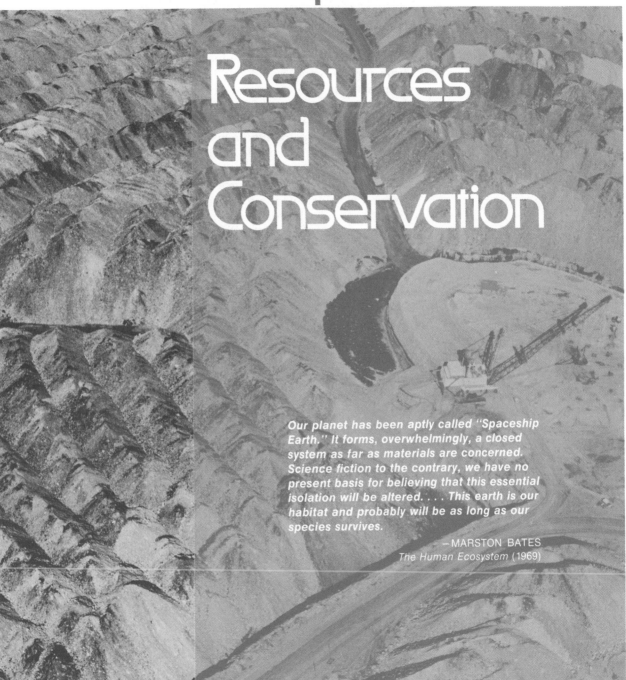

Chapter 8

Resources and Conservation

Our planet has been aptly called "Spaceship Earth." It forms, overwhelmingly, a closed system as far as materials are concerned. Science fiction to the contrary, we have no present basis for believing that this essential isolation will be altered. . . . This earth is our habitat and probably will be as long as our species survives.

—MARSTON BATES
The Human Ecosystem (1969)

In late 1859 the first experimental oil well was drilled at Oil Creek near Titusville in northwestern Pennsylvania. The rig struck oil at a depth of 20 m (about 66 ft). Twenty years earlier the oil had been merely a nuisance as a contaminant in salt wells or had been collected from surface seepages and sold in small bottles as "Rock Oil," a medicine with uncertain curative powers. Twenty years later 30 million barrels were being filled from wells around the world, 80 percent of it from Pennsylvania. A new era in world fuel resources was under way.

The story of petroleum is just one of the dramatic examples of the selective use man has made of the earth's natural resources. We could supplement it with parallel studies — of copper, uranium, even sand — that depict a natural substance rising rapidly in man's estimation to become a valuable resource. Such resources raise various questions that we consider in this chapter. What are natural resources, and how do we measure them? What determines whether or not particular resources will be used? If we do use them, how long will they last? Such issues lead on to the question of conservation, and we look at the issues this in turn raises at the end of the chapter.

8-1 | THE NATURE OF NATURAL RESOURCES

The language in which we talk about the earth's natural resources has become somewhat tangled. In particular, we tend to confuse the concept of potential resources — like the potential hydroelectric power of the Amazon River system — with resources that have actually been developed, such as the electric power produced by the Niagara Falls. It is useful, therefore, to distinguish at the outset between three apparently similar terms: stocks, resources, and reserves.

Stocks, resources, and reserves

The sum total of all the material components of the environment, including both mass and energy, both things biological and things inert, can be described as the *total stock*. In earlier chapters we saw that the prime source of energy on the earth is solar radiation. The fact that the earth receives 17×10^{13} kW per day of solar energy provides some theoretical upper limit to production systems dependent on that source of power. Then too, we can argue that all material goods must be derived ultimately from the 6.6×10^{21} tons of matter that make up the planet Earth. (See Table 8-1.)

In spite of its abundance, the vast proportion of the earth's total stock of matter and energy is of very little interest to man. Either it is wholly inaccessible with our existing technology (e.g., as is the iron and nickel core of the planet) or it is in the form of substances man has not learned to use. Resources are a cultural concept. A stock becomes a

Table 8-1 Earth's most abundant materials

Elements in the earth's crust	Percentage by weight	Metals in sea water	Percentage by weight
Oxygen (O)	46.60	Sodium (Na)	10.60
Silicon (Si)	27.72	Magnesium (Mg)	1.27
Aluminum (Al)	8.13	Calcium (Ca)	0.40
Iron (Fe)	5.00	Potassium (K)	0.38
Calcium (Ca)	3.63	Strontium (Sr)	0.01

resource when it can be of some use to man in meeting his needs for food, shelter, warmth, transportation, and so on. The petroleum stocks of Texas were substantially the same in 1790 and 1890, but in the intervening period man's attitude toward those stocks changed dramatically. Uranium ores provide a more recent example of the stock-to-resource transformation.

The transformation from a stock to a resource is reversible. Figure 8-1 is an aerial photograph of one of the most valuable resources of Neolithic Britain, the flint-axe mines near Brandon. As iron axes replaced flint, around 500 B.C., the resource lost its usefulness and rejoined the unvalued stockpile. Thus, we can define *resources* as that portion of the total stock which could be used under specified tech-

Figure 8-1. The changing status of natural resources. Flint mines produced a key resource for Stone Age cultures but have been abandoned for over 2000 years. Grimes Graves, in eastern England, was one of the most important English sources of flint. Traces of mining activity are evident in the disturbed ground in the upper-middle section of the unforested area. [Royal Air Force photo. Crown copyright reserved.]

nical, economic, and social conditions. Resources as such are determined by human concepts of what is useful, and we can expect *resource estimates* to change with technological and socioeconomic conditions. In this context, *reserves* are the subset of resources available under prevailing technological and socioeconomic conditions. They form the most specific but the smallest of the three categories and are relevant to one period of time only, the present.

Renewable and nonrenewable resources

Geographers classify natural resources in various ways, as Table 8-2 shows. The primary distinction made is between *nonrenewable resources*, which consist of finite masses of material like coal deposits, and *renewable resources*. Nonrenewable resources form so slowly that, from a human viewpoint, the limits of supply can be regarded as fixed. Some, like stocks of coal or metal ores, are unaffected by the passage of time, while others deteriorate. The planet's stocks of refined ore, for example, are reduced by oxidation. The stock of natural gas is reduced by seepages. Renewable, or flow, resources are resources that are recurrent but variable over time; an example would be water power. Flow resources are usually measured in terms of output over a certain time. For instance, the upper limit on world tidal power is about 1.1×10^9 kW per annum.

Renewable resources can be separated further into those whose

Table 8-2 Types of natural resources

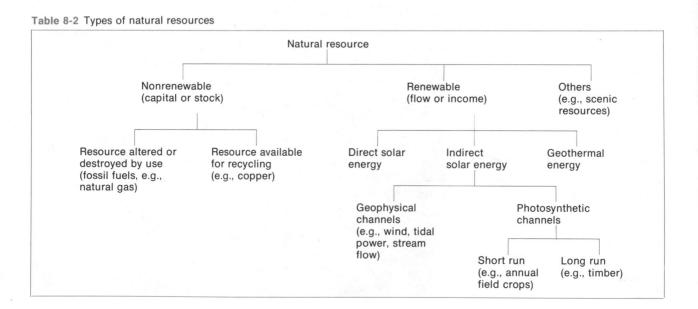

levels of flow are generally unaffected by human action and those demonstrably affected. It is difficult to envision man ever being able to interfere with the world's potential tidal energy. In contrast, the yield of groundwater resources can be permanently reduced. Overpumping may irreversibly close fissures capable of storing water or, as in the coastal valleys of southern California, allow incursions of saline ocean water. Between these two extremes stand resources like forests where a reduced flow (e.g., due to overcutting) can be counterbalanced by subsequent remedial action.

Estimating the size of reserves

How do we go about estimating the size of a particular reserve? We first need to know the distribution of the resource. Figure 8-2 indicates areas in which geologic conditions could produce petroleum. These areas are sedimentary basins where the organic products from which oil is derived were once deposited, compressed under other sediments, and preserved. The probable location of specific fields can be narrowed by geophysical surveys and confirmed by trial bores.

Figure 8-2. Actual and potential resources. Potential petroleum-producing areas in the continental United States are classified according to the likelihood of oil finds. "Possible" areas are those where small commercial quantities of oil can be found. [From U.S. Geological Survey, *Map of the United States Showing Oil Fields and Unproductive Areas* (U.S. Government Printing Office, Washington, D.C. 1960).]

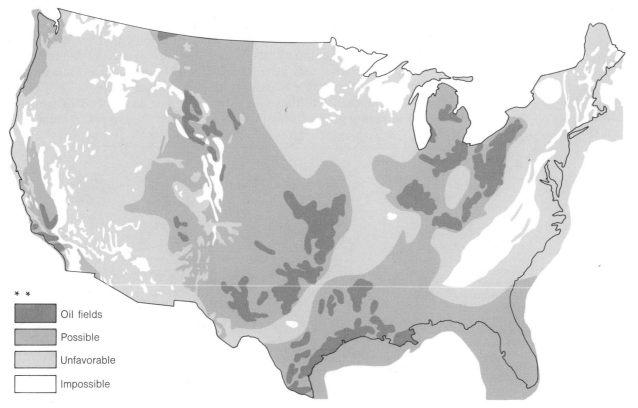

* *
- Oil fields
- Possible
- Unfavorable
- Impossible

Whether a particular oil field will be used depends not just on geologic conditions. As Figure 8-3 indicates, we can consider the size of the reserves the field represents as a joint function of four factors:

1. the quality of the oil, its chemical characteristics, and its freedom from impurities like sulfur;
2. the size of the field and whether it is large enough to justify the capital investment needed to work it;
3. the accessibility of the field, both in a spatial sense (i.e., its distance from refineries or consumers) and in a vertical, geologic sense (its depth); and
4. the relative demand for oil as indicated by the prevailing price level.

By altering any of these four factors the size of the reserve can be changed. Note the effect of a low price in Figure 8-3 in reducing the estimated size of the reserve. We could go on to elaborate the relationship in various ways. For example, we could define our third element, accessibility, as including strategic accessibility (which depends on who owns the field). Potential oil reserves in the North Sea are much more likely to be exploited because they fall within the ownership of countries (like Norway and the United Kingdom) that were wholly dependent on imported oil (see Figure 18-1). In cases like this the high cost of exploitation may be counterbalanced by strategic advantages to a country in being able to control its own sources of oil.

Criteria similar to those for estimating petroleum deposits can be used to gain a general idea of the size of current reserves of other re-

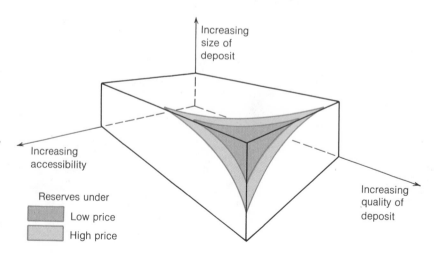

Figure 8-3. Factors affecting size of reserves. Hypothetical relationships between the size of a deposit, its quality, its accessibility, and the prevailing price level are indicated. The block indicates the total size of the world stock of a particular resource, and the colored "corner" the extent of reserves. High prices *increase* the area of reserves, low prices *decrease* the area of reserves.

sources. In the case of stock resources, reserves are expressed as a finite total; reserves of flow resources are described in terms of the potential output in a particular time period. In both cases, estimates of reserves are usually only approximate, refer to a specific time, and must be continuously updated as technical and market conditions change.

8-2 LIMITED RESERVES? THE DILEMMA OF STOCK RESOURCES

The extent of man's exploitation of terrestrial resources is staggering, particularly in the most recent period of human history. We know that the world population doubled between 1800 and 1930 and doubled again between 1930 and 1975. Each human being whose existence is projected in the population models of Chapter 6 is going to require basic necessities such as food, water, shelter, and space (as well as a growing range of nonessential goods). As living standards rise, the pressure on resources created by an exponentially growing population is being increased still further by a rising per capita demand for resources. This demand is being met by a massive consumption of available natural resources.

The combined effect of increased population and resource consumption per capita has been a quintupling in the level of resource extraction between 1870 and 1970. The amount of most metals and ores used since 1930 is in excess of the combined amount used in all previous centuries. A projective study called *Resources in America's Future*, published in the mid-1960s by Resources for the Future, estimates that by A.D. 2000 the world will need a tripling of aggregate food output, a fivefold increase in energy, a fivefold increase in iron alloys, and a tripling of lumber output. If we add to these estimates the new demands only dimly foreseen but certain to occur, then we can expect a massive increase in the use of resources for the remainder of this century.

How long will reserves of the earth's nonrenewable resources last? There appear to be two kinds of answers. The first, which concerns the medium term (about 30 years), is primarily based on economics and is generally optimistic; the second involves a much longer historical period, is based on ecological arguments, and is less hopeful.

An optimistic view

The classic economic test of increasing scarcity is a marked rise in the real cost of a product in comparison with the general price level. How do natural resources meet this kind of test? Changes in price levels for natural resource products since 1870 are shown in Figure 8-4(a) by fluctuations in the price of each commodity. Because the prices are

Figure 8-4. Trends in resource prices. Figure (a) shows relative fluctuations in the deflated price of natural-resource products between 1870 and 1960. Figure (b) shows undeflated-price trends in natural-resource products over a ten-year period. [After H. H. Landsberg *et al., Resources in America's Future* (Johns Hopkins Press, Baltimore, Md., 1963), p. 13, Fig. 4.]

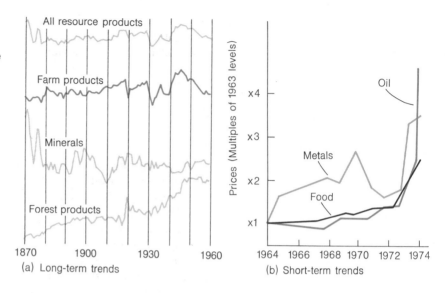

(a) Long-term trends

(b) Short-term trends

plotted in ratio terms, the erratic movements represent considerable fluctuations. For example, real prices of forest products are now over three times what they were in 1870. In comparison, mineral prices have declined and farm products have increased relatively little. Perhaps the most remarkable fact is that the prices for all resource products varied rather little over the past century. According to this conventional economic index of scarcity, natural resources do not appear to have become significantly scarcer since 1870.

It is important to bear this long-term trend in mind in looking at the leaps in prices so alarmingly reported in the press in the 1970s. Figure 8-4(b) shows price trends over the last ten years and emphasizes the surge in oil prices following the Arab-Israeli conflict in 1973. To see these changes in perspective, we need to recall that it is the *relative* price of resources that is at issue in determining scarcity. Figure 8-4(b) also uses a linear rather than a ratio scale on the vertical axis.

To understand the relative stability of natural-resource prices over the long term, we need to think about what happens when a rapid hike occurs in the price of a commodity—as happened in the case of tin in the 1960s (see Figure 8-5) or of oil in 1974.

As the price of a resource rises, a chain of compensating movements occurs. First, high prices bring greater economy in the way resources are used. Supplies of products with high values per ton are carefully metered and their use carefully recorded; conversely, wastage results from low-value products. Water is a classic case of the effect of a change in attitudes toward a resource regarding the way in which it

Figure 8-5. Resource price rises: their spatial impact. Long famous for copper and tin, southwest Cornwall reached the peak of its mining activity about a century ago. Its rapid decline in the late nineteenth and early twentieth centuries was due to falling world prices as overseas deposits brought lower-cost ores into the world market; falling prices were accompanied locally by mine closures and a major emigration of Cornish miners to other, more active mining areas of the world. Recent price rises in the 1960s, especially for tin, have brought a second wave of mining investment to the area. New mines have been sunk and old ones have been reopened. Mine tailings, too, are being reworked. However, West Cornwall has become an important recreational area in the twentieth century, and the new mining investors are running into environmental planning restrictions not faced by their predecessors. See the locations of mines proposed but refused. [From K. Warren, *Mineral Resources* (Penguin, Harmondsworth, 1973), p. 238, Fig. 36.]

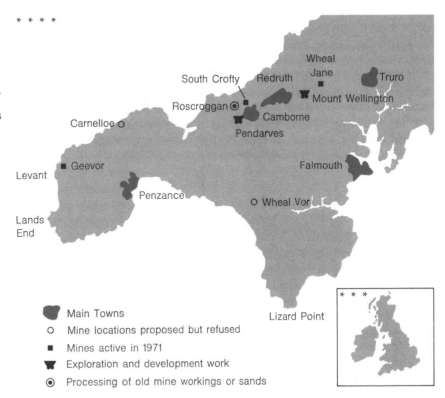

Main Towns
O Mine locations proposed but refused
■ Mines active in 1971
⚓ Exploration and development work
◉ Processing of old mine workings or sands

is used. Once free, its distribution is now metered; and the fact that people must pay for water discourages reckless use of it.

Another reason for stability in the price of resources generally is that various resources may be substituted for one another. An exponential increase in overall demand for one resource may be met by using other resources for the same purpose. The increased demand for textiles is balanced by a switch from natural fibers (such as cotton and wool) to synthetic fibers (such as Dacron, Orlon, and nylon) derived from coal, oil, and even urine. (See Figure 8-6.) Consequently, each resource has a convex consumption curve. Consumption does not fall because of a physical shortage of the old product but because it becomes more expensive in relation to the new substitutes. Ways in which substitution may occur are illustrated by the substitution which occurred in Germany during World War II, when a variety of ersatz products, many based on chemical derivatives of coal, were produced. The changing pattern of world fuel requirements, another example of substitution, is summarized in Table 8-3.

Figure 8-6. Resource consumption and substitution. The graph shows what happens to consumption when a sequence of natural-resource products (1 through 4) is used to meet an exponential increase in the demand for one product. Thus the original need for fuel was originally met by burning wood (1). As fuel consumption has grown exponentially it has been met by using a sequence of other resources (coal, oil, nuclear fuel, and so on). Note that the original resource may show a characteristic rise and fall as it is undercut in price by new substitutes.

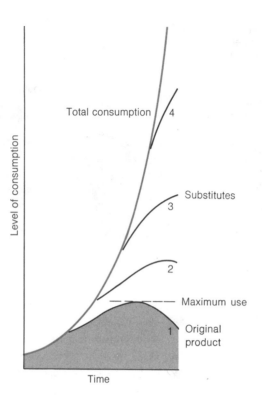

Table 8-3 Changing emphasis on world energy sources[a]

Energy source	Percentage contribution to total energy used					
	1875	1900	1925	1950	1975 (est.)	2000 (est.)
Wood, vegetation	60	39	26	21	13	5
Coal	38	58	61	44	27	21
Oil	2	2	10	25	40	39
Natural gas	<1	1	2	8	15	15
Other sources (mainly hydroelectric and nuclear)	<1	<1	1	2	5	20

[a] Data compiled from U.N. and other sources.

Still another reason for stable resource prices is a switch in methods of extraction. Even when the same natural resources continue to have the same uses, considerable changes in methods of extraction can occur. In copper mining, since the last century there has been a radical switch from the selective mining of high-grade deposits to the mass mining of low-grade deposits. In 1900 ore had to have a copper content of at least 3 percent to be worth mining; by 1970, ore with a copper content of around 0.6 percent was being used. Oil drilling now involves less wasteful extraction methods than in the 1870s, and so on.

This switch to lower grade resources significantly affects the total assessment of the reserves of a stock. The amount of the element in the earth may represent the total stock; in practice, however, the recoverable reserves will be a small fraction of the ultimate reserves. Figure 8-7 presents the theoretical distribution of an element as a frequency distribution diagram; generally, the element is so diffusely located that it is economically unrecoverable. Only when geophysical and geochemical processes concentrate the element in certain locations will recovery normally be possible. Other concentrations may be achieved biologically (e.g., as direct organic concentrations lead to coal, lignite, and petroleum deposits) or by mechanical means (e.g., fluvial sorting of gold into placer deposits).

Note all resources follow the simple arithmetic and geometric model in Figure 8-7. Two specific deposits that fail to show the regular (but inverse) relationship of ore quality to ore volume are the lead-zinc ores found in limestone, and mercury. Here there are abrupt changes in concentration ratios. Although the principle that ore resources increase at a constant geometric rate as ore quality decreases can be a valuable guide to the abundance of some mineral deposits (notably iron, aluminum, magnesium, and copper) in the earth's crust, it would be dangerous to assume that it applies to all minerals, and

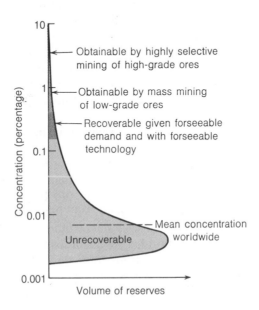

Figure 8-7. Patterns of resource distribution. The curve shows the general concentration and volume of mineral resources. Note that the vertical axis is logarithmic. For the upper part of the curve the quality of ores (concentration) is inversely related to their quantity (volume of reserves).

more dangerous still to assume that it applies to nonmineral resources. Work remains to be done on developing mathematical models of the relationship between quantity and quality for the widest possible range of natural resources. Such models would be valuable guides to the probable future availability of some key resources.

A pessimistic view

If we take a much longer view of resource extraction, then our predictions may be more pessimistic. We noted in Chapter 6 that the present period of rapid population growth and resource exploitation, far from being part of the normal order of events and projectable into the future, is very abnormal. A continuation of the present rate of world population expansion would allow each person a share of only 1 m² of the earth's land surface — Antarctica and Sahara included — in around 525 years.

Population growth is a useful starting point, for we have seen that the consumption of resources is partly a function of the rate of population increase. If we take the rates of energy consumption per capita as an example, the daily minimum needed by primitive man to keep himself alive was equivalent to about 100 watts (for food). As other sources of energy, notably firewood, were added, the level rose to around 1000 watts per capita. Here the rate stayed until the continuous mining of coal (beginning about eight centuries ago) and the production of oil (beginning just over a century ago) brought an exponential increase in energy consumption of around 10,000 watts per capita. We can gain some idea of the recentness of the consumption of fossil fuel deposits by noting that half the cumulative total (or world) production of coal has occurred since 1930 and half the oil consumption since 1952! Given the rates of consumption typical of the present decade, natural resources that took 100 million years to form by sedimentation will be consumed in about 100 years of industrialization.

The problem of dwindling fossil fuels Energy resources are the key to the pace of resource extraction. The current form of world society is highly dependent on energy, and the future availability of other resources, organic and inorganic, is indirectly related to energy resources. Models for estimating the likely volume of fuels available have recently been developed. (See the marginal discussion of projecting future reserves.) These models allow us to project complete cycles of production for the major fossil fuels. Despite some variation, the results of the projections indicate that approximately 80 percent of the petroleum family's resources (crude oil, natural gas, oil from tarsands, and shale oil) will probably be exhausted in about a century. Similar cal-

Projecting future reserves

Assume that the volume of a mineral resource available in any area is finite and that discovery and production follow logistic curves over time. Then the rates of change of discovery, production, and reserves must follow a generally cyclic form. [See diagram (a) below.]

The peak of proved reserves comes where the curves for the rates of discovery and production intersect, with the production rate still rising but the discovery rate already on the decline. This intersection occurs roughly halfway between the two peaks. With similar curves for actual resources, we can project how far their cycle of exploitation has run. As an example, consider the curves for crude oil in the United States, exclusive of Alaska. [See diagram (b) below.]

This figure indicates that reserves probably reached their peak in the 1960s and are now on the decline. Problems in calibrating such curves, their applications, and the reservations needed in interpreting them are discussed in M. King Hubbert, *Energy Resources* (National Academy of Sciences—National Research Council, Washington, D.C., 1962).

culations show that roughly 80 percent of the world's coal reserves will be depleted in about 300 to 400 years. From a historical perspective, the age of fossil fuels will be a limited one, lasting from about A.D. 1500 to A.D. 2800—a very short time even in terms of man's brief occupation of the planet.

Petroleum illustrates well the rapid geographic change in areas of supply which followed the working out of easily accessible fields. We noted earlier that in 1880 world production was 30 million barrels, 80 percent of it coming from Pennsylvania. The succeeding decades have seen a dynamic shift in both the quantity and the location of supplies. By 1910 the 1880 world total of 30 million barrels had increased tenfold, and by 1950 it had grown by over 100 times. New producing areas had sprung up all over the world: the Caucasus in southern Russia and Dutch Indonesia in the 1880s and 1890s, Texas and Oklahoma in the 1900s, Venezuela and the Middle East in the 1930s. Today the pattern continues to change. To the discoveries in Libya and Algeria in the early 1960s, we now must add those in Nigeria, the North Slope of Alaska, and Europe's North Sea. The dominant spatial fact which emerges from the changing historical geography of oil production is the increasingly important role of the Middle East in general and the Persian Gulf in particular. Gulf states now control about three-quarters of the world's petroleum reserves, a share which seems unlikely to change very significantly in the next decade or so. We look at some implications of this for political geography in Chapter 18-1.

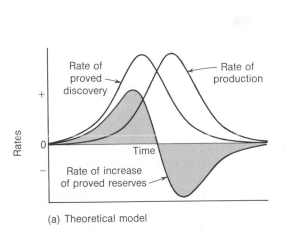

(a) Theoretical model

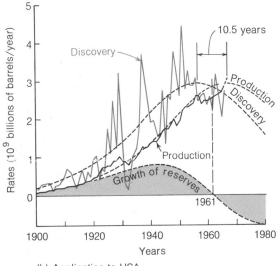

(b) Application to USA

Alternative energy sources? According to current projections, within the next two centuries there will be a need for a reliable source of energy to substitute for the fossil fuels. The possible sources are solar radiation, water power, tidal power, geothermal energy, atomic fission, and atomic fusion. The problem with the first four conventional sources is their scale. As Table 8-4 indicates, only hydroelectric power plants are currently capable of producing power on the kind of scale likely to be needed. To meet the needs of a large city, a solar power plant of 10^{10} thermal watts would have to collect solar power over an area of the earth's surface equivalent to 6.5 km² (2.5 mi²).

Table 8-4 The magnitude of new energy sources[a]

	World annual capacity in 10^3 megawatts				
	Solar	Hydroelectric	Tidal	Geothermal	Nuclear
Current (ca. 1970)	<0.1	152	<0.1	1.12	25
Probable maximum	Small-scale special purpose	2860	64	60	Large, but highly dependent on technological developments

[a] Data from M. King Hubbert in *Resources and Man: A Study and Recommendations* (Committee on Resources and Man of the Division of Earth Sciences—National Research Council, with the cooperation of the Division of Biology and Agriculture. W. H. Freeman and Co. Copyright © 1969), Chap. 8.

The most important sources are therefore the two nuclear sources, atomic fusion and fission. Although the earth's resources of nuclear materials (uranium, thorium, and deuterium) are finite, they are large enough so that, given a reasonable rate of innovation in nuclear-reactor technology, the future scale of supply looks promising. The chief limitation to nuclear fuel may lie less in the adequacy of resources than in the safe disposal of radioactive wastes. (See Figure 8-8.)

The possible environmental threats from nuclear-power production are immense and increase in proportion to the increasing number and size of the power plants. Radiation release, the amount of buried radioactive waste, and even the possibility of nuclear hijacking are likely to get more rather than less important in the next few decades. Yet the importance of success in nuclear-power production is equally vital. The world's greatest potential source of energy is not fossil fuels, but the hydrogen locked up in the waters of the world's oceans. Hydrogen as a liquid or gas may well be the element powering our cars and heating our homes in the 21st century. The electrolysis of water to produce hydrogen is already being undertaken on a small scale where

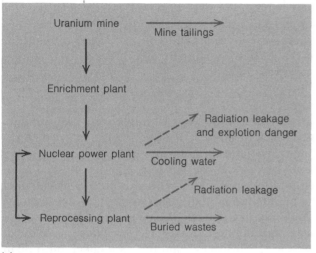

(a)

(b)

Figure 8-8. Nuclear power and environmental hazards. (a) Main stages in the mining and production sequence for the use of fissionable fuel for energy production. The main environmental threats at each stage are indicated. (b) Sites of nuclear power plants (Stage 3 in the production process) are adjacent to cooling-water supplies. Plants are commonly located in rural areas to minimize the population at risk from local leakage problems. The Indian Point nuclear complex at Buchanan, New York, is shown here. The number of such stations is expected to grow exponentially over the globe in the next few decades to meet surging power demands; the associated radiation hazards are likely to pose some major environmental dilemmas. [Con Edison photo.]

cheap power is available; if fossil fuel costs continue to rise and nuclear-power costs continue to fall, the prospects for extracting hydrogen on a mammoth scale should improve.

Energy looks like the main resource bottleneck over the medium term. We could argue that since energy resources are only one of the sets of natural resources used by man, a pessimistic view in this area need not prejudice a more optimistic view of other resources. Unfortunately, this is not so. The main achievements of man in these other areas of resource use have been dependent on an expanding rate of energy consumption. For the last half millennium, we have been drawing on rather readily available sources of fossil fuel energy stored up from geologic processes. The finite nature of these resources is causing some scientists to take a cautious, even gloomy, view of the future.

8-3 | SUSTAINED YIELDS: THE PROBLEM OF FLOW RESOURCES

Renewable resources depend on the great energy cycles of the earth that we have already met in our consideration of global environments in Part One of this book. These are of two kinds: first, *physical* energy cycles related directly to solar energy, and second, *biological* energy cycles indirectly related to solar energy through photosynthesis.

Since we have already met some examples of man's use of earth's renewable resources related to physical cycles—solar power, water power, and tidal power—in our discussion of energy problems, we shall concentrate here on those related to biological cycles. In this category come plant and animal energy sources man taps through agriculture, forestry, and fishing. We shall also look briefly at the special problem of recreational resources, which lie on the border between renewable and nonrenewable resources. The key theme linking the examples chosen is that of how to maintain and increase the sustained yield from a particular resource.

The green revolution

Our first example of renewable resources is drawn from crop production in the agricultural sector. As we saw in Section 7-2, man's disturbing and mixing of plant species to produce new hybrids has been going on for many thousands of years. Plant breeding, albeit of an accidental kind, has been a camp follower of all human agriculture. Since Mendel (1822-1884) laid the foundation of plant genetics with his experiments in heredity, plant breeding has played an increasingly important role in increasing and maintaining agricultural yields; indeed, without it the Malthusian forecast of worldwide famines as a check on human population growth would surely have been fulfilled.

Perhaps the most striking example of the impact of new hybrids is in the recent *green revolution*. The green revolution is an evocative term used to describe the development of extremely high-yielding grain crops that allow major increases in food production, particularly in subtropical areas. In 1953 scientists in Mexico began to develop rust-resistant dwarf *wheats* which doubled Mexico's per-acre production in the next decade. After a major drought in India in 1965, Mexican dwarf wheat was widely planted in the northern part of that country with dramatic results in terms of wheat yield in areas like the Punjab.

Rice development followed a course similar to but more laggardly than that of wheat. The Los Baños research institute in the Philippines was set up with Ford and Rockefeller Foundation backing in 1962 to develop varieties of improved rice (IR). The now famous IR-8 variety was spotted in 1965. Its first harvest, from 60 trial tons of seeds, produced an astonishing sixfold increase of rice under field conditions. From these beginnings, other varieties of IR with important additional characteristics—better resistance to disease, better taste, and (very important in Asia) better appearance after cooking—were developed. Broadly speaking, the new varieties of rice have produced results wherever they were planted in the humid tropics. About 10 percent of India's paddy land is now planted with IR varieties, and the Philippines, once a major rice importer, is now near self-sufficiency in this area.

How far have the "miracle grains" of the green revolution proved an unmitigated blessing? Any seed which can (a) give two to four times the yield of indigenous grains, (b) has such a shortened growing season that two crops per year are often possible, and (c) has a wider tolerance of climatic variations must be welcome. At the same time, some severe side problems have appeared. The high yield is dependent on high applications of fertilizer and insecticides plus, in the case of rice, copious irrigation. Hence, innovation has been most rapid in the most prosperous areas and among the most prosperous farmers, and, in the short run, interregional and social gaps have widened. There is an urgent need for more widespread adoption of the new varieties in poorer sectors, but the fertilizer and water they need are still beyond the financial reach of many of the agricultural peasants of South Asia. At a quite different scale, traditional marketing patterns have been upset. Countries like Thailand and Burma — major exporters of rice — have found their traditional markets disappearing. Japan, normally a great rice importer, has bulging elevators and now looks for export areas.

The successes of the last twenty years have brought to the tropics and subtropics the kind of benefits which plant breeding has been bringing to the midlatitude farmlands for the last half-century. Given the much greater growth potential of the tropics (see Figure 3-4) and the lower yields of indigenous crop varieties, the impact of the revolution has been dramatic. It remains to be seen whether demands for fertilizers and pesticides will take the edge off some of the more promising aspects of this important step in the development of the world's agricultural resources. It may well be that since increased agricultural production in this case depends on fertilizer supplies — much of which come from mineral sources — increased yields from a renewable resource may indirectly depend on a nonrenewable resource.

Sustained-yield forest resources

The view of forests as a renewable resource is a relatively recent one. For most of man's history on the earth, his activities have tended to reduce the world's forest cover. (See Chapter 9.) Heavy inroads were made to clear land, for fuel wood, and for construction materials. Destruction was encouraged by the belief that forest resources were inexhaustible, if not in the immediate neighborhood, then certainly in the world beyond.

Local devastation of forests and the increasing costs of bringing timber from far afield brought gradual changes in this attitude. As early as 450 B.C., official attempts were made to restrict the cutting of the cedars of Lebanon in the eastern Mediterranean. General signs of change did not appear, however, until the 12th and 13th centuries, when drastic restrictions on cutting were imposed in central Europe. Then, gradually, the belief that forests must be protected as a finite and dwindling resource gave way to a belief in new planting and forest-management techniques. By the middle of the 18th century, timber was beginning to be regarded as a slow-growing but renewable crop rather than a nonrenewable source of fuel and timber.

Modern forest management follows two main ecological principles. The first is that of *sustained-yield*—the continuous production of forest products from an area at some appropriate yield level. This is achieved through planned rotational systems (some with cycles of over 100 years), careful species selection, and protection of the timber crop from both fire and disease. There are presently available a wide range of ways of obtaining a continuous flow of forest products. Fast-growing species in the subhumid tropics (e.g., some members of the Eucalyptus family) may be cropped as frequently as every seven years. Douglas fir in the Pacific Northwest may be cropped by *patch cutting* (see Figure 8-9), after which timber from areas around each patch naturally re-covers the cleared areas.

The second ecological principle followed in managing forests as a renewable resource is that of *multiple use*. A forest's yield may be measured in more than wood products, and the objective of multiple-use forestry is to maximize the total flow. Thus timber production may have to be balanced against the role of the forest as a protection against erosion and pollution, as a wilderness or wild-life refuge, or as a recreational area. Not all such uses may be mutually compatible, and we shall look later in the book (in Chapter 19) at the planning problems that the use of land for multiple purposes entails. Indeed, the role of a forest area in recreation may be so important that commercial logging may have to be stopped. The precedent set by the United States in 1872

Figure 8-9. Sustained yield forestry. The early history of lumbering in America shows that forests were treated as a stock resource to be clean-felled. Today, conservation programs ensure that forests are husbanded as a flow resource. Patch cutting is a technique used for Douglas fir lumbering in the Pacific Northwest; it allows natural regrowth of the felled areas from the surrounding untouched ring of forest. [Photo by James H. Karales, DPI.]

when it established Yellowstone National Park has been followed frequently in the last century. The United Nations *List of National Parks* contains the names of some hundreds of parks and refuges in scores of countries around the world.

Sustained-yield recreational resources

As the demand for leisure increases, the role of forests as primarily recreational areas is sure to increase. Sustaining the yield of a forest given an increasing number of picnickers, ramblers, and the like may prove even more difficult than maintaining a sustained timber flow. Most countries have now set aside substantial tracts of country for recreational use (see Figure 8-10), but their management poses increasingly severe problems. A shorter working week, increased leisure time, and falling travel costs have created a surging demand for increased recreational facilities.

Table 8-5 summarizes some of the current trends in the demand for outdoor activities in the United Kingdom. The figures for annual growth rates are based on varying periods of time in the 1960s and should be treated with caution. They indicate, however, the nature of the pressures and the special demands for use of water areas.

We can illustrate the impact of demand on a recreational resource by going back to the lake we used to illustrate ecosystem concepts in Chapter 5. Let us suppose the demand in this case is for sailing. On a weekend day with good sailing conditions a number of folk will bring their sailing dinghys to the lake. Figure 8-11(a) shows this influx of

Table 8-5 The growth in the demand for recreational resources[a]

Growth rate	Urban resources	Countryside resources	Water-area resources
More rapid than population growth	Athletics Golf (8)	Motoring Mountaineering Skiing Camping Horse riding Nature study Gliding (10)	Diving (24) Canoeing (18) Sailing (7) Angling (7)
Similar to population growth	Gardening Major team games Swimming	Walking (3) Hunting	
Slower than population growth	Major spectator sports	Cycling (−2) Youth-hosteling	

[a] General patterns of growth in the United Kingdom. Where figures in parentheses are given they indicate the average annual increase as a percentage and are based on different periods.

SOURCE: J. A. Patmore, *Land and Leisure* (Penguin, Harmondsworth, 1972), p. 49.

National parks

(a)

Areas of outstanding natural beauty

(b)

Areas of great landscape, scientific or historical value

(c)

Green belts

(d)

Figure 8-10. Conservation of recreational resources. Most countries have now introduced legislation conserving areas of land for public recreational uses. These maps show the situation in England and Wales in 1970. Not all the proposed ''green belt'' areas have been formally approved. [From J. A. Patmore, *Land and Leisure* (Penguin, Harmondsworth, 1970), p. 179, Fig. 62.]

people building up over the day simply as a line that rises until a saturation point is reached. How much sailing enjoyment does each family receive? Clearly, at the beginning [point A in Figure 8-11 (b)], the early arrivals have the lake to themselves. But as numbers build up a point is reached (point B) at which boats begin to impinge on each other's territory and the amount of sailing each family can do is reduced. If numbers were to continue to build up, the lake would finally be so jammed with small boats (point C) that no sailing at all would be possible. At this point the enjoyment of each family would presumably have dropped to zero!

How does this crowding affect the "recreational yield" of the lake? We can define the *yield* as "the number of boats × the amount of sailing activity" and plot its curve in Figure 8-11(c). The curve begins at zero (no boats on the lake), rises to a maximum, and then falls back to zero (when the lake is packed with boats but no sailing is possible).

To obtain the maximum yield from the lake, we would like to keep the number of people using it around point D at the top of the yield curve. But how do we do this? A rule of "first-come-first-served" is one

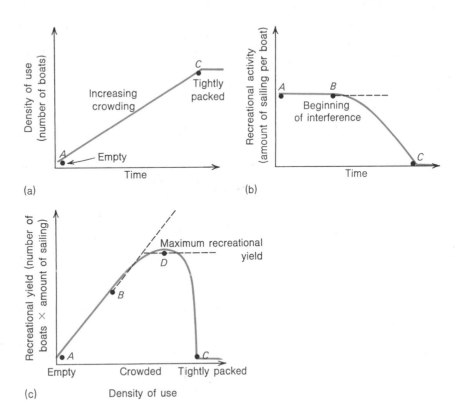

Figure 8-11. Crowding and resource use. The graphs show characteristic curves for the buildup of population pressure on a given recreational resource. In this case, the buildup of sailing boats on a small lake is shown to affect the "recreational yield" of the lake. Details of the relationships in the three graphs are discussed in the text.

way; charging a fee, rationing, or exclusive club memberships are others; but each has undesirable social implications. Of course, no sensible family would launch a dinghy onto an already crowded lake. Thus we might expect feedbacks to operate which would keep the maximum population at around point E. Latecomers would be likely to give up the idea of going sailing or go on to a more distant lake. But if the lake has other uses, if it is used for fishing or as a water supply, then even point D may be too high if it causes too much disturbance for the lake to be used for these other purposes. (Compare this situation to the Chew Valley Lake problem discussed in Chapter 19.)

Once again, the problems of the lake are representative of other and wider problems. Pressures on beaches, on wilderness areas, and on historical sites all raise broadly similar problems of trying to preserve resources—and yet obtaining the maximum yield from them. It is to this theme of resource conservation that we turn in the next and final section of this chapter.

8-4 | CONSERVATION OF NATURAL RESOURCES

The notion of conserving the earth's natural resources is an immediately appealing one. No reader would willingly see Great Plains topsoil eroded, Tennessee hillsides filled with gullies, or redwood groves needlessly felled. But what about a natural gas field in Oklahoma? What are the special merits of leaving the gas in the ground? Who benefits?

Some definitions

Let us begin by trying to define resource conservation. One oft-quoted definition runs as follows: "*Resource conservation* is the scheduling of resource use so as to provide the greatest yield for the greatest number over the longest time period." This definition fits renewable resources nicely and encompasses the targets we met in our discussion of sustained-yield forestry, though it is less easy to apply to nonrenewable resources (e.g., our gas field). Restricting the use of a finite resource means that we are saving for future generations some portion of a resource that would otherwise have been used in the present generation.

Over the short run this idea is appealing, since it encourages the use of other resources and stimulates substitution. However, we have already noted that few natural resources of the past were ever fully worked out (though some natural gas fields may stand as an exception); instead, they were priced out of production. Such is the rapidity of technical substitutions that resources saved by an overcautious conservation policy may never be used at all. Indeed, by limiting the investment of the present generation in them, we may actually be making future generations poorer. Clearly, this is a hard view to accept, but if we really wish to make the best use of nonrenewable resources it makes sense to use up the cheapest and best ones first and then go on to the more inaccessible ones. Clearly, timing is of the essence; and a more useful definition of resource conservation than that above is simply that it is "the optimal timing of the use of natural resources." And along with acceptance of this revised definition goes acceptance of the possibility that the optimal time for the use of some nonrenewable resources could be right now.

The conservation movement

Even if preservation for preservation's sake is untenable for nonrenewable resources, there remains a wide range of resource areas where preservation is both sound ethics and sound economics. The richness of natural recreational areas in the United States owes much to the ethical views of Gifford Pinchot. Head of the U.S. Forest Service from 1898 to 1910, Pinchot caught President Theodore Roosevelt's interest and aroused his enthusiasm. Roosevelt added more acres of the United States to its protected forest lands than all the presidents before and after him.

Presidential concern with conservation has, however, a long history. George Washington was acutely anxious at the loss of soil from his Mount Vernon estate and had his servants bring up mud from the Potomac River to fill in gullies. Nearly two centuries later, in 1970, President Nixon set up a National Environmental Protection Agency. In the intervening period, and particularly since Pinchot, legislation governing the use of all kinds of natural resources has been passed by the Congress. The National Park Service was established in 1916, and the first international treaty relating to resource protection (with Canada, on protecting migrating birds) was signed in the same year. The 1930s were a decade of special concern with soil erosion and water control and saw the passing of the Taylor Grazing Act (1934), the setting up of the Soil Conservation Service (1935), and the passage of the Flood Control Act (1936). In the last thirty years, the emphasis has swung to a concern with conserving waste from the use of mineral resources, to protection from pollution (via the 1970 Clean Air Act), and to general improvements in recreational facilities and improved environmental standards.

Federal legislation on conservation in the United States has been important on a global scale. It not only initiated complementary legislation at the state level *within* that country, but was widely copied in other areas. The Tennessee Valley Authority (TVA) set up by New

Deal legislation in 1933 served as a basinwide "demonstration farm" for watershed-control programs elsewhere. The São Francisco Valley scheme in Brazil, the Gal Oya scheme in Ceylon, and the Snowy Mountain scheme in Australia are examples of the scores of developments that followed, in some measure, the TVA lead.

Currently, interest in conservation is expanding spatially in two ways. First, at the global level, there is increasing concern with international cooperative action on the wise use of resources and on environmental protection. United Nations agencies are now playing a key role in setting the world on the long road toward acceptance of international standards of air pollution control and joint action on the use of and abuse of ocean resources. Second, there is unprecedented conservation activity at the local community level. Locally active groups involved in conservation-minded activities range from Sierra Clubs and Friends of the Earth, right through to local Women's Institutes in English villages valiantly protecting their oaks and elms from the chain-saws. This significant increase in public support for conservation measures at both international and local levels represents a most encouraging shift in man's view of his place in the ecosystem. Whether it is enough of a shift in view of the unprecedented inroads now being made on all our natural resources—both stocks and flows— is still in doubt.

Reflections

1. List the main characteristics that separate *renewable* from *nonrenewable* resources. Can you think of any examples of resources that (a) fall on the borderline between the two types or (b) have changed from one type to the other as human attitudes toward them have altered?

2. Debate the contention that "resource shortages are simply the result of prices that are too low." What happens when the price of a natural resource rises? Does this have geographic implications for the spatial pattern of its production?

3. Review the natural nonrenewable resources of your own state or local area. Look for examples of resources which were once important but are no longer so. What factors account for this?

4. Follow up on the problems of nuclear-energy production by reading some of the material suggested in "One step further . . . ". What are your own views on the balance between the advantages of this fuel source and its possible environmental hazards? Find out if nuclear power plants are planned for your own area and, if they are, where they will be located.

5. Evaluate the model of the recreational yield of the lake in Figure 8-11. How would you go about restricting the use of a wilderness area to ensure that it

was not degraded by overuse? Who benefits from such restrictions? Who pays for them?

6. Think critically about the conservation movement. Would you favor a $10 or a $100,000 fine for shooting members of an endangered bird species (e.g., bald eagles)? Is there a price limit on environmental protection in terms of other social goods? How much conservation can we afford?

7. Review your understanding of the following concepts:
   (a) the planet's total stock   (f) the green revolution
   (b) resources   (g) sustained-yield forestry
   (c) reserves   (h) recreational yield
   (d) resource substitution   (i) resource conservation
   (e) fossil fuels

One step further . . .

A useful way to begin is to browse through readings that give a wide range of approaches to the resource—population relationship, such as

> Burton, I., and R. W. Kates, Eds., *Readings in Resources Management and Conservation* (University of Chicago Press, Chicago, 1965).

The notion of resources, how they are defined, and how we use them is presented in two substantial works,

> Barnett, H. J., and C. Morse, *Scarcity and Growth: The Economics of Natural Resource Availability* (Johns Hopkins Press, Baltimore, Md., 1963), and
> Firey, W. I., *Man, Mind and Land: A Theory of Resource Use* (The Free Press, New York, 1960).

Future patterns of natural-resource availability are discussed at length in

> Fisher, S. C., *Energy Crisis in Perspective* (Wiley, New York, 1974).
> Landsberg, H. H., et al., *Resources in America's Future: Patterns of Requirements and Availabilities 1960–2000* (Johns Hopkins Press, Baltimore, Md., 1964).

The special problems of mineral resources and recreational resources, respectively, are attractively reviewed in two recent paperbacks by geographers:

> Warren, K., *Mineral Resources* (Penguin, Harmondsworth, 1973), esp. Chaps. 10 and 11; and
> Patmore, J. A., *Land and Leisure* (Penguin, Harmondsworth, 1972), esp. Chaps. 1, 2, and 6.

Important papers on the fuel and power problem from *Scientific American* are now available in book form in

> *Energy and Power* (Freeman, San Francisco, 1971).

Geographers have traditionally played a leading role in resource evaluation and in the conservation movement. In addition to the regular geographic serials, you should also browse through one of the more popular resource-oriented journals such as *The Ecologist* (published quarterly).

Chapter 9

Man's Role in Changing the Face of the Earth

Not here for centuries the winds shall sweep
Freely again, for here my tree shall rise
To print leaf-patterns on the empty skies
And fret the sunlight. Here where grasses
    creep
Great roots shall thrust and life run slow and
    deep:
Perhaps strange children, with my children's
    eyes
Shall love it, listening as the daylight dies
To hear its branches singing them to sleep.

— MARGARET ANDERSON
(1950)

Our opening quotation and our opening photograph in this chapter stand in stark contrast. Biogeographer Margaret Anderson catches gentle moments of reflection on the planting of a young tree and picks up the theme of man coaxing environmental change little by little. Our photograph peers down at an altogether cruder scene. Here a copper-rich hillock near Butte, Montana, is being gouged and blasted away in successive strips until it now forms one of the greatest manmade hollows on the planet. Thus the threads of the last three chapters—increasing human numbers, man's interference in the structure of ecosystem, and the accelerating search for resources—begin to intertwine here, as we see the landscape change.

Landscape change and the role of man in changing the face of the earth have been a constant theme in geographic writings across the centuries. In the United States one of the earliest volumes on these topics came from a Vermonter, George P. Marsh. In his *The Earth as Modified by Human Action* (1874), Marsh drew attention to the unsuspected importance of human intervention in shaping what had hitherto been regarded as natural America. In the present century, Marsh's ideas have been developed in two ways: First, by detailed historical reconstruction, geographers have estimated the magnitude of the effects of man's intervention. This reconstruction has been facilitated by modern techniques of measurement and monitoring. Second, academic (and, increasingly, public) concern over the harmful effects of human intervention on environmental quality has created a convergence of disciplines. Contemporary geographic work in this area draws not only on its own long traditions, but on parallel work in such fields as biology and civil engineering.

In this chapter we look at the net effects of all the ways man has influenced his environment in terms of how he has changed the landscape. For it is in the changing face of the land that we can see most clearly the spatial effects of the several processes disentangled in Chapter 7. However, measuring changes in land use turns out to be a more perplexing problem than it appears at first sight. As we noted in Chapter 4, environmental change also can have natural causes, and in any situation these two sources of change—human and natural—must be singled out and identified. We shall also, in this chapter, assess the extent of human intervention on different geographic scales, from the global to the local-township level; and we shall pay special attention to the present and future roles of man in changing the face of the United States.

9-1 DIFFICULTIES IN INTERPRETATION

We have already encountered, in the last two chapters, some dramatic examples of landscape change. Figure 9-1 shows an aerial view of Brazilian area discussed in Chapter 7. Just over a century ago, this rolling country in the Paraíba Valley (midway between the cities of São Paulo and Rio de Janeiro) was covered with a high tropical rain forest. In the 1850s it was caught up in the great swath of forest felling and burning that preceded the coffee boom in this part of Brazil. The boom lasted hardly a generation, and by the 1890s the plantation houses and the slave quarters were being turned into cattle ranches. In the meantime, the coffee frontier was moving into virgin country hundreds of kilometers to the west.

The environment that we see in the photograph has experienced faster and more radical changes in the last century than in the tens of thousands of years humans have lived there. Instead of the original forest cover, we now have a mixture of fire-controlled grassland and scrub forest. Replacing the original soils on the steep hillsides and debris-choked valleys is a thinned and denuded skeletal soil. Molasses grass from South Africa and eucalyptus from Australia are helping to stabilize the vegetational community and to make it of some continuing use to man.

The story of the Paraíba Valley could be retold with suitable changes throughout the rest of eastern Brazil. (See the maps in Figure 9-1.) Indeed, Brazil's story of widespread deforestation and consequent environmental changes is typical throughout the heavily populated areas of both the tropics and the midlatitudes. Over much of these areas, the environment is increasingly a manmade artifact. Some of the effects of man's impact on the changing landscape are immediate and apparent. The suburban sprawl of Los Angeles into the San Fernando Valley, the flooding of Lake Nasser above the Aswan dam on the Egyptian Nile, and the reforestation of Scottish moorlands are plainly visible evidence of ongoing processes. It is when we extend our investigation to eras before our own that we run into more difficult problems in interpreting change.

Here we look at two basic questions. First, how do we establish the facts of past change? Second, once we have established the facts of change, how do we know what parts of the environmental change we have measured to credit (or debit) to the activities of man?

Historical evidence

As we saw in Section 4-1, a variety of methods have been developed to reconstruct the nature of past environments. Table 9-1 summarizes the methods available for reconstructing changes in land use. Note that as we approach the modern period the variety and accuracy of the

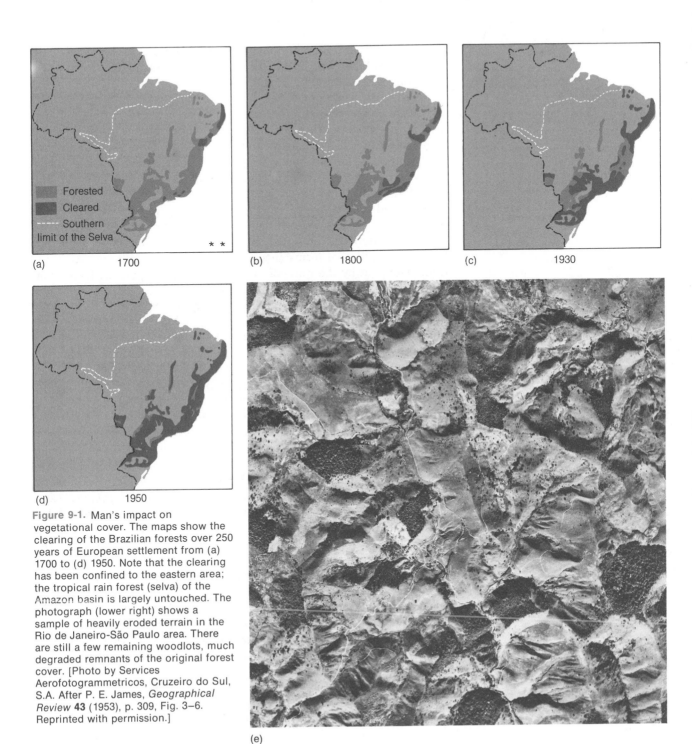

Figure 9-1. Man's impact on vegetational cover. The maps show the clearing of the Brazilian forests over 250 years of European settlement from (a) 1700 to (d) 1950. Note that the clearing has been confined to the eastern area; the tropical rain forest (selva) of the Amazon basin is largely untouched. The photograph (lower right) shows a sample of heavily eroded terrain in the Rio de Janeiro-São Paulo area. There are still a few remaining woodlots, much degraded remnants of the original forest cover. [Photo by Services Aerofotogrammetricos, Cruzeiro do Sul, S.A. After P. E. James, *Geographical Review* **43** (1953), p. 309, Fig. 3–6. Reprinted with permission.]

(a) 1700

(b) 1800

(c) 1930

(d) 1950

Forested
Cleared
Southern limit of the Selva

(e)

evidence tends to increase. How have geographers used these tools, and what results have they found? Let us illustrate the answer by taking a representative case and looking at it in some depth.

Historical geographers, led by H. C. Darby at Cambridge University, have diligently explored manuscript archives from the medieval period in reconstructing the massive changes in the forest cover of Western Europe. The changes can be inferred from an interpretation of such sources as the great survey carried out in Norman England nearly nine centuries ago, during the Domesday Inquest of 1086. For example, one of the questions asked by the Royal Commissioners was "How much wood in this place?" The form of the answer varied among the thousands of replies received from small hamlets, villages, and towns across the country. Some respondents specified the amount of woodland in quantitative terms in one of two ways: by its area or linear dimensions, or by the number of hogs it would support (by feeding on acorns or beech mast). Still other respondents simply stated that there was enough wood for fuel, mending fences, or repairing houses. Darby's patient scrutiny and assembly of detailed and dissimilar information built up a picture of the area of woodlands in early medieval England that shows marked local and regional variations in the pattern of heavily and lightly wooded tracts. (See Figure 9-2.)

Detailed statistical evidence for such an early period in landscape evolution is highly unusual. Even when documentary evidence is missing, however, some land uses can be determined from place names. Figure 9-3 presents the distribution of names of hamlets and villages in an English county with characteristic woodland place names: *leah*, *feld*, *wudu*, and *holt*. The location of these names is well to the north of a band of names (*tun*, *ingham*, and *ham*) indicating early settlement. There is a contrast between early settlement of areas of light gravel and loamy soils in the south and later settlement of areas of intractable heavy London clay to the north. The distribution of this place-name evidence acts as a cross check on and supports the evidence of the Domesday records.

The reconstruction of successive patterns of woodland distribution in medieval Europe indicates that the clearing process was not one of continuous retreat before the axe and the plough. Working at the Würzburg Geographical Institute in Germany, Helmut Jager has found an oscillating pattern in the upper Weser Basin of Germany. As over much of Western Europe, the destruction of war, great plagues, and population declines were reflected in changes in land use; land returned to woodland at times of reduced economic activity and was cleared during prosperous periods when population increased.

The possibilities and limitations of archives as sources of informa-

Figure 9-2. Documentary evidence of change. This reconstruction of the eleventh-century forest cover of eastern England is based on the Domesday survey of 1086. The different symbols reflect the fact that the extent of woodland was recorded in different terms in different parts of the area. For example, over much of the east, woodland was recorded in terms of the number of hogs (swine) it would support, while in the west it was more common to describe its size in terms of now-obsolete measures of length (leagues and furlongs). The exact meaning of certain woodland terms such as "underwood" is uncertain. Despite all these difficulties the main regional contrasts in timber cover stand out. Note that the empty area in the center of the map, now the English Fenland, was largely undrained marshes with some woodland on the few "islands" within the marsh. Some of these islands had settlements and are shown toward the southern end of the marshland. [From H. C. Darby, *The Domesday Geography of Eastern England* (Cambridge University Press, Cambridge, 1952), p. 363, Fig. 106.]

Number of swine for which there was wood
● 500-1000
⊕ 100-500
· Under 100
Acres of wood
◗ Over 250
⊗ 50-250
× Under 50

1 league or 12 furlongs

○ Underwood and miscellaneous

tion exemplify the kinds of characteristics peculiar to a single source of evidence about a past environment. Some of the parallel possibilities and limitations of other sources are summarized in Table 9-1. In practice, the reconstruction process involves using various types of

Figure 9-3. Cross checking of evidence. Geologic factors (a), as well as place-name and Domesday evidence (b-d) for Middlesex county in southeast England, suggest that there was a heavy forest in the northern half of the country but a more broken cover in the south during the eleventh century. [After H. C. Darby, in W. L. Thomas, Jr., Ed., *Man's Role in Changing the Face of the Earth* (University of Chicago Press, Chicago, 1956), p. 192, Fig. 55. Copyright © 1956 by The University of Chicago.]

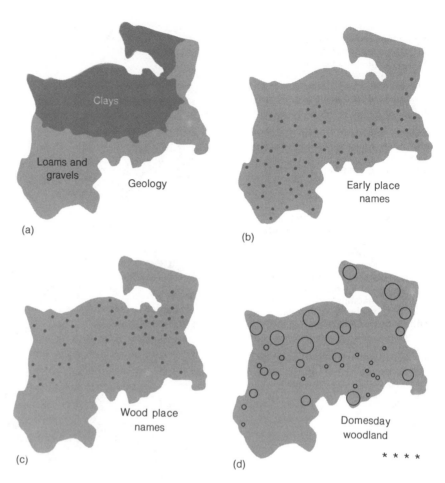

(a) Geology — Clays, Loams and gravels

(b) Early place names

(c) Wood place names

(d) Domesday woodland

★ ★ ★ ★

Table 9-1 Methods of reconstructing land use

Contemporary changes (past 10 years)	Recent historic changes (100 years ago)	Remote historic changes (1,000 years ago)	Prehistoric changes (10,000 years ago)	Postglacial changes (100,000 years ago)
Direct observation			← Radiocarbon dating →	
Aerial photographs		Occasional cross sections (e.g., 1086 Domesday survey)		
← Regular censuses and surveys →				
	Comparison of maps			
	← Written accounts →			
			← Pollen analysis, study of macroscopic remains, lake and bog deposits →	

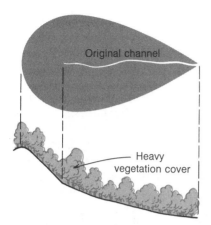

(a) Original morphology

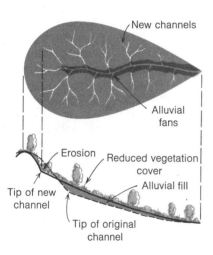

(b) Accelerated erosion

Figure 9-4. Natural changes in the landscape. Environmental changes causing a diminished vegetational cover lead to severe gullying on the slopes and aggradation in the main valley of a small watershed. *Alluvial fans* describe the fan-shaped spreads of eroded materials that form at the lower end of the new channels or gulleys. [From A. N. Strahler, in W. L. Thomas, Jr., Ed., *Man's Role in Changing the Face of the Earth* (University of Chicago Press, Chicago, 1956), p. 635, Fig. 124. Copyright © 1956 by The University of Chicago.]

evidence to verify estimates derived from a single source. It is this cross checking of evidence from various sources that allows a final sequence of landscape changes to be confirmed.

Manmade or natural changes?

Changes brought about by human intervention and changes arising from natural causes may be difficult to separate. Consider, for example, the effect on a small watershed of changes in the rhythm and amount of precipitation (Figure 9-4). Suppose there is a poleward displacement of the present climatic zones that brings semiarid conditions into forested areas. Instead of a year-round pattern of rather light rainfall, we now have a highly irregular pattern of precipitation (occasional large storms separated by droughts). What other changes are likely to follow?

We would expect the droughts and lower precipitation to reduce the moisture available for plant growth and thereby alter the character of the vegetation in the drainage basin. As a result, forest species might be replaced by drought-resistant shrubs. This change would decrease the ability of vegetation to absorb precipitation falling on the soil surface during storms, and increase the amount of material washed downslope. Figure 9-4 shows a sequence of morphological changes within the drainage basin. Increased erosion is followed by severe gullying on the slopes, which in turn is accompanied by the dumping of eroded material in the main valleys (aggradation). The number of streams per square kilometer is increased, and the slopes of both the debris-choked channels and the eroded valleysides are steepened.

Such cycles of aggradation and erosion are well documented for many parts of the world. In the semiarid areas of the southwestern United States, geomorphologists have pieced together a detailed sequence of cycles of gullying (arroyo cycles), although there is considerable dispute over whether these are due to natural variations in rainfall or to alterations in grazing pressures during the post-Columbian period of settlement. More rapid adjustments of the characteristics of channels can be observed over much shorter time periods. Near Ducktown, Tennessee, the denuding of vegetation caused by noxious smelter fumes has transformed a region of few channels and low slopes to one of frequent channels and high slopes. At the same time these gross changes in vegetation and morphology are taking place, more detailed alterations in the structure of the soil and the characteristic pattern of stream flows also occur.

Interpreting environmental change demands, therefore, that we establish *geologic norms*—that is, expected long-term rates of environmental change under natural conditions (discussed in Chapter 4).

Only then can we begin to assess the ways in which human activities have accelerated or distorted these norms.

| 9-2 | THE MAGNITUDE OF PAST LANDSCAPE CHANGES |

How much change has there been in the landscape in the past? To give exact figures on the amount of change induced by the kinds of processes we have described is a much more difficult task than simply itemizing examples of change. Many changes have only recently begun to be viewed as important, and so the desire to measure them is a relatively new phenomenon. However, we can make gross estimates by examining the changes in boundaries between major ecological zones. Within these zones we can get a rough idea of the minor variations that must have accompanied shifts in boundaries by looking at case studies of smaller areas.

On the global level

On the global level, the evidence is particularly difficult to piece together. Table 9-2 lists the assumed distribution of natural vegetation before it was disturbed to any great extent by man. Roughly one-third of the earth's surface was covered by forest; another third was split into polar, mountain, and desert zones; and the remainder was largely open woodland and grassland. The exact estimates given by different authorities vary, but the general proportions given in the table appear to be of the right order of magnitude.

How have these proportions changed under the impact of man? Maps of world population density (see the inside front cover) indicate that well over one-third of the earth—the polar mountains and the dry zones—is unpopulated or very lightly populated. The massive concentrations of population are in environments that originally supported forests or grasslands. The United Nations Food and Agriculture Organization (FAO) estimates that about 10 percent of the world's

Table 9-2 Global changes in land use[a]

Original cover (percentage) (about 10,000 B.C.)		Disturbed cover (percentage) (land use, in 1970s)	
Forest	33	Forest, woodland, and natural rangeland	36
Open woodland and grassland	31		
Desert	20	Desert	19
Arctic and alpine	16	Arctic and alpine	16
		Cropland	10
		Pasture and meadow	19

[a] Figures are averages based on a series of alternative estimates.

land is planted with crops, about 25 percent is forest, and a further 20 percent is grassland (covered by grasses, legumes, herbs, and shrubs). It is difficult to compare the estimates of natural vegetation in Table 9-2 with the FAO figures for current land use because slightly different classifications of land were used. What a comparison suggests, however, is that the natural vegetation has been changed mainly in forest areas. Within these zones, the changes have been highly selective and confined mainly to midlatitude woodlands in eastern North America, Europe, and eastern Asia, and to the monsoon woodlands of South Asia. Large areas of the tropical rain forest and the circumpolar boreal forest remain lightly touched by man.

The reduction in the world's grassland zones has likewise been highly concentrated. The many dramatic changes in midlatitude grasslands, such as the North American prairies or New Zealand's Canterbury Plains, stand in contrast to the less dramatic alterations of the tropical grasslands and savannah zones. Of course, the estimates here refer only to the changes in cover for very general types of vegetation. As we saw in Section 7-2, significant alterations in species composition can occur in areas where the general appearance of the vegetation apparently has been unchanged.

The subcontinental level

At the subcontinental level, the pattern of changes in land use becomes clearer. Figure 9-5 graphs two centuries of figures for categories of land use within the United States. The statistics vary in accuracy from one category to another and over the survey period, but they generally improve in reliability during the last half century. We can see in these figures three types of environmental changes.

First, there was a phase of cropland expansion between 1850 and 1920. The period of this expansion corresponds to the frontier era, when the limits of settlement were extended from the Atlantic seaboard and trans-Appalachian areas to the Midwest, the Great Plains, and the Pacific and Mountain areas. During this period the total amount of cropland increased fourfold, despite the fact that land was going out of cultivation in the East, particularly in New England.

Second, since 1920 the amount of cropland has remained relatively stable. Additions to existing cropland, such as western irrigated areas, have been counterbalanced by the abandonment of farms and the encroachment of urban centers on outlying areas in the East. The expansion in farmland during this period has been due primarily to changes in the ownership of grazing land; that is, land previously in the public domain or owned by the railroad companies has been incorporated into farms. Much of this change has been concentrated in the Great

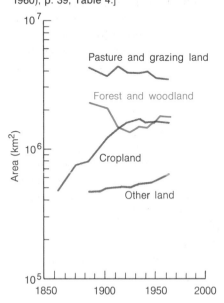

Figure 9-5. Changes in land use in the United States since 1850. Note that the vertical scale is logarithmic. [Data from M. Clawson *et al.*, *Land for the Future* (Johns Hopkins Press, Baltimore, Md., 1960), p. 39, Table 4.]

Plains area. The total area of grazing land as an environmental category has not changed much over this period if we consider both farm pasture and nonfarm grazing land.

Third, the composition of forests has changed drastically. Although the amount of forest land outside the farms has diminished by about one-third, the total area of woodland and forest (despite fluctuations) is much the same now as it was in 1850. Since that time the area of virgin timber has shrunk considerably. But the area devoted to forestry in the sense of managed timber resources has grown, and second-growth timber on abandoned farmland is on the increase. The pattern of change, both in the total area of forest land and in the fraction of the original forest land remaining, has been most noticeable in the eastern half of the country. Here forests were cleared both for lumbering and for cropland. Commercial lumbering showed a distinct east-to-west shift across the country, with New England and the northern Appalachians forming the original center. The first westward movement was into the Great Lakes regions, where white-pine cutting was at its maximum from 1870 to 1890. By 1900 the southern pine region was the chief source of lumber for the national market. The extensive logging of Douglas fir and the shift of emphasis to the Pacific Northwest happened largely after World War I.

The key shifts in land use which began with the general European settlement in the 1700s largely came to an end by 1910 or 1920. Since then, the most significant changes have centered on the small area termed "other" land in Figure 9-5. This land includes urban, industrial, and residential areas (outside farms), parks and wildlife areas, military land, land for roads and transportation, and so on. The rapid growth of urban areas on the one hand and reserved wildlife areas on the other represent the extreme (and compensating) poles of environmental intervention.

On the local level

A small-scale example of changes in land use within the United States is provided by Figure 9-6. This shows the progressive deforestation of a sample area of approximately 10 km² (about 4 mi²) in southwest Wisconsin, Cadiz township, over a 120-year period of European colonization. The map for 1831, before agricultural occupation began, was compiled from the original government land survey. Apart from some prairie and oak-savannah in the southwest corner, the area was covered by upland, deciduous, hardwood forest. By 1882 about 70 percent of the forest area had been cleared for cultivation, and the boundaries of the cleared area reflected the township and range system of land division. (See Section 2-2, esp. Figure 2-9.) By 1902 the forested

Figure 9-6. Changes in land use on the local scale. The maps show changes in the wooded area of Cadiz township, Wisconsin, since the beginning of European settlement. The colored area represents land remaining forested or reverting to forest in each year. [From J. T. Curtis, in W. L. Thomas, Jr., Ed., *Man's Role in Changing the Face of the Earth* (University of Chicago Press, Chicago, 1956), p. 726, Fig. 147. Copyright © 1956 by The University of Chicago.]

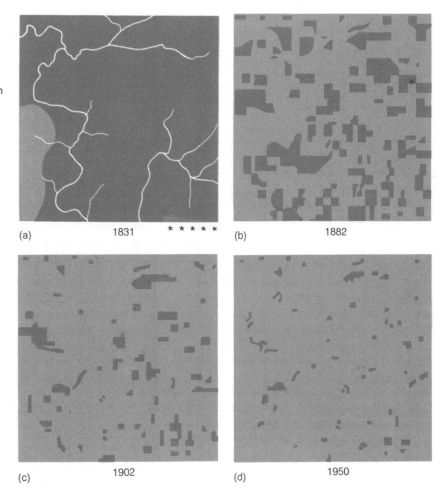

(a) 1831 ★ ★ ★ ★ ★
(b) 1882
(c) 1902
(d) 1950

area had been reduced to less than 10 percent of the total area; it comprised about 60 small woodlots averaging less than 40 acres each. Continued felling of trees for firewood and occasional saw timber, plus heavy grazing by cattle, caused a decrease to less than 4 percent by the 1950s. This diminution in forest land was due largely to a reduction in the size of individual woodlots rather than to the elimination of entire woodlots. The number of these has remained stable in this century.

Detailed investigations by ecologists at the University of Wisconsin have highlighted the significant effects of this environmental transformation. Decreases in the amount of water stored in the subsoil as agricultural fields replaced timber caused springs to dry up; the number of permanently flowing streams had decreased to around a third by 1935. The separation of the forest into isolated blocks reduced widespread burning, while fencing diminished the level of grazing within the woods. Both influences led not only to changes in species composition but to a denser forest cover with more mature trees per acre than under the original, pre-1831 conditions. Animal populations underwent similar changes. Species that were adapted to *edge conditions* (i.e., conditions at the boundary between woodland and open land) benefited from the increased length of the perimeter of the forest caused by its subdivision.

In summary, it is clear that gross changes on the global level conceal subtle environmental adjustments on lower spatial levels. This local case study is one of scores of similar investigations of different ecological zones that illustrate the fine level of adjustment within ecological systems, and the key role that human intervention plays in their shaping. Understanding the interplay of human intervention and environmental adjustments gives us a comprehensive view of the nature and magnitude of changes in land use.

9-3 | ONGOING TRENDS IN LAND USE

What are the probable future trends in the use of land? Our answer must be hedged with caution because the quality of data available to project trends from is generally poor. Even for countries with good statistical records, the information on past land use is unsatisfactory. There is still no comprehensive land-use survey of the United States, and the figures on trends there are culled from a variety of sources, not all compatible. Projections are less hard to make for major land uses, for which the general shape of trends is clear; but trying to measure trends in small but important sectors such as urban lands, land for transportation purposes, and recreational land can be very difficult. Here changes in official classifications may accelerate, dampen, or even reverse trends! Hence, the projections in Table 9-3 should be regarded with more than the usual reservations.

For the United States, the period since World War I has not been one of significant shifts in the primary categories of land use, and the overall pattern of stability is not expected to change greatly in the foreseeable future. As Table 9-3 shows, the overall pattern of land use in the year 2000 will probably not be startlingly dissimilar to that existing now. However, there are likely to be some minor changes.

The regions for which a *marked increase* in area is projected are all essentially urban. They include urban land itself (expected to increase by 141 percent between 1950 and 2000) plus land used by urban dwellers—that is, land for public recreation areas (expected

Table 9-3 Projected changes in land use in the United States, to A.D. 2000[a]

Land use	Projected change, 1950–2000		Probability of change approximately as indicated
	Millions of acres	Percentage change	
Urban purposes	+24	+141	Very high
Public recreation	+49	+107	High
Agriculture	−20	−4	Low
Forestry	−29	−6	High
Grazing	−20	−3	High
Transportation	+5	+20	High
Reservoirs and water management	+10	+100	High
Primarily for wildlife	+6	+43	Moderate
Mineral production Deserts, swamps, mountain tops Unaccounted for	−25	−29	Low

[a] Data from M. Clawson et al., Land for the Future (Johns Hopkins Press, Baltimore, Md., 1960), p. 454, Table 53.

to increase by 107 percent), for reservoirs and water management projects (expected to increase by 100 percent), and for wildlife preserves (expected to increase by 43 percent). We can legitimately regard the last three types of land use as a response to urban demands. Although the exact amount of the increase in each type of land will vary, the direction of change for all these categories seems sure. The area of urban and recreational land will expand to some extent. The total amount of land now being absorbed by all four categories mentioned is small, however, about 18 million acres per decade.

Modest increases in area are projected for land used for managed forestry and transportation (such as roads, rail, airports). The demand for forest products continues to rise, and they are among the few commodities which international price trends indicate are becoming increasingly scarce. Transportation facilities require only a small amount of land, but by the end of the century the amount needed is likely to increase by about one-fifth, a figure which is in line with expected increases in intercity traffic flows. The largest amount of land used for transportation falls within the category of urban land. Streets may occupy up to 40 percent of the land area in the central parts of large cities.

Modest declines are projected for land uses that are essentially agricultural. The area of cropland, unmanaged forests, and grazing land, as well as the area of a residual category (containing swampland, desert, mountain balds, and so forth) is expected to decrease. Exactly what will happen is debatable and will depend greatly on such factors

(a)

(b)

Figure 9-7. The extension and retreat of settlement and its impact on woodland. (a) In the Matanuska valley, Alaska, modern farm settlement is systematically replacing forest by cropland. (b) In southern Illinois, oak and hickory encroach on abandoned fields. Note the relatively uniform texture on the photos shown by the virgin timber stands in Alaska. In contrast the varied texture of the Illinois woodlands shows trees at various stages of regrowth linked to the different times at which fields were abandoned. In both photos the regular checkerboard pattern of fields and roads indicate that land divisions were laid out on the "township and range" system discussed in Chapter 2, Section 2. [U.S. Department of Agriculture photographs.]

as overseas prices for agricultural products and the extent of government support for agriculture. Agricultural surpluses have been an almost regular feature of the American economy, and this plus trends toward still greater productivity per acre and per unit of manpower makes it difficult to foresee any increase in the total amount of land used to grow crops. There are, of course, marked regional variations. The continued absorption of agricultural land in the East for urban purposes is counterbalanced, in part, by a modest expansion of irrigated land in the West. A simple, direct relationship between the intensity of agriculture and changes in land use is not supported by the most recent phase of agricultural production. Industrialized farming (e.g., intensive milk or poultry production) in advanced Western countries is capable of achieving such high yields per acre that the total demand for farmland may be reduced. The increase in woodland, as previously farmed lots in the northeastern and midwestern United States are taken out of cultivation, is related partly to this trend and partly to the changes in the competitive position of farming in this area. (See Figure 9-7.)

Overall, the trend in land use in the United States has matured. The shifts that characterized the frontier era are unlikely to be repeated (at least not until after A.D. 2000). Some changes in land use will take place, but on a local scale. In general, increases or decreases in major categories of land use will probably be less important than changes *within* major categories. Also, the area of land in each category will probably change less than the intensity of use. For example, recreational areas are likely to be much more carefully managed and more heavily used.

How much of what is happening in the United States will be repeated in other parts of the world remains an open question. Many tropical countries appear to be at a stage where rapid shifts in the major categories of land use are still taking place. The remainder of this century is likely to bring greater changes in the landscapes and environments of the less developed countries than in those of North America or Western Europe.

Reflections

1. What have been the main changes in rural land use in your own state or district over the last hundred years? Are (a) similar or (b) different trends likely in the next decade?

2. How do geographers know what changes have occurred in land use? Compare the evidence available on changes over (a) a 10-year period, (b) a 100-year period, and (c) a 1000-year period.

3. Obtain a map which shows the distribution of forested land in your own country. Why are the forested areas located where they are? Is their distribution the result of (a) natural environmental conditions or (b) human intervention? How is the pattern changing?

4. Look at the growth of your own city or community. How large is its population expected to be in 1980, 1990, and 2000 A.D.? Calculate, roughly, the number of new homes that will be required to house these extra people and the additional building land that will be needed. What land will be used for this purpose? What alternatives are there?

5. Review your understanding of the following terms:
   (a) land use   (d) place-name evidence
   (b) Domesday survey   (e) edge conditions
   (c) geologic norms

One step further . . .

The standard geographic account of the topics discussed in this chapter is provided by

> Thomas, W. L., Ed., *Man's Role in Changing the Face of the Earth* (University of Chicago Press, Chicago, 1956).

This is a splendidly produced book with a wealth of talented authors and is strongly recommended for browsing. Another thoughtful book on the same theme is

> Wagner, P. L., *The Human Use of the Earth* (Free Press, New York, 1960).

For a massive survey of Western philosophers' views of man as an agent of terrestrial change, look through

> Glacken, C. J., *Traces on the Rhodian Shore* (University of California Press, Berkeley, 1967).

Detailed accounts of changing land use in specific areas of the United States are provided in

Marschner, F. J., *Land Use and Its Patterns in the United States* (U.S. Department of Agriculture, Washington, D.C., Handbook 153, 1959) and

Clawson, M., *et al., Land for the Future* (Johns Hopkins University Press, Baltimore, Md., 1960).

Descriptions of what is happening in other sample areas are given in

Thomas, M. F., and G. W. Whittington, *Environment and Land Use in Africa* (Methuen, London, 1969) and

Best, R. H., and J. T. Coppock, *The Changing Use of Land in Britain* (Faber, London, 1962).

Research reports on man's impact on environment occur regularly in all the main geographic serials. You should also browse through biological journals like *Ecology* (a quarterly), to see something of the research in neighboring scientific fields, and the *Journal of Historical Geography* (also a quarterly) to see work on past changes in land use.

Part Three

Regional Mosaics

In Part Three we turn from an ecological view of man in relation to his environment to a cultural view of man's organization of the earth's surface. In *Cultural Fission* (Chapter 10) we look at the geographic implications of our nonbiological diversity in terms of such characteristics as religion and language. Just how these affect man's spatial behavior and the way in which they give rise to a highly differentiated system of cultural regions forms the main substance of this chapter. *World Cultural Regions* (Chapter 11) takes the study of cultural regions further by looking back at the ways in which these early spatial forms began to crystallize. Special attention is paid to the explosive growth of Europe since the 1500s, and the dramatic way in which it has shaped the world geography of the present century. To illustrate this theme, the ways in which the main cultural regions of the United States—areas like New England, the Middle West, and California—emerged are studied in more detail. *Spatial Diffusion* (Chapter 12) looks at how information and innovations pass from one cultural region to another, bringing changes both trivial and lethal. We examine the geographic theory and models developed to predict waves of innovation and suggest how the models can be used. Together, these three chapters describe a regional mosaic forged by man's cultural differences but becoming increasingly mixed and muddied by forces overriding older cultural barriers.

Chapter 10

Cultural Fission

TOWARD REGIONAL DIVERGENCE

Although I am unborn, everlasting, and I am the Lord of all, I come to my realm of nature and through my wondrous power I am born.

— BHAGAVADGITA

When the author engaged, usually unsuccessfully, in the schoolyard scuffles beloved of all small boys, the key word to know was "faines." Children, happily, are one of the few remaining uncivilized tribes, and—at least in English village schools in the 1930s—they still retained the local dialect truce words. Shouting "faines" allowed a small boy at the bottom of a pile of bodies a respite or truce. It was important, however, to know your location when you called for time out. In the next county to the west the truce word was "barsy," and in the next county to the east it was "cree"!

Such local differences in schoolyard language over a few tens of miles in as relatively homogenous a country as England are a microcosmic example of the vast cultural differences that shatter the earth's 3.6 billion people into a regional mosaic of immense intricacy and complexity. Jew and Gentile, Moslem and Hindu, WASP and Afro, Maoist and Mennonite, all serve to recall the countless ways in which the single biological species of man seeks to separate and divide itself into different stereotypes.

So in this chapter we move away from the biological and ecological view of man which dominated Part Two of the book and look at the geographic implications of his nonbiological diversity. In this first chapter we try to clear the ground by defining some of the ways in which humanity differs. We begin by asking "What do we mean by cultural differences?" and "How can we define them?" Once these points have been settled, we go on to look at the spatial pattern of some important cultural differences. Here the critical questions are "How does culture vary spatially?" and "Why is it critical for geographers to know about cultural variations?" Finally, we try to use our concepts of culture to build a code by which geographers can separate one part of the earth from another by establishing a system of cultural regions.

In this chapter, more than in any other, the cultural biases of the author must show. No one can shake himself free from deep-rooted and unconscious prejudices that come from being reared in a particular social and cultural setting. Readers are invited to challenge the judgments made here, and to submit for discussion alternate views of the human cultural mosaic based on their own experience.

10-1 | THE NATURE OF CULTURE

In our chapter title and throughout this third part of the book, we use the term "culture." Just what do geographers mean by this term? Since culture turns out to be a complex word, we shall begin with some negatives before going on to a positive definition.

Some negative conclusions

Let us begin by defining what culture is not. First, culture does not mean simply an interest in artistic pursuits. Though the phrase "cultural activities" is commonly used to describe musical, literary, and artistic efforts, this is a special use of the term culture in a very limited sense. Although such activities may have a distinct spatial pattern and may define certain cultural boundaries (e.g., the geographic extent of interest in a virulent variety of bagpipe music may provide a clue to the distribution of folk of Scottish ancestry), this view of "culture" is too narrow for our purposes.

Second, culture does not mean race. *Race* is a biological term used to classify members of the same species who differ in certain secondary characteristics. Man is a single species (*Homo sapiens*) with a common chromosome number (46), and fertile interbreeding among all its billions of members is possible. Nonetheless, specific biological differences do separate men into recognizably distinct subgroups. (See Chapter 11, Section 11-1.) Such differences range from (a) variations in external features (e.g., in skin pigmentation, in the shape of the skull, in hair type, and in eyefolds) to (b) differences in internal features (e.g., blood types).

Blood types may be an important indicator of group differences that have been maintained for a long time. Thus the Basque-speaking population of southwest France and northern Spain has a very high proportion of Rhesus-negative blood types compared to the population throughout most of Europe. Moreover, Rhesus negativism in blood groups is a peculiarly European trait; it is extremely rare among Asians, Africans, and American Indians.

Racial differences of this kind are almost certainly due to long periods of isolation in which genetic variations are accentuated by generations of interbreeding. This hypothesis seems to be supported by the spatial distribution of certain genetically determined diseases. Some of these diseases are highly localized: Kuru, a progressive disorder of the nervous system, is so far known only among the Fore peoples of eastern New Guinea. Other diseases have distinct but intercontinental distributions. Sickle-cell anemia is a disorder of the blood cells found in much of Africa south of the Sahara (and by transfer, in the black population of the Americas) and in another broad belt running from India to Indonesia.

Whether genetic drift (random changes of gene frequencies) permits long-term adaptation of groups to particular environments is not clear. Some differences have been medically established—for example, differences in the sweating capacity of blacks and whites, the fact that Eskimos' skin temperatures remain higher than normal in cold weather, and the lower metabolic rates of Australian aboriginals at night, which allows them to withstand low night temperatures. The cases of genetic adaptability stand in contrast to the extraordinarily widespread ability of people of all kinds to adapt *technically* to extreme environments by devising life-support systems that range from parasols and the fur wrap to spacesuits for moon-walkers.

Some positive conclusions

Considerable efforts have been made by cultural geographers to define culture in a precise and positive way. We can perhaps summarize their views by saying that *culture* describes patterns of learned human behavior that form a durable template by which ideas and images can be transferred from one generation to another, or from one group to another. Three aspects of this definition need further accenting. First, the transfer is not through biological means. The *same* newborn child will grow up with quite *different* sets of cultural characteristics if it is reared in different cultural groups. Second, the main imprinting forces in cultural transfers are symbolic, with language playing a particularly important role. (By "imprinting" is meant the spontaneous acquisition of information, particularly those habits of speech and behavior acquired in the early years of life.) Third, culture has a complexity and durability which make it of an entirely different order from the learned behavior of other, nonhuman animals.

The diversity of human cultures and their innate complexity is staggering. Anthropological research by such scholars as Claude Lévi-Strauss have long since done away with the notion that there are any "simple" cultures. Even in the smallest and most "primitive" cultural groups (e.g., a small Amazonian hunting tribe or the population of a Micronesian village), there is a massive amount of cultural information to acquire. The child growing up in the simplest society slowly acquires millions of pieces of a cultural pattern, which will be duly passed on (albeit in some modified form) to the next generation. We should note (a) that such transfers are independent of formal education (in the Western sense of going to school) and (b) that the transfer process is always incomplete. Thus the culture of the group is always several times larger than the culture of the individual. The most distinguished Harvard professor or the oldest village elder can never hope to acquire, in a lifetime of study, more than some fractional part of the "genetic code" of the culture of which he is a part.

Huxley's model: If culture is so complex and all-embracing a thing, is there any hope of disentangling it into simpler, molecular units which we can study and comprehend? Let us look very briefly at the solution of one man. One of the simplest ways of categorizing culture was proposed by English biologist Julian Huxley in a comparison of cultural and biological evolution. Huxley's model has three components: mentifacts, sociofacts, and artifacts.

Mentifacts are the most central and durable elements of a culture. They include religion, language, magic and folklore, artistic traditions, and the like. They are basically abstract and mental. They relate to man's ability to think and to forge ideas, and they form the ideals and images against which other aspects of culture are measured.

Sociofacts are those aspects of a culture relating to links between individuals and groups. At the individual level they include family structures, reproductive and sexual behavior, and child rearing. At the group level they include political and educational systems.

Artifacts are those aspects of a culture relating to a group's links with its material environment. Sometimes termed "cultural freight," they include those aspects of a group's material technology which allow basic needs for food, shelter, transport, and the like to be filled. Systems of land use and agricultural production are cultural artifacts, as are tools and clothing of a particular design.

Like all such schemes, Huxley's model is only an approximation of reality. In practice, we find aspects of culture in which the three components seem tangled into an intractable knot and others where a strand can be pulled free and studied individually. (See Figure 10.1.)

Culture and ethnicity: One such knot weds culture and biology. Thus, critical readers of this chapter so far will by now be becoming in-

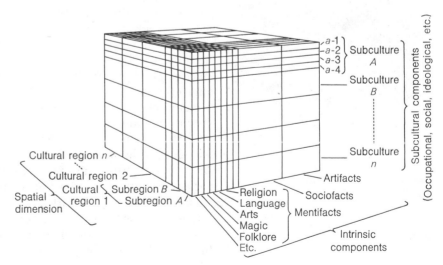

Figure 10-1. The components of culture. Wilbur Zelinsky has created a model of culture using a three-dimensional cube which can be analyzed in terms of (a) its intrinsic components, or (b) the cultural characteristics of a given region, or (c) the cultural characteristics of a distinctive group or subculture. Note how the three approaches interlock, so that we can study small cubes within the master cube, that is, the cube of the spatial distribution of terracing in areas of Chinese culture within the cultural "realm" of Southeast Asia. (See Figure 10-9.) [From W. Zelinsky, *The Cultural Geography of the United States* (Prentice-Hall, Englewood Cliffs, N.J., 1973), p. 73, Fig. 3-1.]

creasingly incensed by an apparent paradox. We saw earlier that race and culture are two entirely different concepts. The first is related to biologically determined and minor differences in the human species which are genetically determined, that is, imprinted on the unborn fetus. The second describes major differences in learned human behavior which are culturally determined, i.e., imprinting begins only after birth. Thus if we take identical twins and bring them up in different areas within the same city (say, Sutton Place and Harlem in New York, or Beverly Hills and Watts in Los Angeles), their biological characteristics will be unaltered but their cultural characteristics will be recognizably different by adulthood.

Despite this fundamental difference in the way cultural and biological characteristics are transferred from one individual to another, we see all around us an apparent association between ethnic divisions and culture. Minor biological differences in skin color are used as visual signals to identify a whole chain of associated cultural characteristics. At the international level, we talk about the voting behavior of "black African states" at the United Nations. Within the United States, scholars like Chicago anthropologist Melville Herskovits, who wrote *The Myth of the Negro Past* (1941), have tried to sort out just what aspects of black culture are directly tied to African cultural heritages. Meantime, sociologists like Gerald Suttles, whose seminal study of the microgeography of a few square blocks of Chicago's Near West Side, *The Social Order of the Slum*, was published in 1968, have shown how sharp are race-related differences at the local level.

Such overlaps between culture and race are themselves mentifacts. That is, they relate to the images human groups associate with minor genetic variations. In studying the geography of the real world, we shall need to be aware of the blurred pattern of human biological variation because of the way it magnifies cultural variations. Figure 10-2 depicts the largest indigenous city in Africa, Ibadan, in the western zone of Nigeria. The city can be divided into distinct zones, each with a separate ethnic origin and a different culture. The largest zone is the original core area of the Yoruba town. It is characterized by densely packed compounds, each housing a large and extended family unit. It has a high population density and an intricate system of special markets. On the northern edge of the Yoruba town lies a small zone of Hausa people from northern Nigeria. They fulfill a special role in marketing cattle from the north. In the third zone is the European community associated with the original British colonial administration. It now has houses of banking and commerce, medical and educational services, and low-density suburban housing. In addition to the three broad zones, there are areas dominated by Lebanese and Syrian

Figure 10-2. Ethnic and cultural divisions within a city. The map shows the main regions within the Nigerian city of Ibadan, which has a population of approximately 700,000. [From A. L. Mabogunje, *Urbanization in Nigeria* (University of London Press, London, and Africana, New York, 1968), p. 206, Fig. 26.]

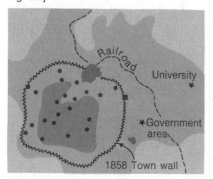

• Yoruba market

■ Hausa market

Core region pre-1850

Western banking and commercial area

Boundary of built-up areas in 1960s

traders; and within the Yoruba area there are important differences in neighborhoods related to the origin, skills, and time of arrival of family groups. The spatial form of the city, its population density, and its economic organization are inseparable from the cultural diversity of its inhabitants. Each ethnic group forms part of a social and spatial mosaic whose components touch but do not penetrate one another.

<table>
<tr><td>10-2</td><td>SPATIAL VARIATIONS
IN CULTURE</td></tr>
</table>

Does culture have a specifically geographic pattern? To answer this basic question, we shall take just one of the many elements in a culture—language—as a proxy for the multitude of other mental, social, and material artifacts that might suggest themselves. But the questions we shall ask will be general ones about spatial stability, order, and change that may provide a basis for your own analysis of other cultural elements that may interest you more.

The scale of cultural variations

Language is the essential linking device in human cultures, enabling members of a group to communicate freely with each other. Just how many different languages are spoken today depends partly on our definition of language. If we leave out minor dialects, there are still around 3000 different languages in current use. At least another 4000 were once spoken but have now gone out of use.

One of the most useful ways of classifying languages is according to the number of people who speak them. *Global languages* are spoken by very many people indeed; local languages are spoken by very few. English, a global language, is the primary tongue of around 350 million people (almost 1 out of every 10 persons in the world) and serves as a second language for many more. However, the most widely spoken language (though those who speak it are more concentrated spatially) is Mandarin Chinese, which, with its many dialects, is the language of about 600 million people in East Asia. Figure 10-3 shows the principal languages spoken on the Indian subcontinent. Were we to rank the world's languages, putting the global ones like English and Chinese at the top, these would come about halfway down the list. At the bottom of the list would be the really local languages. Research in New Guinea has revealed some wholly distinct languages (uncomprehended by neighboring groups) confined to single valleys, spoken and understood by only a few hundred people, and having a spatial extent of less than 65 km² (25 mi²). Actually, a few languages are spoken by a disproportionately large percentage of the human population. The top 14 languages are spoken by 60 percent of the world's people. At the

Figure 10-3. Linguistic differentiation of cultural areas. The map shows the principal languages used in the Indian subcontinent. Darker shading is used for areas in which two or more languages are spoken. [Based on Karta Narodov Indostana. From J. O. M. Broek and J. W. Webb, *A Geography of Mankind* (McGraw-Hill, New York, 1968), p. 109, Fig. 5–3. Copyright © 1968 by McGraw-Hill, Inc. Used with permission of McGraw-Hill Book Company.]

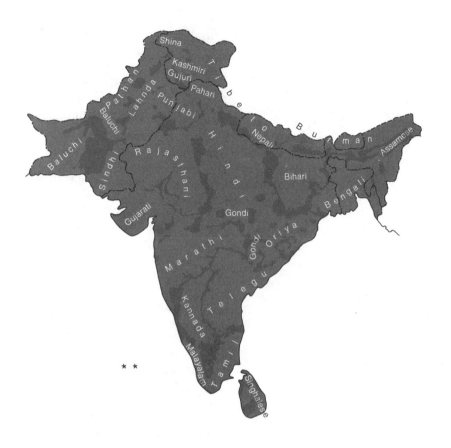

other extreme, the bottom 500 are divided among no more than 1 million people in the remoter parts of Asia, Africa, and Australasia.

The origin and dispersal of cultural elements

Language illustrates clearly a second theme in cultural geography—the origin and dispersal of cultural elements. The questions we must ask to understand this process are about the relation of one language to another.

Extensive linguistic research has revealed that many different languages seem to have emerged from a common stock. For example, it is possible to trace back Indian languages in the northeastern United States like Cayuga, Seneca, and Tuscarora to a common Iroquoian stock which has some linguistic connections with Sioux language groups further west. However, the language group whose evolution we know most about is the Indo-European family of languages. (See Figure 10-4.) A wealth of written records in these languages has allowed us to

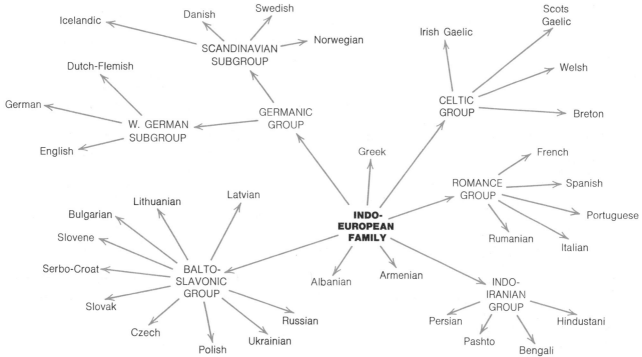

Figure 10-4. Linguistic origins and differentiation. About one-half of the world's 3.6 billion people speak languages in the Indo-European family. The chart shows the main languages in this family and the links between them. Note that some European populations speak languages outside the Indo-European group (e.g., Finnish and Basque).

unravel slow linguistic drifts over the centuries. Despite the fact that these languages are spoken by half the world's people they still have many simple, basic words in common. The English word "mother," for example, is "Mutter" in closely related German. It is also recognizable in quite different subgroups—in the Romance group, in the Spanish "madre"; in the Balto-Slavonic group, in the Russian "mat," and in the Indo-Iranian group, in the Sanskrit "mata." Even in Greek, whose position in the family language tree is still hotly debated, mother is "meter."

The distilling out of the main language groups shown in Figure 10-4 was a slow process, taking over tens of thousands of years. Changes over a much shorter period are noticeable in dialects within a language, but languages themselves are stable enough to provide useful spatial signals of the migrational history of various groups.

Forces of spatial change

Cultural patterns are clearly not static in either time or space. The proportions of the world's population speaking each of the major languages is changing, and their spatial distributions are waxing or waning. Patterns of language are changed not only by the demographic tides of birth and death—which affect mainly our "first," or "native," language—but by the aggressive spread of second languages. Currently, the proportion of English-speaking persons is rapidly increasing in the urbanized and "Westernized" world. Figure 10-5 summarizes the main forces at work hammering out these changing linguistic patterns.

At the same time as global languages are spreading, some small languages are slowly dying out. The Celtic languages of Western Europe, for example, have been losing ground for centuries to the more aggressive English and French tongues. Celtic languages were once spoken in the western parts of the British Isles, the Brittany peninsula in France, and northwest Spain. One of the Celtic languages—Cornish— was confined to the extreme southwest corner of England. Until the 15th century, Cornish was spoken over most of the county of Cornwall; but by 1600 it was heard only in the extreme west. The mining industry brought an increasing number of English-speaking outsiders into the area, and by 1800 the language was virtually dead. The last Cornish-speaking person died in the 1930s. Even much stronger Celtic languages like Welsh are now confined to a part of their original area.

Figure 10-5. Spatial changes in languages. The chart gives a schematic view of some of the main forces causing linguistic changes. Part of eastern Canada is bilingual (speaking French and English). Creole languages (e.g., the French creole in Haiti) have emerged from the mixing of French or Spanish with Caribbean languages. "Pidgin languages" are very basic forms of English spoken as a trading language in many Pacific island communities. "Loanwords" are borrowings due to human communication. (The words *jazz* and *taxi,* for example, are loanwords common to very many languages.)

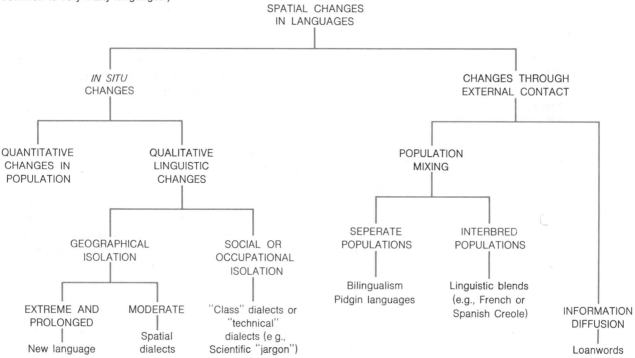

Figure 10-6. Contraction of culture areas. The core of the Welsh language and characteristic Celtic culture has retreated westward and now lies well inside its traditional political boundary with England. Vigorous steps are being taken to halt the decline of the language as part of a resurgent Welsh nationalist movement, but the cultural forces bringing about the decline of local languages are rather pervasive. "Welsh speakers" are defined as those persons with an ability to speak the language; the use of the phrase does not necessarily imply that Welsh is the primary language in an essentially bilingual population. [From E. G. Bowen, *Institute of British Geographers, Publications* **26** (1959), p. 4, Fig. 2.]

* * *

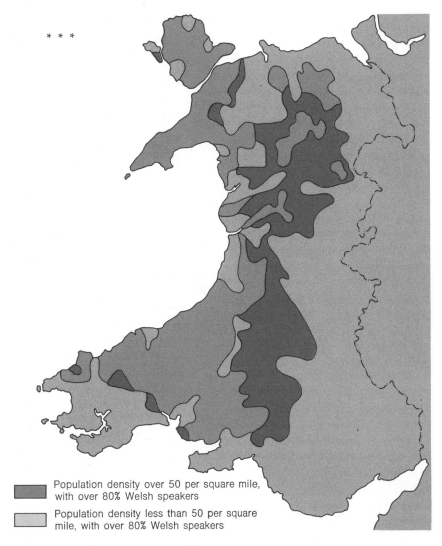

Population density over 50 per square mile, with over 80% Welsh speakers

Population density less than 50 per square mile, with over 80% Welsh speakers

(See Figure 10-6.) Only in Ireland, where Irish (also called Gaelic, or Erse), has been revived and taught in the schools as part of a program to stimulate the national sense of identity, has the language held its ground.

Language and cultural landscapes
We noted in Chapters 2 and 9 that the pattern of place names in a country can also be used by geographers as evidence of cultural influences. Studies like those of the English Place Name Society or lin-

guistic geographer Hans Kurath in the eastern United States allow detailed reconstructions to be attempted.

In England there are major differences in the place names left over by the different invading groups that overran that country. If we take names meaning homestead or settlement as an example, we find the early Celtic names (e.g., those with the Cornish prefix *tre-* as in Tremaine, "the settlement of the stone") confined to the western peninsulas. The period from the fifth to the seventh century brought Anglo-Saxon invasions from the continent, leaving a swathe of names ending in *-ham* or *-tun* (e.g., Aldeham, "old homestead," or Skipton, "sheep farm") in the south and east of England. Danish invasions in the ninth century were confined to the northeast of the country and it is there that *-by* and *-toft* place names are found: hence Normanby ("settlement of the Norsemen") and Wigtoft ("homestead on the creek"). It is ironic that by the time of the Norman-French invasions in the eleventh century the spatial pattern of settlements was wellnigh complete. So despite the near-revolutionary effect of this invasion on the structure and form of the old English language (with its German roots), the impact on the cultural landscape in terms of place names was very small. The French names that do occur (e.g., Beauly) are usually related to large estates.

Kurath's work is based on direct analysis of the speech of rural populations in the eastern United States over the last forty years. He finds that dialect terms show considerable richness traceable to the original admixture of English, German, Dutch, and Scotch-Irish settlers. His study of the words used for farm buildings illustrates this variation in dialects. The loft of a barn is sometimes the *high beam* in New England (except in Connecticut), *overhead* mainly in Pennsylvania, and the *mow* from Virginia to Maine.

10-3 | GEOGRAPHIC IMPACT OF CULTURAL VARIATIONS

Cultural variations have an intrinsic fascination. Since culture is a distinctively human characteristic, any geographer trying to understand the mosaic of world regions must emphasize its cultural variety. In addition, geographers are concerned with cultural variations because of their important secondary impact on a wide range of relations between man and his environment.

We can illustrate this theme by taking a second cultural element (religious beliefs) and tracing its impact on two topics of general interest: (1) attitudes toward the use of resources and (2) attitudes toward innovation.

Spatial variations in religious beliefs

Religion is a central, some would argue *the* central, element in man's cultural differentiation. Figure 10-7 shows variations in religious beliefs on the world scale. Each of the world's main religions has a distinctive geography. Christianity's 1 billion adherents are located largely in Europe and the Near East, the Americas, and Australasia. Islam has diffused from its birthplace in western Arabia through the northern half of Africa, central Asia, and India, and into Indonesia. Hinduism and Buddhism are highly localized, the former being largely confined to the Indian peninsula and the latter to East Asia.

We can, of course, divide each major religion into its various subgroups. If we examine the Christian subgroups within the United States, we find a strong zonal pattern. The Roman Catholics are strongly represented in New England and the industrial Northeast; the Baptists in the southern states and Texas; the Lutherans in Wisconsin, Minnesota, and the Dakotas; the Mormons in Utah. Even within a metropolitan area geographic differences in religion may occur, with Christian Protestant churches most frequent in the wealthy suburbs.

Figure 10-7. Cultural core areas. The map shows Old World core areas for the world's main religions. The figures indicate the approximate number of adherents of each religion, in millions, in the 1960s. The lines showing the expansion of Buddhism (with dates) are approximations only.

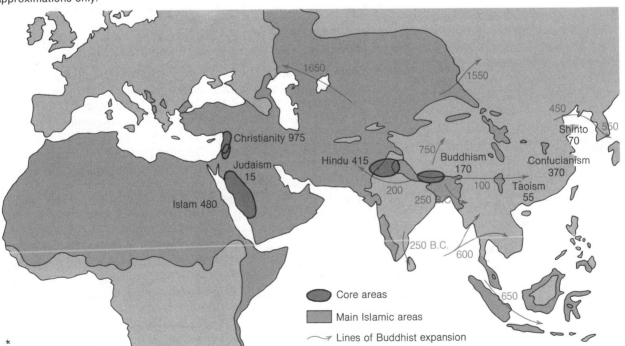

Core areas

Main Islamic areas

Lines of Buddhist expansion

Why are these variations important in determining the cultural mosaic of the world? Religion's role in group organization, its close relationship to politics and the state, and the attitude of churches toward change and development are part of the answer. Most of the great political conflicts of world history have had a religious basis, and lines of conflict (like those between Israel and Egypt or within Northern Ireland) may still be drawn up along religious divides. Some aspects of religion's role in determining geographic divisions will be considered in our treatment of boundary conflicts in Chapter 17. Here we shall confine ourselves to illustrating its impact in two more immediately relevant areas.

Philosophies of resource use Views of natural resources cannot be separated from a group's view of its role and purpose in existing on the earth's surface. It has been argued that Protestant (and particularly Calvinistic) ethics preached the virtues of thrift, the accumulation of capital, and the storing of resources through the accumulation of capital for the benefit of the next generation. Conversely, it has been said that Hindu and Buddhist beliefs encouraged an otherworldly view, and—in Western eyes—the investment of a ruinous amount of capital in ceremonies like funerals linked to life in other worlds.

Chinese civilization underlines the extent of such contrasts. For example, when the newly founded Chinese University of Hong Kong came to lay out its new buildings—including a geography department—at Sha Tin, an unusual locational factor had to be considered. Sites were chosen in which *feng-shui* ("local currents of the cosmic breath") were harmonious. Chinese landscapes continue to have paths, structures, and woodlots designed to blend with the natural landscape rather than dominate it. This response to the environment is not uncommon. For most of the period of man's occupation of the earth, most of its peoples believed that individual natural elements—trees, springs, and hills—had guardian spirits. Before such objects could be used, these spirits had to be raised and mollified.

Historian Lynn White of the University of California has argued that the Judeo-Christian tradition played a major role in replacing this sensitivity to the natural environment with indifference. For example, Christianity fostered the idea of man as an ecologically dominant species of a world whose resources were freely available for the benefit of man. Other historians, not to mention theologians, would strongly dispute White's thesis; to blame Biblical teaching for the whole "cowboy economy" approach to natural resources in the Western world would be going too far. What is important from our viewpoint is that the relationship of man to the environment is not just a matter of demography and economics. Beliefs and fantasies, traditions and taboos, also have their place in any interpretation of man's relations with his environment.

One of the clearest illustrations of this fact is in the *plural societies* and *dual economies* that have grown up within parts of the humid tropics. Figure 10-8 presents a series of maps of Vanua Levu, one of the largest of the Fijian Islands in the southwest Pacific Ocean. The maps underline a significant distinction between the old and new Melanesians. The old Melanesians have a traditional economy characterized by a culture with limited wants, a fatalistic and resigned attitude toward life, and an apparent incapacity or unwillingness to develop large-scale commercial organization. Alongside this culture exists the new Melanesian economy linked to the European discovery of the island and the subsequent introduction of labor from India. This

Figure 10-8. The economic impact of plural societies. The maps show the distribution of (a) the new Melanesian (Indian) and (b) the old Melanesian (Fijian) population of the Fijian island of Vanua Levu in relation to (c) major cash crops. One dot is equal to 40 persons. [From R. G. Ward, *Land Use and Population in Fiji* (Her Majesty's Stationery Office, London, 1965), pp. 87 and 92, Figs. 4–3 and 4–6. The maps originally appeared in *Geographical Review* **49** (1959) and *Journal of the Polynesian Society* **70** (1961), p. 258.]

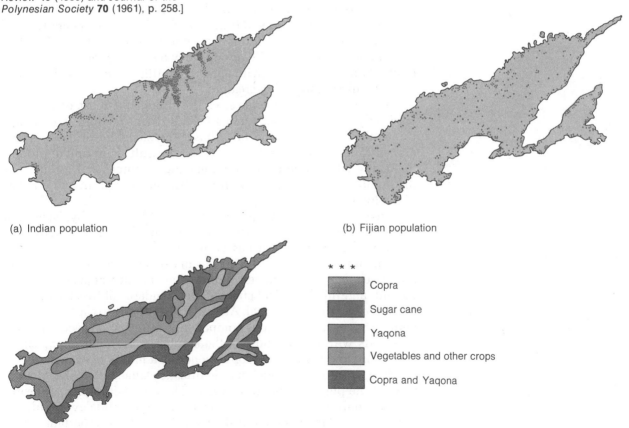

(a) Indian population

(b) Fijian population

(c) Major cash crops

* * *
Copra
Sugar cane
Yaqona
Vegetables and other crops
Copra and Yaqona

economy is characterized by unlimited wants, a strong motivation to get the best possible return on investments, and a facility for large-scale commercial organization. In such plural societies, two cultures may exist side by side, the one restricting its resource-organizing activities to cultivating a small range of subsistence crops, the other introducing new crops, growing them intensively on plantations for export, and exploiting the local mineral resources. Similar situations are encountered throughout the world: There are Chinese, Malay, and European cultures in Malaysia; Singhalese and Tamil cultures in Ceylon; African and Asian cultures in Kenya.

Diet and agricultural resources Religious beliefs affect agricultural development indirectly through constraints on diet and through the symbolic significance given to animal life. For although man has been biologically designed as an omnivorous animal (to judge from our teeth), a major portion of the world's population restricts its diet in some degree. The world's 170 million Buddhists are generally vegetarians, its 415 million Hindus may not eat beef, and its 15 million Jews do not eat pork. Smaller groups may have still more precise rules; India's Jain communities (with about 2.3 million people) are forbidden to kill or injure *any* living creature.

As a result of these views, cattle throughout most of India are used only as draft animals and to some extent for milk production. Under the Hindu doctrine of *ahimsa*, the slaughter of cows is prohibited in many of the Indian states. As a result, aging and unproductive cattle add to the pressure on grazing resources and estimates of the number of "surplus" cows run to between one-third and one-half of the total.

An extreme view of the importance of cattle is taken by the herding tribes of eastern and southern Africa. Among the Pakot of Kenya, the number of cattle a person has is directly related to his prestige and wealth, and cattle serve as the means of exchange, most notably in the purchase of brides. Numbers rather than quality appear to be important, and this has a bad effect on the standards of livestock and the amount of grazing per animal. Although the attitude to cattle among the herding tribes has a religious component (cattle having been said to be "the gods with the wet nose"), the emphasis on numbers appears to have more to do with their convenience as a means of exchange.

More specifically, religious significance attaches to the Moslem view of the pig as an unclean animal. Thus in Malaya pigs are reared for food only in the Chinese enclaves; the native Malay population follows the Islamic code and deprives itself of an important source of food.

Innovation and modernization

Religions of most kinds lay heavy emphasis on continuity, tradition, and strict adherence to long-established patterns of behavior. They have acted, and continue to act, as a vital stabilizing influence, or — depending on one's point of view — an inhibiting drag on change.

Religion is often held to be a major factor inhibiting the spread of family-planning practices. The moral values attached to the human fetus in Roman Catholic doctrine serve as a major barrier to the spread of abortion and certain contraceptive methods. This barrier may operate at the individual and family level for members of the Catholic faith or may become a matter of national policy in countries where there is a strong link between the Catholic Church and the state. Thus contraceptive devices are banned in Ireland, and different attitudes are taken on abortion laws in various states of the United States.

It is difficult to determine the importance of these attitudes from a strictly demographic viewpoint. Population-control practices are clearly described in the Old Testament and in Egyptian wall paintings dating from 5000 B.C. There is ample evidence that human groups throughout history have been able to control family size when this was considered desirable. Attitudes toward what is the most desirable family size are demonstrably more important than which birth-control method is followed. Thus in Europe, a continent with lower birth rates than any area of comparable size in the world (generally about 8 per thousand), there is no major difference between Catholic and non-Catholic populations at the national level. Countries in which contraceptives and birth-control information are banned or restricted have birth rates just as low as those where both are freely available.

In other areas of human behavior, the influence of religious beliefs on the acceptance of innovations is more clear. Let us take a specific regional example. About 20,000 of the world's 370,000 Mennonites follow the Amish religious code. The Mennonites emerged in the early sixteenth century in Switzerland as a nonconformist branch of the Protestant church, with half their numbers in North America today. The Amish represent an extremist breakaway from that Mennonite movement and are today concentrated in farm communities in certain counties of Pennsylvania and Indiana. Located in the midst of the most highly modernized and swiftly changing regions of the world, Amish communities stand out as islands of tradition. Services are conducted in Pennsylvania Dutch (Palatinate German with some English mixed in), traditional plain clothes continue to be worn, telephones and electric lights are shunned, and the horse and buggy continue to serve as a means of transport instead of the all-pervasive automobile. Here, religious beliefs serve as a cement which continues to hold together a human group with a behavior pattern more reminiscent of seventeenth-century rural Europe than America in the mid-1970s. We shall look at more formal models of resistance to change in Chapter 12.

10-4 | CULTURE AS A REGIONAL INDICATOR

In an earlier part of this book, we set out to divide the world into ecological zones. To some extent, the nine zones proposed in Section 5-2 serve as a proxy of the regional mosaic as well. For example, if we compare Figure 5-9 with the world population map on the inside front cover, we see some notable correlations. Three of the zones—the polar, tundra, and arid (comprising over one-quarter of the earth's land surface)—are virtually empty. By contrast, the midlatitude woodlands and the Asian part of the savannah-monsoon zones contain greater numbers of people.

World cultural realms

But population density is itself too scant a guide to the cultural variety found in the human population. Is a general system of actual cultural regions discernible?

If we take the individual cultural elements of social organization, technology, and language and add to them biological characteristics, we have some of the ingredients for a system of social regions. Various proposals for a global system of cultural realms have been put forward. University of Minnesota geographer Jan Broek has suggested four major cultural realms (Occidental, Main Islamic, Indic, and East Asian) and two minor realms (Southeast Asian and Meso-African). The extent of these realms is shown in Figure 10-9. As in the case of environmental regions, the divisions on the global level are only grossly accurate.

Within each realm each cultural element can be used to create finer and finer subdivisions along the lines suggested in Figure 10-10 for the Meso-African realm. Thus, we can divide a Bantu group on the basis of language. Bantu has about 60 distinct component languages, like Venda or Nguui, and roughly 300 tribal dialects. Differences in the distinctive material culture of the Bantu—its land tenure system based on hoe agriculture, cattle, settled villages, and iron, copper, and gold metallurgy—can be used in the same way, to distinguish subgroups within the primary culture. For example, we can recognize three separate types of Sotho pottery. At each stage in the classification process the group of pottery makers we are talking about becomes smaller and, usually, confined to a more limited spatial area.

Cultural traits as regional indicators

The danger here is that we shall end up with a mass of very small units of interest only to very few. The geographer's task is to bring in local

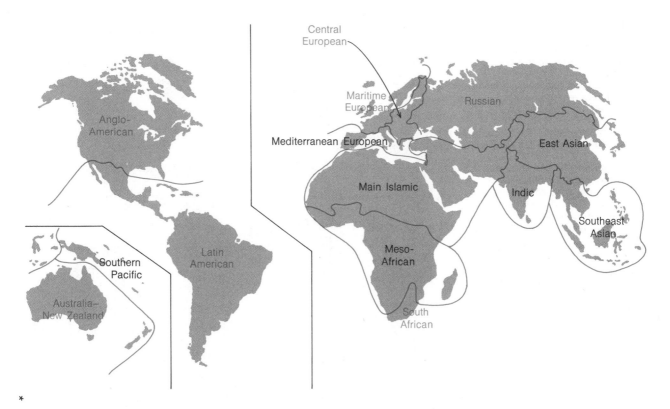

Figure 10-9. Major world cultural realms. The map shows Jan Broek's proposed divisions of the world into cultural realms, with the Western (Occidental) realm split into subregions. [From J. O. M. Broek and J. W. Webb, *A Geography of Mankind* (McGraw-Hill, New York, 1968), p. 189, Fig. 8–3. Copyright © 1968 by McGraw-Hill, Inc. Reproduced with permission.]

details of the culture being analyzed while keeping a firm grip on the overall spatial pattern. Now, in describing cultures some elements are more noteworthy than others. Key elements that seem to provide a guide to other elements are termed *cultural traits*. Among some East African tribes the fact that prestige is gained through owning cattle is a key trait. It helps to explain the group's food, clothing, shelter, and economy and provides a valuable clue to their population density, patterns of settlement, and systems of land use. Such traits can be recognized on all spatial levels. We cannot understand the spatial structure of Swedish towns without being aware of the emphasis that Swedish society places on social welfare, or understand the regional organization of Soviet Russia without recognizing the influence of Marxist-Leninst views on strategies of industrial location (e.g., use of planned heavy industry as the core of economic growth). We can illustrate the geographer's use of culture traits in developing cultural regions by considering, for example, religion and then artifacts.

Religion has been used by American cultural geographer Donald Meinig in delimiting the area of Mormon culture in the western United States. As Figure 10-11 indicates, Meinig recognizes a series of con-

Figure 10-10. Regional indicators. The chart shows alternative criteria for dividing human populations into regional groups. Each of the four groups of overlapping cultural elements spans a large population size range. Thus a *Bantu culture region* could be distinguished in Africa on the grounds of a distinctive set of technologies (here termed the Bantu "technocomplex"), its skein of closely related languages (the Bantu "multilanguage"), or its racial affinities (the Bantu "race group"). Within this general culture region, minor regional boundaries could be drawn down to the level of the different material technologies of different villages and family groups (the "site assemblage"). [Based on D. L. Clarke, *Analytical Archaeology* (Barnes & Noble, New York, and Methuen, London, 1968), p. 361, Fig. 61.]

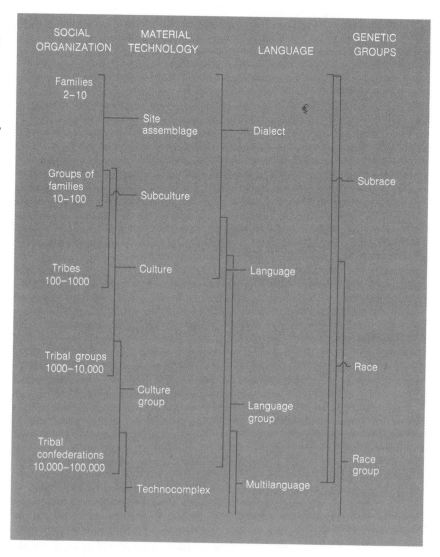

centric regional shells. At the center lies the Mormon *core* area, the key zone where the Mormon population is densest, the religion is strongest, and history of occupation by the group the longest. This core is formed by the Wasatch Oasis, a 210-km (130-mi) strip along the base of the Wasatch Mountains east of Salt Lake City. The area is growing rapidly in population, and for the last half century has contained about 40 percent of the total Mormon population of the United States. It is a principal center of Gentile immigration into the essentially Mormon re-

Core	Sphere
Domain	Metro

* *

Figure 10-11. The Mormon cultural region. The dominance of the Mormon culture decreases as the distance from the Wasatch Oasis core area near Salt Lake City increases. [From D. W. Meinig, *Annals of the Association of American Geographers* **55** (1965), Fig. 7. Reproduced with permission.]

gion, but the Mormon-Gentile ratio nevertheless has tended to remain rather constant.

Areas where the Mormon culture is dominant but with less intensity than in the core, and where local differences in social organization are evident, Meinig terms the Mormon *domain*. The domain extends outside Utah, notably into the river country of southeast Idaho, and has an area more than 20 times the size of the core. It contains a little over a quarter of the total Mormon population, largely in rural settlements, and has a small proportion of Gentiles. Outer zones of influence and peripheral contact, where the Mormons form significant local minorities, are termed the *sphere*. The Mormon sphere forms a fringe all the way from eastern Oregon to northern Mexico, greatly extended in the South and representing the last wave of rural Mormon expansion in the late nineteenth century. About 13 percent of the Mormon population lives in the sphere, where they form a varying minority proportion of the local population. Outside the sphere are small *outliers* (outlying areas) containing the remaining fraction of the Mormon population. The key outliers are in Pacific coast cities, notably Los Angeles. In the last two decades the outliers have been extended outside North America to England, Switzerland, and New Zealand.

Meinig's analysis clarifies the spatial structure of one of the most distinctive regions to emerge on the subnational level within the United States over the last century and a quarter. In the Mormon cultural area, the theological base affects significant aspects of the demography, economic organization, and political viewpoints of the population in the Southwest. Although Meinig has used his method only to analyze other areas in Broek's Occidental realm—notably Texas—it has obvious relevance to other non-Western cultures as well.

Geographers show special interest in the visible impact of mentifacts, sociofacts, and artifacts on the earth's surface. As we saw in Chapter 9, large areas of the world's landscape have been shaped and patterned by human activity. In rural areas, special attention has been paid to the patterns of fields and farms, roads and boundaries. Different cultural groups have had different ways of settling the land and different ways of fixing boundaries, so that strong contrasts in the *cultural landscape* are observable from the air. Figure 10-12 gives typical examples of such contrasts.

Other geographers have taken individual elements in the cultural landscape and traced their origin and spread. Evidence on the spread of covered bridges in America has been carefully assembled by Louisiana geographer Fred Kniffen. Originally developed in Switzerland, Scandinavia, and northern Italy, this type of bridge was first used in the United States in an area running from southern New England to

(a)

(b)

(c)

eastern Pennsylvania. As the datelines in Figure 10-13 show, it spread rapidly through the Midwest and into the southern Piedmont, but only very slowly into northern New England. About 1850, when the number of these bridges was expanding most rapidly in the East, new centers of growth emerged in the western United States—notably in Oregon's Willamette Valley. Thus do artifacts, like bridges and barns, fields and fences, house types and street patterns, provide clues to the limits of cultural regions.

10-5 | REGIONS AS ENIGMAS

When, in Lewis Carroll's *Through the Looking Glass*, Alice objects that "glory" doesn't mean the same as "a nice knockdown argument," she elicits Humpty Dumpty's evasive reply: "When I use a word, it means just what I choose it to mean—neither more nor less." The word *region* has caused centuries of "nice knockdown arguments" among geographers.

Arguments over regional boundaries

We can illustrate the nature of regional arguments by Figure 10-14, which shows how the Great Plains region of the United States has been defined at various times by various scholars. The discrepancies in the three maps have arisen in two ways. First, there is a difference in the criteria used to define the region. The first map describes the limits of the characteristic vegetation of the region and the second the limits of the typical landforms. (You may like to refresh your memory of the character of this part of the United States by turning back to Figure 3-3, which shows a cross section of that region from the Rocky Mountains east to the Mississippi River.) The boundaries of the region in the last map are based on the spatial extent of a particular American Indian culture, that of the Great Plains tribes.

Second, seven different definitions of boundaries, by seven leading scholars, were used for each map. Although the maps may seem to support the theory that "the number of different boundaries for any region is equal to the square of the number of geographers consulted," the area of agreement is considerable. Areas within the shaded zone represent a consensus on what to *include*; areas outside the solid line represent agreement on what to *exclude*. Between lie the areas of disagreement. If we were to trace the boundaries of the region in each of the three maps and superimpose them on one another, we would even find a small area that all twenty-one scholars would be happy to call part of the Great Plains. We may regard such areas (like western Nebraska and eastern Montana) as particularly representative of the special geographic flavor of the Great Plains. As we noted in our

Figure 10-12. Cultural landscapes. Patterns of fields and farms reflect the manner and timing of agricultural settlement. (a) Irregular shapes and sizes of fields in a long-settled area of southwest England. (b) Regular 40-acre fields in an area on the Illinois border settled when the township and range system was used for land surveying. (See Figure 2-9.) (c) A pioneer Canadian settlement in the Lake St. John lowland with farm boundaries perpendicular to a road. [Photos courtesy of United Kingdom Department of the Environment (a), U.S. Department of Agriculture (b), and Canadian Department of Energy, Mines, and Resources (c).]

(a)

Figure 10-13. Elements in the cultural landscape. (a) A typical example of a covered road bridge. (b) Localities with covered road bridges in the United States, with timelines for the bridge's spatial diffusion. [Photo by Hugh Mackey, DPI. Map from F. Kniffen, *Geographical Review* **41** (1951), Copyright by the American Geographical Society of New York.]

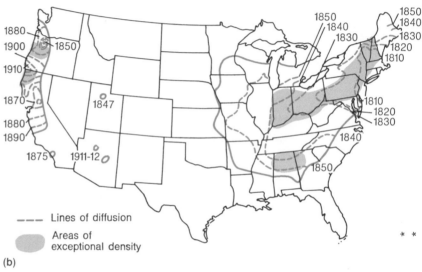

- - - Lines of diffusion

Areas of
exceptional density

(b)

consideration of the Mormon cultural area (Figure 10-11), geographers use the term *core area* to define such regional heartlands, which have a special significance in our study of the regional mosaic of the earth.

Types of regions

We said earlier in this book that a *region* is any tract of the earth's surface with characteristics, either natural or manmade, which make it different from the areas that surround it. In Chapter 5 we saw examples of ecological regions and earlier in this chapter (Section 10-4), examples of cultural regions.

Geographers also draw a distinction between regions on two other grounds. The features which distinguish regions may be singular or plural. Kniffen's region of covered bridges (Figure 10-13) is a *single-feature region*, while the Great Plains core area (Figure 10-14) defined by the overlap of three sets of features, is a *multiple-feature region*. Since culture is such a multifaceted concept, most cultural regions

(a) Region of
distinctive vegetation

(b) Region of
distinctive landforms

(c) Region of distinctive
cultural elements

 Area included within
all definitions

〜 Boundary of area
excluded by all definitions

Figure 10-14. Regional arguments. Alternative definitions are possible even for major geographic regions. The Great American Plains region is defined in terms of three of its characteristics: its distinctive vegetation, landforms, and culture. Each map is based on definitions given by seven leading authorities. [After G. M. Lewis, *Transactions of the Institute for British Geographers,* No. 38 (1966), pp. 142–143, Fig. 11–13.]

tend to have multiple rather than single distinguishing features.

A distinction also is drawn between the degree of spatial organization within regions. *Uniform regions* are defined by the presence or absence of a particular distinguishing feature. A "covered bridge" region is an area in which this element is part of the rural landscape. The boundary of such regions is rather clear-cut compared with that of, say, the Great Plains region, in which we have a clearly defined core and a gradual weakening of "Great Plains" features as we move outward from that core. Regions of this type are termed *nodal* regions. (Some geographers prefer the term "focal" region.) The center is well defined but the regional characteristics die out toward a periphery in a way which makes it very difficult to plot an outer boundary. Perhaps the best examples of nodal regions are the urban regions we shall study in Part Four.

You should not think of the worldwide regional mosaic as a collection of separate, nonoverlapping units with sharp boundaries. Rather, it is a mixture of uniform and nodal regions. An appropriate analogy is not the stained-glass window of your church, but rather the litter of overlapping papers on your desk! The earth's surface is so large and its cultural fissioning so complete that the temptation is always there to break it up into a very fine mesh of regions. But the parts must not be too small. A world system of 999 cultural regions may be an athletic geographic feat, but it is self-defeating if our purpose is to provide a shorthand guide to the world's diversity. Too much information (too many regions) can be as much a problem as too little, and geographers are always searching for the right balance. The ideal set of regions is one which has just enough differentiation to serve our purposes—but not a boundary more.

Regional images

One final reason for the enigmatic quality of regions is that they are cultural assessments. The same qualities may be judged quite differently by different cultures. One of the simplest ways of illustrating this is to show the changing assessment of the same area over time.

Consider Figure 10-15. William Brockedon's engraving of the Val d'Isère in the French Alps was made in 1829. It shows a lyrical

Figure 10-15. The regional image of the alps. Once considered a natural hazard, the European Alps are now regarded as a recreational resource. This early nineteenth-century engraving is symbolic of this change in viewpoint. Its sweeping views comprise a number of well-harmonized elements—the sheep and distant church giving an air of pastoral peace and tranquillity, the two small human figures underlining the smallness of man in relation to the grandeur of nature. [William Brockedon, 1829.]

summer scene with the emphasis on pastoral tranquility in the fore-ground and the majesty of Mont Blanc beyond. These were the Alps which the painter John Ruskin would write of later as "alike beautiful in their snow, and their humanity." This is a view which has persisted until today. It is not, however, the only view of the Alps. Travelers in the eighteenth century, anxious only to cross the Little St. Bernard Pass, would curse "this awful place" in their journals, and hurry on to the welcoming towns of the Italian plains beyond.

Debate over the regional image of newly settled lands runs through the historical geography of European overseas settlement. The grass-lands of North America were variously regarded as the Pastoral Garden of the West and the Great American Desert, and a few early historical descriptions of areas led to mistaken assessments that persisted for decades. In part, the errors were related to changes in semantics over time. The term "desert" originally signified an uninhabited (deserted) place, but it now has a narrower meaning as an arid place without trees or water. In a similar fashion, before the days when height could be determined accurately, the term "mountains" was often applied to hills of modest elevation.

If we examine culturally conditioned appraisals of terrestrial resources and landscapes in the past, we can understand that present viewpoints are similarly affected by our contemporary value systems. For instance, over the past 150 years the climate of California has been successively regarded as unhealthy, ideal, and, most recently, dangerously polluted. Concepts of what is beautiful or healthy have been quite different at various times and have played various roles in the ordering of personal preferences. The notion of the European Alps as places of great beauty and attraction is a Victorian culture concept. Earlier generations regarded them through different cultural eyes.

In the next two chapters, we look at the ways in which the cultural regions and cultural landscapes of the world have evolved and grown. In Chapter 11 we take a historical perspective and trace the slow emergence of culture centers and their spread through population transfers and migration. In Chapter 12 the emphasis is more squarely on the modern period and the exchange of culture through diffusion.

Reflections

1. Consider whether your own county, state, or province can be divided into distinct cultural regions. What bases are there for such divisions? Sketch in some tentative boundaries and compare them with those proposed by other members of the class.

2. Identify the main types of housing in the community around your campus. To what extent are housing characteristics and designs a useful indicator of culture?

3. List (a) the languages you yourself speak and read, (b) those your parents speak and read, and (c) those your grandparents speak and read. How far does your list reflect birthplaces and family migration? Is the range of languages known decreasing, fairly constant, or increasing over the three generations?

4. Consult the Yellow Pages of your local telephone directory and plot the distribution of churches of different denominations in your city or county. Do the different denominations have distinctive spatial patterns? Can you think of any likely causes for these patterns? Check your local findings by

looking at maps of church membership for the whole of your country in its National Atlas or a similar source. (See Appendix C for atlas suggestions.)

5. Use either a place-name dictionary for your own country *or* a local county history to check on the place names of settlements in your own locality. How far do the names of places reflect (a) the origins and language of their founders and (b) the period when they were founded?

6. Review your understanding of the following concepts:

(a) race
(b) mentifacts
(c) sociofacts
(d) artifacts
(e) culture
(f) cultural regions

(g) cultural landscapes
(h) plural societies
(i) cultural traits
(j) regional indicators
(k) uniform regions
(l) nodal regions

One step further . . .

Excellent general introductions to the spatial diversity of human cultural groups are given in two texts:

Broek, J. O. M., and J. W. Webb, *A Geography of Mankind* (McGraw-Hill, New York, 2nd Ed., 1973) and

Spencer, J. E., and W. L. Thomas, *Introducing Cultural Geography* (Wiley, New York, 1973).

Follow these up with a look through some of the papers in a very useful set of readings,

Wagner, P. L., and M. W. Mikesell, Eds., *Readings in Cultural Geography* (University of Chicago Press, Chicago, 1962).

For a further discussion of some of the more specialized topics touched on in this chapter, see

Sopher, D. E., *Geography of Religions* (Prentice-Hall, Englewood Cliffs, N. J., 1967) and

Stewart, G. R., *Names on the Land: A Historical Account of Place-Naming in the United States* (Houghton Mifflin, Boston, Rev. Ed., 1958).

Try especially hard to look at atlases showing the distribution of major cultural elements—patterns of settlement, languages and dialects, religions, voting behavior, ethnic groups, etc.—for your own country. Some of the major national atlases are listed in Appendix C at the end of the book. Few national cultural geographies exist, but for United States students, one is

Zelinsky, W., *The Cultural Geography of the United States* (Prentice-Hall, Englewood Cliffs, N.J., 1973).

This is a most exciting book that leads you into both the highways and the fascinating byways of your country. Regular research tends to be published in the main geographic journals, but look also at *Landscape* (a monthly) for articles on the cultural landscape.

Chapter 11

World Cultural Regions

THE EMERGING MOSAIC

"I will [tell] the story as I go along of small cities no less than of great. Most of those which were great once are small today; and those which in my own lifetime have grown to greatness, were small enough in the old days."

—HERODOTUS
(ca. 440 B.C.)

I n an age of transistor radios, packaged holidays, and earth-circling satellites, the world seems sadly shrunken and homogenized. The same Coca-Cola cans litter the strand lines from Coney Island to the Congo, and the same advertising jingles are crooned on the television sets. The urban-industrial forces that appear to be shrinking and standardizing the world's culture are very powerful ones, and we shall be looking closely at their increasing impact on world geography in Part Four.

At the same time, the resilience and continuity of the variety of human cultures remains impressive. There are few signs that the American melting pot—still less the global melting pot—has reached a temperature in which the kinds of differences we saw in the last chapter will be dissolved. On the contrary, the 1970 Census of the United States recorded a small swing toward ethnicity since 1960, with more individuals wishing to stress their American Indian, Finnish, or Hawaiian forebears. In an age where there are strong trends to "sameness", the scarcity value of some aspects of our differences appears to be rising.

Since it seems that cultural differences and cultural regions are here to stay, it is natural for geographers to want to know how they originated and how they have changed over time. In this chapter we take a historical perspective and look at three basic questions. First, how and where did the great cultural differences we now see originate? Second, how did change occur and some cultural regions spread at the expense of others? Third, how stable is the present pattern of cultures, and how likely is it to persist? The present mosaic of regions reflects not only the pressures and balances between man and his environment today; it also reflects the tens of thousands of years of past population growth, cultural differentiation, and migration.

Of course, it would be impossible, in the span of a single chapter, to cover all these topics. The course we shall follow is a compromise. We shall tackle questions of origin from a global viewpoint and choose one cultural region (Western Europe) for our study of the diffusion of cultural elements. Many areas have been affected by European culture, and we shall choose the situation in the United States as an example of the way various factors affect the persistence of cultural regions. Thus we shall move from a global to a subcontinental scale in our discussion, and finally to a national and regional scale. To use a photographic analogy, we shall change our focus and lens as we examine each of the three major questions raised in this chapter.

11-1 THE QUESTION OF ORIGINS

We can split the question of origins into three parts. Where did the first clusters of the human population form on the earth? Where were the first centers of agricultural innovation? And, finally, where, and at what stage, did urban cultures come into the picture? The order of these questions is a historical one.

The origins of the human population

Let us begin our analysis of this question by recapping some facts we met in Chapter 6. There we noted that the mammal that we call *Homo sapiens* is a very recent arrival on the earth. According to most current estimates, the earth is about 4.5 billion years old. The first living forms, algae and bacteria, originated about 2.2 billion years ago, and the first primitive mammals about 0.22 billion years ago. Just which date is assigned to the origin of the human species depends on which skeletal remains archeologists are prepared to call human. Several manlike species emerged during the last 3.5 million years of the most recent geologic period (the Pleistocene). *Homo sapiens* can be traced back to one of the interglacial periods about 1.5 million years ago. If we consider earth history as a clock measuring out 24 hours, with the great geological periods represented by portions of the various hours and present time represented by midnight, we must place the advent of man at only a few moments before midnight.

The notion that man originated in a single area, let alone the possible location of such an area, has been a matter of acute archeological debate. From current archeological evidence it looks as if man originated in the Old World rather than the New. Recent research is tending to narrow down the location of the source area to tropical Africa in general and to East Africa in particular. Asia is now regarded as a secondary rather than a primary source and Europe is no longer seriously in the running. We shall use the term *hearth* to describe a center of evolution, using it to describe not only centers of biological or genetic evolution (for plant and animal species) but also centers of cultural evolution (e.g., for agricultural methods or city living).

The differentiation of man into the three major races—Caucasoid, Mongoloid, and Negroid—may well have occurred at the same time that man evolved from the hearth areas. Toward the end of the Pleistocene era (around 25,000 B.C.), human groups had spread to most of the land masses, except Antarctica. Figure 11-1 charts the likely sequence of migration from the Old World. Migration across very wide sea areas posed problems for early man and the spread of population is thought to have followed island chains (using them as stepping stones). As we saw in Section 4-1, sea levels fluctuated considerably during the Pleistocene epoch and so-called "land" corridors were opened up during the major ice ages when more of the earth's water was locked up in the ice sheets and ocean levels fell. Certainly the Bering Strait between Siberia and Alaska was replaced by a land corridor at these times. We must stress, however, that the routes shown

Figure 11-1. Early migration of human groups. The probable migration path of the three main racial elements in the human population is shown here schematically.

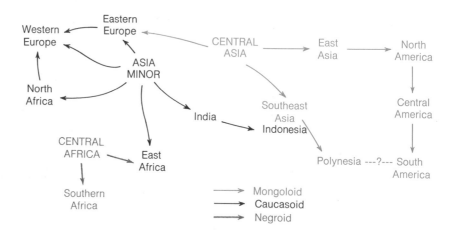

on Figure 11-1 are only tentative, and will, in all probability, have to be substantially revised as more genetic evidence is gradually accumulated. It seems probable that Western Europe, southern Africa, and Australasia were all peripheral areas (i.e., at the end of migration routes) and that the Americas were colonized from East Asia, probably at a late stage.

Many of the smaller and more remote islands in the world's oceans were reached only in the recent past. Radiocarbon dating reveals that the first settlements on the Hawaiian Islands may have been as late as A.D. 1200. Charles Darwin saw the possibilities that island chains offer for research during his visit to the Galapagos in 1835. Since then, an increasing amount of research is being done on the spread of human populations through island chains. (See Section 12-4, which reports research in the Pacific Ocean.) Figure 11-2 throws some light on the question of island colonization by using a model first developed to explain the varying number of plant and animal species found on different islands. Generally, large islands which are close to the continental land masses have a much richer range of fauna and flora than islands which are small and remote. Why should this be? If you follow the figures and caption in Figure 11-2 you'll see one possible explanation. So far as man is concerned, the same kind of rules appear to apply. If we measure human variety by means of genetic terms, it is the small, remote islands which display the smallest range in blood types or biochemistry. Whether the model would also help us to understand cultural variability is more debatable.

Estimates of the total world population in this early period are necessarily vague. If man lived in small food-gathering and hunting communities, we can assume that the average population densities

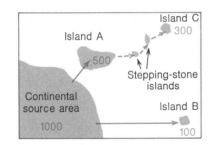

(a)

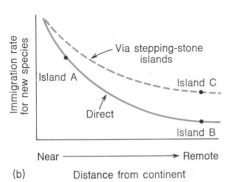

(b)

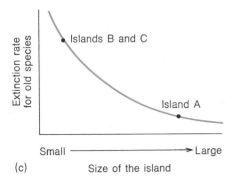

(c)

Figure 11-2. Island colonization models. Since Charles Darwin's *Origin of Species* (1859), islands have held a special fascination for biogeographers. Typical relations between a continental land mass and some offshore islands are shown in (a), where the figures indicate the number of different species of plants or animals found on each. The number of species on each island is determined by the *immigration rate* (the number of new species which arrive from the continent over a given time) and the *extinction rate* (the number of existing species on each island which fail to establish themselves and die out in a given time). Immigration rate is inversely related to the distance of the island from the continent (b), so that nearby islands have more species arriving (whether by wind, wave or animal dispersal) than do remote islands. The upper curve shows that this simple relationship may be modified by "stepping-stone" islands which allow species to migrate by a series of shorter steps. Extinction rate is indirectly related to the size of the island (c) so that more species die out on small than on large islands. This fact is largely due to the narrower range of ecological conditions normal on small islands. If we now combine these two factors—island accessibility and island size— we have a reasonable explanation for the kind of variation in species number so generally observed. Island A is both nearby the continental source and large in area and so has many species, while island B is both small and remote and thus has a low number of species. Island C is also small and remote but is better linked to the continental source via the intermediate chain of islands, so it has a higher number of species than B. [From R. H. MacArthur and E. O. Wilson, *The Theory of Island Biogeography* (Princeton University Press, Princeton, N.J., 1967), p. 22, Fig. 8.]

might have been between around 0.008 and 0.03 people/km² (0.02 and 0.8 people/mi²). Such figures are drawn largely from studies of surviving preagricultural groups like the Australian aborigines, who have an average population density of about 0.02 people/km² (0.5 people/ mi²), or the Haida Indians on the northwest coast of North America, who have a population density of 0.6 people/km² (1.6 people/mi²). These densities, along with the probable environmental limits on human occupation created by deserts and subarctic and alpine areas, suggest a total population of only around 5 million by the end of the preagricultural period.

The origin of agricultural hearths

Geographers commonly divide human culture into four distinct technical stages. These are (1) food-gathering and hunting cultures, (2) herding cultures, (3) agricultural cultures, and (4) urban cultures. Each stage is matched by an increasing complexity of material goods and social organization, by increasing ability to support high population densities, and by ever greater interference with the natural environment (see Chapter 7). The origin of the food-gathering and hunting cultures is the same as that of the human population itself; the first human groups in East Africa supported themselves in this way. Little is known about the earliest domestication of animals and the origin of herding cultures; some authorities regard this as a late, rather than early, stage in human cultural development. Most debate has centered on the origins of the third cultural stage—agriculture—and it is to this controversy that we first turn.

The origin and location of the world's agricultural hearths have been the subject of intense academic debate. Archeological evidence indicates the domestication of plants and animals by 8000 B.C. in the hills of what are now Iraq and Iran. Other finds reveal some similar activity in scattered spots in India, northern China, and central Mexico. It seems likely that wheat and barley were cultivated in the Middle East at an early date, and that the cultivation of corn by the Indians of Central America came later. Little is known of the early beginnings of rice cultivation in Asia, but new archeological finds and new ways of dating finds may yet enable us to revise and rewrite the fragmentary story of the development of agriculture.

The Sauer hypothesis Despite the scarcity of firm evidence, there has been plenty of conjecture on the location of the first agricultural communities. In the sweeping survey *Agricultural Origins and Dispersals*, Berkeley geographer Carl Sauer (Figure 11-3) argued for separate hearths of domestication in both the Old and the New World, outside the conventional hearth areas. As Figure 11-4 illustrates, he places the Old World hearth in South Asia and the later New World hearth in the valleys and lowlands of the northern Andes. Sauer chose these areas on the basis of five criteria. First, the domestication of plants could not occur in areas of chronic food shortages; the domestication of crops and animals implies experimentation, and a sufficient abundance of food so that the experimenters can wait awhile for results. Second, hearths must be in areas where there is a great variety of plants and animals and thus a large enough gene pool for experiment and hybridization to occur. Third, large river valleys are unlikely hearth areas because their settlement and cultivation require rather advanced techniques of water control. Fourth, hearths must be restricted to woodland areas where spaces can readily be cleared by killing and burning trees; grassland sod was probably too tough for primitive cultivators. Finally, the original group of cultivators had to be sedentary to stop crops from being consumed by animals. The main nomadic groups probably did not meet this requirement; nor, probably, did the areas they inhabited (see Figure 11-4) meet the first four requirements.

By combining these criteria, Sauer chose as his hearth areas the most probable environments for agricultural innovation. They had the climatic range to induce diversity and the rivers to provide a regular supply of fish for a sedentary settlement. Here wild plants were de-

Figure 11-3. Carl Ortwin Sauer. For more than half a century, Sauer's research on the early relations of man and plants has placed him at the forefront of cultural geographers. Many of his more speculative and controversial views on the location of agricultural hearths and on the antiquity of man's colonization of the New World have been supported by recent archaeological evidence. As professor of geography at the University of California at Berkeley from 1923 to 1957, he built up one of the most distinctive and distinguished graduate schools in the United States. (See Figure 22-3 on page 577.) [Photograph courtesy of Mrs. C. O. Sauer.]

veloped by centuries of selecting, propagating, and dividing. In Sauer's view the seed agriculture of the Middle East, China, and Central America is a much later, and more sophisticated, outgrowth of the activity at the two earlier centers of this type of agriculture in central Mexico and Asia Minor. The lively and hostile response of many archaeologists to this view suggests that the debate is still wide open.

The spatial impact of the agricultural revolution Whatever the precise location of the first agricultural communities, the impact of permanent agriculture on the spatial organization and density of the human population is clear. It increased the reliability of the food supply as well as the volume of food, so that more people could be supported by a given area. Man no longer needed to concentrate totally on food production and could branch out to nonagricultural crafts. As food surpluses became available, goods began to be exchanged. Pottery, weaving, jewelry, and weapons were bartered and traded over long distances.

The impact of these changes on man's spatial organization was two-

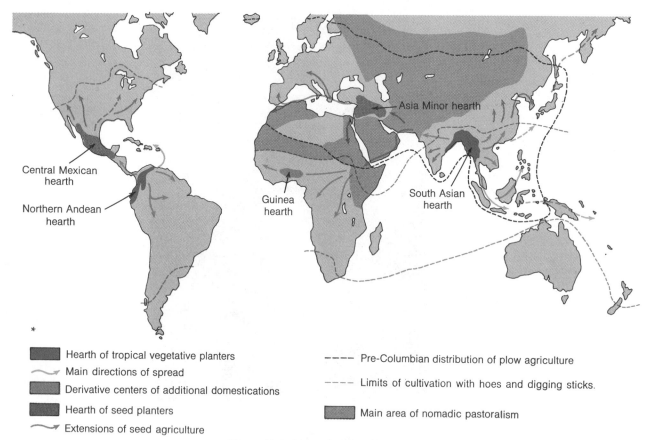

Hearth of tropical vegetative planters

Main directions of spread

Derivative centers of additional domestications

Hearth of seed planters

Extensions of seed agriculture

- - - - Pre-Columbian distribution of plow agriculture

- - - - Limits of cultivation with hoes and digging sticks.

Main area of nomadic pastoralism

Figure 11-4. Agricultural origin and dispersal. This world map shows in a highly generalized form the supposed main features of agricultural diffusion in the pre-Columbian world. It is based on the views of geographers Carl Sauer and Eduard Hahn and geneticists such as N. I. Vavilov. Note that there are two main hearths shown in the Old World: (1) tropical South Asia as a center for agricultural systems based on reproduction by vegetative planting (e.g., subdividing an existing plant into several parts, each of which grows into a new plant); and (2) subtropical Asia Minor as a center for agricultural systems based on reproduction by planting seeds. This fundamental distinction between the two types of plant propagation is repeated in the two centers (again, tropical and subtropical) suggested for the New World. The probable main lines of spread around each hearth are shown together with a secondary hearth area in West Africa. The map also shows the important technological distinction between methods of cultivation based on hoes and digging sticks (found in both Old and New Worlds) and that based on the plow (found only in the Old World in pre-Columbian times). Areas of nomadic pastoralism are those where herding of animals by migratory peoples had been established in the Old World. Compare this map with the detailed table of plant origins given later in this chapter (Table 11–2). [After E. Isaac, *The Geography of Domestication* (Prentice-Hall, Englewood Cliffs, N.J., 1970), p. 41, Fig. 3.]

fold. First, the centrifugal forces that scattered small numbers of people over large areas were weakened; isolation gave way to contact, and some degree of agglomeration into settled agricultural villages became possible. Second, the population densities rose in some areas to levels several hundred times greater than those of the preagricultural communities. For example, we know from archaeological evidence that the hill-farm communities of northern Mesopotamia had densities of approximately 70 people/km² (180 people/mi²) around 8000 B.C.

By 4000 B.C. the total population of the globe had reached around 87 million. The greater part of this population was probably concentrated in areas where village agriculture was intermingling with and slowly replacing food-gathering and hunting. Such areas certainly included a belt stretching from Western Europe and the Mediterranean through the Middle East to western India, northern China, Indonesia, and Central America. Outside this area population changed little from its preagricultural pattern. The extreme zones of the Arctic and Antarctic, together with the more remote ocean islands, remained wholly unoccupied.

Urbanization and its origins

Although the evidence on the origin and early growth of cities is more plentiful, its interpretation has led to academic controversy hardly less acute than that over agricultural hearths. Specific evidence of urban forms is available for several sites in the Tigris-Euphrates Valley for the period 3000 to 2500 B.C. Calculations based on the size of these built-up areas yield probable populations of around 50,000 for Uruk and 80,000 for Baghdad in that period. Less controversy surrounds the time of early urban centers (although new excavations in Asia Minor indicate that they may be older than they were first thought to be) than how they fit into a developmental sequence.

The developmental sequence Figure 11-5(a) shows a highly generalized version of the traditional view of man's developing resource-organizing technology. There are four main stages (primitive food-gathering and hunting, herding, agriculture, and urbanization), linked by three processes (the domestication of animals, the permanent cultivation of crop plants, and the trading of goods). The position of these stages and processes with respect to the time line at the left reflects the pattern of archaeological evidence.

Archaeologists are increasingly divided over the ways in which urbanization fits into this sequence. Figure 11-5(b) shows a conventional "linear" view in which urbanization is a late stage in the developmental sequence, dependent on the buildup of a food surplus

Figure 11-5. The position of cities in the developmental sequence. Figure (a) shows the traditional main stages and processes in the developmental sequence. Figures (b), (c), and (d) provide alternative models of the place of the cities in human development. Of course, the models shown here are highly simplified. Herding, for example, has a very complex origin and probably developed in different ways in various areas.

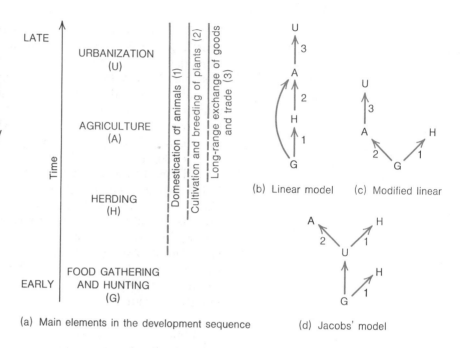

(a) Main elements in the development sequence

(b) Linear model

(c) Modified linear

(d) Jacobs' model

as a result of the increasing production by agricultural communities. More generally, herding is regarded as an incidental side-shoot, contributing little to the sequence. [See Figure 11-5(c).] But is this the correct order? Town planner Jane Jacobs has entered the fray with a controversial book called *The Economy of Cities*. She emphasizes (1) the increasing evidence of highly specialized and long-range trade (e.g., in obsidian axes) among human populations in the basic food-gathering and hunting stage and (2) the increasingly early dates assigned to cities.

Utilizing this evidence, Jacobs questions both the assumptions of the conventional "linear" view. In her own model [Figure 11-5(d)] urbanization is an early response to trade and exchange, and permanent agriculture a byproduct of the food needs and hybridizing environment of the city. There are other scholars who question the assumption that cities originated primarily for *economic* reasons. Leading urban historian Lewis Mumford cites documentary evidence from the cities of ancient Egypt which suggests that they were founded as centers of royal or priestly power. The view of cities as *Zwinburg* (control centers) rather than marketing or manufacturing centers may be a correct interpretation of their role in the preindustrial Near East. However, by the time cities appeared in the eastern Mediterranean area, during the third millennium B.C., their role as control centers was becoming inextricably entwined with their role as centers of interregional trade.

The location of urban hearths If we leave alone the muddy waters of when and how cities began, we are left with the question of where they began. Unfortunately, our notions of the spatial distribution of early cities necessarily reflect the concentration of archaeological activity. The patterns observed depend in part on where archaeologists have chosen to look for evidence of urban centers and on fortuitous factors like the durability of building materials and the preservation of foundations. Most successful cities have experienced so many cycles of building and rebuilding on the same sites that traces of their early outlines are difficult to establish. We know from the available evidence that urban development appears to have begun in four major river valleys: (1) the land lying between the Tigris and Euphrates rivers in the Mesopotamian area of the Middle East, (2) the valley of the Nile in Egypt, (3) the area near the Indus system in western India, and (4) the Hwang Ho (Yellow River) Valley in northern China. Table 11-1 gives the location and dates of the main areas of early urban cultures.

Table 11-1 Main urban core areas

Zone	Location	Early urban culture[a]	Representative city
Middle East (Fertile Crescent)	Nile Valley Tigris-Euphrates Valley	Egyptian (3000 B.C.) Sumerian (2700 B.C.)	Memphis, Thebes, Ur, Uruk
	Indus Valley	Indus (2500 B.C.)	Mohenjo-daro, Harappa
East Asia	Hwang Valley Mekong Valley	Shang (1300 B.C.) Khmer (A.D. 1100)	Anyang Angkor
Southern Europe	Aegean Islands and Peninsula	Aegean (2000 B.C.)	Knossos, Mycenae
	Italian Peninsula	Etruscan (400 B.C.)	Felsina, Rome
America	Yucatan Peninsula Central Mexico Peru	Mayan (A.D. 500) Aztec (A.D. 1400) Inca (A.D. 1500)	Palenque, Tikal Tenochtitlán Cuzco
West Africa	Niger Valley	Yoruba (A.D. 1300)	Ife

[a] The dates are indicative of the middle period of the culture.

Urbanized hearth populations of over 125 people/km² (325 people/mi²) existed in limited sections of agricultural districts as long ago as 4500 B.C. By the beginning of the Christian Era, the earth's total human population had grown from around 133 to 300 million, a substantial but not massive increase since the beginning of urbanization. At this time the main lines of the earth's present pattern of population were beginning to be blocked out.

The overwhelming majority of the world's population was located in three huge clusters. Probably the largest concentration of people was in the Indian subcontinent, which had over 40 percent of the estimated world population. The second largest concentration was that part of China within the Han Empire; perhaps 25 percent of the world's population was concentrated primarily in the delta plains of the Hwang Ho. Outside these two largest clusters lay the ancient Roman Empire, extending from western Europe and the Mediterranean to the Middle East, and including the long-established population concentrations of the Nile Valley and Syria. These three areas contained well over four-fifths of the world's population. In the fertile alluvial parts of these areas, the population density reached levels in excess of 1000 people/km². Outside these areas, population continued to be thinly spread over the land surface, and only a few high-density pockets (e.g., in central Mexico) broke up the pattern of primitive agricultural settlements and food-collecting groups.

11-2 | THE QUESTION OF DIFFUSION

In settling questions of regional origins, the main problem geographers face is lack of evidence. But as the evidence on the recent historical period builds up, the challenge shifts to making a comprehensive and convincing story of known but perplexing events. Each of the centers of urban culture we have recognized grew in population during the Christian Era, and each merits special study. Here, however, we choose to follow the history of only one of them, western Europe.

The reason for selecting western Europe rather than China, for example, is less chauvinistic than it may appear. The growth and spread of European culture is well documented; it provided, in fact, a pattern for much of the non-European world. In some areas like the Americas and Australia the pattern has been followed closely; in others like China, it has had scarcely any effect; in yet others like Africa and South Asia, we still await the final results of contact with the West.

The European hearth

Between the beginning of the Christian Era and A.D. 1500, the world's population doubled to around 500 million. This increase was particularly noticeable in the world's third major population concentration — the area of the former Roman Empire. Here the largest gains were registered in the more recently settled areas of western and east-central Europe, areas which were to emerge in the later stages of the period as modern states like France, Britain, and Poland. For Europe, the improved information on population during the historical period allows us to see much more clearly the pattern of human organization. We can, for example, trace the westward spread of urban institutions in the Roman Empire (Figure 11-6). For the largest city, Rome, housing densities within the known city limits indicate a maximal population approaching 200,000 in A.D. 200. Below Rome extended an urban hierarchy that had the same general form we shall see later in characteristic modern city systems. (See Chapter 14.) The collapse of the Roman Empire and the reduction in the scale of political and commercial organization from a subcontinental to a local level was followed by a breakup of urban and regional links.

With the slow revival of trade in early medieval Europe came the emergence of a fine mesh of small inland cities whose sites were often chosen because of their good defensive positions (e.g., the *bastides* of southwest France). Populations remained surprisingly small. For example, by A.D. 1450 Nuremburg, Germany, an important inland town, still had a population of only 20,000. Even London, which by A.D. 1350 had regained the level of population it had reached in Roman times, contained only around 40,000 inhabitants. Below these principal cities came a regular hierarchy of smaller cities, not unlike that which exists in the modern world.

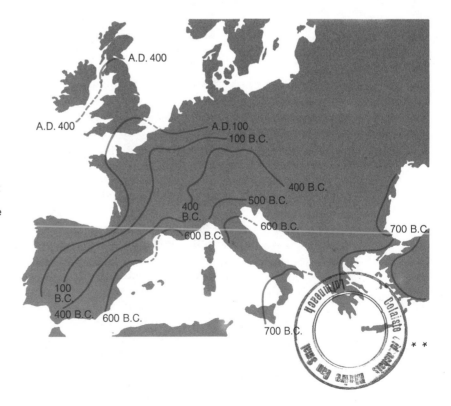

Figure 11-6. The spread of urban organization across Europe. The earliest European cities occur in the extreme southeast corner of the continent, where cities like Knossos in Crete or Mycenae in southern Greece were established by around 2000 B.C. Major extension of city-building in Europe accompanied the spread of the Roman Empire. The generalized contours show cities spreading north and west from the Aegean over an 1100-year period of history. [From N. J. G. *Annals of the Association of American Geographers* **59** (1969), p. 148, Fig. 6.]

At the end of the medieval period, western and central Europe were firmly structured into an organized regional system of cities. At the top of the hierarchy stood the trading cities of emergent industrial areas like London, Flanders, Lombardy, and Catalonia. Below was a network of smaller inland cities like that in Figure 11-7, often playing an important role in trade and administration. The network of cities was growing in two ways. First, by spatial expansion through colonization and the establishment of new cities in eastern Europe. Second, by the growth of smaller cities around fast-growing centers like Venice and Genoa, both booming as trade with the Levant increased.

Outside Europe lay other city hierarchies that were also highly differentiated and expanding. In the rest of the Old World the main areas of urban civilization were in eastern China and northern India. In the New World only central Mexico and the Peruvian valleys were urbanized, and these areas had much smaller populations than the Old World hearths of urban culture. Of the five hearths in existence in 1500, the European one was to experience the most significant expansion in the next 400 years of world history. Three of the four remaining hearths came directly under European influence in that period; only the Chinese hearth remained untouched by the major spatial reorientation of world trade produced by the growth of the European center.

Transoceanic rim settlements

The first phase of European overseas expansions, that of transoceanic *rim*, or coastal, *settlement*, lasted from the original Age of Discovery in the fifteenth century to the early part of the nineteenth century. Different European peoples took the lead at different times—the Spanish and Portuguese earlier, the French, English, and Dutch later—in establishing settlements along the coastal rims of the Americas, Africa, and South Asia. The settlements were of three main kinds: trading stations, plantations, and colonies of farm families.

Coastal trading stations Small trading posts were established widely on the coasts of India and southern China. Ports like Goa, Madras, and Canton served at times as points of exchange through which the products of the two great Asian urban civilizations of India and China could be brought to the growing urban markets of western Europe. Trade was mainly in luxury articles such as spices, tea, and handcrafted products like silks. The number of European settlers at any one time was quite small compared with the size of the indigenous Asian population, and only in India was it possible to exercise any real political control over the large hinterland areas that served the ports. Counterparts to the Indian ports, trading stations, were established on a minor scale in Malaysia, Indonesia, and East and West Africa.

Figure 11-7. The growth of the West European city. St. Gallen is a city in northeast Switzerland. Although the Romans colonized this part of Europe, the founding of St. Gallen is ecclesiastical in origin and later in date. It developed around a Benedictine abbey founded in the early seventh century. For the next four centuries this was to be the most famous educational institution north of the Alps, and it remained an important seat of learning throughout the medieval period. Along with the growth of the abbey came increased political and economic importance for the city. It was walled in the tenth century, became a free city in 1304, and joined the Swiss federation in 1454. An early seventeenth-century print (a) still shows the medieval core of the city with its walls and gateways. But in 1600 its population was probably still below 5000 people. Growth in the period since then to its present 80,000 has been associated with its role as a commercial center for the surrounding area (the canton of St. Gallen), its long-established textile industry, and new industries such as glass and metalworking. With this growth the city has sprawled well outside its original walls and now extends east-west along the Steinach valley. The photo (b) shows the medieval core within the modern city today; forested valleyside slopes north and south of the city show as very dark areas. In comparing (a) and (b) note that north is to the right of the print but to the top of the photo. [From (a) H. Boesch *et al.,* Villes Suisses à Vol d'Oiseau (Kümmerly & Frey, Berne, 1963), pp. 212, 216. (b) Courtesy of Swiss National Tourist Office.]

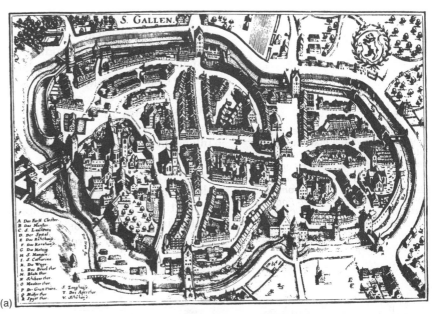

(a)

(b)

Tropical and subtropical plantations Plantations were originally established to grow sugar and spices but the range of crops was later expanded to include the production of various foodstuffs (coffee, cacao, bananas, etc.). The earliest plantation settlements were generally on ocean islands like Madeira, offshore islands like Zanzibar off the east coast of Africa, or coastal strips like the Baixada Fluminense around Rio de Janeiro in Brazil. Inland extensions of these settlements appeared largely in the nineteenth century. Such plantations demanded intensive labor, and European settlers primarily occupied only organizational roles, while the non-European population provided the field labor. When the indigenous population was unable to supply workers, slave laborers were brought in from other areas. (See Figure 11-8.) The current population mix in tropical America, East Africa, and parts of Malaysia and Australasia is largely a legacy of tropical plantation settlements. (See Figure 10-8.)

Figure 11-8. Intercontinental migration. The map shows the main currents of intercontinental migration since the beginning of the sixteenth century. The main flows only are shown, and in a highly generalized form. [From W. S. and E. S. Woytinsky, *World Population and Production* (Twentieth Century Fund, New York, 1953), p. 68, Fig. 27. Copyright © 1953 by The Twentieth Century Fund.]

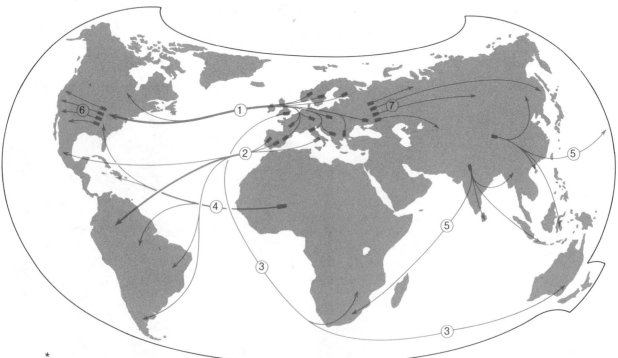

1. Migration from all parts of Europe to North America
2. Migration from southern Europe to Latin America
3. Migration from Britain to Africa and Australasia
4. Shipment of African slaves to the Americas
5. Indian and Chinese movements
6. Westward colonization in Anglo-America
7. Eastward colonization in Russia

Midlatitude farm-family settlements The third type of settlement associated with the period of European expansion is the farm-family colony of European migrants in the middle latitudes. This type included the settlements, mainly of English- and French-speaking groups, on the northeastern seaboard of North America, as well as later settlements in Australia and New Zealand. These settlements were in marked contrast to the tropical plantations because of their dependence on an influx of Europeans. Moreover, their agricultural products were destined for the local market rather than for export to Europe. Different groups from various parts of Europe (Swedes, Germans, Irish, etc.) brought to their overseas settlements some of the distinctive characteristics of their home areas. The layout of farms, villages and towns, the cropping patterns, and the farming technology often reflected traditions and practices in the original homelands. Even today the different ethnic backgrounds of farmers can be traced in the design of their farm buildings; the Pennsylvania barn shows just such a rich variability and has been carefully mapped by cultural geographers.

Continental penetration

The second phase of European expansion, that of *continental penetration*, began early in the nineteenth century and lasted until about World War I. This phase was accelerated by rapid industrialization in the European hearth, the development of innovations in transport like the railroad, the growing overseas migration of Europeans, and a rapid increase in the rate of exploitation and trade in non-European resources. Its chief impacts on the distribution of population were the springing up of industrial cities in the midlatitude colonies and the inland penetration of the agricultural frontier as the rich, midcontinental grassland zones were exploited for grain or stock production.

Midlatitude grassland settlements The nineteenth century witnessed the occupation of the prairies and the pampas in the Americas, the veld in Africa, and the Murray-Darling and Canterbury plains in Australasia. The pattern and timing of settlement was greatly affected by technical innovations like the railroad, refrigeration, and barbed wire. Railroads made it cheap to move agricultural products to ports, refrigeration made it possible to preserve meat for long-distance shipping, and barbed wire resulted in the fencing of open rangelands. Meanwhile, on Europe's eastern continental border the Russian state was expanding its settlement of the steppe grasslands at a comparable pace. In the tropics the demand for plantation products accelerated, and the movement of non-European peoples, first African slaves and later indentured Indians and Chinese, into plantation areas like the Caribbean continued. Trading contacts with the Orient increased and intensified as western countries extended their political control through both treaties and military occupation. More European ports were established on the coasts of China.

Mining and mineral "rushes" The mid- and late-nineteenth-century gold discoveries brought a rapid rise in both the white and the non-white population of mining areas. In some areas with an environment favorable to agriculture, mineral strikes provided the trigger which set off continued migration and long-term settlement. The 1851 gold strikes at Ballarat and Bendigo, in Victoria, Australia, brought in 250,000 prospectors in the next five years. (See Figure 11-9.) By 1855

Figure 11-9. European overseas migration. Gold rushes in the middle of the nineteenth century brought major increments of population to Australia. These contemporary prints show (a) police checking mining licenses on the Ballarat, Victoria, gold field in the 1850s, and (b) a mining camp in the Klondike. [Print (a) from News and Information Bureau, Department of the Interior, Canberra, Australia; print (b) from The New York Public Library Picture Collection.]

(a)

more people lived in Victoria than had lived in the whole of Australia before the discoveries. But in marginal environments like the subarctic zone the population and settlement that followed mining did not last. In 1898 some 30,000 prospectors moved down the Yukon River to the new gold strikes; today the total population of the whole Yukon territory is now less than 15,000. The search for gold in the nineteenth century was followed by a search for oil in the Middle East in the early decades of this century. Mining strikes and oil exploration were part of a phase of expansion that led to a vastly improved system of world communication and a systematic reduction in transport costs over both sea and land.

Consolidation and withdrawal?

We might say that a third phase of economic consolidation but political withdrawal began around World War I and appears to be continuing today. It was marked by a shift of economic power from the original European hearth to the United States and Soviet Russia. In addition, there was a political withdrawal of formal European control of much of Africa and Asia, signaling a virtual end to the British, French, and Dutch overseas empires. Population movements from Europe to mid-

(b)

latitude countries like the United States, Australia, and Argentina have continued, and they have not been balanced by the counter flow of black population to western Europe. Despite lessening political control, the presence of European capital, culture, and means of communication in much of Africa and Southwest and South Asia remains a fact of life. In Latin America Europe's economic role has been largely taken over by the United States. The rise of Japan as a prime industrial and trading power and the increasing role of China are bringing new waves of regional expansion based on East Asian hearths.

What was the effect of the 500 years of European expansion? On the spatial organization of the world community, the effect was enormous. For one thing, the expansion involved a transcontinental movement of around 95 million people. Over two-thirds of these were Europeans moving to temperate latitudes (notably to the United States). Another 20 percent were Africans forcibly transported from their homelands, especially to the American tropics and subtropics. The remaining fraction of over 10 percent were Asiatics. The pattern of Asian population movements is more complex because there are considerable Asiatic populations growing in parts of Africa, in the Caribbean, and in parts of the Pacific.

Along with the changes in population went a massive exchange and mixing of the world's crops. Table 11-2 gives a list of just some of the leading crop plants now used by man and their probable areas of origin. As we saw earlier in this chapter (see especially Figure 11-4), the precise origins of agriculture occurred so far back in human history that we can only make reasoned guesses on their actual location. What is clear is that 500 years of European expansion turned the pre-Columbian pattern of crop distribution upside down. American crops like the potato and tomato were to become regular farm crops in Europe while Old World crops like coffee and wheat were to become major crops in the Americas. Indeed, your own back yard is now likely to contain a variety of plants unquestionably richer and more diverse than seemed possible to our more continent-bound forebears of the pre-Columbian period.

Along with the interchange of population and of crops went a fundamental reorganization of wealth. Figure 11-10 shows in a very generalized form the map of world income today in comparison with that of population. The twin-peaked North Atlantic center symbolizes the extreme degree of financial control exercised by institutions like Wall Street, London's "City," Paris's "Bourse," or the Zurich banks in the organization of much of the world's resource development. Certainly this dominance may now be past its peak with the emergence

Table 11-2 Sixty of man's leading crop plants and their probable areas of origin[a]

Group 1: beverages and drugs	
Cacao	Orinoco basin, S. America
Coffees	E. Africa
Opium poppy	S.W. Asia
Quinoa	Andean America
Teas	S.E. Asia
Tobacco	Plate basin, S. America

Group 2: ornamentals	
Bougainvillea	E. Brazil
Dahlia	Mexico
Marigold	Mexico
(Tagetes)	Tropics

Group 3: root crops	
Cassava	Tropical America
Potatoes	Andes
Sweet potatoes	Meso-America
Taro	S.E. Asia (New Guinea)

Group 4: grains	
Amaranths	Meso-America
Barleys	S.W. Asia
Maize	Meso-America
Oats	Near East
Rice	S.E. Asia
Rye	Asia Minor
Sorghums	E. Africa
Wheats	S.W. Asia

Group 5: sugars	
Sugar cane	S.E. Asia (New Guinea)

Group 6: fibers and oil plants	
Cotton	Tropical America (Caribbean and Ecuador)
Flax	Mediterranean basin
Peanut	E. South America
Sunflower	N. America

Group 7: forage plants	
Alfalfa	S.W. Asia
Bluegrass	S.E. Europe
Cowpea	E. Africa

Group 8: vegetables	
Beets	Mediterranean basin
Broad bean	E. Africa and S.W. Asia
Cabbage	Mediterranean
Carrot	S.W. Asia
Cucumber	India
Gourds	Tropics
Kidney bean	Meso-America
Lima bean	Tropical America
Pea	S.W. Asia
Red peppers	Tropical America
Rhubarb	China
Scarlet runner	Meso-America
Soybean	China
Squashes	Meso-America
Tomato	Andean America

Group 9: fruits	
Apple	Caucasus
Avocado	Meso-America
Banana	Malaysia
Citrus fruits	S.E. Asia
Coconut	S.E. Asia
Date palm	W. India
Fig	S.W. Asia
Grapes	Turkestan
Mango	S.E. Asia
Melon	E. Africa and S. Asia
Papayas	Tropical America
Peach	China
Pear	Caucasus
Pineapple	E. South America
Plum	S.E. Europe
Quince	S.W. Asia
Strawberry	Americas
Watermelon	S. and E. Africa

[a] There are many hundreds of plant species used by man. This highly selective list is designed to illustrate some of the types of plants used and their probable areas of origin. Cultivated plants are very difficult to classify botanically because of the amount of hybridization and the ancestry of only a few has been fully worked out. The source areas should therefore be regarded as reasoned guesses from incomplete evidence in most cases. I am grateful to Professor Jonathan Sauer of the University of California at Los Angeles for commenting on this list, which was largely derived from Edgar Anderson, *Plants, Man, and Life* (Melrose, London, 1954), Chap. X and C. D. Darlington, *Chromosome Botany and the Origins of Cultivated Plants* (George Allen & Unwin, London, 1963).

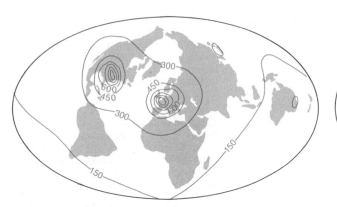

(a) Income

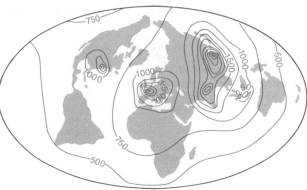

(b) Population

Figure 11-10. The legacy of European spatial organization. These generalized maps show the pattern of world income and world population in the early 1960s. Note the similarities and differences between the two. [From W. Warntz, Macrogeography and income fronts (Regional Science Research Institute, Philadelphia, 1965), pp. 92, 111, Figs. 19, 24.]

of a separate Russian center and the promise of new centers of financial power in Japan—and later China. Nonetheless, much of that control remains and the inequity between world population and world income that persists today is partly a reflection of the superimposition of a world urban system—centered first on Western Europe and then on a combined and widened North Atlantic core—upon much of the remainder of the world.

Geographers have sought to build a spatial model of the diverse pattern of European overseas expansion. We pick up their attempts when we look more closely at models of migration in Chapter 15 and economic growth in Chapter 18.

11-3 | THE QUESTION OF PERSISTENCE

In the first part of this chapter we looked at the origins of cultural regions on the global scale. In the second part we took one culture—that of western Europe—and traced its spread. In this third and final section, we reduce the scale of our investigations again and look at the question of the persistence of cultural elements in one of the areas of European overseas settlement, the United States.

Cultural regions of the United States

To understand something of the cultural complexity of the United States today we must begin with its history of immigration. The relatively small number of Indians, the inferiority of their weapons, and their low resistance to imported infections made it almost certain that they would be progressively overrun by waves of migrants from the Old World.

Sequent migration waves A survey of the period since 1607 (the date of the first permanent European settlement, at Jamestown, Virginia) shows a series of five distinct *sequent waves* of immigration, each associated with a particular set of migration sources. Between 1607 and 1700, there was an initial wave of English and Welsh, along with a small number of African slaves. The period from 1700 to 1775 brought increased migration from the same sources and an influx from Germanic and Scotch-Irish sources as well. The years 1820 to 1870 saw increased numbers arriving from northwest Europe (especially Britain, Ireland, Holland, and Germany), but the influx of Africans came to a halt. The 1870s saw the arrival of a vanguard from southern European countries, plus some Asians, Canadians, and Latin Americans. In the half century from 1870 to 1920, the period of "The Great Deluge," there was a massive increase in the number of immigrants and a widening of the source areas to include eastern and southern Europe and Scandinavia. Since 1920, there has been a drop in the number of immigrants, but the pattern of sources has remained wide and there has been a steady rise in the percentage of immigrants coming from Latin America.

While data for the early period is fragmentary, immigration in the period since independence is well documented. Figure 11-11 shows the pattern of numbers and changing origins over the last 150 years. The timing of the various waves of migration is reflected in the original areas of rural concentrations of migrants from different countries. The "early-wave" Scotch-Irish (the term used to describe both immigrants from Scotland and the Protestant areas of northern Ireland) settled in a belt running west from New England through the Appa-

Figure 11-11. Sequences of migration waves. The graph on the left shows the total number of migrants entering the United States over a 150-year period. The graph on the right shows the changing composition of the migrant stream. The curves indicate the different source areas dominant in the middle of the nineteenth century, the early twentieth century, and at present. Note that the graph describes the percentage of migrants from each source area and not the absolute number.

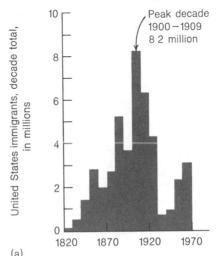

(a)

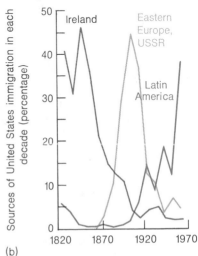

(b)

lachians into the near Midwest, while the "late-wave" Scandinavians were concentrated in the upper Midwest around Minnesota. The initial concentrations of African blacks in the South, of Mexicans in the Southwest, and of Italians in the cities of the Northeast and parts of California are all well known. In each case, the primary pattern of settlement and the original balance of urban and rural concentrations has been muddied and altered by subsequent internal migration, particularly to the growing metropolitan areas. (We shall look at these continuing processes of secondary migration a bit later in the chapter.)

Zelinsky's regions How do geographers make sense of the cultural patchwork that has emerged from these waves of immigration, settlement, and subsequent population growth? Many regional systems have been devised to describe the United States mosaic. Pennsylvania geographer Wilbur Zelinsky's system is shown in Figure 11-12. Zelinsky uses a fivefold classification. His first division, *New England* (Region I), was largely shaped by English migrations over the period from 1620 to 1830, with the development of settlement in northern New

Figure 11-12. Cultural regions of the conterminous United States. The map shows Wilbur Zelinsky's division of the United States into five major cultural regions and various subregions. The interregional boundaries vary in importance, and the status of three regions (Texas, Oklahoma, and peninsular Florida) is uncertain. The dates refer to the approximate limits of settlement and the emergence of a distinctive regional character. [From W. Zelinsky, *The Cultural Geography of the United States* (Prentice-Hall, Englewood Cliffs, N.J., 1973), p. 118, Fig. 4.3.]

England lagging more than a century behind that of the southern nuclear area. The *Midland* (Region II), south of New England, was settled slightly later (between 1624 and 1850), by a wider variety of migrants. To the English element were added important Rhineland and Scotch-Irish populations in Pennsylvania. In the New York region, Dutch and southern European migration was more important, as was in-migration from New England.

The most complex and diffuse of the three original hearths of culture on the eastern seaboard is the *South* (Region III). Beyond the narrow coastal strip of English plantations with their African slave population (settled before 1750), the South is divided into two major regions. Each of these has important subregions: Louisiana (in the Deep South) and the Ozarks and Bluegrass country (in the Upper South). The triangular region of the *Middle West* (Region IV) has more definite boundaries. Settled largely in the century after 1790, it was strongly affected by the westward extension of two existing cultural areas—the Midland and New England. Other cultural elements were superimposed on the existing pattern by new waves of European migration (particularly from Germany and Scandinavia).

An attempt to piece together the cultural dependence of the later regions on the earlier ones and on outside sources is presented in Figure 11-13. Only the main lines of influence are shown. Beyond the four main regions that comprise the eastern half of the country stands the enigma of the *West* (Region V). Here Zelinsky chooses to isolate nine subareas with some claim to distinctive cultural identities and to leave the remainder of the West as something of a cultural vacuum. The archipelago of subregions includes those shaped by particular ethnic groups (e.g., the Mexican element in the upper Rio Grande Valley), by religious beliefs (the Mormon region, also described in Figure 10-11), and by resource-exploitation patterns (e.g., the Colorado Piedmont). Outside the five main regions lie three intriguing areas whose status and affiliation is uncertain. Texas and Oklahoma are distinctive subregions that run across the boundaries of major divisions, while peninsular Florida lies beyond the South yet is not part of it. Some of the links between these three areas and possible sources of cultural influence are shown in Figure 11-13.

Forces of change

How long are the cultural divisions hammered out by nearly four centuries of immigration likely to persist? Urbanization, mass communication, and extreme social and geographic mobility all appear to be reducing regional differences. We now consider how these forces of change are powerfully reshaping the United States.

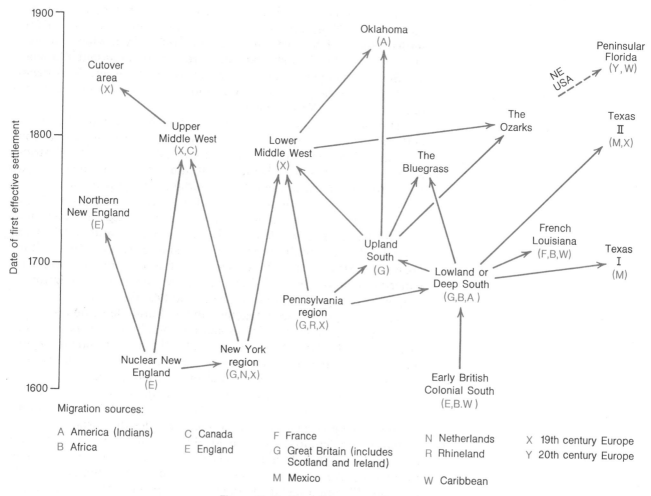

Figure 11-13. The origin of cultural regions. The chart shows the main cultural links between the sixteen cultural regions in Zelinsky's model of the eastern United States. (See Figure 11–12.) The area's approximate geographic location is indicated on a north-to-south (left-to-right) horizontal scale. The date of the first effective settlement appears on the vertical scale. "Texas I" refers to Texas during its Spanish-American period (1690–1821) and "Texas II" to the area incorporated after 1821 into the United States. The arrows indicate major *internal* migrations and cultural contacts within the United States, and the letters indicate major *external* migrations from outside the United States.

Urbanization Urban geographer Brian Berry, of Chicago, sees the change that is occurring in terms of four major trends, the first three of which appear to be reducing the contrast between regions, while the fourth has a contrary effect.

Figure 11-14. Forces changing cultural regions. The map shows some elements in the changing spatial organization of the United States today. [From B. J. L. Berry, *Area* **1** (1970), p. 46, Fig. 1.]

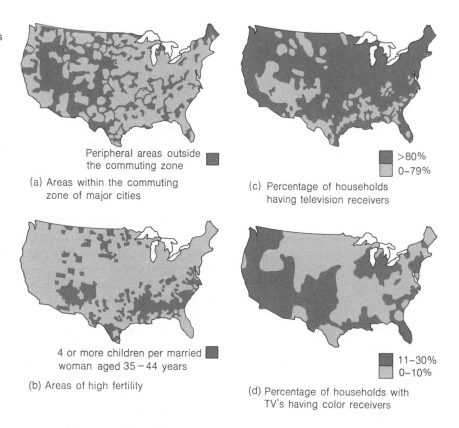

Peripheral areas outside the commuting zone ■

(a) Areas within the commuting zone of major cities

>80% ■
0–79% ▢

(c) Percentage of households having television receivers

4 or more children per married ■ woman aged 35 – 44 years

(b) Areas of high fertility

11–30% ■
0–10% ▢

(d) Percentage of households with TV's having color receivers

First, over the past century the intimate spatial contact between workplace and home has been diminished by advances in rapid transportation. The railway, electric street car, and automobile have each widened the gap. Chicago's daily urban system is now reaching out 150 km (93 mi) from the downtown. As Figure 11-14(a) shows, by 1960 all but 5 percent of the nation's population lived within the bounds of such daily urban systems.

Second, there has been a continuing population decline in the peripheral areas between main cities, due to out-migration and an aging population. Exceptions to this trend occur in peripheral areas that overlap with rural areas of more fertile population.

Third, we are witnessing a gradual coalescence of metropolitan areas. The tendency for transport links to forge urban centers into a megalopolis is already well established. By the year A.D. 2000 the existence of three principal megalopolises—sometimes termed *Boswash*, *Chipitts*, and *Sansan*—will have become more evident. By that date three gargantuan metropolises will probably contain roughly one-half of the total United States population: Boswash, extending from Boston to Washington, should have a population of around 80 million; Chipitts, the lakeshore strip from Chicago to Pittsburgh, should have around 40 million people; and Sansan, stretching from Santa Barbara to San Diego, should have around 20 million. Sansan will probably eventually extend north to include San Francisco. There also may be later megalopolises with lower suburban population densities in Florida and Texas. The three megalopolises will probably contain a large fraction of the scientifically most advanced and most prosperous segments of world population. Even today, the smallest of the megalopolises (Sansan) has a larger total income than all but five or six of the world's nations today.

Against these three trends toward integration on a massive scale can be set a contrary fourth trend toward the continuing social and ethnic differentiation of inner-city districts. In the central residential areas nearest the downtown, employment opportunities are low and population is on the decline. Nevertheless, these districts continue to serve as reception areas for migrants from poor rural areas. The international migrants from Ireland and Italy have been succeeded in the middle part of this century by blacks from the southern states, whites from Appalachia, and a substantial Mexican population in the southwestern cities. The proportion of black people in the cities is rising and, if we assume that present city limits will not change, will probably continue to rise. Washington, D.C., now about 60 percent black, is expected to be 75 percent black by the end of the century. By the same date Cleveland, Ohio, and Newark, New Jersey, will be over 60 percent black, while six other cities (Baltimore, Chicago, Detroit, New York, Philadelphia, and St. Louis) will be 50 percent black. In the suburbs surrounding these cities the black population is expected to expand, but it is unlikely to exceed 25 percent. The highest increases are projected for the San Francisco–Oakland suburbs.

The net effects of these various trends is reflected in the *scale* of cultural contrasts. At the same time that many of the differences between regions are disappearing, the differences between the inner and outer regions of the same cities are sharpening.

The geography of the affluent society All four trends we have discussed suggest that we shall see an intensification of the present metropolitan pattern by the end of the century. Not all evidence points in that direction, however. There are signs of a new urban transformation in the search for better natural amenities. More wealth and leisure have made natural environments (high mountains, persistent sunshine, clear water, extensive forests) a pervasive new influence in the location of population. Of course, in interpreting these signs, we should recall that the total number of people who can respond to this desire is still small and that the area into which the affluent population is moving is largely in the western, southwestern, and southern parts of the country.

Among the first group affected was the retirement population. To these retirees have been added workers employed in research and development projects whose location is not greatly dependent on either natural resources or urban markets. Both groups are likely to be more significant elements in the makeup of the end-of-century populations than now. Already, research has created important new centers in areas that were previously lagging behind the rest of the United States (e.g., Huntsville in Alabama) or that were sparsely populated (e.g., Los Alamos in New Mexico). Urban growth in states like Texas and Florida has also vastly increased.

This transference of affluent groups from the outer fringes of suburbia (the "stockbroker belt") to distant amenity-rich parts of the nation outside the metropolises is one of the latest phenomena in a spatial sorting process. This transfer can be illustrated by the spread of a prime medium of communication, television. For black and white TV, stations were opened in a hierarchic sequence from a few major cities in the early 1940s down to the smallest in the late 1950s. The spatial distribution of receivers followed a similar pattern; there were fewer TV's per capita in the remote interurban areas. [See Figure 11-14(c).]

Color TV has a different spatial pattern. Color television receivers were distributed widely in new settlement areas. [See Figure 11-14(d).] The startling contrast between Figures 11-14(c) and 11-14(d) is an indicator of an important swing toward amenity-resource areas by the country's wealthier population. This movement may well intensify prior to A.D. 2000. If the United States experience is repeated in other areas of the world with high living standards, then some major relocations of population may be in line. The Mediterranean and Alpine parts of western Europe will probably become increasingly attractive relative to other areas, particularly as barriers to the movements of people and their capital are reduced in an enlarged Common Market.

In this chapter we have turned only a few pages in a rich library of research in cultural geography. Yet we have moved in time from eras that saw the earliest crystallizing of cultural differences to the modern age of mass culture in the United States. In our study, the major emphasis has been on cultural change through *migration* — the slow movement and mixing of peoples, each group carrying with it distinctive patterns of language and behavior and other cultural freight. In the next chapter we continue to look at cultural transfers between regions, but the accent is on the *diffusion* of cultural elements by mass communication and the exchange of information.

Reflections

1. Where were the main hearths of agriculture thought to have been located? Why there?

2. List the crops grown on farms and gardens in your own locality. How many of these crops are native to the area, and how many are species introduced by crop exchanges between the Old and New World? (See Table 11-2 for a list of source areas for crops.) Select *one* crop plant and research its spatial spread.

3. Examine the different developmental sequences shown in Figure 11-5. Which do you think is most likely? Can you make a case for any alternative sequences?

4. Assume that the United States was settled, over 350 years, by people of Chinese rather than European stock. What major differences in the cultural landscape would you expect to see?

5. Why are some immigrant groups in the United States concentrated in distinctive cultural areas, while others are widely dispersed throughout the land?

6. How are the cultural regions shown in Figure 11-12 likely to change over the next 30 years? What factors are likely to cause distinctive cultural elements to (a) persist and (b) change.

7. Review your understanding of the following concepts:
   (a) agricultural hearths
   (b) Sauer's hypothesis
   (c) Jacobs' hypothesis
   (d) nodal hierarchies
   (e) rim settlements
   (f) trading stations
   (g) plantation settlements
   (h) continental penetration
   (i) sequent waves of migration
   (j) The Great Deluge
   (k) megalopolises
   (l) daily urban systems
   (m) interurban peripheries
   (n) amenity resources

One step further . . .

The literature describing human evolution and migration and the development of major cultural realms is a very rich one. A useful starting point, in which the growth and spread of population around the world is extensively treated from an ecological viewpoint, is

Thomas, W. L., Jr., Ed., *Man's Role in Changing the Face of the Earth* (University of Chicago Press, Chicago, 1956).

The origins of the early food-producing centers are discussed in
> Isaac, E., *The Geography of Domestication* (Prentice-Hall, Englewood
> Cliffs, N.J., 1970) and
> Sauer, C. O., *Agricultural Origins and Dispersals* (American Geographical
> Society, New York, 1952).

Urbanization and its antecedents are discussed in
> Mumford, L., *The City in History: Its Origins, Its Transformation, and
> Its Prospects* (Harcourt Brace Jovanovich, New York, 1961) and
> Sjoberg, G., *The Preindustrial City* (The Free Press, New York, 1960).

All students should also look at the splendid account of the evolving geography of the Chinese city in
> Wheatley, P., *The Pivot of the Four Quarters* (Aldine, Chicago, and
> Edinburgh University Press, Edinburgh, 1971).

The overseas expansion of Europe and the special case of the United States is described in terms of geographical models of expansion in
> Brown, R. H., *Historical Geography of the United States* (Harcourt Brace
> Jovanovich, New York, 1958) and
> Webb, W. P., *The Great Frontier* (University of Texas Press, Austin, Texas,
> 1964).

For a full treatment of the cultural regions shown in Figure 11-12, see
> Zelinsky, W., *Cultural Geography of the United States* (Prentice-Hall,
> Englewood Cliffs, N.J., 1973), pp. 110-134.

Much relevant research on the origins of cultural regions is published in archaeological and anthropological journals. A useful summary of this work is given from time to time in the "Geographical Record" section of the *Geographical Review* (a quarterly).

Chapter 12

Spatial Diffusion

TOWARD REGIONAL CONVERGENCE

Because I know that time is always time
And place is always and only place
And what is actual is actual only for one time
And only for one place.

—T. S. ELIOT
Ash Wednesday (1930)

In 1905 the El Tor strain of cholera was first identified in the bodies of six Muslim pilgrims at a quarantine station outside Mecca. (El Tor was the name of the quarantine station.) In the 1930s the same strain was recognized as endemic in the Celebes, which has a largely Muslim population. For another 30 years there was little news of El Tor until, in 1961, it began to spread with devastating speed outward from the Celebes. By 1964 it had reached India (replacing the normal cholera strain endemic in the Ganges delta for centuries), and by the early 1970s it was pushing south into central Africa and west into Russia and Europe. (See Figure 12-1.) The seventh of the world's great cholera outbreaks was getting into its stride.

At about the same time an epidemic of a totally different kind was spreading from city to city. Hot pants were introduced as a style of Western female dress in the spring fashion shows in Paris in 1970. In the autumn of that year boutiques from Sydney to San Francisco had caught on to the style, and by the spring of 1971 the first secretary in a conservative British university had been sent home for wearing the shorts to work. Now the style is just part of fashion history.

Waves similar to waves of cholera and hot pants have swept around the world in record time. Among the inconsequential waves were the brief Western passion for all things Japanese in the 1880s, and the 1950s craze for hula hoops among children on both sides of the Atlantic. Things as different as influenza epidemics and oral contraceptives, bank rate charges and computer data banks, Dutch elm disease and fire ants, have one thing in common. They originate in a few places and later spread over a much wider part of the world.

Figure 12-1. Spatial diffusion. The map shows the spread of the El Tor strain of cholera from the Celebes during the decade 1961–1971. This strain of cholera appears to be *endemic* (i.e., permanently present) in the population of the Celebes. From this island it erupts from time to time to spread temporary epidemics in surrounding areas. When such epidemics reach major proportions and span several continents in the manner shown on this map, they are termed *pandemics*. Geographers are interested in the paths followed by epidemics and pandemics through the populations of human settlements because of the insights they yield on other spatial diffusion processes. [Data compiled from press reports and WHO bulletins.]

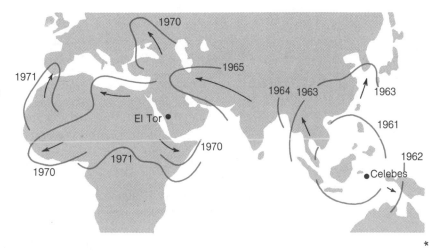

Why are geographers interested in such diverse things? Principally because their spread provides valuable clues to how information is exchanged between regions. Where are the centers of diffusion—and why? At what rates do *diffusion waves* travel, and along what channels? Why do some waves die out rapidly and others persist? Some innovations may move slowly and quietly, like a tide lapping over mud flats. But rapid innovations are studied most frequently, not because of their intrinsic importance (indeed, they sometimes tend to be trivial), but because we can see the whole cycle of diffusion in a relatively brief time period.

The speed with which cultural changes now occur is clearly linked to rapid communication channels. In Chapter 11 we looked at the slow readjustments of the worldwide regional mosaic to the ponderous movements and migrations of human population. In this chapter we turn to the much more rapid changes that can occur through the diffusion of cultural elements, and we see how the ripples from information explosions in one part of the world find their way into another. In this chapter we try to present some of geographers' more recent research on these topics and their use of computer models. (If you prefer to avoid these subjects, read only Sections 12-1 and 12-4). Such models have far more than academic interest. If we wish to speed up the diffusion of certain cultural elements, like the adoption of family-planning methods, a knowledge of precisely how waves of change pass through a regional system may be of help. If we wish to halt or reduce the spread of other cultural patterns of behavior, like drug abuse, or protect certain very fragile cultures from being swamped by Western civilization, this knowledge may be helpful there, too. In all such cases the geographer, as he did in our case history of a beach in Chapter 1, contributes his spatial viewpoint as one way of gaining insight into a many-sided, multidisciplinary problem.

12-1 | THE NATURE OF SPATIAL DIFFUSION

In Chapter 1 we saw how population spreads out over a beach. Again, in the last two chapters, we have come across other cases where a spatial distribution that occurs over time has a distinct diffusion pattern. In everyday language the term "diffusion" means simply to spread out, to disperse, or to intermingle; but for geographers and other scientists, it has acquired more precise meanings.

Types of diffusion

In geographic writing diffusion has two distinct meanings. *Expansion diffusion* is the process by which information, materials, and so on, spread from one region to another. In this expansion process the things being diffused remain, and often intensify, in the originating region; that is, new areas are added between two time periods (time t_1 and time t_2 are both located in a way that alters the spatial pattern as a whole). [See Figure 12-2(a).] A typical example would be the diffusion of an improved crop, such as a new strain of hybrid maize or rice, from one agricultural region to others.

Relocation diffusion is a similar process of spatial spread, but the things being diffused leave the areas where they originated as they move to new areas. The movement of the black population of the United States to the northern cities from the rural South could be viewed as such a relocation process, where members of a population at time t_1 change their location between time t_1 and time t_2. [See Figure 12-2(b).] In a similar manner, an epidemic may pass from one population to the next. Figure 12-2(c) illustrates the two processes and shows how they can be combined. The El Tor outbreak is an example of diffusion by both processes. The strain diffuses by relocation through some areas (e.g., as it did in Spain, where small outbreaks were recorded in 1971), but it also diffuses by expansion because it remains endemic in the Celebes. In this chapter we are discussing interregional interaction and are therefore mainly concerned with expansion processes. Relocation diffusion is treated more extensively in the discussion of regional growth models in Chapter 18.

Expansion diffusion occurs in two ways. *Contagious diffusion* depends on direct contact. It is in this way that contagious diseases like measles pass through a population, from person to person. This process is strongly influenced by distance because nearby individuals or regions have a much higher probability of contact than remote individuals or regions. Therefore, contagious diffusion tends to spread in a rather centrifugal manner from the source region outward. This is clearly shown in Kniffen's study of the spread of the covered bridge over the American cultural landscape, described in Figure 10-13.

Figure 12-2. Types of spatial diffusion. (a) Expansion diffusion. (b) Relocation diffusion. (c) Combined expansion and relocation processes.

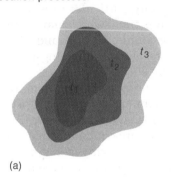

(a)

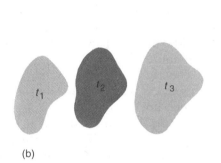

(b)

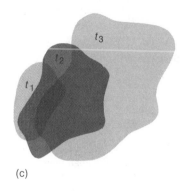

(c)

Cascade diffusion describes transmission through a regular sequence of order, classes, or hierarchies. This process is typified by the diffusion of innovations (such as new styles in women's fashions or new consumer goods like TV) from large metropolitan centers to remote rural villages. Within socially structured populations, innovations may be adopted first on the upper level of the social hierarchy and trickle down to the lower levels. Cascade diffusion is always assumed to be downward, from large centers to smaller ones. When specifying a movement that may be either up or down a hierarchy, geographers generally prefer the term *hierarchic diffusion*. Figure 12-3 demonstrates how diffusion may begin at a lower point in a hierarchy, move slowly upward, and then expand rapidly. We might think of this as a "Beatles pattern." A musical style beginning in a provincial city (Liverpool) moves to the national capital (London), then on to other capitals throughout the world. Finally, it reaches the local music store in small towns thousands of miles from its point of origin.

Elements in the diffusion process

Much geographic interest in diffusion studies stems from the work of the Swedish geographer Torsten Hägerstrand and his colleagues at the University of Lund. Hägerstrand's *Spatial Diffusion as an Innovation Process*, originally published in Sweden in 1953, was concerned with the spread of several agricultural innovations, such as bovine tuberculosis controls and subsidies for the improvement of grazing, in an area of central Sweden. This book was the precursor of various practical studies, particularly in the United States.

Hägerstrand's work is less important for its empirical findings (described in Section 12-4) than for its general analysis of the diffusion process. Hägerstrand describes six essential elements in spatial diffusion. The first element is the *area*, or environment, in which the process occurs. An area may be uniform and *isotropic* (meaning that it is equally easy for diffusion to move in any particular direction), or highly differentiated. The second element is *time*, which may be either continuous or broken into phases. For example, Hägerstrand divided time into discrete periods like days or years, with t_0 indicating a starting point and t_1, t_2, t_3, and so on, representing successive periods.

The third element in the diffusion process is the *item* being diffused. As we have already seen, the item can be material (people, TV sets, hot pants, etc.) or nonmaterial (behavior, messages, illness, etc.). Items vary in the degree to which they are communicable and in their acceptability. For example, measles is highly communicable (or contagious) and has a high acceptance rate; that is, you are likely to

Figure 12-3. Hierarchic diffusion. These diagrams show the spread of an innovation through a hierarchy. The innovation begins on a middle level (e.g., a small county town) and spreads rapidly down to a lower level (e.g., villages in its vicinity) but more slowly to an upper level (e.g., a regional capital). Once there, its downward spread is again rapid. Downward spread through a hierarchy is termed *cascade* diffusion.

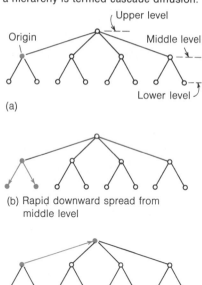

(a)

(b) Rapid downward spread from middle level

(c) Slow upward spread to upper level

(d) Rapid downward spread from upper level

be affected if you have not already had the disease. In contrast, family-planning techniques may be expensive to advertise (communicate) and have a low acceptance rate. The last three elements in Hägerstrand's model relate to the spatial pattern of the item's spread: the different places of *origin*, the *destinations*, and the *paths* along which the item being diffused moves.

Diffusion waves

In one of his early studies of a contagious diffusion process, Hägerstrand suggested a four-stage model for the passage of what he terms "innovation waves" *(innovations-forloppet),* but which are more generally called diffusion waves. From maps of the diffusion of various innovations in Sweden, ranging from bus routes to agricultural methods, Hägerstrand drew a series of cross-sections to show the wave form in profile. Here we discuss the wave in profile and then the wave in time and space.

The wave in profile Diffusion profiles can be broken into four types, each of which describes a distinct stage in the passage of an innovation through an area. Consider Figure 12-4, which shows the relationship between the rate of acceptance of an innovation and the distance from the original center of innovation. The first stage, or *primary stage,* marks the beginning of the diffusion process. Centers of adoption are established, and there is a strong contrast between these centers of innovation and remote areas. The *diffusion stage* signals the start of the actual diffusion process; there is a powerful centrifugal effect shown by the creation of new, rapidly growing centers of innovation in distant areas and by a reduction in the strong regional contrasts typical of the primary stage. In the *condensing stage* the relative increase in the number accepting an item is equal in all locations, regardless of their distance from the innovation center. The final, *saturation stage* is marked by a slowing and eventual cessation of the diffusion process. In this final stage the item being diffused has been accepted throughout the entire country so that there is very little regional variation.

Since Hägerstrand's original work, other Swedish geographers have carried out parallel studies to test the validity of this four-stage process. For instance, Gunnar Tornqvist has traced the spread of televisions in Sweden by observing the growth of TV ownership from 1956 to 1965. Using information obtained from 4000 Swedish post office districts, he demonstrated that television was introduced into Sweden relatively late, yet within 9 years about 70 percent of the country's households had bought their first set. Tornqvist's results broadly

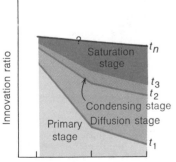

Figure 12-4. Diffusion waves in profile. The graph shows four main stages in the spread of an innovation by diffusion. The *innovation ratio* measures the proportion of a population accepting the item.

confirm Hägerstrand's analysis. The diffusion process slows down, thus indicating the beginning of the saturation phase, at the end of the study period.

The wave in time and space More advanced work on the shape of diffusions in space and time has confirmed their essentially wavelike form. Figure 12-5 is based on American geographer Richard Morrill's work. By fitting generalized contour maps (called *trend* surface maps described in the margin) to the original Swedish data, he showed that a diffusion wave first has a limited height (reflecting a limited rate of acceptance). It increases in both height and extent, and then decreases in height but increases further in total area. The gradual weakening of the wave over time and space is both time-dependent (as the simultaneous slackening of acceptance rates shows) and space-dependent (as the effect when the innovation wave enters inhospitable territory, strikes barriers, or mingles with competing innovation waves shows). The Nature of the medium through which the wave is traveling may cause it to speed up or slow down, and a wave traveling from one center of innovation will lose its identity when it meets a wave coming from another direction.

The exact form of a wave may be difficult to spot when diffusion data are first plotted. There may be an apparently chaotic distribution

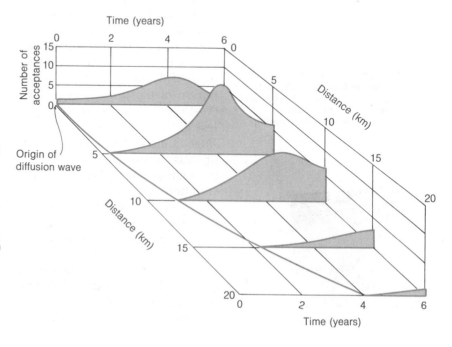

Figure 12-5. Diffusion waves in time and space. Waves of innovation change character with distance from the time and point of origin. In the case shown, the maximum height of the wave (i.e., the largest number of new acceptances of the item being diffused) occurs at a point 5 km from the origin in space and 4 years from the origin in time. Although individual waves will vary, some moving very slowly and some very rapidly, a large number appear to follow the general shape shown here. [After R. L. Morrill, *Economic Geography* **46** (1970), p. 265, Fig. 12.]

of locations and dates. Geographers have experimented with mapping techniques designed to filter out local variations so that the main form of the waves can be observed. For example, the spread of agrarian riots in Czarist Russia from 1905 to 1910 has a very spotty pattern. Using maps to filter out irrelevant data, two broad centers of unrest emerge: the southeastern Ukraine and the Baltic provinces. The location of the two centers is related to a high level of local tension caused by extreme contrasts in prosperity, the size of farms, and conditions of tenancy. Outward diffusion from the two cave areas follows different patterns. As Figure 12-6 indicates, rioting spread more rapidly along the Baltic Coast from the northern hearth, and in gentler, ripplelike movements from the southern hearth. The intersection of waves from other centers may create complicated patterns and make data difficult to interpret.

12-2 | THE BASIC HÄGERSTRAND MODEL

From his empirical studies Hägerstrand went on to suggest how a general operational model of the process of diffusion could be built. We shall look first at how the first and simplest of his models was constructed.

Contact fields

If we take any of the examples of spatial diffusion in the past few pages, we see that the probability that an innovation will spread is related to distance. Distance can be measured in simple geographic terms, as when we measure the number of meters between trees affected with Dutch elm disease around the campus. Alternatively, distance can be measured in terms of a hierarchy: e.g., the lower-level centers in Figure 12-3 are two "steps" away from the upper-level center. Let us take the first method and use it to measure the spread of information through a human population.

Let us begin by making the simple assumption that the probability of contact between any two people (or groups, or regions) will get weaker the farther apart they are. If we call one person as the *sender*, we can say that the probability of any other person receiving a message from the sender is inversely proportional to the distance between them. Near the sender the probability of contact will be strong, but it will become progressively weaker as the distance from the source increases. The exact form of this decline with distance is difficult to judge, but the evidence on telephone calls indicates that it may be *exponential*. That is, it may fall off steeply at first but then ever more slowly. (Check back to Section 6-1, if you wish to refresh your memory on exponential curves.) Thus, we expect the volume of calls to fall off in the ratio 80, 40,

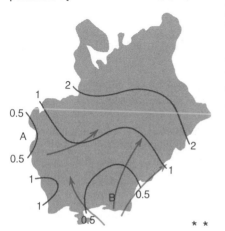

Figure 12-6. Multiple waves. The arrows indicate the spread of agrarian riots in Russia from 1905 to 1907, an example of diffusion from multiple centers. This is termed *polynuclear* diffusion in contrast to *mononuclear* diffusion (where diffusion waves originate from a single source). [From K. R. Cox and G. Demko, *East Lakes Geography* **3** (1967), p. 11, Fig. 2. Reprinted with permission.]

Trend-surface maps

Geographers use trend-surface maps as a device for separating *regional trends* (regular patterns extending over the whole area under study) from *local anomalies* (irregular or spotty variations from the general trend with no regular pattern). Thus trend-surface maps are like filters which cut out "short-wave" irregularities but allow "long-wave" regularities to pass through. Thus the contours in Figure 12-6 show the general form of the spread of riots in Russia but cut out confusing local variations.

20, 10, 5, and so on with the first, second, third, fourth, fifth kilometers. This is, of course, an idealized decline and actual patterns will be less regular. Geographers term this spatial pattern a *contact field*, drawing their language from the use of gravitational and magnetic "fields" in physics.

In models of cascade diffusion, we can retain the exponential contact field but replace geographic distance with economic distance between cities in an urban hierarchy or with social distance in a social-class hierarchy. Distance may not be symmetric in the hierarchic cases; for example, population migration up the hierarchy (from small to large towns) may be easier than migration down the hierarchy. This implies that the socioeconomic distance between levels depends on the direction of movement.

The contact fields in epidemics may be very complex. For instance, studies of measles indicate that the probabilities of contact (and thus infection) *within* a given group like a family or the students at an elementary school may be random. However, the probability of contact *between* such groups may be exponentially related to distance in the way we have already described. For example, research on southwest England showed that the probability of measles outbreaks in an area immediately adjacent to an area that had already reported cases was about 1 in 8. With further distance from the infected area the probabilities of infection fell steadily to about 1 in 30. (See Table 12-1.) The general term for models that assume random probabilities within a unit but exponential change between units is *island-biased* models because each unit is assumed to behave as an isolated island. Some of the cultural groups we met in Chapter 10 clearly illustrate this concept. The Amish groups in North America, for example, form an archipelago of "cultural islands" within the broad sea of Western culture.

Mean information fields

How can we translate the general idea of a contact field into an opera-

Table 12-1 The probability of new measles outbreaks in southwest England, 1969–1970

Location of areas with respect to existing areas:	Number of areas	Number of outbreaks	Probability of new outbreaks
Adjacent	2603	316	1 in 8
1 intervening area	1749	83	1 in 21
2 intervening areas	529	17	1 in 31
3 intervening areas	55	0	nil

SOURCE: P. Haggett, in N. D. McGlashan, Ed., *Medical Geography* (Barnes & Noble, New York, and Methuen, London, 1972).

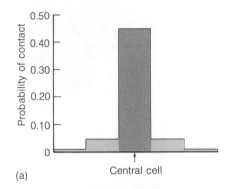

(a)

0.0096	0.0140	0.0168	0.0140	0.0096
0.0140	0.0301	0.0547	0.0301	0.0140
0.0168	0.0547	0.4432	0.0547	0.0168
0.0140	0.0301	0.0547	0.0301	0.0140
0.0096	0.0140	0.0168	0.0140	0.0096

(b)   Mean information

0– 95	96– 235	236– 403	404– 543	544– 639
640– 779	780– 1080	1081– 1627	1628– 1928	1929– 2068
2069– 2236	2237– 2783	2784– 7214	7215– 7761	7762– 7929
7930– 8069	8070– 8370	8371– 8917	8918– 9218	9219– 9358
9359– 9454	9455– 9594	9595– 9762	9763– 9902	9903– 9999

(c)

Figure 12-7. Mean information fields in the Hägerstrand model of diffusion. The probability of contact with distance (a) is superimposed on a square 25-cell grid (b). The probabilities for all the cells in the grid are summed to give (c) a mean information field.

tional model that can be used to predict future patterns of diffusion? Hägerstrand considered the problem in his early research, and he formulated various models to simulate diffusion processes. Figure 12-7 illustrates how he used probabilities of contact to determine a *mean information field* (MIF), that is, an area, or "field," in which contacts could occur. Superimposing of the circular field shown in cross section in Figure 12-7(a), on a square grid of 25 cells, enabled him to assign each cell a probability of being contacted. As Figure 12-7(b) indicates, the probability (P) of contact for the central cells is very high, in fact, over 40 percent (P = 0.4432). For the corner cells at the greatest distance from the center, the probability of contact is less than 1 percent (P = 0.0096).

To make the grid operational we add together the probabilities assigned to the MIF cells. Thus, the upper left cell is assigned the first 96 digits within the range 0 to 95; the next cell in the top row has a higher probability of contact (P = 0.0140) and is assigned the next 140 digits within the range 96 to 235, and so on. Continuing the process gives the last cell the digits 9903–9999, to make a total of 10,000 for the complete MIF [Figure 12-7(c)]. As we shall see shortly, these numbers are important in "steering" messages through our simple distribution of population.

Rules of the Hägerstrand model

We can present the basic structure of the Hägerstrand simulation model in terms of formal rules. The rules given here refer only to the simplest version. They can be relaxed to allow modifications and improvements.

1. We assume that the area over which the diffusion takes place consists of a uniform plain divided into a regular set of cells with an even population distribution of one person per cell.
2. Time intervals are discrete units of equal duration (with the origin of the diffusion set at time t_0). Each interval is termed a generation.
3. Cells with a message (termed "sources" or "transmitters") are specified or "seeded" for time t_0. For instance, a single cell may be given the original message. This provides the starting conditions for the diffusion.
4. Source cells transmit information once in each discrete time period.
5. Transmission is by contact between two cells only; no general or mass media diffusion is considered.
6. The probability of other cells receiving the information from a source cell is related to the distance between them.

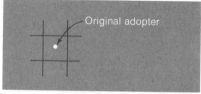

(a)   First generation (t_1)

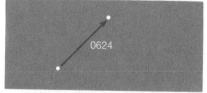

(b)   Second generation (t_2)

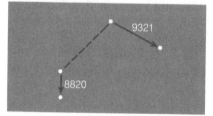

(c)   Third generation (t_3)

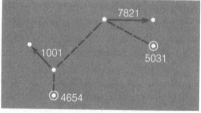

(d)   Fourth generation (t_4)

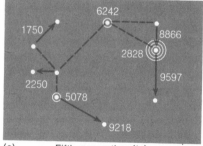

(e)   Fifth generation (t_5)

7. Adoption takes place after a single message has been received. A cell receives a message in time generation t_x from the source cells and, in line with rule 4, transmits the message from time t_{x+1} onward.

8. Messages received by cells that have already adopted the item are considered redundant and have no effect on the situation.

9. Messages received by cells outside the boundaries of the study area are considered lost and have no effect on the situation.

10. In each time interval a mean information field (MIF) is centered over *each* source cell in turn.

11. The location of a cell within the MIF to which a message will be transmitted by the source cell is determined randomly, or by chance.

12. Diffusion can be terminated at any stage. However, once each cell within the boundaries of the study area has received the message, there will be no further change in the situation and the diffusion process will be complete.

Using the model

The key to putting this model into use is in rules 10 and 11. In each time interval the MIF is placed over *each* source cell so that the center cell of the grid corresponds with the source cell. A random number between 0000 and 9999 is drawn and used to direct the message, following rules 4 to 6. *Random numbers* are sets of numbers drawn purely by "chance" (e.g., by rolling a dice). They can be taken from published tables of random numbers, or generated on a computer, or, for small problems, drawn from a hat. We show this process in Figure 12-8. In the first generation the number 0624 is drawn from a table of random numbers and a message is passed to a cell that lies to the northeast of the original adopter, located in the source cell.

Figure 12-8 goes on to present the first few stages in the diffusion process. In each generation the MIF is recentered in turn over each cell that has the message. Because the Hägerstrand model uses a random mechanism, each experiment or trial produces a slightly different geographic pattern. If we ran thousands of such trials (using a computer), we would find that the sum of all the different results matched the probability distribution in the original MIF; that is, we should

Figure 12-8. Simulated diffusion. The opening stages of the Hägerstrand model are illustrated by a mean information field. The numbers refer to contacts determined by drawing random numbers. When contacts are *internal* (i.e., with the cell on which the MIF is centered), a circle is added in that cell.

arrive back at our starting distribution. In order to reap the benefits of the model, it should be applied not to simple, predictable diffusions whose end result is known, but to complicated, unpredictable diffusions whose end result is in doubt.

12-3 | MODIFYING THE HÄGERSTRAND MODEL

If we think about the rules of the basic Hägerstrand model, we can see that they represent a considerable simplification of reality. Areas where diffusions take place are not uniform plains with evenly spread populations; innovations are not adopted the instant a message about them is received; information is not passed solely by contacts between pairs of people; and so on. Hägerstrand was fully aware of these complications, and he used his basic model to provide a logical framework for more realistic versions of the diffusion process. Hägerstrand's later variants of his model contain significant modifications. Others have been added by American researchers.

Some of the modifications introduced into the original model are minor technical improvements in its structure. For instance, regular square cells can be reshaped to fit other regular divisions (hexagonal units have been adopted in some versions), but irregular areas present more of a problem. Adaptation of the contagious diffusion model to cascade processes involves substituting a hierarchy of settlements for an isotropic plain. Probabilities must be assigned to the links between the settlements rather than to cells.

Abandoning the uniform plain

Some of the modifications can be simply made. Let us relax rule 1 and assume that the distribution of population is not uniform and that there are a variable number of people within each cell. The probability of contact is then a function both of the distance between the source and destination cells and of the number of people in each cell. Thus, we can multiply the population in each cell by its original contact probability to find a joint product. The ratio between the joint product of any cell and the sum of the joint products for all 25 cells in the MIF gives us a new contact probability based on both population and distance. (See the marginal discussion of weighting contact probabilities.) We have to buy this added realism at the cost of some tedious arithmetic, particularly because the new probabilities must be recomputed each time we move the MIF grid. On the other hand, such computations can be readily done by a computer.

Although this procedure may seem complicated, we are simply putting *back* into the model the geographic reality which the original assumption of a uniform plain took out. If we were concerned with

Weighting contact probabilities

If we assume that the probability of contact in a diffusion model is a function of both the distance between the source and destination cells and the number of people in each cell, then we can estimate that

$$C_i'' = \frac{C_i' N_i}{\sum\limits_{i=1}^{25} C_i' N_i}$$

where C_i'' = the joint probability of contact with the ith cell based on the MIF and population,

C_i' = the original probability of contact with the ith cell based on the 25-cell MIF,

N_i = the number of people in the ith cell, and

$\sum\limits_{i=1}^{25}$ = the summation of all $C'N$ values for the 25 cells within the MIF, including the ith cell.

These revised values for the probability of contact (C'') must be recomputed each time the MIF grid is moved to allow for spatial variations in the population.

understanding the spread of a cultural artifact (e.g., TV ownership) through a region, one of our first concerns would be the distribution of population and thus of potential purchasers for the product.

Varying resistance to innovations

In discussing the impact of religious beliefs in Chapter 10, we noted their importance in insulating a group against change. One of the examples we used was the persistence of some seventeenth-century cultural traits in the present-day Amish communities in the United States. Can the model be adapted to incorporate factors of this kind?

Well, it can be if we relax another of the original rules, in this case rule 7. The statement that adoption of an item takes place as soon as a message is received by the destination cell is an oversimplification. From research on agricultural innovations we know that there is generally a small group of people who are "early innovators" and another small group of "laggards"; the majority of a population adopts an innovation after the early innovators and before the laggards. In the case of spatial diffusions of population over a territory, this implies that settlements are established sporadically at first. The sporadic phase of settlement is followed by a period when everyone gets on the bandwagon, and eventually by a period of restricted settlement as the number of suitable unsettled locations in the territory diminishes. In the case of spatial diffusions of an innovation throughout a population, there are regional variations in the time of acceptance of the new item or way of doing things. For example, Figure 12-9 shows extreme regional variations in the time of adoption of tuberculin-tested (TT) milk by farmers in the counties of England and Wales. In some southern counties, 50 percent of the milk sold off farms in 1950 was TT milk; the proportion of TT milk in the milk sold in the far southwest had

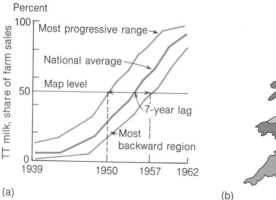

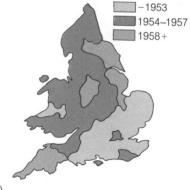

Figure 12-9. Resistance to change. Regional variations in adoption rates are illustrated by (a) the diffusion of tuberculin-tested (TT) milk on farms in England and Wales. The map (b) shows the year by which each county had achieved 50 percent TT milk production. [Data from Milk Marketing Board. After G. E. Jones, *Journal of Agricultural Economics* **15** (1963), pp. 489–490, Figs. 6, 7A.]

(a)

(b)

reached this level eight years before. Another example of varying resistance to change is illustrated in Figure 12-10.

We can approximate the symmetric course of the diffusion process by S-shaped curves. (See the marginal discussion of innovations and

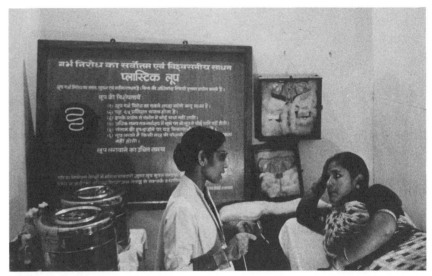

(a)

Figure 12-10. Social factors in resistance to innovations. One practical outcome of geographers' research on spatial diffusion is a concern with ways of speeding up desirable innovations (family-planning clinics) and slowing down or halting undesirable innovations (disease epidemics). The spread of family planning in (a) Rajasthan and (b) Benares, India, illustrates the former concern. [Photographs by (a) Michaud and (b), Paolo Koch, Rapho Guillumette.]

(b)

Innovations and logistic curves

The resistance of a population to adopting an innovation usually follows an S-shaped curve. [See diagram (a).]

This curve can be approximated by a *logistic distribution* given by the equation

$$P = \frac{u}{1 + e^{(a-bt)}}$$

where P = the proportion of the population adopting an innovation,
u = the upper limit of the proportion of adopters,
t = time,
a = the value of P when t is zero,
b = a constant determining the rate at which P increases with t, and
e = the base (2.718) of the natural system of logarithms

Thus, with u = 90 percent, a = 5.0, and b = 1.0, the proportion of adopters will be 4 percent at t = 2, 28 percent at t = 4, 66 percent at t = 6, 85 percent at t = 8, and so on. As diagram (b) shows, the constant b has a critical effect on the form of the innovation curve. Low b values describe smooth innovation curves (curve 1), whereas higher b values describe rates of acceptance that have a slow initial buildup, explode rapidly in a middle period, and enter a final period of slow consolidation (curve 2). [See P. R. Gould, *Spatial Diffusion* (American Association of Geographers, Commission on College Geography, Resource Paper 4, Washington, D.C., 1969).]

logistic curves.) Standardized *resistance curves* of this type were used by Hägerstrand to take into account resistance to innovations. After one message, the probability of acceptance was very low (0.0067); after two messages, it rose to nearly one-third (0.300); and after three, to nearly three-quarters (0.700). From then on the rate of acceptance fell again. The probability of acceptance rose slowly after four messages (to 0.933), and still more slowly thereafter. After five messages, even the worst laggards had accepted the item, and the rate of acceptance was 1.000. Like changing probabilities of contact, varying rates of acceptance (or resistance) can be readily incorporated into a computer simulation of a diffusion process. And so, if we have a community highly resistant to change, like the Amish, we can increase the number of messages sent to any appropriate large number. If a community were wholly resistant to change, the number of messages needed would be infinite!

Boundaries and barriers

In the original model, messages moving outside the boundaries of the study area were considered lost and had no effect on the situation (rule 9). In later models a boundary zone over half the width of the MIF grid was created so diffusion could proceed by way of these external source cells. More important modifications were involved in the introduction of internal barriers that act as a drag on the diffusion process. Like the other modifications we have discussed, such barriers allow observed variations in both the natural environment and the cultural mosaic to be incorporated into the model.

At the University of Michigan, Richard Yuill programmed the Hägerstrand model to stimulate the effect of four types of barriers on the diffusion of information through a matrix of 540 cells within a

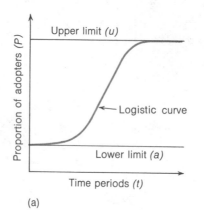

(a)

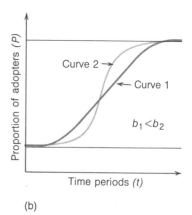

(b)

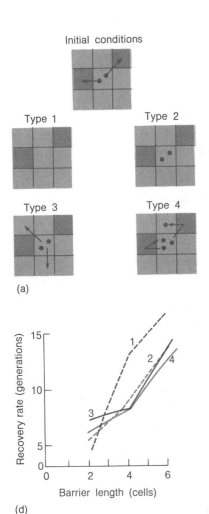

Initial conditions

Type 1   Type 2

Type 3   Type 4

(a)

(d)

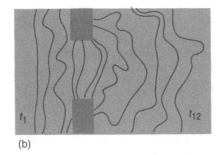

(b)

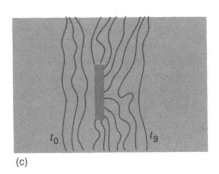

(c)

Figure 12-11. Barriers and diffusion waves. Four types of barrier cells (a) are used in this simulation model. In (b) diffusion waves pass through an opening in a bar barrier. In (c) diffusion waves pass around a bar barrier. The graph (d) shows the recovery rates around bar barriers constructed from the four different types of barrier cells. "Recovery rate" is the time taken for the straight line of the diffusion wave front to re-form. [From R. S. Yuill, *Mich. Inst. Univ. Comm. Math. Geogr. Disc. Papers*, No. 5 (1965), pp. 19, 25, 29.]

9-cell MIF. Figure 12-11(a) shows the 9-cell grid with the barrier cells indicated. Four types of barrier cells that provide a decreasing amount of drag are considered: a *superabsorbing barrier* that absorbs the message but destroys the transmitters; an *absorbing barrier* that absorbs the message but does not affect the transmitters; a *reflecting barrier* that does not absorb the message but allows the transmitter to transmit a new message in the same time period (see the arrows in the figure); a *direct reflecting barrier* that does not absorb the message but deflects it to the available cell nearest to the transmitter.

Each situation was programmed separately and the results plotted. Figure 12-11(b) shows the advance of a linear diffusion wave through an opening in a barrier. The time taken for the original line of the wave to reform determines the *recovery rate*. Varying types of barriers and gaps of varying widths were investigated. In the example shown, the line of the wavefront has recovered by about the eleventh generation (time t_{11}). Another type of barrier is presented in Figure 12-11(c). Here the diffusion wave passes around the barrier and reforms after about nine generations. The recovery rate of a wavefront is directly related to both the type and the length of the barrier it encounters; the curve for a superabsorbing barrier is quite different from the curves for the other three types of barriers. [See Figure 12-11(c).]

Yuill's work expands and develops modifications already begun by Hägerstrand. The original model postulated what was, in effect, a row of absorbing cells around the periphery of the study area. The internal barriers in Hägerstrand's model were represented by the lines between the cells. Such barriers could be adjusted to be absolutely effective (i.e., to allow no messages to get through) or 50 percent effective (i.e., to let one out of two messages cross the barrier). With such *permeable barriers* we can replicate a variety of environments. Thus, the original assumption of isotropic movement can be brought into line with known patterns. In other words, we can build *low-resistance corridors* into the model to allow faster diffusion in certain directions, as well as *high-resistance buffers* to slow down diffusion across barriers.

12-4 | REGIONAL DIFFUSION STUDIES

Many of the applications of Hägerstrand's model stem from his own pioneering work in Sweden. Here we review some of the applications of the model to regions with contrasting environmental conditions. We look first at the spread of cultural attitudes (farmers' attitudes toward farm subsidies); second, at a cultural artifact (irrigation wells); and third, at the spread of a cultural group (the Polynesians).

Farm subsidies in central Sweden

In the late 1920s the Swedish government introduced a scheme to persuade farmers to forego their traditional practice of allowing cattle to graze the open woodlands in summer. Grazing was proving to be a problem because it restricted the growth of young trees. To encourage fencing and improvements in pastureland, the government offered a subsidy. Figure 12-12 presents computer maps of the central part of Sweden and indicates areas where farmers accepted the subsidy during the years 1930 to 1932.

The maps indicate that in 1930 a few farmers accepted the subsidy in the western part of the region but there were scarcely any takers in the east. The next two years brought a rapid increase in the number of acceptors in the west but little change in the east. The sequence of maps suggests a spatial diffusion process in which distance is an important factor. To stimulate this process, Hägerstrand built a model using the 1928–1929 distribution of adopters as a starting point. The basic model was modified in two ways. First, the potential number of adopters (i.e., farmers) in each cell was added; second, barriers that were 100-percent and 50-percent permeable were added to simulate the long north–south lakes that lie across the region. Figure 12-12 compares the simulated diffusion process with the actual one. Because of the random element in the model, we should not expect the simulated pattern to match exactly the actual pattern. But the degree of matching is close, and both the general form of the expansion process and the location of the major clusters of adopters in the western areas are correct.

Irrigation in the Great Plains

An American geographer, Leonard Bowden, recently extended the Hägerstrand model and used it to simulate agricultural changes in the northeastern high plains of Colorado. Droughts had hampered the expansion of cattle farming, and the solution to the problem involved

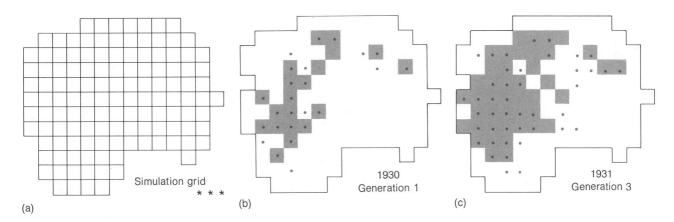

(a) Simulation grid * * *

(b) 1930 Generation 1

(c) 1931 Generation 3

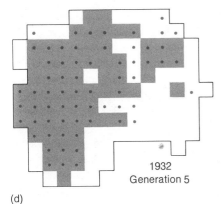

(d) 1932 Generation 5

Actual diffusion
▨ Simulated diffusion

Figure 12-12. Simulating diffusion on a grid. Here we see Hägerstrand's simulated diffusion and the actual diffusion of a decision by farmers in central Sweden to accept a farm subsidy. For simulation purposes the test area was approximated by a regular grid (a). The data for the three trial years shown in (b), (c), and (d) are part of more extended study. [After T. Hägerstrand, *Northw. Univ. Stud. Geogr.*, No. 13 (1967), pp. 17, 23, Figs. 6, 9.]

irrigation by wells and pumps that tapped groundwater. However, the decision to irrigate was very costly and was not made without consultations with successful innovators. How did such discussions take place? Bowden proposed that *mean discussion fields* analogous to MIF's be established by studying telephone calls and social get-togethers like barbecues in the agricultural community.

The actual pattern of irrigation facilities in 1948 included 41 wells. These were concentrated spottily in the eastern borders of the state between two main rivers, the South Platte and Big Sandy Creek. Bowden devised simulation models similar to Hägerstrand's and began his investigation with 1948 as the starting point. Ten trial diffusions were simulated by the computer. As we would anticipate, each one gave a slightly different result and the map in Figure 12-13(b) shows the general pattern as revealed by averaging the results of the ten trials. Comparison of this map with the *actual* distribution in 1962 [Figure 12-13(c)] shows a rather close match. Encouraged by this fact, Bowden went on to run his diffusion model for further "generations" so as to give the projected 1990 pattern. This projection assumes that the same rules for diffusion of irrigation that gave a good match over the past (i.e., 1948 to 1962) will also hold for the future. The only variation introduced was to place a ceiling of 16 on the number of wells per township so as to avoid too great a removal of underground water supplies by irrigation pumping. The location of townships that reached this ceiling, the so-called "saturated" townships, are shown in Figure 12-13(d).

Kon Tiki voyages in the Pacific

When Thor Heyerdahl undertook his now historic voyage on the raft *Kon Tiki* from the coast of Peru to the Tuamotu Islands, he was carrying out a single experiment: He was testing whether it was possible to

Figure 12-13. A diffusion wave for irrigation. The maps show a simulation of decisions to irrigate in eastern Colorado. (a) Actual pattern of wells per township in 1948. (b) Simulated pattern in 1962. (c) Actual pattern in 1962. (d) Projected pattern for 1990. [From L. W. Bowden, *Univ. Chic. Dept. Geogr. Res. Paper,* No. 97 (1965), p. 108, Fig. 21.]

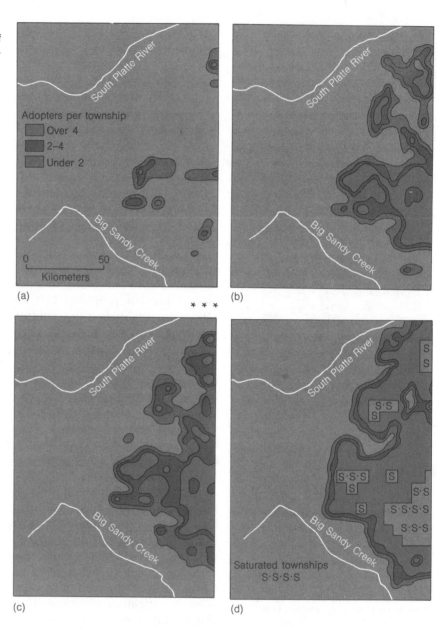

cross the Pacific in such a craft. To analyze thoroughly the probabilities of contact between South America and different island groups by this means would require too many voyages to be feasible. When direct experiments prove too costly, too risky, or unlikely for other reasons, we may be able to turn to computer simulation for answers to our questions. For example, nearly 30 years ago, in Project Manhattan (the name for the atom bomb project), the atomic radiation from bombs was simulated mathematically. Heyerdahl's trans-Pacific migration exemplifies a spatial simulation that can be approached by means of the Hägerstrand model.

The central issue to decide is how the Polynesians came to discover and settle on the islands of the central Pacific. This question has recently attracted considerable attention from anthropologists, navigators, and geographers; but the sparsity of evidence has led to a clash between two schools of thought. The first holds that the colonization process involved intentional two-way voyages and hence a high degree of navigational skill on the part of the Polynesians. (See Figure 12-14.) The alternative view was that colonization was largely accidental, by travelers drifting off course.

To test the probability of interisland contact as a result of accidental

Figure 12-14. Diffusion mechanisms. This eastern Polynesian double canoe was used in interisland voyaging. The settlement of Polynesia by canoes of this kind is a question hotly debated by anthropologists and geographers. Where did the Polynesians come from and how did they move from island to island? Figure 12–15 shows how computer models of diffusion may throw some light on both questions. [Drawn by John Webber, artist on Captain James Cook's third voyage. From A. Sharp, *Ancient Voyagers in the Pacific* (Penguin, London, 1957), Fig. 8.]

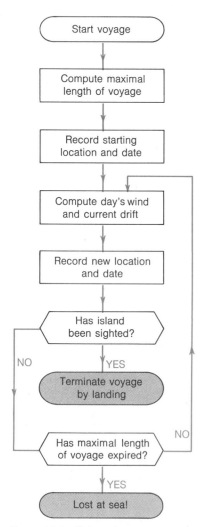

Figure 12-15. Polynesian voyaging. This flow chart shows the main elements in a computer program developed to simulate Pacific "drift" voyages. [After R. G. Ward *et al.*, *Information Processing 68* (1969), p. 1521, Fig. 1. Reproduced with permission of North Holland Publ. Co., Amsterdam.]

drifting about, a group of investigators at London University (Levison, Ward, and Webb) constructed a computer simulation model of the drift process. The stages in their computer program are shown in Figure 12-15. There are four main elements in the model: (1) the relative probabilities of wind strength and direction for each month, and of current strength and direction for each 5° square of latitude and longitude in the Pacific Ocean study area; (2) the positions of all the islands and landmasses in the study area, together with their sighting radius; (3) the estimated distances that would be covered by ships given various combinations of wind and current strength; and (4) the relative probabilities of survival of ships during certain periods at sea. Sighting radius (the distance out to sea from which islands can be seen) was built into the model on the assumption that, once land had been sighted, a landfall could be made. During each daily cycle voyages are started from given hearth areas like the coast of Peru, and a simulated course is followed until it ends in either a landfall or the death of the voyagers. By simulating hundreds of voyages from each starting point, we can map the relative probability of contact with different island groups as a potential contact field.

The program has already been run for various Pacific Island groups and for locations on the coasts of South America and New Zealand. Preliminary results indicate that the probabilities of inter-island links from the accidental drift of ships differ from one area to another. Wind and current patterns create environmental boundaries which make drifting in certain directions highly unlikely. Some of these boundaries coincide with long-standing anthropological breaks in the geographical pattern of ethnicity and culture like that separating the Micronesian people of the Gilbert Islands and the Polynesian inhabitants of the Ellice Islands. Other low probabilities of contact coincide with important linguistic boundaries.

The computer model for this research simulates activities that cannot be observed at first hand and are too complex to be simulated by manual calculations. It confirms that certain existing population distributions are possible purely as a result of voyagers drifting off course. However, there remain certain hard cases, notably the Hawaiian Islands, whose settlement still remains a mystery.

We began this chapter with El Tor and hot pants and ended it with *Kon Tiki*. Between, we have seen how the general notions of spatial diffusion can be simulated by probabilistic models—most of them developed from the work of Swedish geographers. These models help to throw some light on the process of diffusion by which past cultural changes have occurred. Modern mass-communication media have made the power and significance of the forces of change we have studied immense. The TV masts in the opening photograph of this chapter signify the trend toward a global village where change no longer requires mass movements of people but spreads far more rapidly through the subtle osmosis of messages carried by the mass media. The long-term implications of the waves of innovation generated and reinforced by mass media for the persistence of the cultural variety of man on the planet earth may be immense.

Reflections

1. Gather data for your local area on (a) the location and (b) the date of foundation of any one sort of public institution (churches, banks, schools, colleges, and so on). Map the data and try to identify the kind of spatial diffusion processes that appear to be operating. Do contagious or hierarchic processes seem to be more important?

2. Imagine you are opening a new chain of motels or hamburger restaurants (e.g., Holiday Inns or McDonalds) in your state. Where would you try to locate the first five establishments? Why would you pick these locations? Compare your results with those of others in your class and identify any common locations you all wish (a) to adopt or (b) to avoid.

3. List the factors which affect the spread of family-planning information in a developing country like India. How many of these factors could you incorporate into the Hägerstrand model of innovation diffusion?

4. Check how information spreads in a small group by planting a rumor (e.g., that your instructor has just become the father of twins) with one other member of the class. At the beginning of the next class check (a) who now knows, (b) from whom the information was obtained, and (c) when the "telling" took place. Try to construct a tree like that in Figure 12-3, showing the way in which the rumor spread through the class.

5. Look carefully at the first four phases of Figure 11-8. What would the pattern of diffusion have looked like if the first seven random numbers had been 1920, 8520, 1567, 7803, 3223, 5059, and 2483?

6. Trace the loops in the simplified flow chart of Polynesian voyages in Figure 12-15. How accurate is such a model likely to be? How might this type of model be used to trace other cultural diffusion processes?

7. Review your understanding of the following concepts:
   (a) expansion diffusion
   (b) relocation diffusion
   (c) contagious diffusion
   (d) cascade diffusion
   (e) isotropic areas
   (f) innovation profiles
   (g) contact fields
   (h) mean information fields (MIF)

(i) random numbers   (k) barrier effects
(j) resistance curves   (l) simulation models

One step further . . .

An excellent introductory review of theories of spatial diffusion and their use by geographers is given in

Gould, P., *Spatial Diffusion* (Association of American Geographers, Washington, D.C., 1969).

The classic Swedish work on diffusion is by Torsten Hägerstrand. This has now been translated into English as

Hägerstrand, T., *Innovation Diffusion as a Spatial Process* (University of Chicago Press, Chicago, 1968).

Examples of more recent work on diffusion in the United States are

Bowden, L. W., *Diffusion of the Decision to Irrigate* (University of Chicago, Department of Geography, Research Paper 97, Chicago, 1965) and

Berry, B. J., and F. Horton, *Geographic Perspectives on Urban Systems* (Prentice-Hall, Englewood Cliffs, N.J., 1970), pp. 419–434.

The "Polynesian drift" model discussed in the chapter and the application of diffusion models to the reconstruction of past cultural distributions is discussed in

Levison, M., R. G. Ward, and J. W. Webb, *The Settlement of Polynesia: A Computer Simulation* (University of Minnesota Press, Minneapolis, 1973), Chaps. 1 and 5.

A comprehensive review of diffusion theory, including its sociological and biophysical applications, is given in

Brown, L. A., *Diffusion Processes and Location* (Regional Science Research Institute, Philadelphia, Bibliography Series 4, 1968).

Although research in spatial diffusion is reported in the standard American geographic journals, it is worth paying particular attention to Swedish serials. The *Lund Publications in Geography* (An occasional publication) and *Geografiska Annaler, Series B* (a quarterly) are of especial significance.

Part Four

Regional Hierarchies

Part Four of the book turns from the study of regional contrasts—the mosaic of different cultural regions—to similarities between regions. We begin with the most powerful common force shaping regional structures in *Urbanization* (Chapter 13). We see how geographers are attempting to unravel the forces that cause more and more of the world's population to cluster into large cities. We look at the past and future trends in city growth. We find that cities and other population clusters grow not in a random fashion, but in a regular and rather orderly way. In *City Chains and Hierarchies* (Chapter 14) this delicate structure is dissected, and the beautiful symmetry and balance that underlies it is exposed. Man is shown to be a surprisingly orderly species, sharing with other populations an almost feudal structure, ranging from world city down to local village and farm. The last two chapters (15 and 16) look to the *Worlds Beyond the City*. Cities are found to be intimately linked to the fortunes of surrounding rural areas, and the two together are found to form integrated city-regions. The mechanisms of this linkage and the way in which agricultural and industrial zones develop around cities is explored at a series of spatial levels. Finally, we examine the flows within a region and the delicate filigree of networks that binds city and region together into an inter-dependent whole.

Chapter 13

Urbanization

It isn't size that counts so much as the way things are arranged.

—E. M. FORSTER
Howards End (1910)

If you had to think of a definition of a city, what would it be? Let us begin with a very simple one. "A *city* is a large number of people living together at very high densities in a compact swarm." Perhaps the easiest way to visualize what our definition describes is to think of a major rock or pop festival. During the summer weekends these may draw crowds of 100,000 or more to green fields around North America and Western Europe. Although newspapers and promoters vary in their estimates of the total attending, it is possible to utilize aerial photographs to measure the densities per square kilometer of these Woodstock-style events. If we put the densities, conservatively, at about 150,000 per square kilometer, then we need only some routine geometry (and a vivid imagination) to compute that, at these densities, the total world population could be corralled in a ring with a radius of only 87 km (54 mi). If we packed people together at the densities encountered on the New York subway, we could get everyone into a ring with a radius of only a little over 11 km (6.8 mi).

The notion of such a seething swarm of humanity might repel you, and the ways in which such a swarm would be fed and watered, let alone sanitized, are wholly unknown. Actually, the world's biggest high-density clusters like Tokyo and New York are several orders of magnitude smaller than the hypothetical world city we have just described. Even so, these *megacities* of around 10 million or so people demand the most complex of life-support and communication systems and pose some of our largest social- and economic-control problems.

In this chapter and the three that follow, we shall be looking at the city as a dominating force in the organization of the human population. Geographers must try to unravel the forces that lead to this dominance. Why do populations cluster into cities, and what keeps them from clustering even more strongly? What are the trends over time in this regard, and how urbanized is the world becoming? We turn, in the second half of the chapter, from questions of urbanization at the global, or macrogeographic, level to the micro level. What is the internal structure of a city? Do Western cities provide a reliable guide for understanding cities in other parts of the world? We can only sketch in the most basic elements of the answers to these questions in this book. For those of you who go on to more advanced geography courses, urban geography will form one of the most important and interesting areas of further study.

13-1 | WORLD URBANIZATION

Cities are the world's most crowded places. In New York City, about 55,000 people live on each square kilometer; in Montreal and Moscow, the corresponding population densities are about 52,000 and 49,000.

These densities are several orders of magnitude higher than the overall densities for the countries that contain these cities. The United States has, on the average, only 15 people living on each square kilometer. For Canada and Soviet Russia the national averages are lower still—about 1½ and 8 persons per square kilometer, respectively.

Trends in urbanization

Whatever the precise yardstick we use to measure a city, or urban place, the measurements show that the proportion of the world's urban dwellers is on the increase. If we use a threshold population of 20,000 for our definition of an urban place, then in 1800 only about one in forty (2½ percent) of the world's people were urbanites. By 1970 the proportion had jumped to about one in four (25 percent), and it is expected to reach one in two (50 percent) by the year 2000.

Trends in the rate of urbanization vary from country to country. For example, the United States is well ahead of the world as a whole. In 1800, about 5 percent of its population were classifiable as urban (using the same 20,000 threshold to define an urban place). By 1970 the proportion of urban dwellers in the population had risen to 70 percent, and by the end of the century it is expected to reach about 80 percent.

Can we detect any general pattern in the urbanization rates for different countries? Figure 13-1 shows a typical curve for a Western country. This S-shaped *urbanization curve* has the same kind of logistic form which we have observed in other curves earlier in this book. (See Section 12-3.) Slow rates of growth in the early part of the nineteenth century are followed by sharp rises in the growth rate in the second half of the century and a progressive slowing down thereafter. During

Figure 13-1. Urbanization curves over time. On the left in the graph is an idealized S-shaped urbanization curve for a Western country. Geographers are uncertain how reliable a guide this is for projecting the growth of cities in developing countries today, where natural population increase plays a more important role than inward migration.

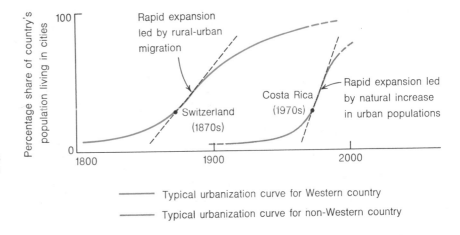

the most rapid period of growth, the key factor in population change was migration from rural to urban areas. Not only were birth rates lower in the cities, but the higher risk of epidemics and degenerative diseases there meant that death rates were higher too. Even when the figures are modified to take differences in the age structure of the populations in the two areas into account, London death rates in 1900 were one-third higher than rates in the surrounding rural counties.

It would be tempting to regard the S-shaped Western curve as a model for urbanization rates in the developing countries of the world. As Figure 13-1 shows, however, these countries do differ from the already-industrialized countries in two important respects. First, the process of industrialization not only started later, but is proceeding much more rapidly in the developing countries. The figure for the average annual gain in urban population for 34 countries in Africa, Asia, and Latin America over the last two decades is around $4\frac{1}{2}$ percent. By contrast, nine European countries, during *their* period of *fastest* growth (for most, the latter half of the nineteenth century), had average gains just above 2 percent. Curiously, the United States, Canada, and Australia, which were hit by huge waves of European immigration, had urbanization rates closer to those of today's developing countries.

There is, however, a more important, second difference between the two urbanization curves in Figure 13-1. Nineteenth-century urbanization was essentially "migration-led"; that is, the great proportion of the new urban dwellers were from rural areas. This is not true in today's developing countries. Despite the popular images of the rural poor streaming into the shanty towns on the edges of Latin American or African cities, the relative share of migration in the growth of urban populations is small in proportion to natural births. In Switzerland in the late nineteenth century, about 70 percent of the urban growth was due to rural-urban migration. In Costa Rica, with a roughly similar proportion of urban dwellers in its population now (as compared to our Swiss example), such migration accounts directly for only about 20 percent of urban growth. In much of the developing world today, the new urbanites are predominantly home grown.

Why do cities grow?

What are the forces fueling the present rapid growth of world cities? If we leave aside growth by the natural increase of existing urban populations, we are left with the puzzle of why man has become increasingly concentrated in clusters so dense that the environment cannot support him harmoniously. Let us take the familiar example of the United States. As Table 13-1 indicates, most big cities have severe social problems — soaring crime rates, drug abuse, poverty, and increasing pollution. In general the problems appear to get worse the larger the city. But, as the table also shows, the rewards (as measured by median income) also increase with city size.

Urbanization continues to increase because the benefits to be gained from crowding greatly exceed the costs. The major benefits gained from high-density crowding are the so-called agglomeration economies. *Agglomeration economies* are the savings that can be made by serving an increasingly large market distributed over a small, compact geographic area. Economies of scale make production costs for each unit low, while the short distance separating buyer and seller in cities cuts back the costs of transporting goods. Clearly production efficiency is directly related to how much is produced. A large output commonly results in lower unit costs because machinery and plants are used more fully and labor can become more specialized. Research and development costs and fixed costs can be spread over more units of production.

Table 13-1 Some implications of city size[a]

Characteristic	10,000 Inhabitants	100,000 Inhabitants	1,000,000 Inhabitants
Median income	90	100	120
Crime rates			
Murder	37	100	310
Rape	38	100	260
Auto theft	30	100	320
Air pollution	82	100	155

[a] United States data for the 1960s standardized to give characteristics of cities of 100,000 an index value of 100. Note that median income is probably underestimated due to the "suburbanization" effect (the fact that those who earn large incomes in the city may reside outside its legal or statistical boundaries).

Economist Adam Smith's dictum that "specialization depends on the size of the market" is clearly proven in the modern city. You need go no further than the Yellow Pages of New York's or London's telephone directories to be aware of the intricate degree of specialization the large metropolitan city permits.

Geographers have dissected the links between extreme specialization and extreme accessibility by examining the flows of information within a city. The persistence of highly specialized office complexes like New York's Wall Street and London's "City" district depends on face-to-face contacts as well as other forms of information flow and exchange. Each individual's activity is dependent on ready access to others, and the actions of the whole group form a complex, or knot, of specialized activities.

Limits to urban growth?

It is clear from Figure 13-1 that the rates of urbanization have slowed down in the Western, industrialized world. Indeed, in some Western countries an equilibrium position of low overall population growth and very little continuing urban growth has been reached. Some major cities, like London, show little overall change in population and even absolute declines in the central areas.

Can we then talk about some limits to the eventual size of cities? We've already seen that a major factor in urban growth is the *centripetal* force of agglomeration economies. But these advantages do not accrue indefinitely as size increases. Lower production costs may be outweighed by increased transport costs as urban areas grow to such a size that raw materials must be imported and finished goods exported over greater distances. Cities may become increasingly congested, internal transport costs rise, and public health dangers from infectious diseases or antisocial behavior increase. These factors contribute to the *centrifugal* forces that tend to halt concentration and to scatter population. As the cost of overcoming spatial separation has fallen, so the relative strength of centripetal forces has grown. Water pipelines, supertankers, and refrigerated ships are symbols of a continuing transport revolution which enables centralized population agglomerations to draw on widely distributed resources; equally important are the sewage lines and disposal services which allow a city to dispose of its massive daily burden of wastes. It is difficult to measure the exact balance of costs of cities, but the probable form of the cost curves as the size of the cities increases is given in Figure 13-2. This figure suggests that there is a threshold over which further increases in size are uneconomic and growth will slow down. At any one time, one particular bottleneck may be critical. For the eighteenth-century European city, the water supply and infectious diseases were significant barriers to growth; for today's cities, crime and finance may be higher on the agenda. In Section 21-3 we take a longer-term view of the future of the city and review Professor Jay Forrester's MIT model of urban stagnation. (See especially Figure 21-14.)

Urban multipliers

The export base of a city, like that of a country, is the activity that lets it sell goods or services or investments beyond its immediate boundaries. Although cities don't have precise boundaries like states, we can nonetheless think of them as having a balance of trade—importing and exporting. Of course, the distinction between the export sector and the rest of the activity of the city is not always clear-cut. Nonetheless, a useful distinction can be drawn between activities like manufacturing

Figure 13-2. Urban economies of scale. The general cost curve for all activities at the top of the graph suggests that middle-sized cities may be more efficient than either very large metropolises or small cities and towns. This does not necessarily mean that they are preferable, however, as comparable curves for welfare or satisfaction levels show. [From R. L. Morrill, *The Spatial Organization of Society* (Wadsworth Publishing, Calif., 1970), p. 157, Fig. 8.01. Copyright © 1970 by Wadsworth Publishing Co., Belmont, Calif. Reprinted with permission.]

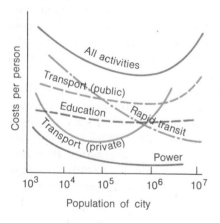

aircraft engines (to be "exported" to aircraft assembly plants in other cities) and those like baking bread (to be consumed in the city itself). The *urban base ratio* is used by geographers to describe the ratio of jobs in "export" sectors of the city economy to the total population. If a city of 60,000 people has 10,000 jobs in its export sector, the urban base ratio is 1:6.

Let us look at the effects of an increase in the export base in a highly simplified situation. Figure 13-3 shows a small Minnesota city whose export base is iron-ore processing. The ore is not consumed by the local community directly. But the inflow of funds from exporting the ore helps balance the "import" of food, gasoline, and so on, enabling the city to survive as an economic unit even though it is not self-sufficient. Let us suppose that a new mine is opened, bringing an increase of 100 jobs in this basic export sector. What will be the effect on the local city?

A typical cycle of events is shown in Figure 13-4. First, as we have noted, the new mine creates a number of entirely new jobs in the local area. If the average family size is four, we can say that every 100 jobs will lead to 400 more people in the household sector. Since these people will demand a set of service facilities—schools, churches,

Figure 13-3. Resource processing and urban growth. The new town of Silver Bay (background) has grown up in this wilderness area of northern Minnesota in direct response to the location of an iron ore processing plant (foreground). Low-grade iron ore is hauled 70 km (43 mi) by railroad from the Mesabi range to this coastal site on Lake Superior. After processing, the ore is shipped a further 1000 km (over 600 mi) to steelmaking plants around the Great Lakes. [Photography courtesy of Reserve Mining Company.]

Figure 13-4. The impact of changes in the basic sector on urban growth. An increase in jobs in the basic sector of the urban economy has a multiplicative effect on other sectors within the city. In this highly simplified model, every basic job creates an increase of four in the population of the household sector (i.e., the worker plus a hypothetical family). Every increase of eight people in the household sector creates one extra job in the service sector (for teachers, gas-station attendants, doctors, etc.). The arrows show successive cycles of expansion, since each person in the service sector is also assumed to have a hypothetical family, which also demands services. And so on! Actual multipliers may vary from those shown here (4⅛).

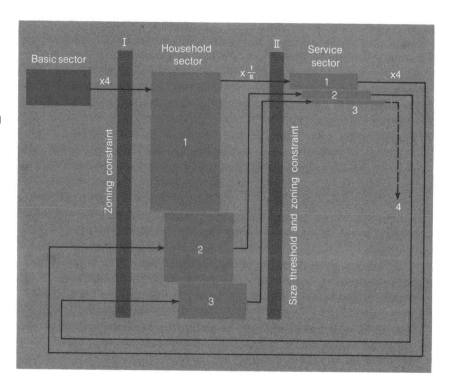

shops, hospitals—the households will, in turn, create a set of service jobs. Moreover, the workers in the service jobs will have families and create a second, smaller cycle of more households and more service jobs. As Figure 13-4 indicates, these diminishing cycles can be repeated any number of times. If you follow the calculations through, you will find that the 100 new mining jobs eventually cause an increase of 800 people in the household sector, plus a further 100 jobs in the service sector.

This model of the cascading effect of one activity on another was developed by an economist, I. S. Lowry, and is generally termed a *Lowry model.* Two checks may be built into the model to modulate the cycles and thus get a "truer" picture. First, we can assume that land for residential development cannot be available in the same zone as the mine, and therefore housing must be located elsewhere. Second, we can stipulate a minimum size for service sectors. For instance, we can stipulate that a hospital may be created only when the population in the household sector exceeds an appropriate threshold level. Otherwise, the local population will have to go elsewhere for medical treatment.

Constraints

Constraints act as checks on the numbers in the household and service sectors. A *zoning constraint* limits numbers by stipulating a maximum density in any area within the city; when that is reached a new housing tract or a new area of service activities must be started. The *size-threshold constraint* stipulates a minimum level of demand (i.e., numbers in the household sector) before a particular service facility (e.g., a hospital) can be established.

The impact of the mine may not be confined to the household and service sectors of the economy. Mining may lead to associated industrial activities involving the refining, smelting, or processing of mineral ore, or the manufacture of equipment for the mining process. This will cause a proportional drop in jobs in the *primary sector* (a term used to describe jobs in agriculture, mining, fishing, etc.) and a relative increase in jobs in the manufacturing, or *secondary* sector. Further growth will lead to an expansion in the wide range of service jobs (in shops, education, hospitals, transport, etc.) in the *tertiary* sector. Large cities usually have a larger proportion of their work force in the tertiary sector. In the last half century, the most rapidly growing job sector has been jobs in research and administration, the so-called *quaternary* sector. "Office" and "research lab" jobs account for an increasing share of employment opportunities in major metropolitan areas. One way of distinguishing between the four sectors is to take a given natural product and consider which jobs are related to it. For example, timber production gives jobs to a chainsaw gang (in the primary sector), a lathe operator in a furniture-manufacturing plant (in the secondary sector), a furniture-store operator (in the tertiary sector), and a researcher in wood technology (in the quaternary sector).

Whether or not the extraction of a given resource will trigger a chain of associated growth depends on a great variety of factors. To take one example, consider the enormous copper pit in Butte, Montana; it has had a fairly modest and short-lived impact on the surrounding area. The population of Butte reached a peak of 41,000 in 1920. It has since been slowly dwindling in size. Increased mechanization reduced the labor force needed in the mining industry; and because of its rather remote location, Butte failed to attract new industry to employ surplus labor. Conversely, in many of the coal mining areas of northwest Europe that attracted large populations in the early nineteenth century, the major cities have continued to expand long after the original mining activity has ceased to be significant. The extraction of resources located in central, highly accessible zones (that may already have a dense population) is much more likely to trigger prolonged urban–industrial growth than the extraction of similar resources set in remote, peripheral locations. Figure 13-5 gives a typical curve of the changing proportion of jobs in ex-mining centers like Denver, Colorado.

13-2 | URBANIZATION AS A SPATIAL PROCESS

One aspect of the urbanization process of great interest to geographers is the role of changing accessibility. For agglomeration economies to occur, markets have to be not only large but highly accessible. Over the

Figure 13-5. Urbanization and changing job structures. As a city grows, the proportion of people in the three or four major job sectors (see text for definitions) will change. One way of showing this is to plot the relative percentage of jobs in each sector on (a) a triangular graph for three sectors, or (b) a tetrahedral graph for four sectors. In the three-sector case, a city with exactly one-third of its jobs in each sector would be located at point 1, while one with 90 percent of its jobs in the manufacturing sector and 5 percent in each of the other two sectors would be located at point 2. The more a city specializes in employment within a single sector, the more it moves toward an apex of the triangle or tetrahedron. The red lines indicate what happens when a city which begins as a mining town evolves into a major regional capital with many jobs in the quaternary sector.

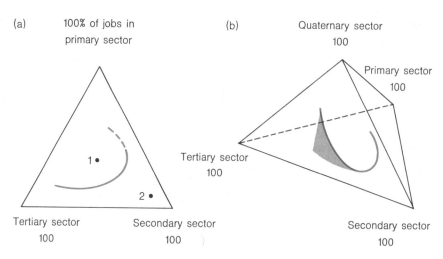

last two centuries, in which time the average level of world urbanization has increased tenfold, overall travel times and transport costs have *fallen* in real terms by an even greater amount. We look here at the impact of this improved accessibility both at the intercity and at the within-city levels.

Accessibility between cities: The urban implosion

Table 13-2 summarizes the major transport innovations of the last 200 years. To make comparisons easier, the average speed for each mode is given in terms of the most basic form of transport—man's walking speed. To the items in the table could be added similar improvements in the movement of energy sources at increasing speeds and volumes and decreasing costs through innovations such as pipelines and electric cables.

Changing travel times The impact of these space-shrinking innovations on cities varied greatly. It was between existing cities that the need for improved transport was greatest, and it was here that the major innovations came first. New transport technology—whether it is a railroad, Telex, shuttle jet, or monorail—tends first to connect key cities, thereby increasing their locational advantage in relation to important cities. We can illustrate this "ratchet effect" by measuring changes in the average cost of movement between centers of different sizes. If we use time as a proxy for cost, between the years 1850 and 1900 the distance from New York to California was cut from 24 days (3 days by railroad plus 21 days by overland coach) to 4 days (by direct train). The next half-century reduced this distance to 8 hours by DC-6, and further

Table 13-2 Ranges in travel times

Innovations	Period of introduction	Maximum distance coverable in 12 hours[a]	Speed relative to man's walking speed
In land transport			
Horse-drawn coaches	} 1770s	200 km (124 mi)	× 2.5
Canal and river transport systems			
Railway systems	1830s	800 km (497 mi)	× 10
Multilane highways for automobile	1930s	1,300 km (808 mi)	× 17.5
High-speed rail systems	1950s	2,000 km (1,243 mi)	× 25
In sea transport			
Sailing vessels	1770s	200 km (124 mi)	× 2.5
Early steamships and clippers	1830s 1850s	400 km (249 mi)	× 5
Diesel-powered vessels	1920s	600 km (373 mi)	× 7.5
In air transport			
Balloons	1850s	200 km (124 mi)	× 2.5
Propeller-driven aircraft	1900s	3,000 km (1,864 mi)	× 37.5
Jet-engined aircraft	1950s	15,000 km (9321 mi)	× 187
In communications			
Organized postal services	1840s	—	—
Telephone	1870s	—	—
Telegraph	1890s	—	—
Radio	1920s	—	—
Television	1940s	—	—
Satellite links	1960s	—	—

[a] Distances are averages based on the fastest mode of public transport available at each date and are approximations only.

Figure 13-6. Shrinking worlds. The graphs show changes in the average amount of time needed to go from New York City to Los Angeles since 1850, using the fastest means of public transport. [From R. E. G. Davies, *History of the World's Airlines* (Oxford University Press, London, 1964), p. 509, Fig. 91.]

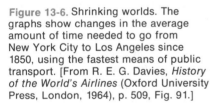
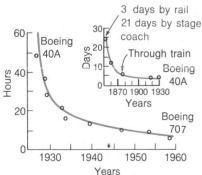

reductions have followed. (See Figure 13-6.) Thus the relative cost of the trip (in hours) has been reduced by over one-half since the 1930s.

Yet these changes have not been uniform and the bigger cities gain more from improvements in services over time. As a result, the cost in time of traveling between them is reduced, and they move closer together in relative terms.

We can borrow a term from astronomer Fred Hoyle and describe the effects of this differential spatial shrinkage as an urban *implosion* — that is, the inverse of an explosion. The ways in which large cities converge on each other is shown by the diagram in Figure 13-7. Note that since large cities converge rapidly and the smaller cities converge

Figure 13-7. Urban implosions. Transport innovations are usually first introduced between pairs of large cities between which important flows of people and goods already occur. Such innovations make links between such pairs easier, setting off a cycle of increased contacts and flows. As Figure 13–6 shows, the net effect is to bring larger cities relatively closer to each other. Meanwhile smaller centers become relatively (and sometimes absolutely) more remote. This figure illustrates this process of "implosion." Note the convergent movement of the three largest cities (the size of the spheres is proportional to the size of the cities) in comparison with the divergent movement of the three smaller ones.

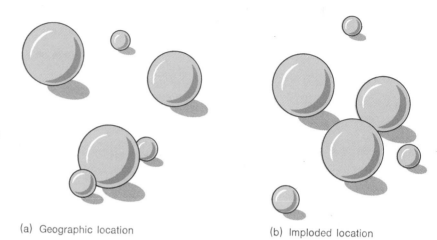

(a) Geographic location   (b) Imploded location

less rapidly, the larger and smaller cities diverge in relative terms. The speed and self-reinforcing nature of this spatial implosion are principal factors in the current rapid growth of large cities. At Bristol University, Philip Forer has traced the implosion of New Zealand cities over the last thirty years. Basing his research on changing air travel times, he was able to map New Zealand's changing structure; as larger cities were dragged closer together, smaller cities pushed outwards into less central locations. For the South Pacific as a whole, Forer's time maps show a curious pattern in which the largest Australian city (Sydney) is closer to the largest New Zealand city (Auckland) than are many small New Zealand cities (Figure 13-8).

Contact patterns Another approach to the urbanization process has been taken by a group of Swedish geographers headed by Torsten Hägerstrand and Gunnar Tornqvist. They note the growing importance of the quaternary sector in the urban growth pattern of the present century. Despite the growth of telecommunications of all kinds, the need for direct person-to-person contacts is critical in this sector. In research planning, for example, a large number of people may need to come together in informal groups. Note, however, that such groups are likely to meet occasionally and irregularly (rather than on a fixed daily basis) and that the membership of the group may be flexible.

In this situation the ease of contact between urban areas may become critical. Using detailed diaries of personal movements of business executives, Swedish geographers have been able (a) to track down the degree of interdependence of different industries and types of activity in different geographic locations and (b) to rank Swedish

Figure 13-8. Time maps of urban implosions. (a) A conventional geographic map of New Zealand shows the location of major airports. (b) A time map shows the "distance" in minutes between the same urban airports. The map is based on travel times and service frequencies in 1970. Note the way in which the three leading cities—Auckland, Wellington, and Christchurch—are clustered near the center of the map. In contrast, the smaller city of Tauranga is displaced toward the periphery. (c) A time map shows the relations between the main New Zealand airports and other major urban airports in the South Pacific. It too is based on air travel times and service frequencies in 1970. Note how large numbers of high-speed flights between Sydney and the main New Zealand cities draw these locations together. These unfamiliar time maps are produced by a computer analysis of a mass of information on flight times and frequencies between all the pairs of locations shown on the map. [From P. Forer, *Changes in the Spatial Structure of the New Zealand Internal Airline Network,* doctoral dissertation, Bristol University.]

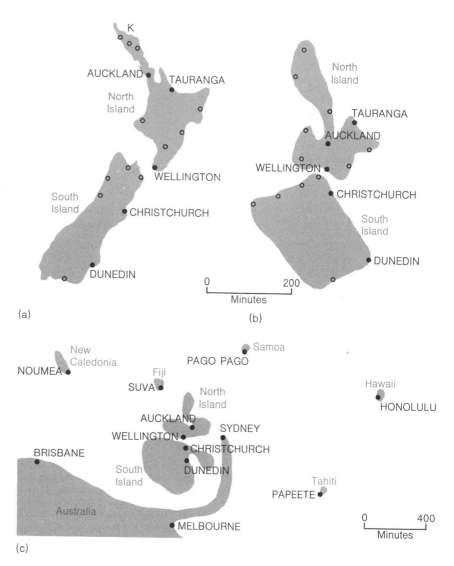

cities on a *contactability scale*. To do this they use journeys made by private automobile or the fastest form of public transport. (See Figure 13-9.) Only journeys made between 6 a.m. and 11 p.m. are counted, and a "contact" is defined as a meeting of not less than four hours during the "normal" working day (8 a.m. to 6 p.m.). When journeys are weighted to include the cost of contacts, a very detailed regional picture can be built up. If we give the "most contactable" city—the

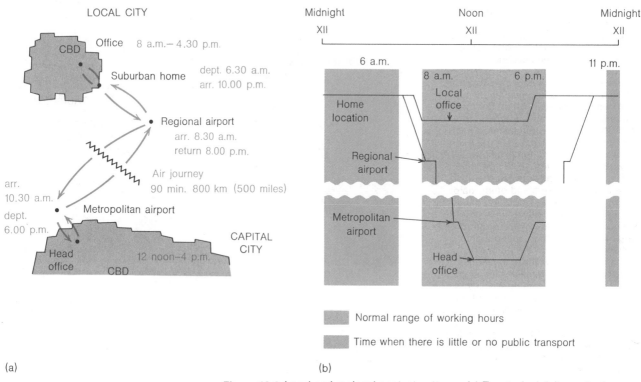

Figure 13-9. Local and regional contact patterns. (a) Two typical daily contact cycles for a suburban-living executive: first, the regular daily cycle between home and local office, with a short commuting period (about 40 minutes by automobile), and second, the occasional visit to the head office in the capital city, with a long commuting period (about 12 hours, using an automobile, plane, and taxi) and a short (four-hour) conference period. These contact patterns may be plotted on space-time diagrams, as in (b). In this diagram the horizontal lines indicate a static location and the diagonal lines show travel between locations. The more a line diverges from the horizontal, the greater the speed of travel. Using detailed travel diaries, a group of Swedish researchers led by Torsten Hägerstrand and Gunnar Tornqvist have been able to show the importance of contact possibilities between Swedish cities in the growth of the quaternary sector. Clearly, increased transport speeds allow lower-density living patterns within the daily commuting range and an increased meshing of cities within the occasional daily contact range.

capital, Stockholm—a score of 100, we can then rate the other cities by comparing them to the capital. Some smaller cities in the vicinity of Stockholm have scores in the 80s and 90s. Gothenburg and Malmo, the second- and third-largest cities, have scores of 78 and 70, respectively. The "least contactable" city is the iron-mining city of Kiruna north of the Arctic Circle, which has a score of only 36.

These results are useful in picking out the likely centers of further urban growth in the quaternary sector. Planned changes in transport services (e.g., the introduction of new airline schedules or the opening of new highway links) can be checked for their effect on the contactability rating of different cities. Which cities will benefit? Which will lose out? By repeating studies over time, the changing access of one city to another can be plotted and the course of the urban implosion charted and projected into the future.

Accessibility within cities: Urban sprawl

To understand the second major impact of transport innovations on urbanization, we must recall that man has evolved biologically in an environment with a regular cycle of lightness and darkness, and most basic human activities — feeding, sleeping, working, procreating — have a daily rhythm. In a recent study, the pattern for the use of time in the United States [Figure 13-10(a)] can be broadly compared to that of 40 other countries in North America and Europe. Minor differences between the patterns were matters of culture (e.g., the later time of dining in southern Europe) or of technology (e.g., the higher proportion of TV watching in the United States). Even on the campus, the same kind of regular rhythms can be shown [Figure 13-10(b)]. Not only is the basic biological clockwork that controls human activity important to agricultural communities, but it also controls the vast tides of traffic that flow into our great cities each morning and out again each evening.

How do these time rhythms affect spatial organization? Basically, they operate through the dominant need of most of the human species for a fixed, or relatively fixed, home base. This home base may vary from an apartment in New York City to a sampan in Hong Kong, but it has the same essential economic, social, and biological role. At the very least the home base provides a place for a stock of household goods. It may also provide a site for the production and rearing of children, and a retreat for the satisfaction of sexual and emotional needs. The need to return regularly to a home base puts a severe constraint on the number of hours that can be devoted to travel to work, and therefore on the distance that separates the elements in any household economy. Even when the adult male breaks away from this pattern and works away from the home base for long periods, the constraint usually remains for females, children, and older folk in the household.

Over what distance does this constraint operate? Before Stephenson's *Rocket* rolled along the rails in October 1829, the only regular ways of moving over land were by manpower or horsepower. (See Table 13-2.) Horses and carriages could attain a speed of over 50 km/

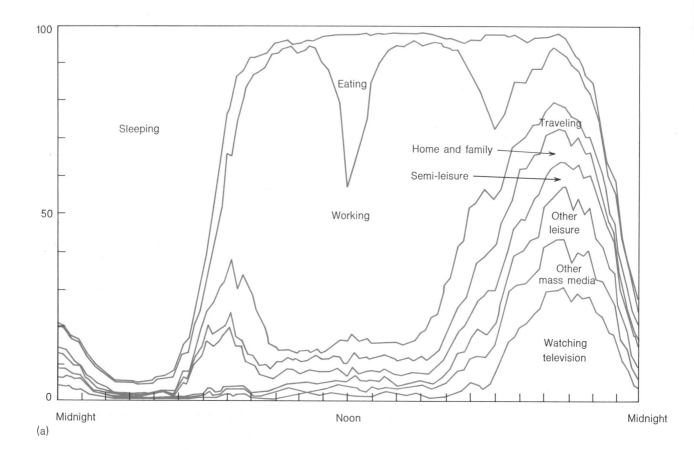

Sleeping

Eating

Traveling

Home and family

Semi-leisure

Working

Other leisure

Other mass media

Watching television

Midnight

Noon

Midnight

(a)

hour (over 30 mi/hour), although in practice the rate of travel was reduced by poor or congested roads to less than a quarter of this figure. How fast or far man moved on his own legs is more difficult to estimate. Modern measurements of our walking speeds on city sidewalks indicate that the average adult pedestrian walks at 5.5 km, or 3.4 mi, an hour. The values vary with age and sex. Adolescents gallop along at a steady 6.4 km, or 4 mi, but mothers with young children walk only 2.6 km/hour (1.6 mi/hour). A slope of only 6° cuts speeds by over a fifth; a slope of 12° cuts speeds in half.

It is not surprising that most communities in the prerailway period were quite compact, usually with a diameter of not more than a kilometer or two. The average walking time from one end of town to another would thus be about 11 to 22 minutes. This compactness was directly related to the need for communication between different parts of the city and to the limitation imposed by walking or carriage

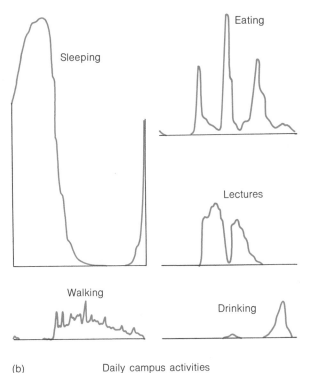

Sleeping

Eating

Lectures

Walking

Drinking

(b)   Daily campus activities

Figure 13-10. Time rhythms in human activity. (a) Monday-through-Friday activity patterns for a sample of 375 employed men from a sample of 44 United States cities. (b) Comparable activity patterns for a sample of students at Reading University, England: the axes in these graphs are the same as those in (a). The concentration of drinking reflects English licencing laws. [From A. Szalai, Ed., *The Use of Time* (Mouton, The Hague, 1970), p. 736, Fig. 5.1–11A; and L. March, *Architectural Design* **41** (1971), p. 302, Fig. 14.]

speeds. The communication links between home and work, merchant and scribe, banker and businessman, magistrate and prostitute all had to be spatially short if transactions were to be completed within an acceptably small part of the day.

The technological revolutions in land transport during the last 150 years have successively reduced the need for compact cities. As steam locomotive speeds increased, and the electric streetcar, the omnibus, the suburban electric railway, and the private automobile succeeded each other, so the links between home and workplace widened. Living above the shop or workshop was replaced by daily commuting over increasingly long distances. We can see the dramatic effects of these changes in the exploding size of London (in Figure 13-11). In 1800 London, Europe's largest city, could still be crossed on foot in little more than an hour. Even at its widest, the built-up area was only 10 km (about 6 mi) across. As the sequence of maps shows, this area expanded so rapidly that by 1914 its diameter had reached 35 km (about 22 mi), and it approached 70 km (about 43 mi) by 1958.

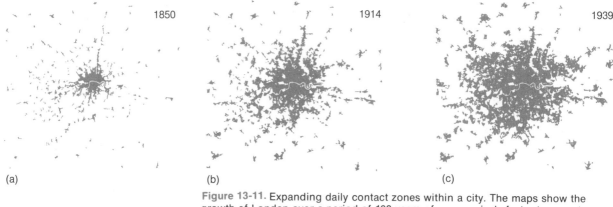

1850   (a)

1914   (b)

1939   (c)

Figure 13-11. Expanding daily contact zones within a city. The maps show the growth of London over a period of 108 years of progressively faster transport. [From D. J. Sinclair, in K. M. Clayton, Ed., *Guide to London Excursions* (Twentieth International IGU Conference, London, 1964), p. 12, Fig. 9.]

1958

(d)

* * * *

Since then it has shown few changes due to a vigorous policy of protecting further extension into rural areas (the green belt policy).

London is modest in size by international standards. Los Angeles (Figure 13-12) already exceeds 80 km (50 mi), and Greater Tokyo 120 km (75 mi) in diameter. If we accept Oxford University geographer Jean Gottmann's idea that the whole network of cities along the East Coast between Boston and Washington, D.C., is being fused by factors such as shuttle jets into one vast megalopolis, then we have a world city of mammoth dimensions, over 600 km (373 mi) in diameter. Each new transport breakthrough expands the area within which the key daily transactions of a city are conducted. Expansions in this "transaction zone" are shown as a three-stage process in Figure 13-13.

13-3 | SPATIAL PATTERNS WITHIN THE CITY

The basic geometry of the North American city is well known to most readers. At the center is the downtown shopping district, the banks and offices, the hotels and theaters. Surrounding this is an area of run-down housing, mixed with some industry. Beyond this we run into modest residential areas, typically houses and apartments for office and blue-collar workers. Still further out come the family homes of middle-class suburbanites, thinning out until we come to the golf courses and the rolling acres of the very rich.

In this section we shall probe the familiar pattern to try and see what factors lie behind it. We shall look at the value of land within the city, the use to which it is put, and the population density it supports. Secondly, we shall take Chicago as an illustration of a

Figure 13-12. Urban sprawl. Los Angeles is often seen as the classic example of urban sprawl and the prototype of the twentieth-century city. It is now the third-largest city in the United States, with an urbanized area that sprawls outside its legal limits to include over 6 million inhabitants in the metropolitan area. Although smaller in population than Chicago, it has twice its area, and it is nearly ten times as large as San Francisco. This low-density, automobile-dependent metropolis has had the fastest growth of any major city in North America. In 1870 it had a population of only 5728 inhabitants; since then, the population has grown to 100 times that figure. This photo shows us the view across the Los Angeles basin to the San Gabriel Mountains, with the civic center and high-rise buildings around the CBD in the middle distance. [Photo by William A. Garnett.]

Western city and examine its pattern in relation to existing models of urban structure. Finally, we raise a preliminary question regarding how far cities in other parts of the world resemble the North American pattern.

The geometry of land values

One measure of the value an urban community places on different sites within the city is provided by land values. As an example, Figure 13-14 shows a three-dimensional representation of land values in Topeka, Kansas. Despite minor variations in peripheral areas, the dominating elements in this and in most Western cities is the extremely high values attached to land in the central part of the city and the

Figure 13-13. A simplified, three-stage model of the sprawling expansion of two cities. In the first stage (a), jobs and homes overlap in a small high-density cluster (like the medieval walled city in Figure 11-7). (b) Transport innovations allow the links between jobs and homes to be loosened. Densities in the central areas of cities appear to fall as less land is available for homes and long-distance commuting creates a daily tidal flow of workers into and out of the city. (c) Continuation of this process leads to further evacuation of the central areas of cities, the destruction of the intervening rural land, and suburban sprawl. Corridors of cities develop like those along the German Ruhr and the Boston-Washington axis.

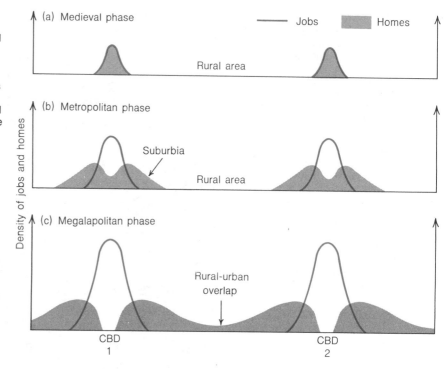

Figure 13-14. Land values in an urban community. This three-dimensional map shows land values in the city of Topeka, Kansas. Note the extremely high values in the city center and the smaller secondary centers. [From D. S. Knos, *Distribution and Land Values in Topeka, Kansas* (University of Kansas, Bureau of Business and Economic Research, Lawrence, Kansas, 1962), Fig. 2.]

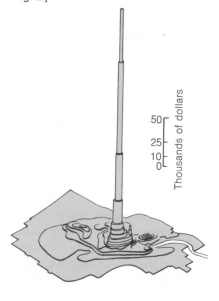

generally steep decline of land values with distance from the center. The high point in values is found somewhere near the center of the central business district (CBD), usually marked by tall buildings, an extremely high daytime population, and high traffic densities.

If we take a series of such cities, we can draw a general picture of average land values (Figure 13-15). The value of land is highest in the city center and decreases toward the periphery, but the pattern is modified by two additional elements: main traffic arteries and intersections of main arteries with secondary centers at regular distances from the CBD. When we superimpose these three effects on one another on a *three-dimensional model*, the result is a conic hill whose flanks are disturbed by ridges, depressions, and small peaks. This land-value surface directly reflects the different accessibility of parts of the city and shows where the most intense competition for space occurs (and hence where the higher land values occur).

Land values and land use

What will the effect of these variations in land values be on the distribution of land uses within the city? Let us assume that a city wishes

Figure 13-15. A general pattern of city land values. This idealized land-value surface emphasizes the high value ridges that run outward from the CBD along the main arterial roads. Secondary centers on these roads cause local peaks.

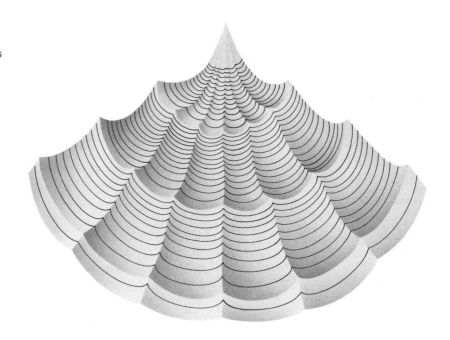

to establish a new university somewhere within its limits. If the university is constructed near the city center, it will be most accessible by public transport to all students, but it will be taking up valuable land that could be leased at high rents to commercial firms. Conversely, if we choose a green field on the edge of the city as the site of the campus, the land values there will be low and the university grounds can be more spacious. This advantage, however, will be offset by the school's eccentric location. Most students will have to travel farther, some right across the city from the far side.

All types of urban land use entail this type of quandary. As businesses move farther from the city center, they gain from the cheaper land prices or rents; but they are increasingly divorced from their potential market and stand to lose money from increased transport costs. We can describe this trade-off between land prices and transport costs by a series of nonintersecting lines, termed by the economist *bid-price curves* [Figure 13-16(a)]. Each line represents the rent values that exactly balance the increased transport costs due to locating away from the city center. If we assume that transport costs increase regularly with distance, then the bid-price curves will be parallel straight lines sloping downward with distance from the city center. Note that the lower lines within the family of parallel lines represent lower rent levels and therefore are always preferred over higher lines.

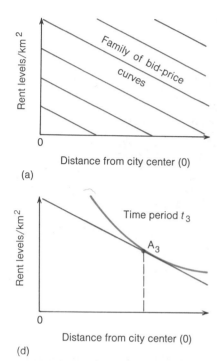

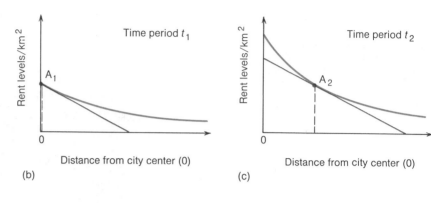

Figure 13-16. Land-use models of the city. For any urban land use—be it a cemetary, a motel, or a college—a location represents a trade-off between the convenience of being near the accessible city center and the high rent levels we must pay to locate there. If we describe our assessment of these two conflicting goals by a family of bid-price curves (a), we can compare these with the actual rent levels (colored line). Note how the equilibrium position (A_1) indicating the best trade-off (and thus the best location) may shift (to A_2 or A_3) as the city grows and land values rise (in time periods t_1 through t_3). Thus certain types of urban land use (e.g., single family homes) may be squeezed out toward the margin of the city as it grows.

By superimposing bid-price curves on the actual rent curve for a given city, as in Figure 13-16(b), we can determine both the point where the best trade-off is reached and the corresponding rent level. The best location will be where the actual rent curve just grazes the lowest possible bid-price curve. At this point (A_1) the values of both the actual rent curve and the bid-price curve are equal. If we think of this diagram as three dimensional, the bid price curves become a series of cones centered on the CBD, and we see that A_1 is not a point but a ring of possible locations around the city center.

As a city grows larger, land values increase, especially at the center, so the land-value curve becomes both higher and more concave, as in Figure 13-16(c). Hence, the ring of optimal locations for land-using activities is forced outward from near the city center (A_1 in time period t_1) to successively more peripheral locations at A_2 and A_3. We can trace this effect in the displacement of activities like manufacturing from inner city to suburban locations.

Each type of land-use activity in a city has a distinct pattern of bid-price curves. Those activities which gain greatly from locating "where the action is" near the city center will have steep curves. Theaters, insurance brokers, publishers, are all examples of activities which depend on a high degree of contactability and need accessible locations. Conversely, other activities may not be greatly affected by their location but may be very anxious to avoid high rentals: These will have bid-price curves with a gentle slope. Activities with steep bid-price curves will be able to cling to the steep slopes in land values around the CBD; others are gently angled and find a foothold only in the remoter parts of the city. Figure 13-17(a) illustrates hypothetical curves for banking and insurance offices (which typically cling tenaciously to downtown locations) and golf courses, which are usually located on the fringes of cities and sensitive to rises in land values.

Figure 13-17. Multiple land uses. An extension of the arguments in Figure 13–16 to two types of land use. (a) Characteristic bid-price curves for an intensive high-rent land use (banking) and an extensive low-rent land use (golf courses). Note how the equilibrium position for the first is in the downtown area, while that for the second is on the outskirts of the city. (b) This shows how a varied land-values gradient with secondary peaks may allow banking to find a series of equilibrium positions at varying distances from the CBD.

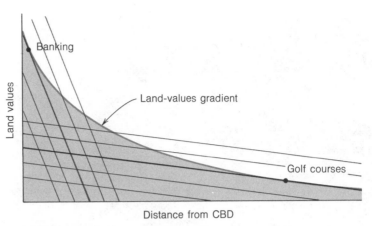

(a) Multiple land uses

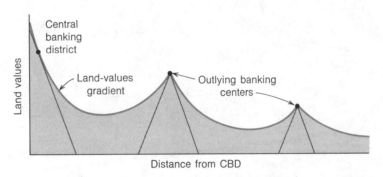

(b) Multiple centers for single land use

Note that the cross section shown here is a very simple one; if we add outlying secondary centers to the figure we can see that banking may be carried on there too, albeit on a smaller scale. [See Figure 13-17(b).]

Land-use mosaics, Chicago style

Land use in urban areas is a major research area for the urban geographer. Probably the most thoroughly dissected city is Chicago, where social scientists like Robert Park and E. W. Burgess began a trail of research in the 1920s that is still being followed a half-century later by geographers like Harold Mayer and Brian Berry. If we take an aerial photograph of Chicago from the south, looking toward the CBD in the Loop (Figure 13-18), we are viewing a city whose structure has formed a touchstone for studies of large metropolises elsewhere.

Can we make any sense of the complex mosaic of land uses shown by Chicago? Figure 13-19 shows one guide to the pattern provided by

Figure 13-18. Chicago. Poet Carl Sandburg's "City of the Big Shoulders" has had a special fascination for geographers. From the 1920s on, its spatial structure has been studied and dissected by a series of distinguished researchers. The key to its semicircular structure is the CBD, whose high-rise buildings form the famous "Loop" area (named from the loop of elevated railroad tracks that encircle it). The zones and sectors that flare out from the CBD are summarized in Figure 13–19. The transportation lines that are so prominent in the photograph have had an important effect on the city. Chicago grew up in the 1830s at the strategic southwest corner of Lake Michigan, and the canal connecting the Chicago River to the Illinois River (constructed in 1948) brought a rapid rise in the city's role as a freight-hauling center. In the same year, the first railroad

link was completed and the stage was set for a period of remarkable economic growth linked to Chicago's dominant position on the expanding railroad network of the Middle West. Its population grew from around 30,000 in 1850 to past the one-million mark by 1890. Today, with over 7 million people in its metropolitan area, Chicago is America's second largest city and the tenth-largest city in the world. To meet the vast transit needs of the metropolis, an intricate system of urban highways like the Ryan Expressway (in the foreground) have been constructed. [Photograph by William A. Garnett.]

geographers. Look first at the top row of the diagram. This shows an idealized Chicago as a circle centered on the CBD with Lake Michigan to the right. The CBD is the key feature, marked out by a score of factors (its high land values, its soaring daytime population, its high buildings, its age, and so on). Around this CBD the city's housing has developed as a series of wedge-shaped *income sectors* (upper left) which fan out as we move outwards. A drive north or south from the CBD along the lakeshore will take you through much higher-income housing than a drive to, say, the northwest. Once housing of a particular type had been established in a sector, it tended to persist as new housing of similar quality was built. Thus the city expanded outwards in this wedge-shaped fashion.

A second factor is a series of ring-shaped *age and family-size zones* centered on the CBD. The three zones shown mark a regular progression from the apartment-dwelling population of Zone I (typically, people in their twenties or an older generation in their fifties or beyond) to the suburban family-home population of Zone III (typically, families with young children). On the right of the top row is the third factor, *ethnic segregation*. This shows the presence of two distinctive black wedges to the south and west, in the inner part of an otherwise white city.

If you follow the arrows from the first three factors they lead to a fourth diagram of Chicago. This is made by overlapping the first three factors to give a mixture of wedges and zones. Thus we get a white, high-income, Zone III in the north of Chicago (the Evanston area). Note how the two black zones distort the simple model: Each black zone is itself subdivided into three zones related to age and family size, and each is an area of relatively low income despite its occurring in middle-income sectors in terms of our first diagram (upper left). Indeed, an area like the black South Side of Chicago is in some ways a detailed replica of the whole city, finely divided into a complex mosaic of income, age, and family-size units.

Of course, Chicago is not the simple ring shown in the upper half of Figure 13-19. So, we introduce a more accurate map of the city showing its *differential growth* between 1840 and 1960. This fourth factor is also blended into the model. Note how the zones within the various sectors are now displaced in our map showing differential

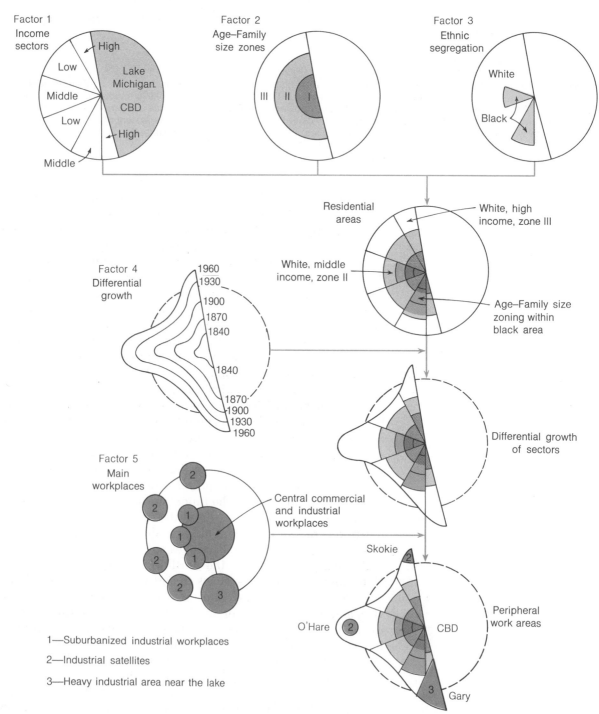

Figure 13-19. The regional structure of Chicago. Much of the variety within Chicago's semicircular urban structure (itself a product of the lakeshore location) can be explained by the combined effect of five factors. The arrows show the ways in which factors are crossbred together in this simplified model of the city. For a description of each factor and its impact on the city, see the text discussion. [From P. H. Rees, University of Chicago, unpublished master's thesis, 1968.]

growth of sectors. Finally, a fifth factor, the *main workplaces*, can be added to the model: The main commercial center plus three types of outlying industrial area are shown. Since to add all these to the model would be too complicated for our final diagram of the city (lower right), we include just some examples of the way in which industry helps to shape the land-use mosaic of the city. Two industrial satellites are shown, the Skokie area in the north and the area near O'Hare airport in the west. South of the city the heavy industrial area of Gary is also shown.

The patterns shown in Figure 13-19 show how various types of spatial segregation interact in shaping the mosaic of land use within a city like Chicago.

Spatial segregation affects not only the clustering of social groups as in the black ghetto on the south side of Chicago or the Italian quarter to the northwest; it also plays a role in decisions to group certain land uses, like light industry, in particular sections of the city. The forces causing segregation are external and internal, public and private, malignant and benign. They include not only discriminatory housing policies but also antipollution restrictions.

Population density patterns

If we take a single Western city and map its changing population density over time, we find that the population tends to spread out like a slowly melting ice-cream cone, covering ever wider areas but at ever lower densities. Figure 13-20 shows the pattern of population change for

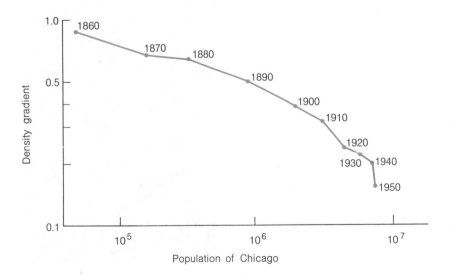

Figure 13-20. Changing urban density gradients over time. The average slope of Chicago's population density away from the city center was steep in 1860 but has declined steadily as the city expanded. [From P. H. Rees, University of Chicago, unpublished master's thesis, 1968.]

Figure 13-21. Density and urban structure. The photos show a cross section of changing urban structures as we move from the center to the periphery of the city. These changes in land use are paralleled by contrasts in land values and the density of the residential population. [Photographs (a) and (b) courtesy of American Airlines; (c) and (d) by George W. Gardner.]

(a)

(b)

(c)

(d)

Urban density functions

Study of scores of urban areas throughout the world led Colin Clark to suggest a general model for the decline of urban population densities away from the CBD. He proposed a *negative exponential* form in which population decreases at a decreasing rate with distance; that is,

$$Z_d = Z_0 e^{-bd}$$

where Z_d = the population density at distance d from the CBD;

Z_0 = a constant indicating the extrapolated population density at distance zero, that is, at the center of the city;

e = the base of the natural logarithms (2.718);

b = a constant indicating the rate of decrease of population density with distance; and

d = the variable distance.

Thus, with a central density of 1000 people/km² at the center and $b = -1.0$, we should expect a density of 368 people/km² at 1 km from the CBD, 135 people/km² at 2 km, 50 people/km² at 3 km, and so on. A comparison of Z_0 and b values allows us to compare different urban structures easily. Clark's work has been extended by others to allow more complex density functions to be matched and compared. [See M. H. Yeates and B. J. Garner, *The North American City* (Harper & Row, New York, 1971), Chap. 10.]

Chicago during a 100-year period by a series of population-density gradients that steadily decline in steepness from around 1860 to 1920. The change over the last half-century has been less spectacular but has included a marked reduction in density at the center of the city.

How typical of the spatial structure of cities is Chicago? Cross-cultural checks on population declines with distance from the city center have been intensively studied, and investigators have reached some general conclusions about the shape of these slopes. For example, economist Colin Clark studied population-density gradients for a group of 36 cities from Los Angeles to Budapest from 1807 to 1950. He found that all the curves could be described as *negatively exponential*— that is, as decreasing sharply at first with distance from the center and then getting progressively flatter. (See the marginal discussion of urban density functions.) In Western cities, the decline in population density is reflected in a familiar sequence of housing types, with high-rise apartment blocks near the CBD and low-density housing on semirural tracts of land on the suburban periphery. (See Figure 13-21.)

Extension of these cross-cultural studies to non-Western cities shows that we still have a lot to learn about the way cities evolve. For example, one striking feature of an Indian city like Calcutta is the continued increase in densities near the center and the stability of the density gradient with urban growth. The falling degree of compactness and crowding that characterizes the growth of Western cities is not seen in Calcutta, where both tend to remain constant. Given the same increase in population, the periphery of the non-Western city expands less than the periphery of the Western city. Figure 13-22 summarizes the variations in both time and space that characterize the two types of city. As the transport revolution which shaped the Western city makes its impact elsewhere, more cities may conform to this pattern.

In this chapter we have chosen to stress broad themes and common elements, the things that give urbanization its special character and cities their general flavor. But each city is also unique and special. Cincinnati is not Columbus, Ohio; neither is Denver a Detroit. Outside the United States the pattern of diversity gets wider. Clearly the cultural mosaic of a Montreal, the vitality of a Hong Kong, the formality of a Vienna, or the poverty of a Calcutta cannot be squeezed into a simplified model. These are cities to be studied at leisure and to enjoy or weep over. Like the learning of a foreign language, the study of geography involves rules of grammar and syntax that provide a guide to the structure of the thing under study, even though the rules may not always be followed exactly. Only by seeing how cities have evolved can we hope to avoid mistakes in the search for new urban futures.

Figure 13-22. Western and non-Western cities. Cross-cultural contrasts are evident in these urban-density gradients for a sample of Western and non-Western cities. [From B. J. L. Berry *et al.*, *Geographical Review* **53** (1963), p. 403. Reprinted with permission.]

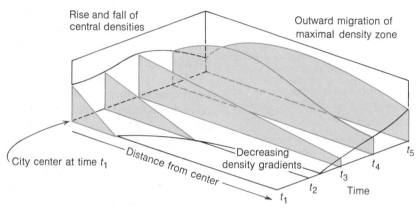

(a) Western cities

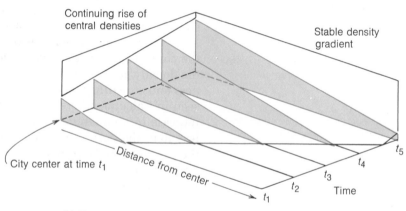

(b) Nonwestern cities

Reflections

1. Why do cities grow? List (a) the benefits and (b) the costs of high-density crowding of the human population into large cities. Is the balance between the two changing? If so, in what direction is it shifting?

2. What is going on in Figure 13-4? Substitute alternative multipliers for those in the diagram (say, 5 for 4, and $\frac{1}{10}$ for $\frac{1}{8}$), and recalculate the effect of 100 new jobs in the basic sector, ignoring the effects of the constraints. How would you improve this model to make it more realistic?

3. How does the biological need to sleep for around eight hours a day affect the organization of cities? What would happen if we needed sleep once every six hours, or once every six days?

4. Gather data on the distribution of different types of land use within your local urban community. (Maps are usually available from the city planning office). How far do the patterns you find (a) resemble or (b) differ from the patterns described in this chapter? Suggest some reasons for the differences, if any.

5. Is your local city's central business district (CBD) growing, stabilizing, or declining in importance? Give reasons for your opinion. What evidence would you need to assemble to test your views?

6. Choose any one city in a country other than your own. Using the resources of your college library, try to build up a picture of how the city evolved. List ways in which it (a) fits and (b) fails to fit the pattern shown in Figure 13-19.

7. Review your understanding of the following concepts:
   (a) urbanization curves
   (b) agglomeration economies
   (c) centripetal forces
   (d) centrifugal forces
   (e) urban base ratios
   (f) urban implosions
   (g) the quaternary sector
   (h) bid-price curves
   (i) population density gradients
   (j) CBDs

One step further . . .

For a general introduction to urbanization and the problems it poses, see
   Scientific American, *Cities* (Knopf, New York, 1966), esp. Chaps. 1 and 2, and
   Berry, B. J. L., *The Human Consequences of Urbanisation* (Macmillan, London, 1973), esp. Chaps. 2 and 3.

Excellent brief case studies of seven of the world's largest cities, together with thoughts on the future form of the metropolis, are given in
   Hall, P. G., *The World Cities* (Weidenfeld & Nicholson, London, 1966).

The classic geographic study of urban growth in the Boston-Washington corridor is
   Gottmann, J., *Megalopolis* (MIT Press, Cambridge, Mass., 1964).

For a general study of North American cities and the special problems of the ghetto, see
   Yeates, M. and B. J. Garner, *The North American City* (Harper & Row, New York, 1971).
   Rose, H. M., *The Black Ghetto: A Spatial Behavioral Perspective* (McGraw-Hill, New York, 1971).

Much of the contemporary geographic work on Chicago, together with reviews of earlier classic studies of the city, is given in

Berry, B. J. L. and F. E. Horton, *Geographic Perspectives on Urban Systems* (Prentice-Hall, Englewood Cliffs, N.J., 1970).

Urban geography now dominates many geographic journals, and you will find something of relevance and interest in most issues. Keep a special eye on the book review section to check the growing literature in this area. Those really enthusiastic about urban studies can keep up to date with journals such as *Urban Studies* (quarterly) and the *Journal of the American Institute of Planners* (monthly).

Chapter 14

City Chains and Hierarchies

For some minutes Alice stood without speaking, looking out in all directions over the country. . . . "I declare it's marked out just like a large chess-board . . . all over the world—if this is the world at all."

—LEWIS CARROLL
*Through the Looking-Glass
and What Alice Found There* (1872)

A night flight over any densely populated part of the world presents a unique opportunity to see the intricate spatial structure of the human ant's nest. Aboard an intercontinental jet flying at an altitude of 7000 m (23,000 ft), our concern with cultural complexity slips away. Large cities are visible only as faint clusters of lights spaced many kilometers apart; as the plane loses height, and smaller settlements and isolated farms come into view, sporadic pinpoints of light appear. At night the earth viewed from above looks much like the heavens from the earth. The great galaxies visible to the naked eye dissolve into a host of stars when we look at them through a telescope.

Geographers have long been fascinated by the galactic patterns of human settlement. What forms do they take? Are their forms random and chaotic—or can patterns and processes be discerned? If there are regularities, what lies behind them? We look here at some of the answers to these questions, at the models of settlement geographers have built, and at how we can use these models to predict changes and to plan more efficient and attractive patterns of settlement.

14-1 | DEFINING URBAN SETTLEMENTS

To answer questions about settlements, we must first puzzle out how to define them. We must try to find ways of describing them and comparing the characteristics of one region with those of another. Perhaps the simplest solution is to begin by asking how big human settlements are. For if we can define their size, we can go on to compare their magnitudes and relate them to other findings.

Questions of size

Let us consider the various definitions of urban settlements. The definitions used in legal and administrative documents will tell us precisely what we mean by Topeka, Kansas, or Melbourne, Australia. Unfortunately, the legal and administrative borders of cities are often a historical or constitutional legacy. Typically, the legal city has fixed boundaries that survive long after urban development has exceeded those bounds. Thus the legal city is often "underbounded" [Figure 14-1(c)]. Parts of the urban area may remain legally outside the city though but share a common boundary with it. Beverly Hills, completely surrounded by the city of Los Angeles, is a case in point. In England, where administrative regions continue to have a strong historic quality, some boroughs still have a municipal status that is a legacy of their former importance and is out of line with their present small size. The number of inhabitants an area must have to be considered urban also varies from country to country. In Iceland places

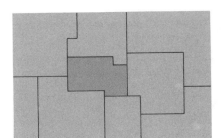

(a) Matched boundaries

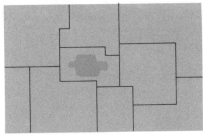

(b) Overbounded city

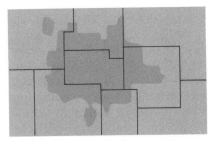

(c) Underbounded city

Figure 14-1. Difficulties in defining cities. The census limits of cities rarely coincide with their actual built-up area (shaded); matched boundaries are rare, underbounded cities the most common.

with a few hundred people are termed urban, whereas in the Netherlands a population of 20,000 is needed.

A second approach to defining urban settlements is to ignore the legal boundaries and try to define each settlement in terms of its physical structure. For example, we might define a settlement on the basis of a continuous distribution of housing, or population above a certain density, or the intensity of traffic. But there are difficulties here too. What do we mean by "continuous" housing, and what happens when different definitions don't all give the same answer? Figure 14-2 presents some different definitions of New York based on both its legal boundaries and its physical structure. Note that New York City itself (Manhattan, Staten Island, Brooklyn, Queens, and the Bronx) is only a small part of the continuous urban sprawl that is greater New York.

The "mismatch" between the legal and the physical city becomes vitally important when the legal city, with its static or declining population and limited tax base, has to provide public services like transport and police for the millions of commuters who cross its legal boundaries to work each day. As the discrepancy between the legal and economic boundaries of the city becomes worse, the pressure for some form of revenue sharing or boundary adjustment grows. This discrepancy also affects our ability to answer even the simplest questions about the size of the city. To take an extreme case, the "legal city" of Sydney, Australia, in 1955 had a population of only 193,000, while the "built-up area" of Sydney had a population of 1,869,000. This difference of over nine times in size is unusual, but important enough to make the definition of settlements a matter of concern.

Some possible solutions

As a result of this problem, international, and indeed intranational, definitions of urban settlements are being standardized. One definition of world metropolitan areas, by demographer Kingsley Davis, runs to 12 pages, including 2 pages on difficult cases. In the United States, the concept of a Standard Metropolitan Statistical Area (SMSA) was introduced in 1960 so that metropolitan areas could be defined realistically by using three criteria. First, a population criterion: Each SMSA must include one central city with 50,000 or more inhabitants. Special rules allow contiguous cities (i.e., those directly adjoining each other) and nearby cities (within 32 km, or 20 mi, of each other) to be combined. Second, the metropolitan character of an area is taken into account. At least 75 percent of the labor force of the county must be employed by nonagricultural industries. Other criteria for SMSAs

Figure 14-2. Varying definitions of a metropolis. The map shows three alternative boundaries for New York, none of which coincides exactly with the limits of the built-up area.

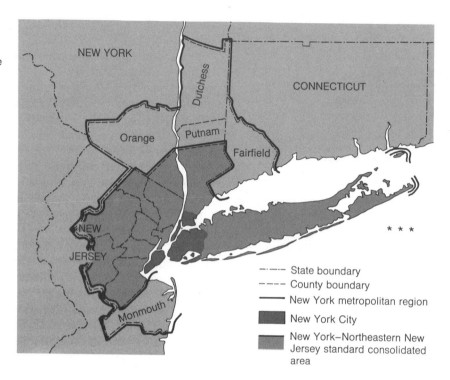

--·-- State boundary
---- County boundary
——— New York metropolitan region
New York City
New York–Northeastern New Jersey standard consolidated area

relate to population density, the contiguity of townships, and ratios between the nonagricultural labor forces of the counties making up the unit. Finally, the integration of the areas that constitute the SMSA is considered. Counties are integrated with the county containing a central city if 15 percent of the workers living in the county commute to the city, or if 25 percent of the workers in the county live in the city. This measure of integration can be supplemented by other measures based on the market area for newspaper subscriptions, retail trade, public transport, and the like.

Despite their apparent comprehensiveness, the SMSA definitions have still not fully solved the problem of urban boundaries. An improved definition using county blocks and commuting data has been suggested by a team of Chicago geographers.

Their recommendations were threefold. First, counties or equivalent units were to be retained as the basic building blocks in any system of classifying areas of the United States. Second, counties were to be classified into *functional economic areas* (FEAs) on the basis of county-to-county commuting data. An FEA would consist of all the counties in which the proportion of resident workers who commuted to a given *central* county (usually containing a city with 50,000 or more inhabitants) exceeded the proportion who commuted to alternative central counties. Third, FEAs were to be grouped into *consolidated urban regions* (CURs) when two or more FEAs sent at least 5 percent of the workers in the central county of one commuting area to the central county of the others.

Not only metropolitan areas but small towns and villages as well can be difficult to define. The smaller settlements, however, unlike urban areas, are usually overbounded [Figure 14-1(b)]. Similar sets of dull, but necessary, rules must be worked out for these cases too.

Measuring patterns of settlement

If we mark the location of cities, towns, and villages on a map, we can see the overall pattern of settlement. Let us look at some examples. Figure 14-3 shows four sample areas from different parts of the United

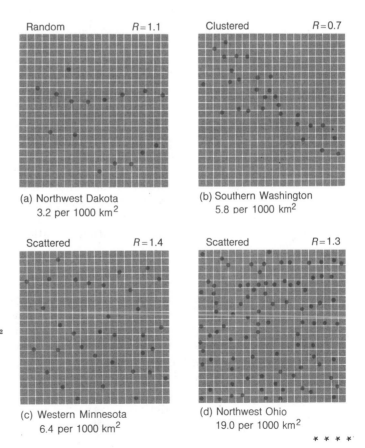

Figure 14-3. Patterns of settlement. The diagrams show the density and spacing of towns in four sample 5000-km² (1930-mi²) areas in the northern United States. The classification of the spatial patterns as "clustered," "random," or "scattered" is determined by the value of the nearest-neighbor index, *R*. (See marginal discussion.) [From L. J. King, *Tijdschrift voor Economiske en Sociale Geografie* **53** (1962), pp. 4–6, Fig. 3–7.]

Random   R=1.1
(a) Northwest Dakota
3.2 per 1000 km²

Clustered   R=0.7
(b) Southern Washington
5.8 per 1000 km²

Scattered   R=1.4
(c) Western Minnesota
6.4 per 1000 km²

Scattered   R=1.3
(d) Northwest Ohio
19.0 per 1000 km²

* * * *

States; each square has an area of 5000 km² (2000 mi²), and each dot represents an "urban settlement" as listed in the United States Census. We can distinguish among the four areas in two ways. First, they differ in density (i.e., in the number of urban settlements per square kilometer). For example, the North Dakota area has only 16 towns (3.2 towns/1000 km²) while the Ohio area has 98 (19.0 towns/1000 km²). Measurement is no problem, and it is a simple matter to arrange the four areas in a sequence of increasing density, as in Figure 14-3.

The patterns of settlement also differ in a second characteristic which is less easily measured. Compare the patterns of the Washington and Minnesota areas in Figure 14-3. Both have similar densities (5.8 and 6.4 towns per 1000 km², respectively) but strikingly different arrangements of the towns in space. The Washington towns are clustered, whereas the Minnesota towns are scattered. To measure this second property, geographers have adopted a spacing index (see the marginal discussion of the nearest-neighbor index) that enables them to rank patterns of settlement along a scale from "highly clustered" to "highly dispersed." The values of the index range from a theoretical low of zero, when all the settlements are concentrated at a single point, to a maximal value (2.15) when the pattern of settlement is triangular.

Most of the patterns of settlement examined thus far would have values on this scale between around 0.5 and 1.5. These values hover around the value we would assign to a group of randomly generated points (1.0), thereby implying that there is no strong pattern-forming forces deciding how settlements are arranged. In relatively uniform environments, the index values for settlements drift toward 2.15. at the uniform end of the spacing scale; conversely, in environments with greater contrasts the values drift toward the clustered end (0). With this index, geographers can compare patterns of settlement with different spacing and estimate the probable amount of environmental influence on the location of settlements.

14-2 | SETTLEMENTS AS CHAINS

Once we find a commonly accepted method of defining cities, then analysis of their comparative size and importance can begin. One of the first steps in such analysis is to arrange them in order of population size. Table 14-1 ranks the 20 largest cities in three areas of decreasing size: the world, the United States, and the state of Texas. At first sight, this looks like a dull collection of statistics. But look at Figure 14-4. This plots the size of each of these "top 20" cities against its rank. Geographers have repeated this process for large and small areas; in each case they have looked for a repetitive pattern in the array of sizes.

The nearest-neighbor index

Assume a spatial distribution of towns as in Figure 14-3. Using measures first developed by plant ecologists, geographers can define a *spacing index* by comparing the observed pattern of settlements in an area with a theoretical random distribution; that is,

$$R = \frac{D_{obs}}{D_{exp}}$$

where R = the nearest-neighbor index of spacing,

D_{obs} = the average of the observed distances between each town and its nearest neighbor in kilometers, and

D_{exp} = the expected average difference between each town and its nearest neighbor in kilometers.

The expected average distance is given by

$$D_{exp} = \frac{1}{2\sqrt{A}}$$

where A = the density of towns per square kilometer. Thus, in an area with an observed nearest-neighbor distance of 3.46 km (2.15 mi) and an observed density of 0.0243 towns/km², the nearest-neighbor index of spacing (R) will be 1.08. Values of 1.00 indicate a random pattern. Dispersed or scattered patterns of settlement have values greater than unity, and clustered patterns of settlement have values less than unity. Nearest-neighbor indices are discussed at greater length in L. J. King, *Statistical Analysis in Geography* (Prentice-Hall, Englewood Cliffs, N.J., 1969), Chap. 5.

Table 14-1 The twenty largest urban settlements for areas of different magnitude[a]

World metropolitan areas		United States cities		Texas metropolitan areas	
New York	14.76	New York City	7.78	Houston	1.41
Tokyo	11.31	Chicago	3.55	Dallas	1.12
London	7.88	Los Angeles	2.48	San Antonio	0.72
Paris	7.74	Philadelphia	2.00	Fort Worth	0.57
Buenos Aires	7.00	Detroit	1.67	El Paso	0.31
Shanghai	6.90	Baltimore	0.94	Beaumont–Port Arthur–Orange	0.31
Chicago	6.79	Houston	0.94	Corpus Christi	0.27
Moscow	6.43	Cleveland	0.81	Austin	0.21
Los Angeles–Long Beach	6.04	Washington, D.C.	0.76	McAllen–Pharr–Edinburg	0.18
São Paulo	5.38	St. Louis	0.75	Lubbock	0.16
Bombay	4.90	Milwaukee	0.74	Brownsville–Harlingen–San Benito	0.15
Calcutta	4.76	San Francisco	0.74	Waco	0.15
Philadelphia	4.34	Boston	0.70	Amarillo	0.15
Rio de Janeiro	4.03	Dallas	0.68	Galveston–Texas City	0.15
Peking	4.01	New Orleans	0.63	Wichita Falls	0.13
Detroit	3.76	Pittsburgh	0.60	Abilene	0.12
Leningrad	3.61	San Antonio	0.59	Texarkana	0.09
Cairo	3.52	San Diego	0.57	Odessa	0.09
Seoul	3.47	Seattle	0.56	Tyler	0.09
Berlin (East and West)	3.28	Memphis	0.54	Sherman–Denison	0.07

[a] The figures refer to the estimated population in millions for the mid-1960s according to the United Nations *Demographic Yearbook* and the United States Census Bureau.

Have any rules been discovered? What do they tell us about the way such "chains" of city sizes are linked together.

A rule for the size distribution of settlements

Although some nineteenth-century investigators sought for patterns, one of the first to actually find a significant one was a German geographer, Felix Auerbach, in 1913. He noticed that if we arrange settlements in order of size (1st, 2nd, 3rd, 4th, . . ., nth), the population sizes for some regions are related. Auerbach found the simplest relationship to be that the population of the nth city was 1/n the size of the largest city's population. Thus the 4th ranking city was found to have approximately ¼ the population of the largest. This inverse relationship between the population of a city and its rank within a set of cities is termed the *rank–size rule*. (See the marginal discussion of the rank-size rule on page 358.) If we apply this rule to the United States and look at Table 14-1, we should expect Chicago (ranked second) to have

Figure 14-4. City chains. The twenty largest cities in three areas (see Table 12–1) are arranged on a population size-rank diagram. The largest city in all three series—the "primate city"—has a relative population size of 100. Actual population sizes are shown on the vertical scales to the left. The continuous line is an idealized rank-size curve.

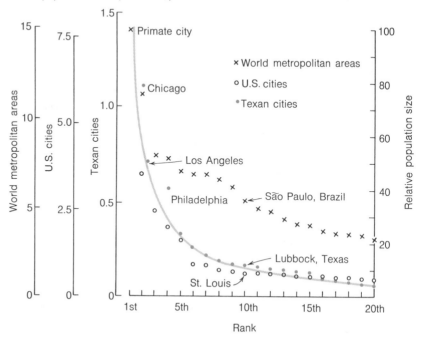

Actual population size (in millions)

The rank-size rule

Assume a set of cities ranked according to size from the largest (I) downward. In its simplest form, the rank–size rule states that the population of a given city tends to be equal to the population of the largest city divided by the rank of the given city; that is,

$$P_r = \frac{P_I}{R}$$

where P_r = the population of the rth city,
P_I = the population of the largest city, and
R = the rank of the rth city in the set.

This basic formula is often modified by a constant (b) to allow variations from the strict rank–size rule—for example,

$$P_r = \frac{P_I}{R^b}$$

Thus, when the largest city has a population of 1,000,000 and $b = 0.5$ (a low-angled slope in Figure 14-5), we should expect the population of the fourth-largest city to be 500,000. If b is raised to 2.0, then we should expect the population of the fourth city to be smaller—that is, only 62,500. For an extensive critical study of rank–size rules in a specific regional context, see C. D. Harris, *Cities of the Soviet Union* (Rand McNally, Skokie, Ill., 1970).

one-half of the population of New York City (ranked first). Thus, in the mid-1960s, Chicago should have had a population of 3.89 million; in fact, its population was less than this (3.55 million). Los Angeles, ranked third, had a population of 2.48 million at that time as against a predicted 2.59 million. Philadelphia, ranked fourth, had a population of 2.00 million as against a predicted 1.94 million. But despite this lack of agreement in the case of individual cities, the over- and under-estimates tend to cancel each other out to produce an overall picture loosely in line with the ideal rank-size rule (the curve in Figure 14-4). The match with the idealized pattern is also fairly good for the Texan cities, but rather poor for the world metropolitan areas.

We can more easily compare the fit between distributions of real cities and the idealized distributions predicted by the rule if we make the axes on which we plot the cities' size and rank nonlinear. The 20 hypothetical cities in Figure 14-5(a) all conform exactly to the rank-size rule and make an awkward, J-shaped curve. If we transform the values on the axes to a logarithmic scale, the curve becomes a straight line, as in Figure 14-5(b). The simple rank–size relationships for the United States and Texas can also be described by a straight line, with

Figure 14-5. Hypothetical relations between city size and rank. Twenty cities arranged in an idealized rank-size pattern are plotted on (a) a graph with arithmetic axes and (b) a graph with logarithmic axes. Figure (c) shows three alternative patterns of city sizes. Figure (d) shows the evolution of idealized city chains as population size increases through three time periods. (See also Figure 14–16.)

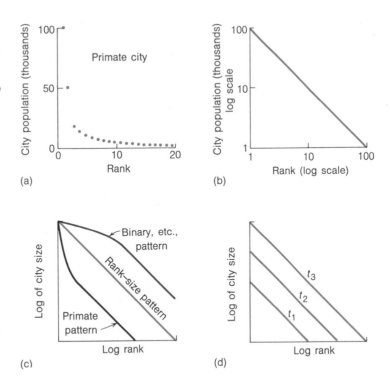

a slope of 45° to the horizontal; that is, they have the form predicted by the rank–size rule and shown in Figure 14-5(b). Yet other lines that are equally regular but have *different* slopes have been found. For example, the curve for the cities of Switzerland in 1960 has a much gentler slope, and that for urban areas in India in 1921 has a much steeper one. Low slopes imply that the decrease in a city's population with rank is extremely slow; high-value slopes imply a sharp falloff in size with rank.

Regional variations in the rule

Since the size of any one city appears to be linked to the size of all other settlements in a region, we can regard the whole set of cities as forming an interlinked chain running from the largest city to the smallest. These are termed *rank–size chains.* When considering these rank–size chains, some geographers have found it more meaningful to subdivide the distribution of settlements into distinct segments. Australia in 1961 had a distinct pattern of urban areas which when plotted on a graph showed a flat upper section and a steeper lower section, and the critical break occurred at a population of 65,000. This convex Australian pattern is in contrast to the concave distribution of Soviet Russian cities in 1956, in which a steep upper section is followed by a flatter lower section. Here the break between the two sections comes much higher than in the Australian pattern; it happens at a population level of around 500,000. In Figure 14-5(c), we distinguish between these two patterns of settlement. The Australian pattern is dominated by a few large cities that are roughly equal in size to a "tail" of smaller cities (with populations below 65,000) conforming to the rank–size rule. We call such patterns *binary, trinary, quaternary, quinary,* and so on, depending on the number of cities on the upper slope. Conversely, in the Russian pattern the falloff in population among the first few cities below Moscow is sharper than the falloff predicted by the simple rank–size rule. We call this a *primate* pattern to indicate the dominant role of the first, or primate, city. (The term primate is taken from ecclesiastical language where the primate of a church is its superior bishop.)

When historical census figures are available, geographers can trace the changes in rank–size relationships over time. If the total population of a region grew and its cities remained distributed in a simple rank–size sequence, we should expect it to change through time periods t_1 to t_3 in the way described in Figure 14-5(d). The figures for the United States over one and a half centuries are rather stable. The curves for 1790 to 1950 are generally parallel, and there is some evidence of increasing regularity (manifested by the straightening of the curves over time), as we shall see later in Figure 14-16. The statistics for Sweden over the same period reveal a reverse effect; here an S-shaped curve is retained and even accentuated by an overall growth in the population.

The logic of rank–size chains

Enough evidence is now available to prove that regular rank–size chains are recognizable for settlements in many types of regions during different time periods. Why are many towns and cities arranged in this regular fashion?

Geographers are not the only ones to be puzzled by these size distributions. Rank–size rules are not confined to human settlements. Similar distributions have been observed by botanists studying the number of plant species, and by linguists studying the frequency with which different words are used in our speech. The pervasiveness of this type of distribution has led organization theorist Herbert Simon to postulate rank–size rules as the equilibrium slope of a general growth process. We can visualize this process as one in which each unit, such as a city, initially has a random size and thereafter grows in an exponential manner that is proportional to size. (See Section 6-1, especially page 147.) Simon points out that extremely general processes of this kind tend to produce distributions that approximate a regular rank–size form.

Simon's hypothesis has been translated into urban terms by Brian Berry. Berry studied the rank–size distribution of towns with populations of 20,000 or more in 38 countries. Of all the countries, only 13 had rank–size distributions like the ones postulated in the Simon model. These countries were among the largest in the group (e.g., the United States), had a long history of urbanization (e.g., India), and were economically and politically complex (e.g., South Africa). By contrast, 15 countries had primate distributions in which one or more large cities dominated the size distribution and were much larger than we might have expected them to be on the basis of rank–size rules.

In contrast to regular rank–size distributions, primate distributions appear to be products of urbanization processes in countries that are smaller than average, have a short history of urbanization, and have simple economic and political structures. Hence, primate distributions typify the impact of a few rather strong forces. For instance, the impact of imperial status on the large cities in Austria, the Netherlands, or Portugal, each the hub of former empires, is certainly potent. Another powerful force is the superimposition of outside influences on an existing hierarchy. Examples would be the institution of a dual economy (such as a peasant and plantation sector in Ceylon) or the influence of a Westernized city such as Bangkok on the Thai system of cities.

Whatever the specific contributions of Simon's model prove to be, it has two principal advantages: It introduces time into the rank–size model by taking into account an urban system's history of development, and it emphasizes the effects of numerous small forces in producing regular structures. Variations from the rank–size model may be caused by the distorting effect of a few powerful forces.

Departure of small settlements from the rule

Most evidence on which rank–size rules are based has come from investigations of the size distribution of cities and towns. Geographers have paid rather little attention to settlements at the lower end of the population spectrum, that is, to small villages and hamlets.

Now, one of the implications of the rank–size rule is that the number of settlements is inversely proportional to their size; we should expect, therefore, a large number of small settlements and a very large number of very small settlements. Yet this may not be the case. We know from quantitative investigations of small settlements in Ceylon that the distribution of small settlements may have a different shape.

Figure 14-6. Frequency distributions of settlements. These curves indicate the general pattern of settlements in two different environmental zones of southern Ceylon. Both the vertical and horizontal axes are logarithmic. [From K. Gunawardena, Cambridge University, unpublished doctoral dissertation (1964), p. 167, Fig. 8.]

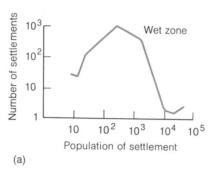

(a)

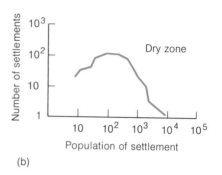

(b)

(See Figure 14-6.) The distribution of settlements with a population above about 1000 seems to match the distribution predicted by the rank–size rule, but the distribution of smaller settlements is completely different. Indeed, the rule seems to work *in reverse* for small settlements with a population below about 100. Although we know that Ceylon's dry and wet zones have different population thresholds at which this reversal sets in, not enough comparable evidence is at hand to make international comparisons. However, it looks as if there is a minimal level of population where the regular chain relationships of size and rank break down. This level may change significantly over time.

14-3 | THE CHRISTALLER CENTRAL-PLACE MODEL

Two-dimensional patterns of settlement on maps of population distribution invite questions similar to those provoked by one-dimensional patterns. Is any order discernible? If so, what forces lie behind it? Although this problem had been stated by German geographers in the nineteenth century and some tentative hypotheses were made, the main breakthrough did not occur until 1933, when Walter Christaller published his now-famous doctoral thesis on *Central Places of Southern Germany.*

Walter Christaller (1893–1969) was a German scholar who, in 1932, submitted to the University of Erlangen a dissertation on the structure of settlements in southern Germany. His ideas were based on those of locational theorists like J. G. Kohl, Johann von Thünen, Alfred Weber (Christaller's former teacher), and the German settlement geographer Robert Gradmann. Christaller's works had little impact in contemporary Germany, and it was not until their introduction into the United States in the 1940s and 1950s that their value was realized. Since then, Christaller's ideas have been verified and extended, and used to analyze the pattern of "central places" within cities, such as urban shopping districts, as well as patterns of cities themselves.

Central places

The terminology of Christaller's model is straightforward. *Central places* are broadly synonymous with towns that serve as centers for regional communities by providing them with *central goods* like tractors and *central services* like hospital treatment. Central places vary in importance. Higher-order centers stock a wide array of goods and services; lower-order centers stock a smaller range of goods and services—that is, some limited part of the range offered by the higher center. *Complementary regions* are areas served by a central place. Those for the higher-order centers are large and overlap the small complementary regions of the lower-order centers.

Schools provide a good example of a central-place organization. The local elementary school provides a lower-order center (to use Christaller's terms) which serves a small part of a city or a single rural community. There are a large number of such schools in any state and they teach children drawn from only a few square miles (i.e., they have small complementary regions). Above the elementary schools come the higher-order services provided by the junior high schools, the high schools, and colleges of various kinds. As we move higher up the educational ladder, the number of centers becomes smaller and their complementary regions become larger. At the top of the ladder stands the state university, often a single institution serving students drawn from the whole state, its complementary region. Education is just one range of central goods and central services that give character to central place organization and help to distinguish the central-place functions of one settlement from those of another.

Christaller defined the *centrality* of an urban center as the ratio between all the services provided there (for both its own residents and for visitors from its complementary region) and the services needed just for its own residents. Towns with high centrality supplied many services per resident, and those with low centrality few services per resident. Christaller found that, in Germany in the 1920s, the number of telephones offered a useful indicator of the range of central goods available in a town. Using telephone data, he defined the centrality of a town as equal to the number of telephones in the town, minus the town's population multiplied by the average number of telephones per population in the town's complementary region. A town of 25,000 with 5,000 telephones in a region with only 1 telephone for every 50 people would have an index of 5,000−25,000 $(\frac{1}{50})$, or 4,500. Thus, the index basically measures the difference between the expected level of services (i.e., that needed by a town to serve its *own* inhabitants) and the level of services actually measured within the center.

Contemporary researchers have revised Christaller's terminology

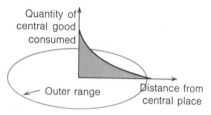

Figure 14-7. Idealized demand zones in the Christaller model. With uniform transport costs, the demand for central goods falls with distance from the central place and the market range (the area within which the goods will be bought) forms a circle.

to include two simple concepts. The first is a *market-size threshold*, below which a place will be unable to supply a market good. That is, below the threshold, sales will be too few for firms to earn acceptable profits. The second concept added was that of the *range of a central good*, the limits of the market area for the good (Figure 14-7). The market area's lower limit is determined by the threshold market size, and its upper limit is defined by the distance beyond which the central place no longer is able to sell the good. If we assume that travel is equally easy in all directions, the range of a good will be a perfect circle. This circle is the outer limit of a *demand cone* in which the quantity of a central good consumed decreases with distance from the central place because of increased transport costs.

Complementary regions

Given a circular demand cone for central goods, Christaller demonstrated that a group of similar central places will have hexagonal, complementary regions with the central places arranged in a regular triangular lattice. Figure 14-8 depicts the stages by which such a pattern might emerge as population colonized a new area and central places were established. (Cf. Section 14-5.) The final hexagonal patterns follow directly from five simplified assumptions.

1. There must be an unbounded isotropic plain with a homogeneous distribution of purchasing power.
2. Central goods must be purchased from the nearest central place.
3. All parts of the plain must be served by a central place; that is, the complementary areas must completely fill the plain.
4. Consumer movement must be minimized.
5. No excess profits may be earned by any central place.

Figure 14-8. Hexagon formation. The overlapping of circular demand cones, along with the close packing of centers, gives a network of hexagonal territories. (Compare with Figure 14–14.)

The hexagons result from our attempt to pack as many circular demand cones as possible onto the plain. If we require all parts of the plain to be served by a central place (assumption 3), the circles will

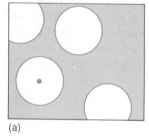

(a)

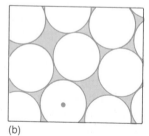

(b)

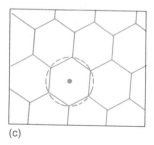

(c)

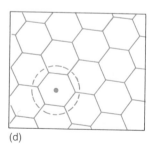

(d)

overlap. But because of our second assumption, that consumers will shop at the closest central place, the areas of overlap will be bisected. A perfect competitive situation will be achieved only when the plain is served by the maximal number of central places offering identical central goods at identical prices to hexagonal complementary regions of identical size. Only this arrangement ensures that consumers travel the least distance to central places.

Christaller was able to vary the level of central places within a settlement hierarchy by varying the size of the complementary regions, as in Figure 14-9. He discusses three cases.

The first is a *market-optimizing* case, in which the supply of goods from central places is as near as possible to the places supplied. A higher-order central place will serve *two* of its lower-order neighbors. It may do this by serving only two of its six equidistant nearest neighbors and thus having an asymmetric complementary region. Alternatively, a higher-order central place may share the same neighbors with two others, for instance, competing neighbors. Note in Figure 14-9(a) how settlement 2 lies on the edge of three complementary regions (those of centers 1, 3, and 4). This arrangement is termed a $K = 3$ system, where K refers to the number of places served, that is, the central place plus two nearest neighbors or the central place plus one-third of each of its six nearest neighbors.

The second case involves a *traffic optimizing* situation in which the boundaries of the complementary regions are rearranged to allow a more efficient highway pattern than in the first case. As Figure 14-9(b) shows, as many places as possible now lie on traffic routes between the larger towns; for example, the direct route from center 1 to 5 goes right through center 2. This situation is represented by the $K = 4$ hierarchy, where a higher-order place serves three adjacent lower-order places. It may do this by dominating three of its six nearest neighbors or by sharing them with another central place of the same order.

The third case Christaller discusses is an *administration-optimizing* situation, in which there is a clearcut separation of the higher-order place and its neighboring lower-order centers. That is, each lower-order center falls clearly within the trade area of a single central place; in Figure 14-9(c), for example, center 2 falls within the area of center 1. Such arrangements are likely to be economically and politically more stable than divided settlements. This relationship produces the $K = 7$ hierarchy.

All three cases assume that relationships established for one level (e.g., between villages and small towns) will also apply to other and higher levels (e.g., between small towns and larger cities). They are usually called *fixed-K hierarchies* because the same fixed relationships

Figure 14-9. Alternative principles of organization in the Christaller model. Settlements can be partitioned in one of three basic ways, (a), (b), or (c), by enlarging and rotating the hexagonal cells. The cells can then be grouped hierarchically to give tiers of higher-order centers; for example, (d) shows higher-order centers in a traffic-optimizing ($K = 4$) hierarchy. Note the way in which lower-order centers "nest" within the market areas of higher-order centers in a manner reminiscent of sets of Russian dolls.

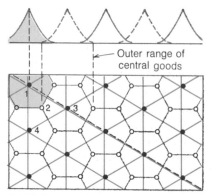

Outer range of central goods

(a) Market optimizing $K = 3$

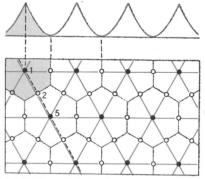

(b) Traffic optimizing $K = 4$

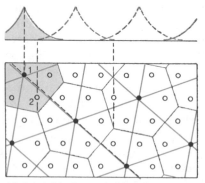

(c) Administration optimizing $K = 7$

● Central place
○ Dependent place
—— Boundary of complementary region
===== Highways between central places

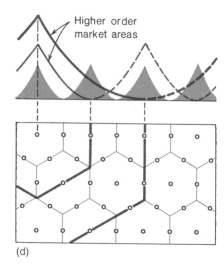

Higher order market areas

(d)

hold at *all* levels of the settlement hierarchy. This means that we can expand each of Christaller's three central-place variants by building higher and higher levels on top of the basic framework. Consider the situation in Figure 14-9(d), where a second and third $K = 4$ central place is superimposed on the first. As we add each successive upper level the size of the hexagonal regions increases and the number of places is reduced by a quarter. Thus if a region had 2000 central places on the lowest level, it would have 500 on the next level, and 125 on the next higher level again. If we start at the top, we can put this a simpler way. In an idealized $K = 4$ school system with three levels or tiers, 1 junior college would draw students from 4 high schools, each of which drew children from 4 elementary schools (i.e., 16 in all). For each of the three cases considered by Christaller, typical sequences would be 1, 3, 9, 27 for the $K = 3$ network, 1, 4, 16, 64 for the $K = 4$ network, and 1, 7, 49, 343 for the $K = 7$ network.

Southern Germany

As we have noted, Christaller developed his basic ideas using southern Germany as his original test area. The theoretical distribution of the status and location of towns in this area, according to his market-optimizing principle, is summarized in Table 14-2. Christaller postulated seven levels of the hierarchy from the level of the hamlet (Figure 14-10) to that of the city, each level showing an increase in the area of the complementary region. Approximate populations have been added to the table by extrapolating from Christaller's detailed work on southern Germany. On the upper level of the hierarchy are the *Landstadt* cities with populations of around 500,000 — Munich, Frankfurt, Stuttgart, and Nuremburg, together with the border cities of Zürich in Switzerland and Strasbourg in France. The lowest market center at the base of the hierarchy has a service radius of a little more than 3 km

Table 14-2 Status of towns in Christaller's system

Type of town	Order	Approximate population	Distance from other towns km	mi	Service area km²	mi²
Landstadt (L)	Upper	500,000	187	116	35,000	13,514
Provinzstadt (P)		100,000	109	68	11,650	4,498
Gaustadt (G)		30,000	63	29	3,880	1,498
Bezirkstadt (B)		10,000	36	22	1,243	478
Kreisstadt (Kr)		4,000	21	13	414	160
Amtsort (A)		2,000	13	8	140	54
Marktort (M)	Lower	1,000	7	4	47	18

[a] Values are based on a study of southern Germany (Figure 14-11).

SOURCE: R. E. Dickinson, *City and Region* (Humanities, New York, and Routledge & Kegan Paul, London, 1964), p. 76.

(2 mi). The application of the Christaller's theoretical model to southern Germany is shown in Figure 14-11.

Despite the general agreement between the model and reality, Christaller found several specialized centers—mining towns, border towns, and so on—that deviated from the general pattern. The resources in a particular region or subregion may cause a general increase in the density of settlement, resulting in a closer spacing of centers.

Figure 14-10. South Germany. South Germany was used by Christaller as a test area for his central-place model. Settlements like this second-tier Amtsort (at the bottom of the photo) were seen as part of an intricate structure linking lower-order rural settlements to higher-order urban settlements. [Photograph courtesy of the German Information Center.]

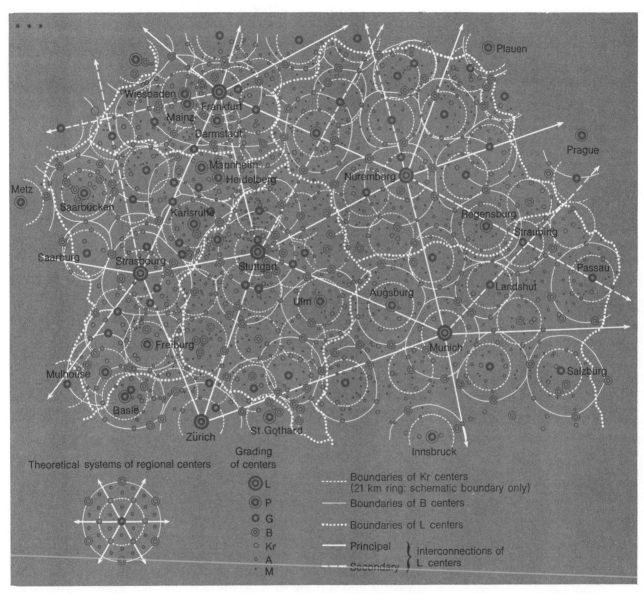

Figure 14-11. South Germany. Distribution of cities, towns, and villages in southern Germany shown by means of a seven-level hierarchy. (See Table 14–2.) The map also shows the boundaries of the complementary regions about the four highest levels of centers (drawn as ellipses or circles) and the main routes interconnecting the highest-level centers (drawn as straight lines). The boundary of the area is simply the limit of Christaller's study area; border cities like Zürich and Plauen lock into a continuing continentwide hierarchic system.

<table>
<tr><td>14-4</td><td>EXTENSIONS OF THE
CHRISTALLER MODEL</td></tr>
</table>

14-4 | EXTENSIONS OF THE CHRISTALLER MODEL

Since its publication, the Christaller model has provoked two main reactions from fellow geographers. First, there are those who have accepted the general argument of his model. Their reaction has been to extend and refine it. Second, there are those who found the Christaller model too rigid and static. They have reacted by trying to build alternative models with a stronger time dimension and a closer correspondence to actual settlement history. In the last two sections of this chapter we look at these two approaches.

Lösch's modifications

The prime theoretical extension of the Christaller model was created by a fellow German, August Lösch (1906–1945), in his *Die raumliche Ordnung der Wirtschaft*. He clarified the ways in which spatial demand cones are derived and verified the optimal hexagonal shapes of complementary regions where the population served was uniformly distributed. However, Lösch's main contribution was to extend the notion of fixed-K hierarchies.

Lösch took all the hexagonal networks in Figure 14-9 and extended them to higher orders by superimposing them on a common central place. This common central place is the hub of the settlement system, its single most important city dominating the trade and services in the whole of the surrounding region. Each of the networks was then rotated about this common central city until as many as possible of the higher-order services coincided in the same centers. Such an arrangement ensures that the sum of the minimal distances between settlements is small and that not only shipments, but transport lines, are reduced to a minimum.

You can envisage this process by imagining that the fixed $K = 3$ network is drawn on a map. The $K = 4$ network is now drawn on an overlay of transparent tracing paper and pinned to the $K = 3$ map by a single thumb-tack through the common central place. By rotating the overlay, many major places on both the $K = 3$ and the $K = 4$ paper are made to coincide. For example, if we have a $K = 3$ school system and a $K = 4$ hospital system, we try to rotate the overlay so that the high school and the doctor's hospital both coincide in the *same* locations rather than being split between two. Lösch went on to add the $K = 7$ and still higher K networks to the map, always trying to get as many services as possible to overlap in the same locations.

A simplified version of his final result is shown in Figure 14-12. It shows that the resulting central-place system changed with distance away from the common central hub and was arranged, like a wagon wheel, with alternating sectors. Twelve sectors are produced, six with many production sites and six with few (called by Lösch "city-rich"

Figure 14-12. The Löschian landscape. City-rich and city-poor sectors in the Löschian landscape. (a) Twelve sectors. (b) Centers with the largest number of functions. (c) An enlargement of a pair of adjacent sectors to show the underlying regular hexagonal pattern; the size of the dot is proportional to the number of functions. [From A. Lösch, *The Economics of Location* (Yale University Press, New Haven, Conn., and Fischer Verlag, Stuttgart, 1954), p. 127, Fig. 32.]

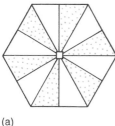

(a)

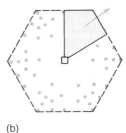

(b)

(c)

and "city-poor" sectors). In Figure 14-12, the metropolitan center is the center of 150 separate fields.

Thus, using the same basic hexagonal unit and the same K concept as Christaller, Lösch evolved a markedly different hierarchy. Christaller's hierarchy consists of several fixed tiers in which all places in a particular tier have the same size and function and all higher-order places perform all the functions of the smaller central places. In contrast, the Löschian hierarchy is far less rigid. It consists of a nearly continuous sequence of centers rather than distinct tiers. So settlements of the same size need not have the same function (e.g., a center serving seven settlements may be either a $K = 7$ central place or a center where both a $K = 3$ and $K = 4$ central place coincide), and larger places need not perform all the functions of the smaller central places.

Lösch's model represents a logical extension of the Christaller model. It is based on the same hexagonal unit and hence suffers from the same rigidity, but it yields a relationship between the size and function of central places that is continuous rather than stepped, and therefore more in accord with the observed distributions described in Section 14-2.

Periodic variations

The permanent provision of central goods implies a high and continuous level of demand. In most peasant societies central goods are provided by markets that are not open every day, but only once every few days on a regular basis. Although periodic markets are now only a small element in the central-place structure of Western society and generally sell only agricultural products, they continue to be important in most peasant societies and are vital to the exchange structure of two-thirds of the world.

The relevance of Christaller's scheme to periodic markets is demonstrated by the central-place network of rural China. Figure 14-13 shows a portion of the Szechwan province, southeast of Chengtu, which is simplified to a basic $K = 3$ system of nesting. [Cf. Figure 14-9(a).] Two levels of the hierarchy are shown: an upper level (Chung-ho-chen) and a lower level (Hsin-tien-tzu) which has a market area about one-third the size. Periodic markets are superimposed on the system as indicated in Figure 14-13. A 10-day cycle is usually divided into three units of 3 days. No business is transacted on the tenth day. With synchronized cycles, central goods can be circulated around several markets on a regular schedule and firms can accumulate enough trade to remain profitable. Hence, a central-place system is maintained by *rotating* rather than *fixed* central-place functions.

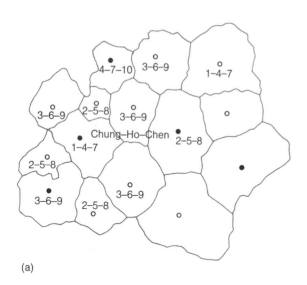

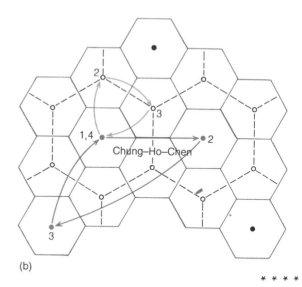

(a)

(b)

* * * *

Figure 14-13. Periodic central places.
(a) A map of rural centers in the Chinese
province of Szechwan, showing the days
on which markets are held. (b) The rural
centers as part of a $K = 3$ Christaller
network. The cyclic movements of
traders around a market ring are
examples of the space-time meshing of
central-place functions. [After G. W.
Skinner, *Journal of Asian Studies* **34**
(1964), pp. 25, 26, Figs. 3, 4.]

Market cycles vary widely in length. In tropical Africa, the market
week varies from 3 to 7 days. For example, Yourubaland in western
Nigeria works on an interlocking system of 4-day circuits. Generally
the higher the population and per capita income, the greater the total
trade and the shorter the length of the cycle. Where demand for goods
is high enough, the market opens every day (and thus is a permanent
central-place function); where demand is low enough, the cycles become so long that a service ceases to be provided within the region.

Historical research reveals instances of space–time interlocking
in medieval Europe, where cattle might be sold or cloth exchanged
at large spring and autumn fairs. In the modern international economy
the time period may widen to years and the region broaden to involve
all the major capitals of the world. The World's Fairs and the Olympic
Games, with their 4-year cycle, could be viewed as extreme extensions
of the periodic case of the Christaller model.

Toward a world model

There have been many attempts to extend the Christaller–Lösch models
to the world level. American geographer Allen Philbrick proposed a
system of 22 regions divided into four main types. (See Table 14-3.)
At one extreme is the modern urban-industrial world, dominated by
the core area around the North Atlantic (Anglo-America, Western
Europe, and the western part of the Soviet Union), together with the
peripheral area of the southern hemisphere (southern South America,
South Africa, and Australia–New Zealand) and Japan. This is termed

Table 14-3 A system of world regions

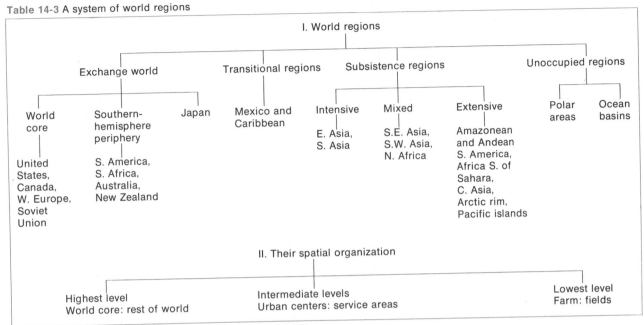

SOURCE: A. K. Philbrick, *This Human World* (Wiley, New York, 1963), p. 30, Table 1.

by Philbrick the "exchange" world to indicate the importance of trade in this zone. At the other extreme stand the unoccupied polar and ocean regions. Between these extremes lie the transitional worlds divided between extensive and intensive regions of land occupation. Within this global pattern of nodal concentration is concealed the lower order of the hierarchy. Within one of the principal core regions, the eastern United States, is a complex mosaic of cores and connections that runs from New York City (a seventh-order world city in Philbrick's scheme) to small, rural, second-order Michigan towns. These rural towns serve, in turn, as local retail centers for the first-order farm settlements.

Philbrick sees human spatial organization as fundamentally nodal. In a nodal system the pattern of regional organization is marked by horizontal or spatial linkages between populations. But there are areas of the earth's surface where the human population is drawn away from its nodal pattern by the attraction of natural resources. There the emphasis is on vertical or ecological relations between populations and immobile or fixed resources. Both concepts have been integrated into Philbrick's model. He suggests that the structure of world regions is composed of many levels where human activities are arranged about nodes of varying size and character. One particularly noteworthy feature of this model is the alternating shifts from uniform to nodal regions (see Section 10-5) as the size and complexity of the organizing units increase. Thus, the smallest unit in the Philbrick hierarchy, the farm (level 1), is essentially a nodal region. On the next level, farms are grouped into adjoining blocks to form homogenous agricultural zones (level 2), essentially a uniform region. The small trading centers that serve the agricultural regions on level 3, however, are nodal regions. This alternation of nodal and uniform regions extends through all the levels of the hierarchy.

14-5 | THE SEARCH FOR ALTERNATIVE MODELS

Not all geographers are happy with the Christaller model, even in its more refined forms. They regard it as essentially a special case in two important regards. First, it is a special case in *conception* in that it describes a closed system where change can occur only from the bottom upwards (i.e., with increased rural productivity in the lowest level of settlement leading to an enlarged hierarchy and therefore more higher-order centers). Second, it is a special case in *reality* since it emerged from the study of a particular area (southern Germany, a mid-continental location) with a particular history of settlement in which the "feudal" organization of agriculture had played a notable part. Geographers who accept this criticism as valid do not see how the Christaller model can help in studying settlement patterns where the main forces of change are from the outside (i.e., the system is open rather than closed). In these patterns, the hierarchy may evolve from the top downwards with large seaboard cities, like those on the east coast of North America in the nineteenth century, acting as centers of innovation for external commercial forces.

There have been a number of attempts to build alternatives in which the time dimension and historical reality play a larger part. Space only permits us to look at one simple example of this, a model based on analogies with plant colonization. For an elegant and carefully-argued alternative which throws light on the history of the American city, browse through Berkeley geographer James Vance's book, *The Merchant's World* (see "One step further . . ." at the end of this chapter).

Competition between settlements: A biological analogy

If we leave a piece of ground untilled—maybe our back yard or an abandoned field on a farm—then the bare ground will slowly fill up with plants. We noted other examples of this plant-succession process earlier (see Section 4-1). Botanists have been able to work out various stages through which the vegetation evolves from the rapid change of

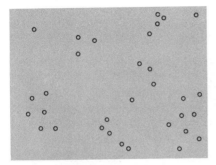

(a) Initial colonization

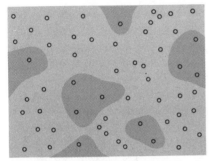

(b) Infilling

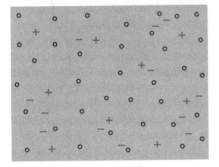

+ Growth centers   − Declining centers

(c) Urban competition

Figure 14-14. Idealized colonization sequence. (a) Low density colonization in the initial phase with a random settlement pattern. (b) Build-up of population in the original colonization areas with secondary infilling of previously empty areas (shaded). (c) Urban competition with the differential growth and decline of centers leading to a more uniform, hierarchically structured settlement pattern.

the early pioneer stage through equilibrium or climax vegetation. Using these biological analogies, geographers have constructed some general models of the spatial organization of settlement that develops as population increases. The sequence of changes on the world level and within individual regions has enabled workers to build general models of population growth. Figure 14-14 presents a synopsis of prevailing views, which see various distinct phases of the spread of population. In the initial *colonization phase*, population growth is accompanied by an extension of low-density settlements into previously unoccupied areas (Figure 14-14(a)). Settlements in the new area are comparable in size and structure to existing settlements. Alternatively, new technologies for using resources may allow a denser pattern of settlements to develop in areas that have or have not been settled before. An example of the first sequence of events is the extension of farm settlements in the northeastern United States during the eighteenth century; mining settlements exploiting new mineral resources typify the second process.

The second stage of population expansion is an *infilling phase* when the boundaries of a settled area remain unchanged but the population continues to grow. Increasing population is followed by short-distance migration that fills in the gaps in the settlement distribution formed by the original colonizers [Figure 14-14(b)]. In the third phase the proportion of the population engaged in geographically dispersed activities (such as agriculture, forestry, and mining) falls while the proportion of concentrated urban activities (such as manufacturing, commerce, or the provision of services) increases. In this *urban competition* phase larger centers grow more rapidly than smaller ones; those in the best locations capture an increasing share of commercial activity in the region, and those less well placed tend to stagnate and decline [Figure 14-14(c)]. The end of this phase is marked by the evolution of the highly structured, regular lattice of settlements we encounter in central-place theory.

In this settlement sequence, the phases of colonization and infilling parallel the stages of hexagon-formation in Figure 14-8. However, the accord between the urban-competition phase and the idealized lattices in Christaller's model will be "disturbed" by separate historical and environmental factors.

The evolution of the U.S. settlement hierarchy

How far does the actual evolution of urban systems parallel this biological analogy? Chicago geographer Brian Berry recognizes a four-stage process in the emergence of cities in the United States (Figure 14-15).

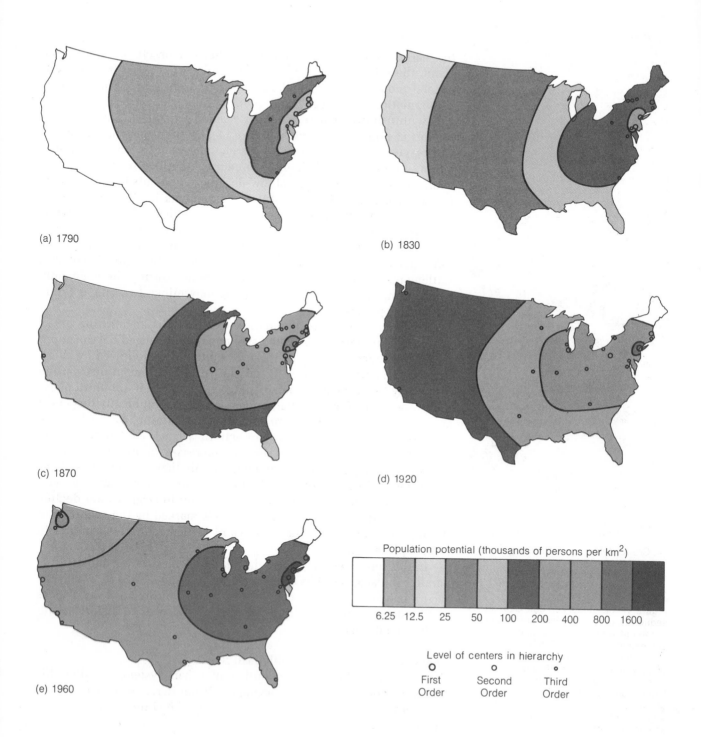

(a) 1790

(b) 1830

(c) 1870

(d) 1920

(e) 1960

Population potential (thousands of persons per km²)

6.25 12.5 25 50 100 200 400 800 1600

Level of centers in hierarchy

○   ○   ○
First   Second   Third
Order   Order   Order

Figure 14-15. The changing urban
organization of the United States. The
maps show the changing status of urban
centers in the U.S. hierarchy and
population potentials at approximately
40-year periods from (a) 1790 to (e)
1960. [From B. J. Berry and F. E. Horton,
*Geographic Perspectives on Urban
Systems* (Prentice-Hall, Englewood
Cliffs, N.J., 1970), p. 23, Fig. 2–1.]

The first stage recognized was a *mercantile phase* beginning with the growth of Atlantic seaboard towns in the eighteenth century. Such towns were generally deepwater ports serving as a nucleus of communication and export center for agricultural hinterlands that produced staples for the world's markets. The fact that the hinterlands of Boston, Philadelphia, and Charleston were physically more limited than those of New York, and the relative separation of New Orleans from the main domestic market, allowed New York City—with its middle location along the Atlantic strip and its good internal communications—to move into a dominant position that it retained in succeeding decades [Figure 14-15(a)]. The increase and spread of population inland from these coastal cities followed natural corridors, reinforced by later canal and railroad links, toward the heart of the agricultural-processing regions. With such expansion came the growth of a second generation of inland rail and processing centers like Cincinnati, Chicago, and St. Louis.

A subsequent *industrial phase*, dating from around 1840 to 1850, took place because of the rapid expansion of manufacturing. The growing demand for iron, and later steel, thrust into locational prominence those areas with (1) appropriate resource combinations (iron ores and coal) and (2) central cities already established during the mercantile phase. Buffalo, Cleveland, Detroit, and Pittsburgh shared these advantages, while other peripheral locations with natural resources (but without ready access to a market) did not. The industrial phase further strengthened the position of New York City and saw the emergence of a major heavy-industry "ridge" running westwards into the heartland of the United States [Figure 14-15(b)].

This heartland is usually delineated as the area within the Boston-Washington-St. Louis-Chicago rectangle. It had the initial advantages of excellent agricultural resources and a strategic location with respect to mineral resources. We could characterize the period after 1870 as a *heartland–periphery phase*, when contrasts between this core region and other parts of the United States were strengthened [Figure 14-15(c)]. The processes of circular and cumulative causation that increased the wealth of the core area are part of the same processes of regional evolution that brought wealth to the European cultural hearth (Section 11-2). We can regard even the spatial changes outside the heartland—that is, the emergence of new peripheral centers in the Far West, the South, and Texas—as direct responses to the needs of the heartland. The resource demands of the peripheral areas fostered regional specialization there, but, at least until World War II, this peripheral growth was essentially dependent on central growth.

Since around 1950 we can detect a *decentralized phase*. In this phase the location of amenity resources (e.g., a sunny climate or unpolluted environment) has become more important [Figure 14-15(e)]. These resources have stimulated interregional movements of population, bringing rapid urban growth to Arizona and the Southwest. They have been responsible for the intraregional rise of small- and medium-sized urban centers with above-average housing, schools, amenities, and so forth. And they have affected local populations by encouraging the suburbanization of manufacturing. These population shifts away from established urban centers are partly related to a rapid overall rise in real incomes. Other factors include an increase in the number of retired persons who move to milder climates, the growth of quaternary institutions and industries such as universities and research and development centers, and an expansion of footloose industries like the aerospace industry, which has few links to either existing markets or existing resources.

The shifts in growth locations over the last two centuries may be partly traced to the urban population's changing definitions of natural resources. In the agricultural period the most valued resource was agricultural land, which relied on basic distributions of climate, water supply, and soil. With industrialization the location of mineral resources—particularly coal—became a dominant factor in growth. During the growth of the heartland and peripheral areas, location and good communications were stressed; the current emphasis on leisure activities makes amenity resources an important locational factor. As different amenity resources become important, they provide new directions for the expansion process.

Postscript: Toward applications

We have left until last the important question of the applications of a settlement model. How can geographers use ideas of city chains and central-place structure? Well, we can use these ideas in two ways. First, the relationship between the size and pattern of settlements summed up by the rank-size rule is stable enough over time for us to use it to project future patterns of settlement sizes." Consider, for example, the regular progression of curves for the United States in Figure 14-16, and try to think what forces we would have to bring in to distort or disturb the pattern. Of course, it is only the whole system of cities that is stable. Individual cities may have widely varying paths of growth. Compare, for example, the constant lead position of New York to the faltering pattern of a city like Savannah, Georgia. Still greater contrasts separate the rapid rise of Los Angeles with the dropout pattern of Hudson, N.Y.

A second use of the central-place model is in regional planning.

Figure 14-16. Hierarchic evolution. The diagrams show changes in the role of individual cities with a hierarchy. The regular growth of the U.S. urban system over 150 years contrasts with the varying growth trajectories of individual cities. [From C. H. Madden, *Economic Development and Cultural Change* (University of Chicago Press, Chicago, 1956), Vol. 4, p. 239, Fig. 1. Copyright © 1956 by The University of Chicago.]

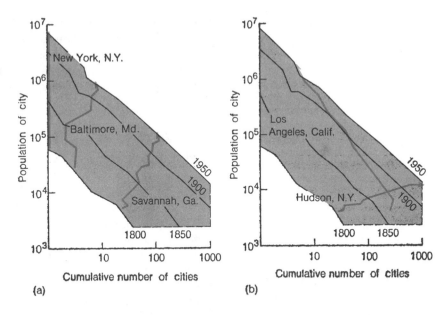

(a)

(b)

Beckmann's hierarchic model

The idea of integrating the concepts of one-dimensional city chains and two-dimensional city hierarchies has been proposed by mathematician Martin Beckmann [*Economic Development and Cultural Change* 6 (1958), pp. 243–248]. In his model, the population of a city of a given order is given by

$$P_r = \frac{LC_1 K^{r-1}}{(1 - L)^r}$$

where P_r = the population of a central place with order r in the hierarchy,

L = the proportion of the total population served by a central place located in that central place,

C_1 = the rural population served by the place with the lowest order in the hierarchy, and

K = the number of places of the next lowest order $(r - 1)$ served by places of order r.

Thus, in a Christaller hierarchy with $K = 4$, $L = 0.5$, and $C_1 = 100$, a second-order central place will have a population of 800, a third-order place a population of 6400, a fourth-order one a population of 51,200, and so on. The more orders there are in the hierarchy, the more closely city sizes will conform to a continuous size distribution. Moreover, if L is small relative to 1, then the product of the rank of a place and its population size will approximate a constant, as the rank–size rule requires.

The settlement hierarchy in a newly settled area often tends to move from a primate to a rank–size form as population increases and the separate settlements are more closely integrated. For instance, if we were to design such a system for the settlement of the middle-west plateau of Brazil around the new primate city of Brasilia, the later readjustment of the hierarchy could be anticipated and investment in infrastructure (roads, power stations, schools, hospitals, etc.) adjusted accordingly. Central-place theories also have played a role in the designing of hierarchies of shopping and service centers within cities. Increasing use is also being made of hierarchic concepts in the design of key service sectors like hospital systems.

In beginning this chapter we looked first at the simple chain models of settlements in one dimension and went on to consider hierarchic models in two dimensions. Both types of models are closely linked. Indeed, mathematician Martin Beckmann has shown that the more levels there are in a system the closer the array of city sizes will come to a continuous distribution. By contrast, regions with only a few units in the hierarchy will have sharply stepped rank–size distributions. (See the marginal discussion of the Beckmann model.) Rank–size distributions must be logical byproducts of the central-place system.

Like Alice in our opening quotation, the geographer finds the world laid out like a large chessboard. In this chapter we have been able to cover only the simplest and most basic moves in the complicated chess game by which cities "capture" and organize one another's territories into a kind of feudal hierarchy of metropolis, city, and village. Like all models, our hexagonal chess set is an oversimplification of reality. To follow the actual "moves" by which any individual city develops would tax the skill of a Bobby Fischer or a Boris Spassky.

Reflections

1. What is the size of your own community? Look very carefully at your answer to see how it is affected by the bounding problem illustrated in Figure 14-1. Have you over- or underestimated your community's real size? How might you make your estimate more accurate?

2. Gather data for the size of cities in your state or province. Plot their position on a population size–rank diagram like that in Figure 14-4. How closely does the resulting distribution correspond to the one predicted by the rank-size rule? Suggest reasons for any departures from this rule you observe.

3. Why do so many small towns in Western countries appear to be losing their central-place functions? Can you suggest any ways in which their decline might be arrested?

4. What periodic central functions are still found in Western countries? List any you can find for your own area, and try to map their tracks. Would you consider college football an appropriate example?

5. Debate the relevance of the very formal settlement models presented in this chapter. Do they (a) fail to reflect the true complexity of city settlements or (b) provide a unique insight into their structure?

6. To what extent does the distribution of schools and colleges in your own city, state, or province have a regular spatial order? Identify some of the factors that you think might be "disturbing" this order.

7. Review your understanding of the following concepts:
   (a) bounding problems
   (b) SMSAs
   (c) rank-size rule
   (d) primate patterns
   (e) central places
   (f) complementary regions
   (g) market-size thresholds
   (h) fixed-K hierarchies
   (i) periodic market cycles
   (j) urban competition

One step further . . .

Reviews of the basic theories of settlement hierarchies and their structures are provided in most texts on human geography. See, for example,

Bunge, W., *Theoretical Geography* (Lund Studies in Geography, Series C, No. 1, 1962) and

Haggett, P., *Locational Analysis in Human Geography* (Edward Arnold, London, 1965), Chaps. 4 and 5.

The classic work in central-place theory is Walter Christaller's study of southern Germany, published in 1933. It should certainly be dipped into and is available now in translation. See

Christaller, W., *Central Places in Southern Germany* (Prentice-Hall, Englewood Cliffs, N.J., 1966, transl.).

Another classic German work that has more ideas in its footnotes than many books have in their text is

Lösch, A., *The Economics of Location* (Yale University Press, New Haven, Conn., 1954).

Some modern theoretical departures are authoritatively presented in

Berry, B. L., *Geography of Market Centers and Retail Distribution* (Prentice-Hall, Englewood Cliffs, N.J., 1967), Chap. 4.

For a more critical approach to settlement theory with emphasis on the historical evidence and dynamic models, the outstanding book is

Vance, J. E., Jr., *The Merchant's World: The Geography of Wholesaling* (Prentice-Hall, Englewood Cliffs, N.J., 1970).

Current research is reported in the major geographic journals. Look especially at the *University of Chicago Department of Geography Research Papers* (published occasionally) for applications of central-place concepts.

Chapter 15

Worlds Beyond the City, 1:

AGRICULTURAL ZONE AND INDUSTRIAL CENTERS

The last rural threads of American society are being woven into the national urban fabric.

—MELVIN M. WEBBER
Cities and Space (1963)

W here does a city end? Think about the answer by recalling the last time you drove along a highway out of a major city. At one mile out from the CBD, the landscape was probably still urban, dominated by a built-up environment of houses and streets. At ten miles out, the landscape was probably semirural; but in this suburban range there would still be massive daily commuting to the city. At fifty miles out, the landscape may have been truly rural; but the car radio would still carry, faintly, the voice of city radio stations, the farmers would still purchase city newspapers from their local general store, and their wives would still frequent the city to carry out their major shopping.

Cities, like old soldiers, don't die—but simply fade away. Their spatial dominance lessens as we move further from their centers but is never entirely eliminated. Some minute proportion of a city's mail or telephone calls will reach out even to the far side of the world; its billboard advertising ("Only 1050 miles to Harry's Place") may stretch across a continent. It is this city-centered world that we look at in this chapter. We begin by looking at the effect of the city on how the rural land around it is used. We then go on to look at the parallel impact of the city on the processing of natural resources and study the see-saw of forces that shape the spatial pattern of industry. In both these sections we shall be concerned with the locational theory that geographers have built up to explain the city-oriented world. Locational theory is one of the three major elements in geographers' research, and we shall be asking questions about why certain activities are located where they are. Since some of the answers turn out to be complicated ones, you may need to move rather slowly through this and the following chapter.

15-1 | THÜNEN AND LAND-USE ZONING

The earliest attempt to correlate land-use patterns with the spatial relationship of a city to its surrounding region was made by a German, Johann Heinrich von Thünen (Figure 15-1). In his classic work on the location of agricultural land-use zones, *Der Isolierte Staat in Beziehung auf Landwirtschaft*, first published in 1826, he not only laid the foundation for refined analysis of the location of agriculture, but stimulated interest in a much broader area of locational analysis. In 1810 Thünen, at the age of 27, acquired his own agricultural estate, Tellow, near the town of Rostock in Mecklenburg on the Baltic coast of Germany. For the next 40 years, until his death in 1850, he supervised the cultivation of the Tellow estate and amassed a remarkable set of records and accounts that provided the empirical basis for his published theories.

Figure 15-1. A pioneer locational theorist. Johann Heinrich von Thünen (1783–1850) laid the foundations for agricultural location theory in a book published in 1826, called *The Isolated State.* Basing his ideas on observations from his own farms, Thünen extended them to encompass worldwide patterns of land-use variation. Although agriculture has changed dramatically in the last 150 years, his models still give useful insights into the geographical patterns observable today. [Photo courtesy of The Mansell Collection.]

The isolated state

Thünen's isolated state is modeled on the agricultural patterns of nineteenth-century Mecklenburg. The basic form of the land-use patterns he envisaged is shown in Figure 15-2, and the characteristics of each of the zones are presented in Table 15-1. The patterns are a series of concentric shells ranging from narrow bands of intensive farming and forest to a broad band of extensive agriculture and ranching to an outer "waste."

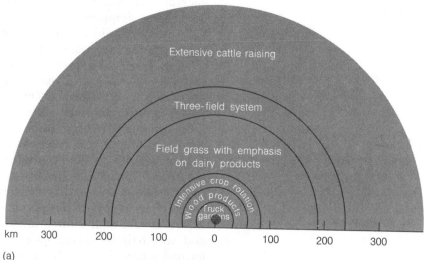

(a)

Figure 15-2. Land-use zones in Thünen's *Isolated State* (1826). (a) Thünen's original land-use rings. (b) Thünen's rings modified by a navigable river bringing cheaper transport costs. The distance figures were added to Thünen's original model by L. Waibel in 1933, in his *Probleme der Landwirtschaftsgeographie.*

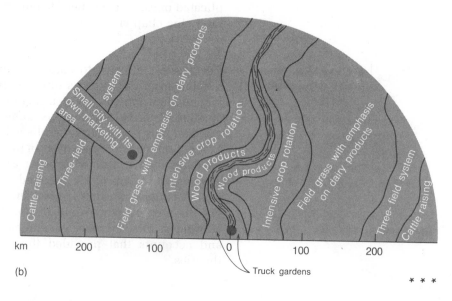

(b)

＊ ＊ ＊

Table 15-1 Thünen's land-use rings

Zone	Percentage of state area	Relative distance from central city	Land-use	Major product marketed	Production system
0					
0	<0.1	—0.1	Urban–industrial	Manufactured goods	Urban trade center of state; near iron and coal mines
1	1	0.1–0.6	Intensive agriculture	Milk, vegetables	Intensive dairying and trucking; heavy manuring; no fallow period
2	3	0.6–3.5	Forest	Firewood	Sustained-yield forestry
3a	3	3.6–4.6		Rye, potatoes	6-year intensive crop rotation: rye (2), potatoes (1), clover (1), barley (1), vetch (1); no fallow period; cattle stall-fed in winter
3b	30	4.7–34	Extensive agriculture	Rye, animal products	7-year rotation system: field grass with an emphasis on dairy products; pasture (3), rye (1), barley (1), oats (1), fallow period (1)
3c	25	34–44		Rye, animal products	3-field system: rye, etc. (1), pasture (1), fallow period (1)
4	38	45–100	Ranching	Animal products	Mainly extensive stock-raising; some rye for on-farm consumption
5	—	Beyond 100	Waste	None	None

SOURCE: P. Haggett, *Locational Analysis in Human Geography* (St. Martin's Press, New York, and Edward Arnold, London, 1965), p. 165, Table 6-4.

To understand the form of these spatial patterns, we need to review the six assumed conditions that dictated it:

1. the existence of an isolated state cut off from the rest of the world;
2. the domination of this state by a single large city that served as the sole urban market;
3. the setting of the city in a broad, featureless plain that was equal everywhere in fertility and in which the ease of movement was the same everywhere so that production and transport costs were the same everywhere; *isotropic plain*.
4. the supplying of the city by farmers who shipped agricultural goods there in return for industrial produce;
5. the transport of farm produce by the farmer himself, who hauled it to the central market along a close, dense trail of converging roads of equal quality at a cost directly proportional to the distance covered; and
6. the maximizing of profits by all farmers, who automatically adjust the output of crops to meet the needs of the central market.

These assumptions are, of course, unrepresentative of actual conditions either in the early 1800s or, indeed, now. Why, then, did Thünen make them? To understand this, we have to go back to our initial discussion of the role of models in science (Section 1-5). The objective of models is to simplify the real world in order to understand some of its characteristics. Thünen's model permits us to do just this by sorting out some of the key factors (albeit in a simplified form) that cause land-use rings to form.

Given these assumptions, Thünen was able to demonstrate that rural land values would decline away from the central city in the same way urban land values do, though at much lower rates and with gentler slopes. (See Figure 15-3.) Like each urban land use, each agricultural land use has a characteristic set of bid-price curves and finds an appropriate location with respect to distance from the city. Thünen stated his model in terms somewhat different from those of urban models (see the marginal discussion of zoning mechanisms in the Thünen model), but the processes by which eventual land-uses are determined are identical. A product that is bulky (i.e., has a large

Figure 15-3. Ring formation in the Thünen model. Figure (a) shows how the bid-price curves for four types of land use have slopes of different steepness. (To understand this figure more easily, you may find it useful to check back to Figure 13–17 where we first looked at the idea of bid-price curves.) Use of land for dairying has a steeper curve than that for wheat farming, indicating that dairying is a more intensive way of using the land and one which stands more to gain from locating near the city. But dairying is itself displaced by land used for residential housing on the fringe of the city. The broken lines in (a) mark the breaks where one type of land use "outbids" the other in terms of the rent levels it can afford to pay for a convenient location near the city center at *X*. If you follow these broken lines down to Figure (b) you will see that four distinct land-use zones are formed. ("Waste" indicates land not in use because it lies too remote from the city.) If you complete the arcs of the circles shown in (b), a series of land-use rings are formed around the city.

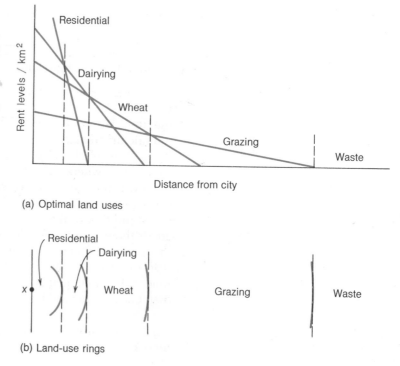

(a) Optimal land uses

(b) Land-use rings

Zoning mechanisms in the Thünen model

Thünen regarded locational rent (*Bodenrente*) as the key factor sorting the uniform area of his isolated state into distinct land-use zones. The location is given by

$$L = Y(P - C) - YD(F)$$

where L = the locational rent (in $/km²),
Y = the crop yield (in tons/km²),
P = the market price of the crop (in $/ton),
C = the production cost of the crop (in $/ton),
D = the distance to the central market (in km), and
F = the transport rate (in $/ton/km)

Thus, for a crop yielding 1000 tons/km², fetching $100/ton at the central market, and costing $50/ton to produce and $1/ton/km to haul, the locational rent at the city center would be $50,000/km²; at 10 km distant it would be only $40,000/km², and at 20 km distant it would be down to $30,000/km². Beyond 50 km production would be at a loss. The competition of two crops (i and j) for the same area depends on their yield (Y) and relative profitability ($P - C$). When the condition

$$1 < \frac{Y(P-C)_i}{Y(P-C)_j} < \frac{Y_i}{Y_j}$$

obtains for crops i and j, they form two distinct spatial zones; crop i dominates a circular area adjacent to the city, and crop j occupies a ring-shaped zone immediately outside it. The symbol < means "is less than." Any other relation of the terms in the equation above results in the two crops' being reversed so that j occupies the inner ring, *or* one crop dominating all available land to the complete exclusion of the other, *or* both crops being grown side by side with no spatially differentiated zoning. [See E. S. Dunn, *The Location of Agricultural Production* (University of Florida Press, Gainesville, Fla., 1954), Chap. 1.]

tonnage per unit of area) or is difficult to transport will have steep bid-price curves and will be quite sensitive to displacement from the market. Conversely, one that is lighter (i.e., has a low tonnage per unit of area) or easy to transport will be less sensitive to displacement. As Figure 15-3 indicates, land-use boundaries occur at the intersections of bid-price curves. Land use in the remoter areas is adapted to take into account the poor accessibility. A smaller tonnage of a given crop may be grown on each unit of area by using extensive farming methods; for instance, long fallow periods may be used to restore the land's fertility rather than fertilizers. Alternatively, a product may be shipped in a more compressed form (e.g., as cheese rather than liquid milk) or animals may be used to "concentrate" a crop (as when hogs raised on corn are marketed in place of corn).

We can extend the basic Thünen model to explain situations quite unlike the ones it was first proposed to explain. Although the original study was of ring formation about a single node, this is merely one geometric case. If we substitute a linear market for a central one, the model still explains the formation of zones, but in straight parallel bands. Land-use zones along a coastal strip or transport axis are also common variants of the conventional Thünen rings. Several alternatives of the ringed model were discussed by Thünen himself. By introducing a navigable river on which transport is speedier and costs are only one-tenth that of land transport, a minor market center with its own trading area, and spatial variations in the productivity of the plain, he was able to explain considerable variations in land-use patterns. [See Figure 15-2(b).] Once we allow these kinds of variations and add to them the wider rings that the technological advances in transport make possible, then the Thünen model can provide insights into contemporary patterns of land use on larger spatial scales.

City-centered zoning in the United States

Attempts to compare the Thünen model with the real world are hindered by the difficulties geographers have had in drawing up unambiguous definitions of land use. Different crops are grown in different environmental zones, so the Thünen rings are disturbed by other zones related to the ecology of crops rather than the accessibility of cities. Geographers have tended, therefore, to use population density as a proxy for land use and to look for gradual changes in density instead of sharp discontinuities between distinct land-use zones.

Demographer Donald Bogue has investigated how population distributed itself around 67 major cities in the United States. Using the census figures for counties, he analyzed the changes in population density with respect to distance as far as 800 km (500 mi) from the city.

Bogue's general conclusion is that the main metropolitan centers dominate the spatial arrangement of population in the United States. If we take the average population density at various locations and plot it against the distance to the nearest city, we find a rapid falloff; if, however, we transform both the axes of the graph to a logarithmic form, the decrease appears as a simple linear rate of decay (Figure 15-4). For example, 40 km (25 mi) outside the city the density exceeds 500 people /km² (1250 people/mi²); 400 km (250 mi) out, the density is only 10 people/km² (25 people/mi²). The detailed pattern of the decay curves reflects the size of the central city. Large metropolitan cities with over half a million people have densities much higher than smaller cities at similar distances. Farther out, these differences diminish.

We can also detect strong regional differences within the United States. The Northeast has curves that are higher and steeper, reflecting its dense network of cities separated by rural pockets of relatively low population density such as northern New England and Appalachia. The South has lower population densities and greatly irregular curves, reflecting its fewer and smaller cities and a more uneven pattern of rural population. In the West there is a sharp decline in population density with distance from the city. [See Figure 15-4(c).]

We can relate these differences in population density more directly to land use by sorting the population into employment categories. Bogue showed that for farm populations there was a gentle decline in density from the metropolitan centers out to about 150 km (93 mi) and a sharp decline at 500 km (300 mi). Again, the national trend concealed

Figure 15-4. Density gradients around cities. Variations in population density with distance from the nearest large city (metropolis) are shown for three regions of the United States in 1940. Both axes of the graphs are logarithmic to allow details of the middle parts of the density curves to be shown more clearly. This also allows easier comparison between the three regions. If plotted on arithmetic scales, all the curves would appear to be ''L'' shaped, emphasizing only the dramatic drop in population density within the first few miles of distance from the metropolis. [From D. J. Bogue, *Structure of the Metropolitan Community* (Scripps Population Institute, Ann Arbor, Mich., 1949), p. 58.]

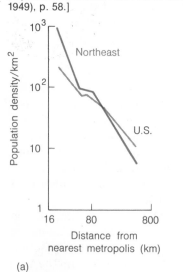

(a)

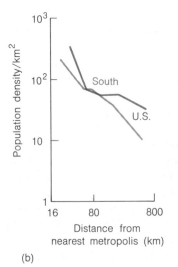

(b)

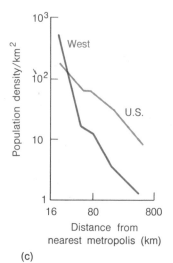

(c)

strong regional contrasts. The density in the South, for example, changed little with distance. Industrial land use as approximated by industrial employment was much more dependent on distance. Employment declines very sharply with distance from the major metropolises, but the decline is arrested around 50–100 km (30-60 mi) out before dropping sharply again. The brief halt in the decline probably represents a concentration of specialized manufacturing towns at this distance.

One important feature of Bogue's study, which has been picked up in later work, is that population density depends not only on distance from the metropolis but also on the direction in which other cities lie. If we divide the area around the city into wedge-shaped sectors, then sectors with routes running to other cities represent ridges of high population density compared with those not containing such routes.

Similar research in different countries but on comparable geographic scales indicates that the findings for the United States are reasonably typical. Population density the world round is sensitive to both the distance from and the direction to the centers of cities.

Zoning in farm communities

Geographic studies of land-use zones at the micro level of the village and farm (Figure 15-5) have been done as well as studies of sub-

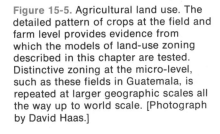

Figure 15-5. Agricultural land use. The detailed pattern of crops at the field and farm level provides evidence from which the models of land-use zoning described in this chapter are tested. Distinctive zoning at the micro-level, such as these fields in Guatemala, is repeated at larger geographic scales all the way up to world scale. [Photograph by David Haas.]

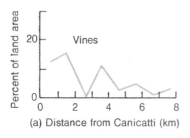

(a) Distance from Canicatti (km)

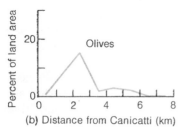

(b) Distance from Canicatti (km)

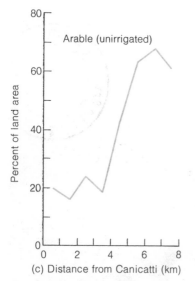

(c) Distance from Canicatti (km)

Figure 15-6. Accessibility and rural land use. The diagrams show changing crop patterns with increasing distance from the Sicilian village of Canicatti. [From M. D. I. Chisholm, *Rural Settlement and Land Use* (Hutchinson, London, 1966), p. 57, Table 6.]

continental zoning. Here, too, the effort required to use an area of land is going to increase with the distance from the center of the community. If we take a small community, the individual family farm, the time taken to reach the most distant fields will be greater than the time needed to visit the home paddock adjacent to the farmstead. By tracking individual farmers from dawn-to-dusk, comprehensive diaries of their movements in space over time can be compiled. What do these show? In Holland arable plots only 0.5 km (.3 mi) from farmsteads receive about 400 man-hours of care per hectare annually. At a distance of 2 km (1.2 mi), the care level drops to 300 man-hours; at 5 km (3.1 mi), it dwindles to only 150 man-hours. We can convert such figures into costs by adding information on the jobs carried out in these movements between farmstead and field. In Punjab villages in Pakistan the cost of ploughing increases by around 5 percent, the cost of hauling manure rises 10–25 percent, and the cost of transporting crops rises 15–32 percent everytime we move an additional half kilometer from the village.

How rapidly costs increase with distance is a matter for debate; different rates have been reported in different studies. What does seem clear is that locational adjustments begin to occur at distances as low as 1 km (.6 mi) from the community center and that the costs of operation rise sharply beyond about 3 or 4 km (2 or 2½ mi). We can see the form of these locational adjustments in the distinctive zones of land use around villages. Figure 15-6 shows the sequence of crops extending a distance of 8 km (5 mi) from the Sicilian village of Canicatti. Note that the growing of olives and vines falls off rapidly with distance; beyond 4 km (2½ mi) the open, arable land is cultivated mostly for wheat and barley. Some clue to the "sorting" process by which the decision is made to grow certain crops on certain locations is provided by the average number of man-days per hectare expended. This energy expenditure begins at 52 man-days in fields near the village but declines to less than 40 in remote fields at distances of 8 km (5 mi) or more.

Geographer Mansell Prothero has described a similar zoning around villages in northern Nigeria. He distinguishes four zones. The first is an inner garden zone with close interplanting, a continuous sequence of crops, and intensive care. A second zone at 0.8 to 1.2 km (.5 to .7 mi) out is continuously used (mainly planted with Guinea corn, cotton, tobacco, and groundnuts) and fertilized. A third zone with an outer boundary at 1.6 km (1 mi) is used for rotation farming; that is, the land is cultivated for 3 to 4 years and then allowed to return to bush for at least 5 years to get back its fertility. Finally comes the fourth zone of heavy bush. Within this zone are isolated clearings, in which the three-zone sequence of the main villages is reproduced.

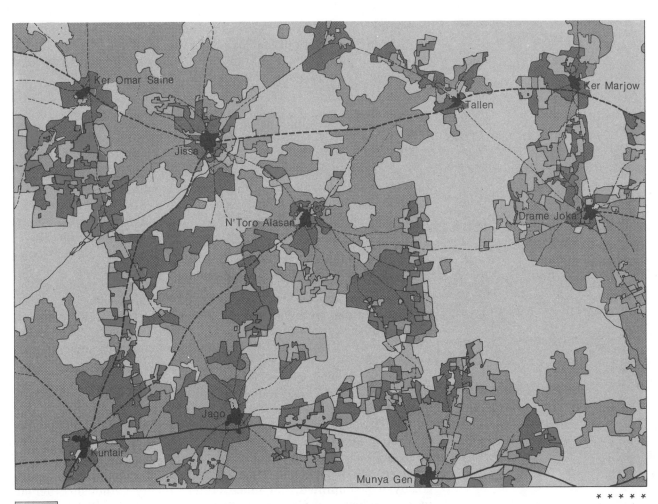

* * * * *

Woodland

Fallow bush

Grass

Groundnuts

Groundnuts, sorghums, and millets

Sorghums, millets, and other food crops

Villages

Figure 15-7. Rural land-use banding in the African tropics. The map shows areas of permanent and shifting cultivation in the Kuntair area of Gambia, West Africa. Note the relations of the different zones to the location of the nine villages named: the total distance across the map is about 5 km (3 mi). Woodland occurs as a residual zone at the margins of the cultivated land and along boundary areas between villages. Fallow bush is woodland which has previously been used for cultivation and the land is now lying idle while recovering its fertility. Groundnuts (peanuts) are an important cash crop, while sorghum and millets (both grain crops) are consumed locally. These crop areas (see color key) tend to be located on the woodland margins in newly cleared patches of land with relatively high fertility. [From the Directorate of Overseas Surveys, Gambia, Land Use Sheet 3/111, 1:25,000, 1958.]

Studies similar to those in Sicily and Nigeria reveal a variety of responses by farmers to distance. Sometimes the response leads to sharp land-use zoning, as in Figure 15-7; in other instances farmers may grow the same crops over a wide range of distances but plant them in different combinations and give them varying degrees of care. Moreover, the simple symmetry of the zones is usually interrupted by local variations in terrain, soils, patterns of ownership, and the like.

15-2 | WEBER AND INDUSTRIAL LOCATION

We have seen how cities, as centers of agricultural consumption, play a dominant role in shaping rural land uses. But cities are also the great consuming centers for all natural resources. What influence does the city as a marketplace have on the location of resource processing and thus on industrial patterns in the world beyond the city?

This question of the location of resource processing, with all its population-multiplying implications (Figure 13-3), has long attracted investigators. One of the simplest but most penetrating analyses of this subject came from the German spatial economist Alfred Weber, who published his original study of industrial location in 1909. Weber paid particular attention to the loss of weight or bulk involved in resource processing. He demonstrated that this weight loss played a significant part in the location of certain industries. Industries with manufacturing processes that involved a large weight loss were found to be *resource-oriented* (i.e., located near the natural resource that was to be processed). To convert wood to pulp and paper involves a 60-percent weight loss, and so pulp and paper mills tend to be located near major forested areas rather than in the heart of the large cities in which the bulk of their products is eventually consumed. Conversely, an industry like brewing is *market-oriented*, since its product is very weighty in relation to the malt, hops, and other materials that go into it. (Water is assumed to be generally available; hence a need for it is assumed not to affect location decisions. This analysis is clearly greatly oversimplified; but much current theory, although different in terminology, sophistication, and method, follows Weber's basic line of thinking and is therefore termed *Weberian*. We first look at Weberian analysis in a single spatial dimension and then go on to more realistic two-dimensional analyses.

Location on a line

Let us assume that a single resource is supplied at location R and a single urban market for that resource is located at city M as in Figure 15-8. We further assume that transport costs increase regularly with distance from the supply point and that all other costs (labor, power, taxes, etc.) are fixed and equal everywhere. Figure 15-8(a) shows a simplified situation with a single resource supply point (R) and a single

Figure 15-8. Least-cost locations. Simplified examples of the least costly location for a resource-processing facility are shown as a "tug-of-war" between the resources located at R and the markets located at M. Cost on the vertical axis is plotted against distance along a line between R and M on the horizontal axis. The three different cases shown are discussed in the text.

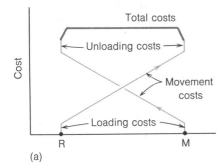

(a)

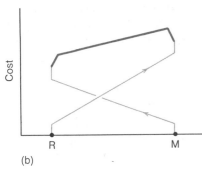

(b)

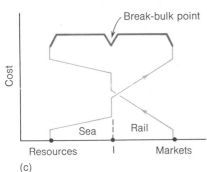

(c)

city (M) located on the horizontal axis; costs are marked on the vertical axis. If the loading costs are equal, and delivery costs from R to M are identical functions of distance, then the total transportation costs have the form shown by the heavy line. Transportation costs may be at a minimum at either the supply site or the market site. At all other possible locations on the line of haul between the two (RM), the transportation costs are equal, but higher, because they involve an additional loading or unloading charge.

The processing of resources commonly involves changes in their mass, weight, or value that affect unit transportation costs. The inequality in transit costs for raw and processed goods is illustrated in Figure 15-8(b) by the steeper slope of the delivery-cost curve from R. Transportation costs are lowest from the supply point, R. When transportation costs are higher for the finished product, the reverse situation obtains and the lowest-cost point moves to location M. To make the situation a little more realistic, we can introduce two zones (land and sea) with different transport costs. Because the commodity being moved must go by two modes of transportation, the costs of loading, unloading, and reloading at the intermediate point [I in Figure 15-8(c)] may be substantial. Thus all three points (R, I, and M) share the status of *least-cost transport points*. By modifying some of the assumptions in Figure 15-8, we could go on to introduce further and more complex least-cost patterns.

We can see something of the tug of war between these supply and demand points by considering the location of oil refineries. In the early period of oil production (to 1920, approximately), there were advantages to be gained from locating refineries on or near the oil fields themselves. Transport costs were high and the demand was mainly for kerosene, with about half the crude oil being dumped or burned as waste. As the advent of larger pipelines and supertankers reduced crude-oil transport rates, and as the range of petroleum products and the capacity of refineries to use more of the oil grew, it became more advantageous to locate refineries near the point of consumption. The vastly increased refinery capacity on the seaboards of Western European countries reflects both the changing economics of location in the industry and the changing political situation. Market-based refineries can deal with a wider selection of crude-oil suppliers, thus reducing the risk of supplies being cut off from a single source; and they can be massive enough for the economies of scale since they can draw on various small oil fields for supplies. The jump from the simple diagram in Figure 15-8 to the complexities of refinery location shows how large is the gap that more advanced locational theory must try to bridge.

Location in a plane

We can move a small way forward if we change the focus of our analysis from a one-dimensional line to a two-dimensional plane. Figure 15-9(a) presents a simple situation in which we have two resource supply sites (R_1 and R_2) and a single market city M. If we begin by assuming equal transport costs per unit of weight, the costs from each of the three points can be represented by a series of equally spaced, concentric, and circular contours called *isotims*. Each isotim describes the locus of points about each source where delivery or procurement costs are equal. Here the isotims are circular about each of the three points because transport costs across the isotropic plane are everywhere the same. Total transportation costs can be computed by summing the values of intersecting isotims. Lines connecting points with equal total transportation costs are termed *isodapanes* and, like isotims, may be regarded as contours showing the cost terrain of a particular region. In Figure 15-9(a), the lowest point on the isodapane is equidistant from each of the three resource supply points.

If we relax the assumption of equal transport costs per unit of weight, then the isotims around each point may vary. Figure 15-9(b) shows a situation in which the movement costs from R_2 are twice those from the other two sites. This situation distorts the shape of the isodapane surface and displaces the point of minimal total transport costs toward the supply point with the higher transport costs (i.e.,

Figure 15-9. *Isodapanes.* Contours of total transportation costs on a plane. In these simplified situations, there are two resources (R_1 and R_2), a single market (M), and a uniform plane. Transport costs around each of the three points (R_1, R_2, and M) are shown as concentric circles (*isotims*) Equal isotims about each point gives a midway compromise point of lowest transport cost in (a). Doubled costs from R_2 draw this compromise point toward it in (b).

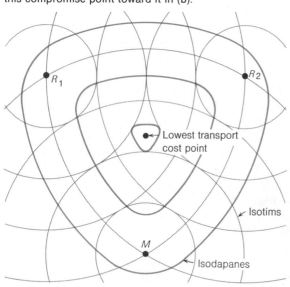

(a) Equal transport costs

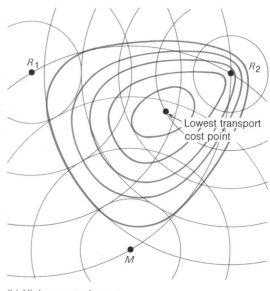

(b) Higher costs from R_2

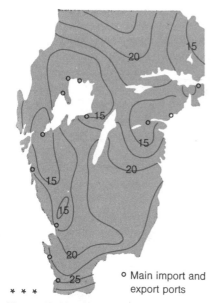

○ Main import and
export ports

* * *

Figure 15-10. Isodapanes in Sweden.
Contours of total transportation costs
for paper making in southern Sweden.
The figures include the total cost of
transporting pulpwood, coal, sulfur,
limestone, and the paper itself in kroner
per ton of output. The lowest-cost zone
runs across the middle of the map and
to the northwest. [From O. Lindberg,
Geographische Annalen **35** (1953), p. 39,
Fig. 20.]

toward R_2). Weber compared this displacement process to a set of
scales that are weighted too heavily in one scale pan. He suggested
that this process might be illustrated by a simple physical model in
which the pull of alternative sites is represented by physical weights
attached to a circular disk by a system of cords and pulleys. Although
clumsy in a practical sense, the idea of locational weights helps us to
visualize the many competing forces involved in locational decisions.

Isodapane maps prove useful in practice, and a Swedish example
of their use is given in Figure 15-10. This map shows the total costs of
transporting pulp wood from various assumed manufacturing centers
in southern Sweden. The map emphasizes the relatively advantageous
location of the central lakes lowland and the coastal strip at its two
ends. Costs increase sharply toward the interior and to the north. Of
course, isodapane maps indicate only transport costs; to get a more
complete picture we need to consider other costs not related to the
transport of goods.

Space-cost curves

So far we have analyzed the location of resource-processing sites
simply in terms of transport costs. This is an oversimplification, as
Weber recognized; for location is affected by three other kinds of
costs. First, there are labor and power costs that vary over space and
are dependent on location. Second, we have costs that are largely independent of location. That is, there may be advantages to be gained
from producers joining together to share the overhead costs of marketing and research or to encourage local suppliers to specialize in certain
goods. Third, government legislation may cause spatial variations in
costs through subsidies or taxation.

We can combine these nontransport costs with transport costs in
locational analyses of resource processing. In Figure 15-11 space–
cost curves are used to create a two-dimensional profile of the distribution of both types of costs. In the first case [Figure 15-11(a)] nontransport costs are considered uniform irrespective of location, whereas
transport costs vary systematically about a central location (A). If we
assume a horizontal demand curve at a common market price, then
the area of profitable production forms a shallow, saucerlike depression where we can locate the point of least-cost production (A)
and the spatial margins of profitability. When we allow factors such
as labor or power costs to vary, the area of profitable production can
be altered. Thus, we can have a second production outlier with low
nontransport costs forming a geographically isolated pocket, as in
Figure 15-11(b). Changes in the third group of nontransport factors,
government intervention, are introduced in Figure 15-11(c). By placing

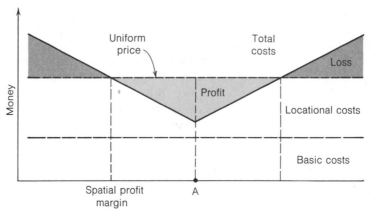

(a) Uniform basic costs

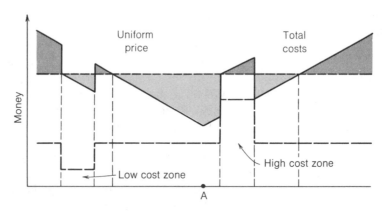

(b) Variable basic costs

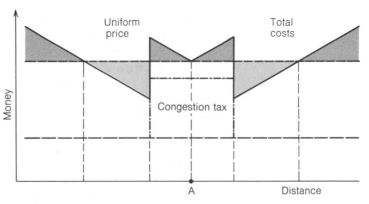

(c) Government intervention

Figure 15-11. The effect of nontransport costs on location. All three diagrams show a cross section through an economic landscape in which transport costs vary systematically with distance from a central location (A), but the basic (or nontransport) costs either (a) are uniform, (b) vary because of local production conditions, or (c) are modified by government intervention. In practice, basic costs are also likely to vary over space, being lower (because of greater economies of scale) near the market center (A).

heavy taxes on locations near A and subsidizing peripheral areas, we ensure considerable alterations in the spatial distribution of profitable locations.

These simple cross sections give some idea of how variations in nontransport costs can be introduced into locational analysis. The cross sections can be supplemented by contour maps whose cost elements are plotted on a plane rather than on a line. Again, these combinations are only first steps in unraveling the complexities of real-world locational patterns and intricate locational criteria.

We can remind ourselves of how sensitive to change are real-world resource-processing locations by considering the decisions faced by a ski-tow operator in Pennsylvania during the 1960s (Figure 15-12). If he moved closer to the natural resource (i.e., areas of heavy and reliable winter snowfall), then he would be further from his main urban markets—the skiers in Philadelphia, Baltimore, and Washington. If he moved toward these sources of skiers, he would have to locate in areas of lower and less reliable snowfall. He also had to keep an eye on other ski-tow operators and evaluate the benefits of locating near or far from them. Thus, in this resource-processing situation, a subtle game was being played in which both nature and the individual processors had a role. Artificial snowmaking would vary this locational

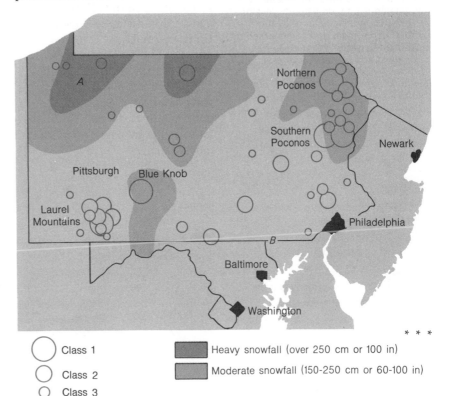

Figure 15-12. Locational decisions in a resource-use framework. This figure illustrates the problem of finding the best location for a ski resort given variations in snowfall, major urban markets, and the competition of other ski resorts. The sizes of the circles indicate the sizes of the existing ski resorts. A ski resort at A in northwest Pennsylvania would be fine for snow conditions and lack of competition, but short on skiers. Location at B would have lots of potential skiers in nearby Baltimore and Philadelphia but be short on snow. [From J. V. Langdale, Pennsylvania State University, unpublished master's thesis, 1968.]

○ Class 1
○ Class 2
○ Class 3
○ Class 4

Heavy snowfall (over 250 cm or 100 in)

Moderate snowfall (150-250 cm or 60-100 in)

* * *

game still further. It is complications of this kind that modern locational theory, derived from Weber's theory but now vastly extended, is designed to explain.

15-3 | COMPLEXITIES OF SPACE

In our discussion of both the Thünen and the Weber families of models, we have stressed geographic distance from the city as a dominating theme. But geographic distance (in the sense of the length of routes in miles or kilometers) is often only a crude measure of the costs of movement. Consider Figure 15-13, which shows actual freight costs per ton from six ports in eastern New Guinea to other points within the territory. The costs form an intricate patchwork related to the volume and type of goods to be carried, the mode of transport (truck, aircraft, or ship), and the degree of competition, as well as the geographic distance to the six ports. The situation in New Guinea is by no means atypical. At any point in time, the exact cost of moving resources from one part of the earth's surface to another is a function of a dozen or more different considerations. Think of the different rates quoted for a passenger fare from Boston to London—from scheduled air fares, through reduced air charter rates, to variable shipping rates.

Notwithstanding these complications, we can still make out some rough order in the relationship of costs and distance. First, we find that geographic distance does play a role in determining most rates. Other things being equal, a longer haul costs more than a shorter haul. There are of course exceptions. Within many countries the rate for sending a parcel varies with its weight but not with the distance it is sent. Similarly, companies may charge a uniform delivery rate throughout their sales area. Such blanket rates do not mean that costs do not increase with distance. The blanket rate subsidizes the longer distance movements by charging more than is necessary for the shorter distance movements.

Figure 15-13. Spatial variations in transport costs. Freight rates per ton are shown for general cargo from the six main ports in eastern New Guinea to a selection of inland and coastal locations, using the cheapest mode of transport available. Note the contrasts between the low unit costs but restricted operating area of coastal shipping (a) in contrast to the very high unit costs but flexible operating area of air freight (c). The fact that the road network into the mountainous interior is poorly developed is reflected in the cluster of high costs in (b). The six main ports are marked on each map by stars. [From H. C. Brookfield and D. Hart, *Melanesia* (*Barnes & Noble, New York, and Methuen, London, 1971*), p. 357, Fig. 14–9.]

Freight rates ($/ton)

- •<8   •8–16   •16–48
- •48–96   •96–144   •>144

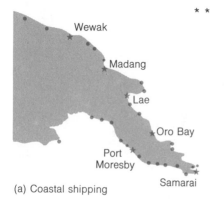

(a) Coastal shipping

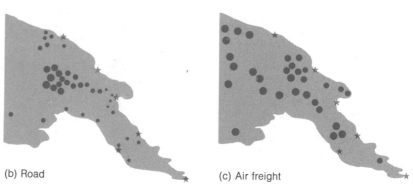

(b) Road

(c) Air freight

Spatial pricing

Three alternative pricing strategies are commonly used to recover the costs involved in transporting a product.

1. *Source pricing*. The price is established at the production point and the customer pays the transfer costs of moving the product. This system is also termed *f.o.b.* (free-on-board) pricing. Customers may be charged freight rates on the basis of the actual distance covered [Figure 15-14(b)], blanket zones [Figure 15-14(c)], or a uniform *postage stamp* rate in which the charge levied is unrelated to the distance involved.

2. *Uniform delivered pricing*. The price is the same for all customers regardless of their location. The producers pay all the transport costs involved in shipping their product but recover this by taking the average transport costs into consideration when deciding on the price at which they offer the product. This system is also termed *c.i.f.* (cost-insurance-freight) pricing.

3. *Basing-point pricing*. All production of a given commodity is regarded as originating from a single point, a uniform price is established for all sources regardless of their location, and customers pay the transfer costs from the basing point regardless of the actual location of the producer from which they purchased a product. The most famous case of basing-point pricing was the *Pittsburgh plus* system that operated for some time in the United States steel industry. Here customers were charged freight costs from Pittsburgh regardless of the location of the plant from which they actually purchased steel.

Pricing policies and their locational implications are discussed in D. M. Smith, *Industrial Location* (Wiley, New York, 1971), Chaps. 4, 5.

Second, we find that total costs have two elements: a terminal, or handling, element (unrelated to distance) and a delivery, or haulage, element directly related to distance. These two cost elements may vary for different modes of transport, as Figure 15-14(a) shows. A more realistic curve for distance costs is convex and nonlinear, indicating that transport costs increase, but at a decreasing rate, with distance [Figure 15-14(b)]; that is, the cost of moving the first 10 km may be much higher than the cost of moving a similar 10-km stretch between 150 and 160 km away. We can add further realism to our model by breaking up the continuous cost curve in Figure 15-14(b) into a series of steps, each related to a particular level of freight charge, as in Figure 15-14(c). An extreme case of this kind of blanket zoning is provided by the postal service, which charges uniform rates over very large areas, regardless of the distance a letter or parcel is transported. The costs of transport may be absorbed by either the producer or the consumer. (See the marginal discussion of spatial pricing.)

Thus it remains broadly true that the economic costs of connecting a city to agricultural zones and industrial-processing centers are a function of geographic distance. The twisting and blurring of freight rates, competition between modes of transport, and the like may be important for individual activities in particular locations. But the

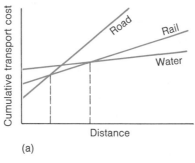

(a)

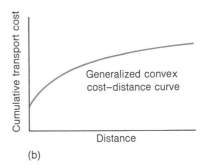

(b)

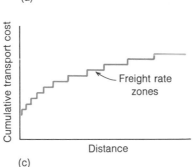

(c)

Figure 15-14. Transport costs and distance. Some of the factors that distort a simple linear relationship between costs and distance. (a) Competition between modes of transport. (b) Length-of-haul economies. (c) Freight-rate zoning.

overall picture of a city organizing the space around it—but with its influence gradually fading as we move further away from it—remains. In the next chapter we go on to explore the implications of this city-centered organization for other aspects of the world beyond the city.

Reflections

1. Examine the sequence of land-use zones in Thünen's isolated state (Figure 15-2). List the reasons why (a) truck farming is close to the city and (b) extensive cattle raising is carried out on the far periphery of the region. Do any of these factors affect location decisions in agriculture today?

2. Transport costs for the bulk movement of agricultural products continue to decline in real terms. What effect do you think this has on the changing importance of environmental resources (e.g., climate and soils) as a factor in the location of crop areas?

3. Give examples from your local area of any industrial concerns which appear to be (a) resource-oriented and (b) market oriented. Is there an "in-between" category? How can the firms in this in-between category be fitted into a locational theory?

4. To what extent do you agree with the statement "Transport costs are a function of distance"? Why?

5. Assume that a major oil field is discovered in your community. Where do you think the oil would be refined, and why there?

6. Review your understanding of the following concepts:
   (a) land-use rings   (f) Weberian analysis
   (b) the von Thünen model   (g) isotims
   (c) resource orientation   (h) isodapanes
   (d) market orientation   (i) space-cost curves
   (e) weight loss

One step further . . .

The classic theory of ring formation by J. H. von Thünen has been translated in
   Hall, P. G., Ed., *Von Thünen's Isolated State* (Pergamon, London, 1966).

Good accounts of its role in the structuring of land use in rural areas is given in
   Chisholm, M. D. I., *Rural Settlement and Land Use* (Pergamon, London, 1966), Chaps. 4 and 7.

Changing patterns of agricultural land use in intercity areas are described in
   Gregor, H. F., *Geography of Agriculture* (Prentice-Hall, Englewood Cliffs, N.J., 1970) and
   Morgan, W. B., and R. C. Munton, *Agricultural Geography* (Methuen, London, 1971).

A modern view of industrial location is given in

Smith, D. M., *Industrial Location: An Economic Geographical Analysis* (Wiley, New York, 1971) and

Estall, R. C., and R. O. Buchanan, *Industrial Activity and Economic Geography* (Humanities, New York, 1961),

and a useful selection of background readings is provided in

Karaska, G. J., and D. F. Bramhall, Eds., *Locational Analysis for Manufacturing: A Selection of Readings* (M.I.T. Press, Cambridge, Mass., 1969).

For an excellent discussion of locational decision-making in the context of resource use, see

Abler, R., J. S. Adams, and P. Gould, *Spatial Organization* (Prentice-Hall, Englewood Cliffs, N.J., 1971), Chap. 12.

Research on agricultural and industrial geography is generally reported in the regular geographic journals, especially *Economic Geography* (a quarterly). Understanding developments in locational theory now demands considerable mathematical competence; browse through the *Journal of Regional Science* (a quarterly) to get some idea of what lies beyond the highly simplified models presented in this chapter.

Chapter 16

Worlds Beyond the City, II:

MOVEMENTS AND PATHWAYS

To arrive at a clear decision on these questions, let us take familiar examples, but set them out in geometrical fashion.

—JOHANNES KEPLER
The Six-Cornered Snowflake (1611)

In this second chapter on the worlds beyond the city, we move away from agricultural zones and industrial centers to the lifelines that hold all three together. Transport links of all kinds—from TV transmissions to crude-oil supertankers—form an essential system by which cities and regions are maintained and grow.

This dependence of the city on links to the world beyond has roots which run back well beyond the modern period. Even in the most primitive of early urban settlements, trade and exchange were the distinctive features which allowed urbanization to emerge. What distinguishes primitive from modern systems of exchange is the intensity of exchange, its product range, and its spatial reach, rather than its essential nature. In this chapter we consider two aspects of this exchange. We look first at the questions of how and why exchanges between city and region take place and report on the regularities that geographers have discerned in these exchanges. Second, we examine the spatial structure of exchanges and the networks of routes that have been erected to facilitate them. As in Chapter 15, we link our discussion to the names of individuals whose concepts have thrown light on these geographic problems.

16-1 | NEWTON AND INTERCITY FLOWS

To raise the name of Sir Isaac Newton, the seventeenth-century English mathematician, in a geography text requires some explanation. Like many men of genius, his ideas in one field have been later found to be productive in others. Specifically, Newton's laws of gravitation have been found to throw light on geographers' understanding of the way in which flows occur between cities. This does not mean that you and I are swept along between cities like molecules in an "urban gravitational field." It does signify that the trillions of telephone messages, billions of freight-car journeys, or millions of aircraft movements that link the world's galaxy of settlements show a tendency, taken as a whole, to move in a way not unlike Newton's physical laws would predict.

Types of flow

We have already mentioned three different kinds of intercity flows: telephone calls, freight cars, and aircraft movements. Just what sort of flows do we have in mind? (See Table 16-1.) We can draw a broad distinction between *transport* flows and *communication* flows. Transport involves the physical movement of something, be it people or iron ore, between two places. This movement can take place through a series of different transport modes, each with a specific set of advantages and disadvantages. These are summarized in Table 16-1. Thus if

Table 16-1 The relative advantages of different modes of transport

Mode of transport	Principal technical advantages	Use
Railroads	Minimum resistance to movement, general flexibility, dependability, and safety	Bulk-commodity and general-cargo transport, intercity; of minimum value for short-haul traffic
Highways	Flexibility, especially of routes; speed and ease of movement in intraterminal and local service	Individual transport; also transport of merchandise and general cargo of medium size and quantity; pickup and delivery service; short-to-medium intercity transport; feeder service
Waterways	High productivity at low horsepower per ton	Slow-speed movement of bulk, and low-grade freight where waterways are available; general-cargo transport where speed is not a factor or where other means of transport are not available
Airways	High speed	Movement of any traffic where time is a factor—over medium and long distances; traffic with a high value in relation to its weight and bulk
Pipelines	Continuous flow; maximum dependability and safety	Transport of liquids where total and daily volume are high and continuity of delivery is required; potential future use in movement of suspended solids

SOURCE: W. H. Hay, *An Introduction to Transportation Engineering* (Wiley, New York, 1961), p. 283, Table 8-5.

we wish to move large cargoes of a bulky commodity, then slow-moving barges are very cheap (on a per tonnage basis) and may be preferable. Conversely, aircraft make up for their very high costs per ton by their very great speed and freedom from the environmental barriers set by mountain, ocean, or icecap. The relative advantages of the various modes have not remained constant; indeed, only two of the five modes listed in Table 16-1 would have been available for intercity flow 150 years ago. The current trends in both passenger and freight flows are toward more emphasis on highways and airways.

More striking than the changes in individual transport modes has been the much faster increase in communication flows. Unlike transport, communication flows do not involve the physical movement of an element between places. Communication is the sharing of information. Although short-range communication is as old as man himself, most of the mass communication now flowing between cities is the product of the last two centuries of technology. The nineteenth century was characterized by the inventions and rapid spread of "wire" communication systems. Thus Washington and Boston were the first two cities to be linked by a commercial telegraph line (1844); Europe and America were linked by submarine telegraph cable (1858). After Bell's Boston experiments in 1876, the telephone system spread slowly. By the century's end, Chicago and New York had been linked. The twentieth century saw breakthroughs in "wire-less" communication systems—radio in the 1910s, television in the 1930s, satellite communications in the 1960s. Intercity communication using these new modes has been increasing exponentially at rates which make the expansion of world population look slow. For example, the volume of intercity telephone calls is now doubling every decade over most of the world.

We have already commented in this section of the book on the way in which the changing costs of flows have led to two contrary spatial movements—an *implosion* of major cities at the intercity level, but an *explosion* of suburban sprawl at the intracity level. (Compare Figure 13-7 with Figure 13-13). This drastic change in the geography of the world's leading cities is closely linked to the innovations in transport and communications reviewed above. New airline services or telex links tend to be first established between pairs of cities where the demand is greatest, so reinforcing the already commanding position of the leading cities. To understand the pattern of urbanization it is therefore critical to look first at the spatial pattern of existing flows.

Spatial patterns of flows

If we map the origin and destination of flows, we find that most moves are over a short distance. As an example, let us take the hundreds of heavy trucks that roll down the freeways from Chicago bound for other parts of the United States. Most unload their contents a few kilometers from the city; relatively few move a long distance. In Figure 16-1(a) we plot the decrease in traffic with distance from Chicago out to about 650 km (400 mi). Notice that the diagram is drawn so that both the amount and the distance of the flow are plotted not on a linear scale, but on a logarithmic one.

Plots of similar flows on much larger geographic scales have a rather similar pattern. Figure 16-1(b) presents the pattern of rail-freight flows within the United States, which fall off regularly out to about 2,400 km (1500 mi). Figure 16-1(c) shows the pattern of world shipping out to a distance of 20,000 km (12,500 mi). Similar patterns can be found on geographic scales from the world level right down to the level of local kindergarten districts.

We can generalize about these *distance-decay,* or *distance lapse-rate, curves* by simply stating that the degree of spatial interaction (flows between regions) is inversely related to distance; that is, near regions interact more intensely than distant regions. The general form of this rule has been firmly established since the 1880s. However, the exact form of the relationship between distance and interaction has been difficult to pin down. On a graph with arithmetic scales, plotting distance against interaction produces a J-shaped curve in which flows decrease rapidly over shorter distances and more slowly over longer distances. Using logarithmic scales on both axes commonly yields an approximately linear relationship, and various alternative mathematical functions can be used to describe such forms. The results of Swedish work on migration indicate that spatial interaction is inversely related to the square of the distance between *settlements,* but this is an approximation of variable empirical findings. (See the marginal discussion of distance-decay curves.)

The gravity model

As early as 1850 observers of social interaction had noted that flows of migrants between cities appeared to be directly related to the size of the cities involved and inversely proportional to the distance separating them. By 1885 the British demographer E. G. Ravenstein had incorporated similar ideas into elementary "laws" of migration. Although the

Figure 16-1. The decay in spatial interactions with distance. (a) Truck trips around Chicago. (b) Railway shipments in the United States. (c) World oceangoing freight. [Data from M. Helvig and G. Zipf. Adapted from P. Haggett, *Locational Analysis in Human Geography* (St. Martin's Press, New York, and Edward Arnold, London, 1965), p. 34, Fig. 2–2.]

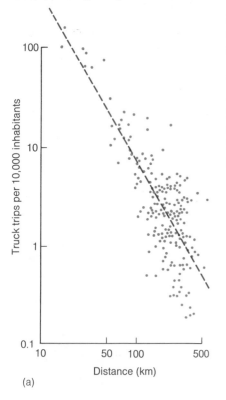

(a)

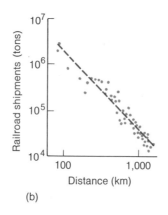

(b)

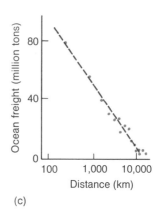

(c)

Distance-decay curves

Consider Figure 16-1, in which spatial interaction falls off with distance. One of the simplest and most common ways of describing curves that relate flows and distance is with a *Pareto function*:

$$F = aD^{-b}$$

where F = the flow, D = the distance, and a and b are constants. Geographers are especially interested in the value for the constant b. Low b values indicate a curve with a gentle slope with flows extending over a wide area, whereas high b values indicate a sharp decrease with distance so that flows are confined to a limited area. This formula was used extensively by a group of Swedish geographers in studies of migration between regions on a large variety of geographic scales as far back as the nineteenth century. Their findings showed b values going from as low as -0.4 to as high as -3.3. The mean value for all the studies was just below -2 (in fact, -1.94). This figure would suggest that

$$F = aD^{-2}$$

which we can rewrite as

$$F = a \frac{1}{D^2}$$

Apparently, spatial interaction falls off inversely with the square of distance. That is, the size of a flow at 20 km (12 mi) is likely to be only one-quarter of that at a 10-km (6-mi) distance. This *inverse-square* relationship is analogous to that used by physicists in estimating gravitational attraction.

specific term *gravity model* did not appear until the 1920s, it is clear that nineteenth-century workers were drawing on the relationships formalized by Sir Isaac Newton in his law of universal gravitation (1687), which states that two bodies in the universe attract each other in proportion to the product of their masses and inversely with the square of their distance. Gravitational concepts were specifically introduced by W. J. Reilly in 1929 in discussing the ways in which trade areas are formed. Reilly's ideas were subsequently expanded by researchers concerned with predicting flows in applied fields like highway design or retail marketing studies.

We can roughly estimate the size of flows between two regions by multiplying the mass of the two regions and dividing the result by the distance separating them. Thus a flow of 6 units would be produced by two regions with masses of 4 and 3 units, respectively, that are 2 units of distance apart. But what do we mean by "units" of mass and "units" of distance?

Mass has been equated with population size in many gravity studies. Information on population is easy to find, and we can readily estimate the size of most population clusters from census figures. However, population data may conceal significant differences between regions that affect the probability of spatial interaction, and the use of some system of weighting has been urged to take these differences into account. Economist Walter Isard has argued that just as the weights of molecules of different elements are unequal, so too the weights assigned to different groups of people should vary. Weights of 0.8 for population in the Deep South, 2.0 for population in the Far West, and 1.0 for population in other areas of the United States (reflecting regional differences in travel patterns) have been suggested. Multiplication of the population of each area by its mean per capita income is another possible way of refining our measuring stick for mass.

As we have already seen in Section 15-3, *distance* can be measured in several ways. The conventional measure in gravity models is simply the straight-line or cross-country distance between two points. In commuting studies, travel time rather than miles or kilometers may be an appropriate measure, as it may take as long to go a short way in urban areas as to go a long way in rural areas. When different forms of transport are available, distance may be measured in terms of the ease and cost of movement also. Fares (for people) and terminal costs and delivery charges (for goods) may be taken into account.

A simple example of the use of the gravity model to estimate flows between four cities is given in Figure 16-2. If you follow the sequence

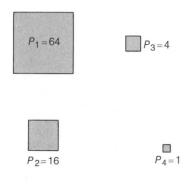

(a) Measure of population (P)

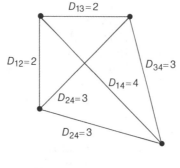

(b) Measure of distance (D)

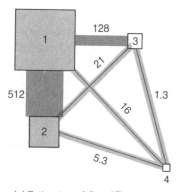

(c) Estimates of flow (F)
using distance

$$F_{ij} = \frac{P_i P_j}{D_{ij}}$$

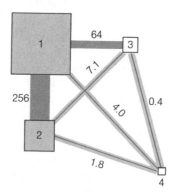

(d) Estimates of flow (F)
using distance squared

$$F_{ij} = \frac{P_i P_j}{D_{ij}^2}$$

Figure 16-2. A gravity model of flows between centers. Population size (a) and distance (b) are used to estimate the spatial interaction between four centers. In (c) the actual distance between cities and in (d) the square of the distance between cities is used to estimate the flow. Note in (b) that D_{12} refers to the distance between city 1 and city 2, and so on. Since the formula in (c) and (d) is general and refers to flows (F) between *any* two pairs of cities, we use the general term F_{ij} to refer to flows between city *i* and city *j*. The same general terms *i* and *j* are used in the rest of the formula so that the population of the two cities is P_i and P_j and the distance between them is D_{ij}. Check to see that you understand how the figures in (c) and (d) were achieved. For example, flow F_{14} in (d) is 4.0. This is given by multiplying the population of cities 1 and 4 and dividing this product by the square of the distance between them, i.e., $(64 \times 1) \div (4 \times 4) = 4.0$.

of maps carefully and refer to the caption to check on terms used, then you should see how simple the arithmetic is. You may be puzzled that two separate estimates of flow are given, one using distance [Figure 16-2(c)] and one using distance squared [Figure 16-2(d)]. Much of the work of geographers has been concerned with estimating just how distance should be measured and how it should be blended into the formulas. It turns out that, although distance squared is the better of the two, we have to use considerably more sophisticated formulas if we are to make estimates of acceptable accuracy. These lie outside the scope of this book but will form an important part of any future courses you select in this field.

The Ullman model

A different approach to the study of flows between regions is to ask why they occur. We can begin to answer this question by inverting it, or by trying to define the conditions under which flows would *not* occur. For example, if travel between the regions were expensive or if each region were highly self-sufficient, then we would expect rather

Gravity models: spatial interaction

Assume two regions, region 1 and region 2, spatially separated by an intervening distance. Spatial interaction between the two regions can be estimated by a gravity model of the form

$$F_{12} = a \frac{M_1 M_2}{D_{12}{}^b}$$

where
F_{12} = the flows between region 1 and region 2,
M_1, M_2 = the mass of the two regions (mass may be equated with population),
D_{12} = the distance between the two regions,
a = an empirical constant, and
b = a distance exponent (assumed to have a value of 2.0 in the original gravity model).

Thus, if we assume two cities of 1000 people each, 10 km (6 mi) apart, with $a = 1$ and $b = 2$, then the total flow between the two cities would be 10,000 units; at 20 km (12 mi) apart, the flow would fall to 2500 units. The units of flow are arbitrarily determined by the definitions of M and D.

Values for the constants a and b can be estimated empirically by studying regional situations in which flows (F) as well as population (M) and distance (D) values are known. For an extension of gravity models see W. Isard, *Methods of Regional Analysis* (The M.I.T. Press, Cambridge, Mass., 1960), Chap. 11.

The concept of gravity models can be extended to the calculation of potential population surfaces. Assume a set of towns (1, 2, 3, . . . , n) each of a known population size ($M_1, M_2, M_3, . . . , M_n$). Then the population potential for the first town is

$$P_1 = a \frac{M_1}{(\frac{1}{2}D_{1.})^b} + a \frac{M_2}{D_{12}{}^b} + \cdots + a \frac{M_n}{D_{1n}{}^b}$$

for the second town is

$$P_2 = a \frac{M_1}{D_{21}{}^b} + a \frac{M_2}{(\frac{1}{2}D_{2.})^b} + \cdots + a \frac{M_n}{D_{2n}{}^b}$$

for the third town is

$$P_3 = a \frac{M_1}{D_{31}{}^b} + a \frac{M_2}{D_{32}{}^b}$$
$$+ a \frac{M_3}{(\frac{1}{2}D_{3.})^b} + \cdots + a \frac{M_n}{D_{3n}{}^b}$$

and so on, where P_1 = the population potential of town 1
D_{12} = the distance between town 1 and town 2
$D_{1.}$ = the distance between town 1 and its nearest town
a = an empirical constant
b = a distance exponent (assumed to have a value of 2 in the original gravity model)

The nearest-neighbor component (*) is introduced so we can take into account the effect of the population of each town on itself. The potential values ($P_1, P_2, P_3, . . . , P_n$) determine the flow potential for each town. Potential maps are usually presented by drawing contour lines of equal potential expressed in percentages of the town with the highest potential. For examples of their use in geographic research see W. Warntz, *Macrogeography and Income Fronts* (Regional Science Research Institute, Philadelphia, 1965).

little to be exchanged. American geographer Edward Ullman has systematized these notions in a useful model of spatial interaction based on three factors: regional complementarity, intervening opportunity, and spatial transferability. We can illustrate Ullman's model using his study of interstate flows of lumber in the United States (Figure 16-3).

The first factor on which his model is based, *regional complementarity*, is a function of the resources available in any particular region. In order for regions to interact, there must be a supply or surplus of resources in one region and a demand or deficit in another. Thus, in Figure 16-3 shipments of forest products from Washington to the southeast states are low partly because of the easy availability of forest products in each; conversely, flows of forest products to New York and Pennsylvania are heavy, despite the long distance, because of the high demand there and the small size of their own forest area.

Complementarity will, however, generate flows between pairs of regions only if no *intervening opportunity* for a flow occurs—that is, if there are no intervening regions in a position to serve as alternative sources of supply or demand. Seventy years ago little lumber moved from Washington to the Northeast because the Great Lakes region provided an alternative and intervening source of supply.

The third factor in Ullman's model, *transferability*, refers to the possibility of moving a product. Transferability is a function of distance measured in real costs or time, as well as of the specific characteristics of the product. Table 16-2 outlines the relationship between

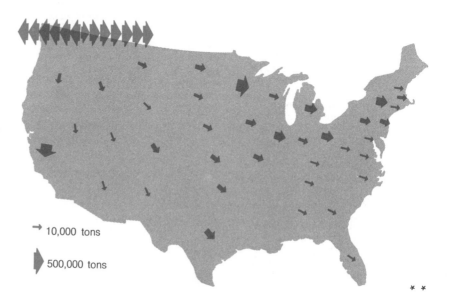

Figure 16-3. Interregional freight flows. Width of the arrows is used to show the volume of forest products moved by rail from Washington to other states in a single year. The row of arrows centered over the state of Washington itself (upper left) shows that most forest products moved very short distances. [From E. L. Ullman, in W. L. Thomas, Jr., Ed., *Man's Role in Changing the Face of the Earth* (University of Chicago Press, Chicago, 1956), p. 869, Fig. 162. Copyright © 1956 by The University of Chicago.]

→ 10,000 tons

500,000 tons

Table 16-2 The relative transportability of three lumber products

	High-value veneer logs	Medium-value pulpwood	Low-value mine props
Value (dollars/ton)	150	20	5
Average length of longest haul by railroad (km)	640	32	8

SOURCE: W. A. Duerr, *Fundamentals of Forestry Economics* (Copyright © 1960 by McGraw-Hill, New York), p. 167, Table 21. Reproduced with permission.

the specific value of three types of lumber products (measured in dollars per ton) and the length of shipment. Local products may be substituted for products that are difficult to transfer in the same way that intervening areas of supply or demand substitute for more distant ones.

16-2 | KOHL AND INTERCITY NETWORKS

In the previous section we assumed that flows between cities and between a city and its region occurred across a uniform plane. For wave transmissions via radio and TV, this is broadly true. Other flows, however, are generally confined to a series of channels, or routes (Figure 16-4). Geographers see this delicate filigree of transport networks as the arterial and nervous system of regional organization, along which flow signals, freight, people, and all the other essential elements that allow the structure to be maintained. Do these networks have a characteristic spatial structure? What governs their location? Can they be improved?

Networks as regional lifelines

Although transport systems form an essential and permanent feature of the economic landscape, the early locational theorists like Johann von Thünen and Alfred Weber had little to say about them. Yet as early as 1850 a German geographer, J. G. Kohl, created a series of branching networks to serve the settlements in his idealized city-region [Figure 16-5(a)]. His ideas were taken up by Walter Christaller nearly a century later in his own scheme for a system of cities [Figure 16-5(b)]; since then, these ideas have been extended by other workers.

Some features of both the Kohl and Christaller schemes deserve notice. First, the transport networks are *hierarchic* in that they consist of a few heavily used channels and many lightly used feeders, or tributary channels. Like the city systems they serve, the segments of the transport system form an inverse distribution of size with frequency.

A second feature also is analogous to river systems, for the transport network has a branching structure in which the angle of branch-

(a)

Figure 16-4. Contrasting modes of spatial interaction. [Photographs (a) from the Bettmann Archive, (b) Rotkin, P. F. I., and (c) courtesy of Port of N.Y. Authority.]

(b)

ing is intimately related to the flow. The rule governing this phenomenon is familiar. The angle of departure between the branch and the main stem is inversely related to the size of the branch. As the flow on the branch diminishes in relation to the main stem flow, the angle of departure becomes bigger. There is a precise interaction between the shape of the system and the work it has to do.

The number of major routes emerging from each city also has been the subject of research. For inland centers the most frequent number of key routes radiating from a city is six; few cities have less than three or more than eight. These figures are about what we would expect from our earlier discussion of the spacing and location of cities.

Mathematician Martin Beckmann has shown that if a region has a uniform population density and the costs of building a route are everywhere the same, the ideal transport system has a hexagonal honeycomb pattern [Figure 16-6(a)]. It is assumed in this system that both the origins and the destinations of flows are evenly spread over the region. Beckmann's system is a complicated one and involves some advanced theory that need not concern us here. What is interesting is that the basic honeycomb pattern can be modified to be more realistic.

For example, suppose we retain the idea of a uniform population but assume that we are concerned simply with linking this population with a single source (say, a central place in terms of Christaller's model). In this instance, all the destinations are uniformly spread

(c)

(a) Kohl 1850

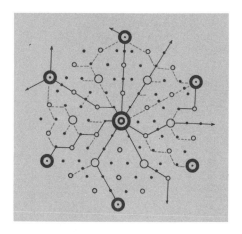

(b) Christaller 1933

Figure 16-5. Transport networks for theoretical settlement systems. Three alternative schemes proposed between 1850 and 1956. Kohl's network (a) serves a Thünen-like isolated state. Only the upper half of the circular state is shown and the broken lines subdivide it into identical segments. The Christaller scheme (b) shows traffic routes in a $K = 3$ landscape. (Check back to Figure 14-9 (a) to refresh your memory.) In this case the networks are not symmetrical about the central city: compare the upper right quadrant (a wealthy region in which long-haul traffic prevails) with that of the lower left quadrant (a poor region in which short-haul traffic prevails). Note the contrasts between the straight routes of the former and the zigzag routes of the latter. Minor routes are shown by broken lines. Isard's network (c) has two centers each of which is surrounded by zones of decreasing population density. Boundaries of the complementary regions are shown as polygons which increase in size with distance from the two centers. [From P. Haggett and R. J. Chorley, *Network Analysis in Geography* (St. Martin's Press, New York, and Edward Arnold, London, 1969), p. 125, Fig. 3–11.]

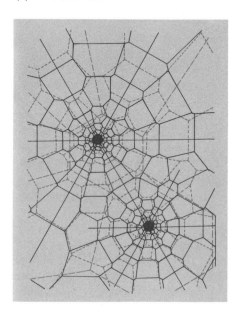

(c) Isard 1956

over the whole region, but the origin is a single point. The best type of transport network is a symmetric honeycomb with holes placed in such a way that the system remains simply connected, as in Figure 16-6(b). By "simply connected" we mean there are no loops in the system and it still has a basic branching form like a tree. This tree-like form is important because it helps us to bridge the gap between

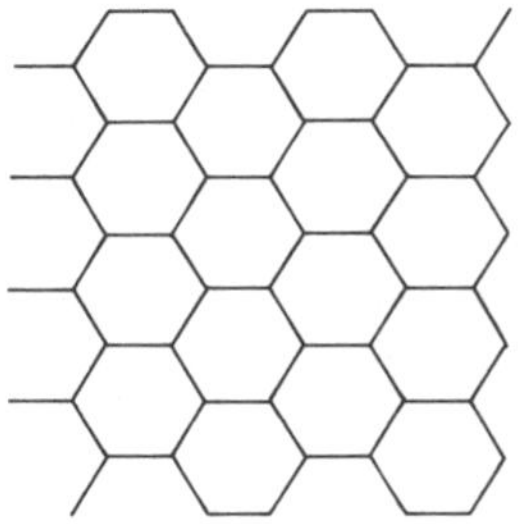

(a) Uniform origins

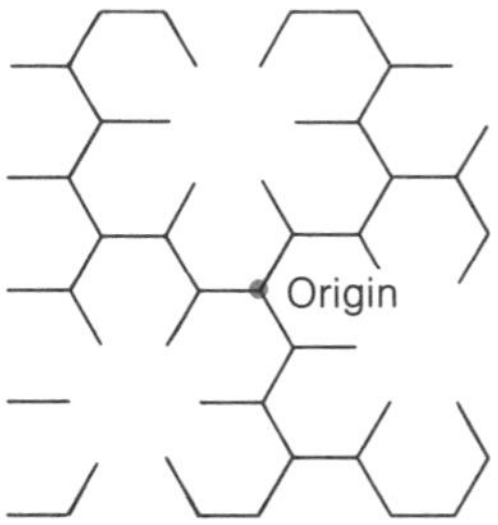

Origin

(b) Single origin

Figure 16-6. Optimal transport networks. These networks assume a uniform population and homogeneous regions. In (a) there are multiple origins and multiple terminals for flows. In (b) there are a single origin and multiple terminals.

Beckmann's honeycombs and Kohl's branching patterns [Figure 16-5(a)]. The two schemes differ in two ways: Kohl's region is bounded (in fact, it is circular) and has a higher density of settlements in the center than at the periphery, whereas Beckmann's region is continuous (no boundary is specified) and the population density is uniform throughout. If we modify the Beckmann model by allowing a higher population density (and thus, smaller honeycombs) near the center and adding a circular boundary the model has a spatial form very similar to that of Kohl's network. One network is then simply a special case of the other. The missing link between the two is provided by Isard's landscape [Figure 16-5(c)] where a honeycomb system of transport links is developed about a pair of centers.

Can geographers find regions in which to test their models of transport networks? Certainly the number of areas of entirely new settlements is fairly small. The pattern of settlement in a Dutch polder (i.e., major area of land reclaimed from the sea) may be the nearest approximations we can find to the pattern that would result if a new, empty, and rather uniform landscape were settled all at once rather than slowly over a historical period. The scheme actually adopted there is rectangular, not unlike the pattern of roads laid out in the new territories of the United States during the 1800s. Other small agricultural areas like plantations also have been designed according to the principle of the Beckmann model.

Network design is important when a new system of roads is superimposed on existing ones. The interstate highway system in the United States and the new highway system of the United Kingdom typify the locational compromise that must be made between the cost of building networks compared with the costs of using them. We can illustrate this need for compromise by considering Figure 16-7, which presents a simple network designed to connect only five towns. If we design the network to minimize the costs to the user, we will make as many of the links as direct as possible. [See direct link AB in Figure 16-7(a).] But if we design the network to minimize building costs, we will have a different kind of pattern. In this second case the link AB will be much longer, but the *total* length of the network will be much smaller [Figure 16-7(b)]. Comparisons with actual transport systems reveal an evolution from the first to the second type as flows increase. Look at a historical atlas, and compare the changing structure of the United States railroad network during different periods of its growth. Indeed, there is still a strong contrast between the sparse rail network in the West, where the costs to the builder were very important and the denser rail network in the East, where greater intercity flows made the costs to the user more significant.

Figure 16-7. The shortest networks connecting five urban centers. Two alternative solutions are shown. In (a) the costs of the system to the user are most important and the interchanges are located within the cities. In (b) the cost of building the network is more important and the interchanges are located in nonurban areas. [After W. Bunge, *Lund Studies in Geography C,* No. 1 (1962), p. 183, Fig. 7-10.]

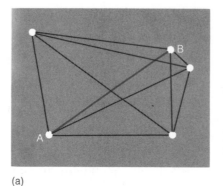

(a)

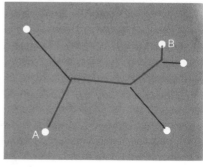

(b)

Networks as graphs

In our discussion of urbanization processes we saw how changes in the relative accessibility of cities had important repercussions on their relative growth. (Take another look at Section 13-1 to refresh your memory on this point.) So, if we add a new transport link to a regional network, we should expect it to affect the relative accessibility of all the cities that are connected to it. Let us consider a specific example. Figure 16-8 shows the road network of the northeastern part of Ontario, Canada. Each of the 37 nodes represents either a settlement with at least 300 inhabitants or a main highway intersection. The road network, extending from Sudbury to Sault Sainte Marie, is connected to road systems in the remainder of Canada, but these external links are so few that we can treat the network as a closed system. Now let us suppose that seven new road links are proposed (links *a* through g in Figure 16-8). Which links will have the greatest impact on the total accessibility of points within the area to one another? Also, what local effect will each new link have on the *relative* accessibility of individual nodes within the network?

One way of answering these questions is to turn to a branch of mathematics that treats networks on the most primitive topological level. Topology is a branch of geometry that is concerned with the quality of "connectivity," that is, whether objects are or are not connected in some way. Its earliest and most famous geographical application is shown in Figure 16-9. Note that Euler was not concerned with distance or direction (the usual concern of geometry) but simply with whether a particular path through a network—the seven bridges connecting the different parts of the city of Königsberg—was possible or not. Euler's puzzling over this problem was to grow into *graph theory,* a branch of mathematics whose concepts are proving of increasing in-

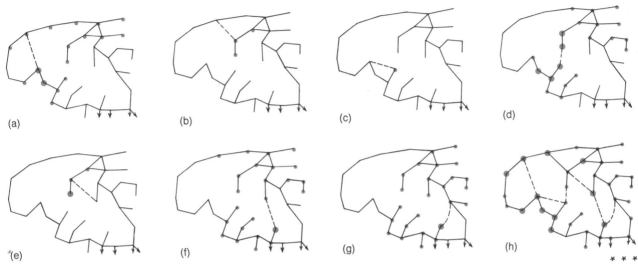

(a)   (b)   (c)   (d)

(e)   (f)   (g)   (h)

* * *

Figure 16-8. The impact of new links on accessibility. The maps show the impact of projected road links (a through g) on the accessibility of nodes in eastern Ontario, Canada. Points on the network which are expected to benefit by the improved connections are shown by dots: large dots indicate major improvements, small dots only minor improvements. The final map (h) shows the combined effect of all new links. [After I. Burton, *Accessibility in Northern Ontario,* unpublished paper, 1963, Fig. 1.]

terest to geographers studying regional networks. To use graph theory, we must reduce networks to graphs. This reduction involves throwing away a great deal of information about flows and characteristics of routes, but retaining the essential spatial factors of networks, nodes, and links. *Nodes* are the termination or intersection points of a graph.

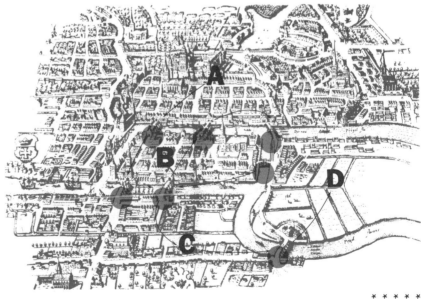

Figure 16-9. The Königsberg problem. Sometimes apparently trivial spatial problems can have profoundly important implications. Mathematician Leonhard Euler (1707–1783) puzzled over why it was impossible for the citizens of this Prussian city to visit four areas (A, B, C, and D) and cross all its seven bridges (marked by colored dots) without recrossing at least one bridge. His later studies of the structure of networks laid the groundwork for a major branch of mathematics, graph theory, which has proved of importance in designing computer circuits and is of direct use to geographers analyzing the spatial structure of networks. [Drawing courtesy of The New York Public Library, The Astor, Lennox, and Tilden Foundations.]

* * * * *

Connectivity in graphs

Consider a simple graph consisting of five nodes (A, B, C, D, and E) connected by five links (shown by solid lines).

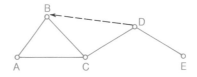

We can summarize the information in the graph in a connectivity matrix. Here the distance between pairs of nodes is expressed in terms of the number of intervening links along the shortest path connecting them, as follows.

	To:	A	B	C	D	E	Row sum	Average path length
From:								
A		0	1	1	2	3	7	1.75
B		1	0	1	2	3	7	1.75
C		1	1	0	1	2	5	1.25
D		2	2	1	0	1	6	1.50
E		3	3	2	1	0	9	2.25
						Total	34	1.70

The row sum for each node provides a measure of its relative accessibility. Thus node C is the most accessible and node E is the least accessible. The grand total (termed the *dispersion value* of the graph) provides a measure of the graph's size in terms of all the paths within it. Dividing the row sums and the dispersion value by the number of positive values provides a measure of the average path length, which can be used in comparing one network with another.

The connectivity matrix in this simple case is symmetric about the diagonal of zeros because all five links are two-way. If we introduce a one-way link from D to B, it will disturb the symmetry. That is, DB will have a link value of only 1, but BD will have a link value of 2. The effect of the new one-way link is to slightly improve the network's connectivity. (The dispersion value falls to 33.) If we were to reverse the direction of the same link from B to D, the improvement in connectivity would be greater. For a further discussion of graphs including weighted as well as directed links, see K. J. Kansky, *Structure of Transportation Networks* (Department of Geography, Research Paper 84, University of Chicago, Chicago, Ill., 1963), Chap. 1.

They may be assigned values denoting their location, size, the traffic they can handle, and so on. Depending on the varying scale of the analysis, nodes may be whole cities or street intersections. *Links* are connections or routes within a network. Links also can be assigned values relating to their location, length, size, and capacity. Some information on the connectivity of graphs can be obtained by measuring the *average path length*. Path lengths are determined by the number of steps (or "hops") between pairs of nodes, moving one link at a time along the shortest path through the network. (See the marginal discussion of connectivity in graphs.)

Armed with this simple ruler, we can now go back to our Ontario road network and measure the impact of the proposed new links on accessibility within the system. For while each of the new links must bring some improvement and cut average path lengths in the network, some do this more effectively than others.

If we build the new links into the network one at a time, we can test just how much each one improves the connectivity of the system. Link *d* (from Folyet to Chapleau) does the most, cutting average path lengths in the network by 9.5 percent. Next best is link *a*. Note that both *a* and *d* short-circuit the network by joining the northern and southern halves. Other links, like *c*, are on the periphery of the network and their building has little overall effect.

How do individual cities benefit from the new links? This is measured by calculating the changes in the average path length for *each* city. Figure 16-8 shows the spatial pattern of improvements in connectivity, city by city, from each link proposed. Note that some links have a very localized benefit (e.g., *c* and *e*). In contrast, *d* brings the biggest gains in improved access to the center of the network; *g* brings less striking improvements, but spreads them rather equitably over almost all the eastern nodes. Some projects which are desirable on local grounds may not have the most beneficial effect on the whole network.

Graph theory provides only a first step in the analysis of transport systems. Links must be weighted to reflect the traffic they carry. We also must relax our implicit assumption that each link is just as costly to construct as all of the other ones. The tools needed for a full analysis of transport systems—cost-benefit ratios, spatial allocation models, and the like—are a bit complicated for an introductory course and are generally reserved for courses in transport geography. Graph theory does, however, serve to highlight the delicate balance between the urban system of cities and the regional transport net that links them and binds them together. Each new airline link, pipeline, highway, or shipping schedule tilts the balance of locational advantage this way and that, in a manner geographers must continue to monitor carefully if they are to be able to track—and eventually forecast (see Chapter 21)—changes in the world about them.

<table>
<tr><td>16-3</td><td>**THE CITY REGION AS AN ECOLOGICAL UNIT**</td><td>The real world is made up of an immensely complex mosaic of regions. As we have seen, geographers have attempted to make sense of this by devising formal systems of regions. There is, however, no single or generally accepted set of world regions. Rather, there are alternatives, each with certain strengths and weaknesses. The most successful types of regional units appear to be those whose spatial boundaries coincide most closely with ecological or socioeconomic systems.</td></tr>
</table>

We have already seen, in Part Three, that the cultural area is a possible candidate for the basic unit of regional organization. The cultural region is attractive for this purpose because it has a variety of spatial levels. It stresses the unique qualities of particular groups in

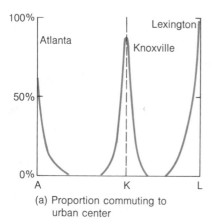

(a) Proportion commuting to
urban center

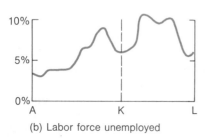

(b) Labor force unemployed

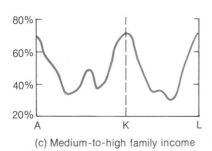

(c) Medium-to-high family income

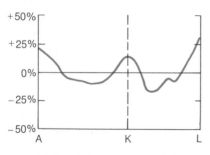

(d) Population change 1950–1960

certain environments and comes close to providing an accurate picture of the immense diversity of human groups. The cultural region suffers from three drawbacks, however, as a geographical module. First, the accurate definition of cultural regions depends on extremely detailed data, if we are to include in the cultural complex a full range of cultural elements. Often these data are simply not available. Second, there is considerable dispute about which indicators should be used to define cultural boundaries. Third, the cultural region emphasizes the unique qualities of human groups at a time when group mixing and the erosion of local character by a rather uniform urban–industrial technology is proceeding rapidly.

Many geographers have chosen to reject the cultural area as a basic unit and replace it with the idea of a "city region." By a *city region* we mean the area that surrounds a human settlement and is tied to it in terms of its spatial organization. (These central settlements may be smaller than cities measured by conventional standards, but the term "city region" is arbitrarily extended to include smaller nodal regions when it is appropriate to do so.) The spatial ties between the city and its surrounding area are essentially movements of people, goods, finance, information, and influence. Figure 16-10, a cross section of links between three cities in the eastern United States, shows how variations in one element of the city region (commuting levels) are reflected in other elements. Like an orchestral piece, the various elements amplify and modulate, but do not destroy, the city-centered pattern of organization.

The arguments for adopting the city region as the basic spatial unit are persuasive. A growing proportion of the world's population is concentrated in cities, and consequently man's organization of the globe is increasingly city-centered. Cities form easily identifiable and mappable regional units for which a flood of reasonably uniform statistical data has been available for the last century and a half. Further, city regions stress the comparability of different parts of the world, and thus encourage the search for general theories of human spatial organization. Finally, city regions are hierarchic. Like watersheds, they nest inside one another, and the city-region concept can be enlarged up to the world level or reduced down to the level of the smallest hamlet.

We have already seen in Chapter 5 the emphasis geographers place

Figure 16-10. City regions. Cross profiles from the city of Atlanta to Knoxville and Lexington, in the southeastern United States, use selected socioeconomic indicators to show the concentration of development about the three large urban centers. [From B. J. L. Berry, *Metropolitan Area Definition* (Bureau of the Census, Washington, D.C., 1968), Fig. 7.]

on the ecosystem in studying the natural world. Can ecosystems be extended to include men? Cambridge geographer David Stoddart thinks they can. He sees several advantages in extending ecosystems concepts, which are usually confined to plant and animal communities, to human communities. Ecosystems bring together man and his physical environment into a single framework, encouraging a monistic rather than a dualistic view of a region. Ecosystems are studied in terms of structure and function, structural cohesion being seen as a logical response to the cycling of material and energy. Finally, ecosystems have certain features in common with other systems, and the networks used to construct models of those systems, say by engineers or physicists, can be extended to ecosystems. Experiments in using electric circuits to construct models of biological systems are an example of how analogies might be developed.

Can we think of the city region as an ecosystem? Like watersheds, city regions need a constant flow of energy to maintain themselves. If we cut off the movement of people, freight, or funds into a city, it will stagnate; if we increase those flows, it will respond by growing in size. The city region is, like the watershed, in a state of balance with the forces that maintain and mold it. We have used the city region as a basic organizational unit in the last four chapters, but it is only one way of ordering the complexities of the real world. A complete integration of such regions with ecological and cultural regions still lies in the future.

Reflections

1. Define the three basic conditions needed for flows to occur in the Ullman model. Illustrate each, using flows into or out of your own state or province as an example.

2. What changes would you expect to occur in the flows between two cities if (a) both doubled in population or (b) the travel time between them was halved (e.g., by a bridge built to replace a slow ferry link)? Use Figure 16-2 to guide your calculations.

3. Sketch a map of the campus and the surrounding community and plot on it the paths followed by members of the class in getting to the building each day. Compare your results with Figure 16-5. How far does your network of paths resemble Kohl's branching tree? How are the paths of walkers, cyclists, bus riders, etc., different?

4. Gather data on any one pattern of spatial interaction (e.g., the number of guests in a local hotel, the number of visitors to a national park, or the number of students to a state university). Plot the observed number coming from each "source" area on the vertical axis of a graph and the expected number on the horizontal axis. For the expected number use a gravity model

with the relevant population of each source area divided by its distance from the hotel, park, or university campus. Comment on your findings.

5. Consider the three modes of transport in Figure 16-4. Sketch typical networks for each mode and convert to a graph. What differences can you see in the connectivity of the three types of networks?

6. Review your understanding of the following concepts:
   (a) distance-decay curves   (f) nodes and links
   (b) gravity models   (g) connectivity
   (c) regional complementarity   (h) average path length
   (d) intervening opportunity   (i) city region
   (e) transferability

One step further . . .

For an excellent but brief introduction to the topics covered in this chapter, see
> Taaffe, E. J., and H. Gauthier, *Geography of Transportation* (Prentice-Hall, Englewood Cliffs, N.J., 1973).

One outstanding example of empirical work by a geographer on the structure of interregional flows is
> Ullman, E. L., *American Commodity Flow: A Geographic Interpretation of Rail and Water Traffic Based on Principles of Spatial Interchange* (University of Washington Press, Seattle, Wash., 1957).

An authoritative review of both classical and modern work is provided in
> Olsson, G., *Distance and Human Interaction* (Regional Science Research Institute, Philadelphia, 1965).

For a geographic study that uses models of spatial interaction based on statistical mechanics and demands some knowledge of advanced mathematics on the part of the reader, see
> Wilson, A. G., *Urban and Regional Models in Geography and Planning* (Wiley, New York, 1974), esp. chap. 9.

The spatial structure of transport networks and the locational principles that determine their form are discussed in
> Haggett, P., and R. J. Chorley, *Network Analysis in Geography* (St. Martin's Press, New York, and Edward Arnold, London, 1969) and
> Kansky, K. J., *Structure and Transportation Networks* (University of Chicago, Department of Geography, Research Paper 84, Chicago, 1963).

For examples of the study of individual transport modes and the transport network of an individual region, see
> Sealey, K. R., *Geography of Air Transport* (Holt, Rinehart & Winston, New York, 1962), 2nd ed., and
> Gould, P. R., *The Development of Transportation Patterns in Ghana* (Northwestern University, Studies in Geography, 5, Evanston, Ill., 1960).

Recent applied geographic research on urban transport networks is presented in

> Horton, F., Ed., *Geographic Studies of Urban Transportation and Network Analysis* (Northwestern University, Studies in Geography, 16, Evanston, Ill., 1968).

Research on both spatial interaction and the structure of networks is reported regularly in the major geographic journals. The more advanced quantitative work is often presented in *Geographical Analysis* (a quarterly) and in the *Journal of Transport Economics and Policy* (also a quarterly).

Part Five

Interregional Stresses

In Part Five we examine the stresses that the regional structures described in Parts Three and Four place on contacts between regions. *Territories and Boundaries* (Chapter 17) probes the barriers man has thrown up to separate and delimit areas of ownership. We look at the basis of territorial organization, the factors that provide stability and instability, and the ways in which conflicts can be reduced. In *Rich Countries and Poor* (Chapter 18) we look at the most important territorial units—the modern nation states—and explore the vast differences in present prosperity and future prospects between them. Special attention is paid to trends over time and the disputed question of whether poorer countries are catching up with the rich, or whether inequalities are growing. Finally, in *Inequalities Within Countries* (Chapter 19) we follow the problem down the spatial scale. We note how governments intervene to try to adjust imbalances in the regional mosaic of wealth and poverty, and consider the issues raised by regional planning and the role of geographic concepts in guiding spatial policy decisions.

Chapter 17

Territories and Boundaries

My apple trees will never get across
And eat the cones under his pines, I tell him.
He only says, "Good fences make good
neighbors."

—ROBERT FROST
Mending Wall (1914)

The word "territory" has several alternative meanings in the English language, so it is useful at the outset to define its interpretation here. Territory is used in a legal sense to mean land belonging to a given sovereign state, such as the Northwest Territory of Canada. More specifically, it describes areas not yet granted full status in comparison with the other parts of the state. For example, Alaska was a territory of the United States, with an appointed governor, not an elected governor, until it became the 49th state of the Union in 1959.

Geographers, like biologists, use the term territory in a much more general sense, to indicate an area over which rights of ownership are exercised and which can be delimited or bounded in some way. We use the word "boundaries" to describe the limits of such territories. Sometimes ownership is formal and may be legally enforced. A householder may legally own the lot on which his house is built, and a nation-state may legally own its lands. At other times ownership may be unstable, and territories may be precariously held on to by displays of strength at the borders. Ornithologists studying the robin singing on the garden post at the edge of its territory and sociologists noting the gang member painting obscenities on a wall at the boundary of his turf are both recording displays of unstable territorial ownership.

Boundaries may be marked by things as fleeting or trivial as bird-songs or graffiti or by boundaries as fixed and patrolled as the Berlin Wall. The earth's surface is laced with an intricate network of boundaries. In this chapter we look at how geographers interpret such territorial limits, and how they work to draw up more equitable, efficient, and stable systems of territorial organization. We look first at the concept of territoriality and ask why territories occur. Secondly, we turn to some of the issues that are raised in partitioning the earth's surface at local, regional, and national levels. Thirdly, we ask how boundaries are actually demarcated on the ground.

17-1 CONCEPTS OF TERRITORIALITY

Territoriality is not a purely human trait. In this opening section of this chapter, we look at the evidence of this trait in other species and compare their behavior with the ways in which man himself stakes out his "home range" on the earth's surface.

Animal territories: Biological evidence

In Chapter 1, we drew attention to the way in which groups on the beach spaced themselves out, creating, in effect, local "family territories." (See the discussion of interpersonal space, Section 1-2.) At many points in this book we have had to recall that, for all his cul-

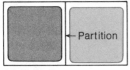

(a) Stage 1

(b) Stage 2

(c) Stage 3

Figure 17-1. Formal and informal limits. Experiments with mice in separate cages (a) show that when the partition is removed (b) the territory of each animal may change, but informal boundaries remain (c). The shaded areas indicate territories.

tural and technical uniqueness, man is still an animal species. To understand the intricate and formal way in which man organizes his territory, we need to begin by looking over the species fence at the simpler, territorial behavior of other animals. In doing so we shall need to draw on the findings of *ethology*, that branch of the study of animal behavior in a natural environment pioneered by European zoologists like Konrad Lorenz and Niko Tinbergen.

Let us begin with a simple experiment. If we keep a group of white mice in a single cage, but place wire partitions between each mouse, their movements are necessarily confined to their own sections. The partitions obviously act as physical barriers that prohibit movement outside individual sections. If we eliminate the internal partitions some days later, we find that, although the barriers have been removed, the mice usually continue to occupy their own sections (Figure 17-1). Eventually, the more active and dominant animals may increase their sections of the cage at the expense of the less aggressive animals.

The study of other animals in their natural, field setting confirms the cage findings. Many animal species "stake out" a specific area of space for their activities (e.g., feeding, breeding, or nesting). Territories will be defended very aggressively against other animals of the same species—sometimes by real fighting, but more often through highly ritualized "displays." Studies of birdsongs by Charles Hartshorne (not to be confused with his brother, geographer Richard Hartshorne, whose work is discussed in Chapter 22) show how this highly varied language plays an important marking role in territories.

Detailed maps of bird territories show distinctive spatial patterns (Figure 17-2). Note the distinction between discrete and overlapping territories for both isolated and gregarious species. Gregarious species of birds live in colonies, and the territories are those of the whole colony rather than the individual breeding pair. Figure 17-2(d) shows a special case, sea birds whose colonies may be confined to a few traditional nesting sites on islands and whose territories are overlapping areas of the sea.

While the facts that animals have territories is not in doubt, there is considerable debate over their meaning. Just why do territories occur? Two reasons appear the most likely. First, territories help to regulate population density and thus preserve an ecological balance with food supply. Those animals that cannot secure a territory for themselves are forced to migrate or starve. Second, territories ensure that the strongest members of a population (i.e., those able to obtain and to hold a territory) are the ones that breed and perpetuate the group. Because it forces out the weaker members of a population, territoriality may be an important mechanism in the natural-selection

Figure 17-2. Territories of animal populations. The maps show four types of "home ranges" for territorial animals. (a) Solitary animals with sharply bounded and discrete territories. (b) Solitary animals with somewhat overlapping ranges. Core areas around a nesting site for bird populations will be discrete and defended. (c) Gregarious animals in colonies with slightly overlapping group territories. (d) Gregarious animals with highly overlapping ranges—in this case, seabird populations with island nesting sites. [From N. C. Wynne-Edwards, *Animal Dispersion in Relation to Social Behavior.* Copyright © 1962 by Oliver & Boyd, Ltd., Edinburgh.]

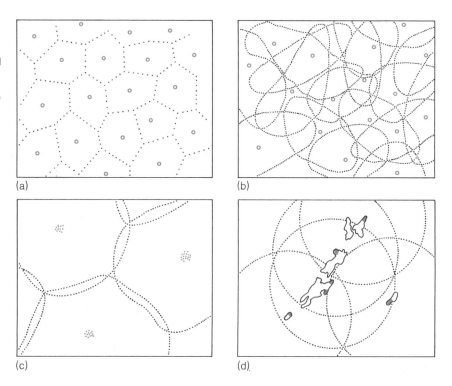

(a)

(b)

(c)

(d)

process. Note that in the special case of seabird colonies [Figure 17-2(d)], the competition is not for the marine feeding areas but for the few square inches of rock ledges in the traditional nesting areas.

Human territories: Similarities

But what have the seabird colonies to do with man? It would be as easy as it would be dangerous to draw facile comparisons between mice in cages and gangs in ghettos, or sororities in suburbia. In approaching the territorial behavior of the human animal, we choose to begin by looking at some points of similarity with the behavior of other animals, and then go on to review the differences.

Median-line models Let us replace our white mice with a scatter of pioneer farmsteads like those which existed during the colonial period of European overseas expansion. Each farm family clears the virgin forests around the homestead, tilling the most accessible land first and then gradually moving further afield as the family grows and more help is available. If we make simplifying assumptions which ensure that all families are the same size and have the same resources and that the land is everywhere of the same quality, what territorial pattern of farm boundaries will emerge?

Figure 17-3 shows the probable course of events. In the earliest stages, each farm can expand in isolation and its territory forms a circle. The untamed forest outside the homestead looks like pastry rolled out on a kitchen table, from which round biscuits have been cut. But as the closer farms expand, their boundaries meet and fence lines are staked halfway between them. In the later stages of farm expansion, only the last vestiges of forest remain and the farm boundaries look like those in Figure 17-3(c). Polygonal territories formed in this way are termed *Dirichlet polygons,* after the German mathematician who studied their geometric properties in the last century. They have the unique quality of containing within them areas that are nearer to the point around which they are constructed (in this case, the farmstead) than to any other points. Essentially, each side of the polygon is a median line, drawn at right angles to the line joining two farmsteads at its halfway point.

Clearly our two simple assumptions — that there is a random scatter of farmsteads and each farmer clears the land nearest to him — produce a territorial division of some complexity. It bears some resemblance

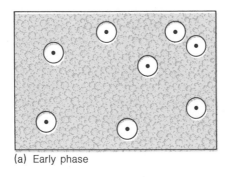
(a) Early phase

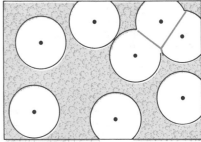
(b) Middle phase

Figure 17-3. The evolution of farm boundaries. The maps show stages in the idealized settlement and colonization of a forest area. The method of drawing boundaries is as follows. First, lines join a given farmstead to each adjacent farmstead; second, each of these interfarm lines is bisected to obtain the median (or midpoint) of the line; third, from this median point a boundary line is drawn at right angles to produce a series of polygons. This type of territorial division is based on two assumptions — that there is a random scatter of farmsteads and that each farmer clears only the land nearest to him.

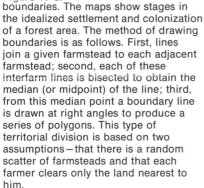
(c) Late phase

 Uncleared forest

• Farmstead

———— Farm boundaries

Estimating boundaries by gravity models

We can estimate the location of a boundary line between the market area of two centers by using a gravity model like those encountered in Section 16-1. Assume that we have two cities (city 1 and city 2), each with a specific market area (M_1 and M_2) and separated by distance D_{12}. We can estimate a break-point (B_2) in units of distance from the second city as

$$B_2 = \frac{D_{12}}{1 + \sqrt{\dfrac{M_1}{M_2}}}$$

If we now assume a simple case in which the two cities are 12 km (7.5 mi) apart and both have the same size market ($M_1 = M_2 = 10$), then

$$B_2 = \frac{12}{1 + \sqrt{\dfrac{10}{10}}} = 6 \text{ km (3.7 mi)}$$

That is, the boundary line occurs halfway between the two equal-sized centers. If we make the two cities unequal in size ($M_1 = 20$; $M_2 = 5$), then

$$B_2 = \frac{12}{1 + \sqrt{\dfrac{20}{5}}} = 4 \text{ km (2.5 mi)}$$

In this second case the boundary is displaced away from the halfway position in the direction of the smaller center.

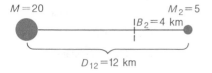

$M = 20$ $M_2 = 5$

$|B_2 = 4$ km

$D_{12} = 12$ km

to the patterns of villages and forests in Gambia, West Africa, that we met in our study of the Thünen model of regional divisions. (See Figure 15-7.) If the farmsteads had been arranged in a regular triangular lattice, then the farm boundaries would have been hexagonal, just like those of Christaller's complementary regions.

Thus we could regard some human territories simply as median-line divisions of space. Each animal, or gang, or farmer, or nation expands its territory outward until it meets its neighbors. The boundary is fixed at the halfway point between an individual and its neighbors. Unfortunately, this simple geometric view of how territories are formed ignores two significant complications. First, the animals, gangs, and so on may not be uniform but may vary in strength and aggressiveness. Second, the space over which the distances are measured may be not simple and uniform (as in Figure 17-3) but complex and highly differentiated.

Unequal competition for space In Figure 17-3(c) the inequalities in farm areas come from the initial irregular location of the farms and the fact the farmers are assumed to be a homogenous group. Let us suppose, however, that the farmers were quite different—that some had large families and some had small ones, that some had more resources than others, were more aggressive, and so on. Can geographers incorporate the effect of such differences into a territorial model?

One approach is to use the gravity models we met in Section 16-1. When we have two farmsteads of equal size, we expect the boundary to be exactly half way between them, that is, at the median point. If they are of different sizes, however, we expect the boundary to be displaced away from the median point in the direction of the smaller farmstead. Just how big the displacement will be can be estimated from a gravity model. (See the marginal discussion of estimating boundaries by gravity models.)

Working with models of competition, regional economists have provided another perspective on how space can be partitioned into territories. Consider the position of the two sellers at center 1 and center 2 in Figure 17-4(a). Both produce homogenous goods and both have to pay the same freight charges, which are proportional to the linear distance the goods are shipped. The costs form an inverted cone about each center. These are shown as circular contours on the left of the diagram and as V-shaped cross sections on the right. The boundary (B) is located where the cones intersect; here the cost of goods from both centers is exactly equal.

In the first case the boundary is a straight line, so that a set of centers would form territories in the same way Dirichlet polygons were

Figure 17-4. Competition for a market territory. The diagram shows the hypothetical effect of variations in transport costs and production costs on the location of boundaries between retail centers. There are two centers (1 and 2) producing identical goods and located on a uniform plane where freight charges are proportional to the straight-line distance across the plane. The boundary line between the two market areas forms an *indifference curve* along which prices from both centers are equal and a potential consumer is indifferent to which center sells him the goods. These curves form hypercircles determined by the equation $P_1 + T_{1x}D_{1x} = P_2 + T_{2x}D_{2x}$, where P is the market price at the point of production, T_x is the freight rate between a production point and a given consumption point, x, and D_x is the distance between a production point and a given consumption point, x. Three alternative boundaries are shown here. The original market partition model was developed in the 1920s by economist F. A. Fetter and is known as Fetter's model. [From H. W. Richardson, *Regional Economics* (Praeger, New York, and Weidenfeld & Nicolson, London, 1969), p. 27, Fig. 2–3.]

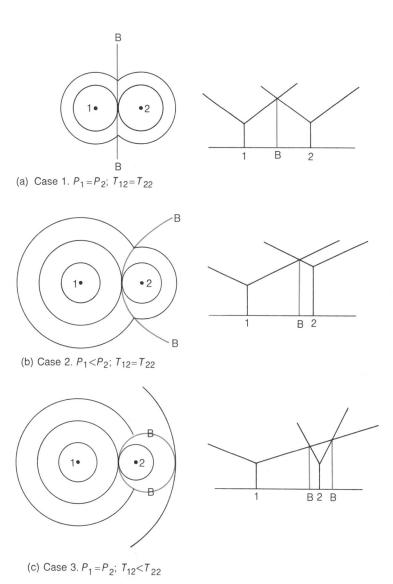

(a) Case 1. $P_1 = P_2$; $T_{12} = T_{22}$

(b) Case 2. $P_1 < P_2$; $T_{12} = T_{22}$

(c) Case 3. $P_1 = P_2$; $T_{12} < T_{22}$

formed. But we can go beyond the simple polygonal model. In the second case, the freight rates remain the same; but the production costs in center 2 are higher than those in center 1. The intersection of the two cones is now a curve, and the boundary forms a hyperbola [Figure 17-4(b)]. In the third case the position is reversed; that is, there are equal production costs but unequal freight rates. In this last case the boundary forms a circle, and the market area of seller 2 becomes an enclave within the much wider territory of seller 1. The circle is displaced eccentrically from the location of seller 1 [Figure 17-4(c)]. The formal structure of the model can incorporate even more complicated variations in both production and transport costs.

So far we have concentrated on certain rough analogies between the spatial form of animal and human territories. What about their purpose? Do human territories also have a role in the control of population density, regulating numbers by limiting the area from which resources are drawn? Certainly in some cultural groups farms cannot be subdivided if the number of potential farmers increases; the legal practice of *primogeniture*, by which the eldest son inherits property and the younger sons leave the estate to make their own fortunes, might, by a long stretch, be seen as ritual ecological behavior. The second reason for animal territories, survival of the fittest, may also have analogies in the division of market territories by business corporations, with small firms "going to the wall."

Human territories: Differences

What about the ways in which animal and human territories differ? Here we shall emphasize one very major difference—the relative permanence of human boundaries.

When we are dealing with unstable or short-lived territories like the market area of a seller or the domain of a small mammal, we can expect boundaries to be well adjusted to the forces that create them. Thus, in Figure 17-4 if seller 1 reduces its production costs more than other sellers, we would expect its territory to increase. But boundaries that are legally defined may persist long after the forces which created them have changed. We have already noted in Section 14-1 the gap between the legal limits of cities and the actual limits of built-up areas or commuting zones. International boundaries also reflect the political balance of forces at the time of their creation. The present boundaries between East Germany and West Germany and North Korea and South Korea relate to the military situations that existed in 1945 and 1953, respectively. Different segments of a country's boundaries may date from different periods. In the United States, the boundary of Maine with Canada has a 1782 vintage, whereas that of Arizona with Mexico dates from 1853. Figure 17-5 shows the extent to which the boundaries of nation–states in tropical Africa are legacies of European colonial expansion and bear little relation to present cultural and economic realities.

Can geographers build this historical inertia into their general models of territorial evolution? Geographers traditionally have taken

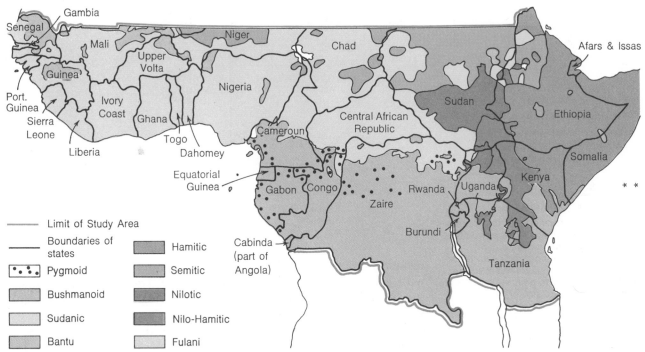

Figure 17-5. State boundaries and ethnic groups. The decolonization of tropical Africa in the decade 1955 to 1965 saw the emergence of many new independent states. These states have, however, retained the boundaries of the original British, French, and Belgian (formerly German) colonies, boundaries which were drawn where opposing forces met during "the scramble for Africa" in the last two decades of the nineteenth century. As the map shows, the boundaries of the new states run across major ethnic boundaries rather than enclosing homogeneous ethnic groups. [From H. R. J. Davies, *Tropical Africa: An Atlas for Rural Development,* University of Wales Press, Cardiff, p. 23, Plate 10.]

classic areas of changing boundaries—Silesia, the Trieste region, the Saarland—and used historical documents and evidence from maps both to reconstruct patterns of spatial change and to unravel underlying shifts in population, political affiliations, and the like. The relatively small number of international boundary changes, their cataclysmic nature, and the richness of historical evidence attending such shifts make detailed case studies an appropriate tool for analyzing the evolution of boundaries. In the case of internal rather than interstate boundaries, the number of cases is substantially greater, and a historically less restricted model can be built.

Figure 17-6 shows a two-stage process of boundary formation. There is an initial stage when a skeleton of primary divisions is established. We would expect the boundary lines of such divisions to be closely related to natural barriers like rivers, mountain ranges, and unpopulated zones [Figure 17-6(a)], and that the areas so bounded would be irregular in shape and size [Figure 17-6(b)]. A later stage in the evolution of boundaries occurs when the primary divisions are broken into smaller but more regular secondary units [Figure 17-6(c)].

The degree of secondary partition is related to external factors. For

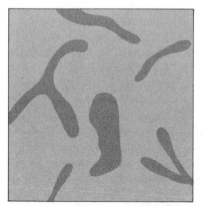

(a) Natural discontinuities

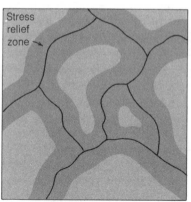

(b) Initial boundaries

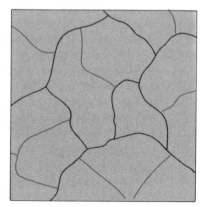

(c) Secondary partitions

Figure 17-6. The inertia of territorial boundaries. The maps show a two-stage model of boundary formation and persistence. Initial boundaries (b) tend to reflect natural discontinuities in the landscape (a) and to be irregular in both their shape and the size of the areas they bound. Secondary boundaries (c) tend to be more regular in both their shape and the size of the areas they bound. Note that the secondary boundaries tend to run through the stress relief zone (and thus intersect the initial boundaries) at right angles.

instance, large units with few subdivisions tend to be typical of zones with a low population density; conversely, small units with dense subdivisions are predominant in areas of high population density. We can relate the degree of regional subdivision to population density and conceptualize this subdivision as resulting from an increasing stress on the administration of larger and larger population units. This stress is relieved by partition. Once partition occurs, the need for division is removed. Furthermore, each boundary creates around it a zone of relieved stress, making it unlikely that new boundaries will be drawn along or near an existing boundary [Figure 17-6(b)].

The zone of relieved stress around a boundary is not a uniform band. Normally, we should expect the amount of relief to die out systematically with distance from the boundary line—to be at a maximum at right angles to the boundary and at a minimum parallel to the boundary. This would explain why a secondary boundary tends to meet a primary boundary at a right angle, and why low-angled intersections are less common than high-angled ones. Historical geographers use this notion in reconstructing the growth of boundary systems. Because the intersection of a secondary boundary with an earlier one tends to be at a right angle, the presence of such intersections in a pattern implies that one of the lines predates the other. In a similar manner, primary boundaries are likely to be less regular than secondary ones and related to natural discontinuities in the region.

Let us summarize as we conclude the first part of this chapter. We have seen that territoriality is strongly developed in certain animals other than man and that it is important in population control and selective breeding. Territories with a similar spatial form arise in human communities, although whether they serve a similar purpose remains a matter of debate. Finally, we have noted the distinctive institutionalized quality of many human territories that sets them firmly aside from their biological counterparts. It is this third type of territory that we shall be concerned with in the remainder of this chapter.

17-2 PARTITION PROBLEMS: LOCAL, REGIONAL, NATIONAL, GLOBAL

Geographers are concerned with territories and boundaries not only as spatial objects with an intrinsic academic interest but also as applied problems. As we noted in Section 1-7, geographers have been busy drawing up new boundaries on various spatial levels. Here we look at some of their work.

Local problems: The search for equity

On the lowest spatial level, geographers are concerned with how boundaries of local districts are arranged. Figure 17-7 shows the choices faced by legislators in drawing up boundaries. Let us assume that we must establish three new school districts with roughly equal numbers of residents in three ethnic groups (A, B, and C) in a city. Two alternative policies, maximal segregation and maximal integration, will produce two unlike spatial arrangements of boundaries. Moreover, these two solutions are only a fraction of the literally billions of alternative ways of partitioning the area. Can we ensure that the solution chosen is a fair one?

Gerrymanders: The problem In 1812 Governor Elbridge Gerry of Massachusetts established a curiously bow-shaped electoral district north

Figure 17-7. Alternative region-building strategies. In (a) the boundaries drawn create the greatest possible segregation of three hypothetical groups (A, B, and C). In (b) each area has a fair percentage of people from each group. Choices between these two types of zoning often face administrators drawing up school district lines in ethnically diverse areas.

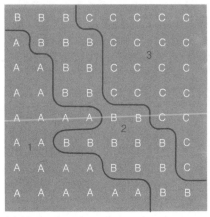

(a) Maximal segregation

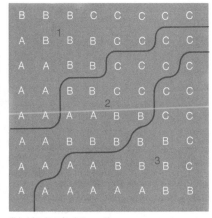

(b) Maximal integration

of Boston in order to favor his own party. In his memory the term "gerrymander" is used to describe any method of arranging electoral districts in such a way that one political party can elect more representatives than they could if the district boundaries were fairly drawn. Since 1812 the number of examples of gerrymandering has grown rapidly, and Figure 17-8 presents two of the most extreme ones.

If we wish to rig boundaries in an unfair way, we have two possible strategies. The first is to *contain* our opponents, to lump all the electors in the opposing party within a single boundary so that a representative receives an unnecessarily large majority of votes. The second possible strategy is to disperse our opponents, to split the electors of an opposing party between many districts so that nowhere are they numerous enough to elect one of their own representatives.

Within the United States the location of equitable political boundaries has been a subject of greater interest since the 1962 *Baker vs. Carr* decision of the United States Supreme Court. In this case the court ruled that legislative seats must be apportioned and states divided so the number of inhabitants per legislator in one district is approximately equal to the number of inhabitants per legislator in another district. The case was triggered by a group of voters in Tennessee who claimed that their votes were debased by inequities in the state's electoral districts. The situation arose because Tennessee had continued to elect its state legislators on the basis of an appor-

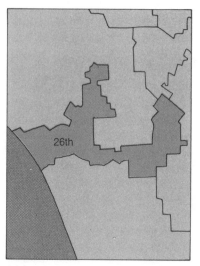

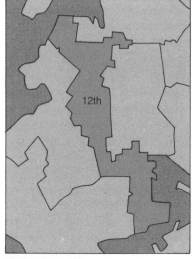

Figure 17-8. Gerrymandering in the United States. (a) California's 26th electoral district in 1960. (b) Brooklyn's 12th electoral district in 1960. In both cases, the boundaries were drawn to give a major electoral advantage to a political party.

(a) Los Angeles, Calif. ★ ★ ★ ★

(b) Brooklyn, N.Y. ★ ★ ★ ★

tionment adopted in 1901. In the intervening 60 years, however, there had been radical shifts in the distribution of the population from rural areas to cities and suburbs. This led to a situation in which one vote in rural Moore County was equal to nineteen votes in urban Hamilton County (containing the city of Chattanooga). Parallel situations existed in other American states. In Vermont the most populous district had 987 times more voters than the least populous district, and in Texas one district elected a representative to Congress with less than a quarter of the votes received by other representatives in the same state.

A computer program for setting electoral-district boundaries

Consider the problem of dividing an area into a number of electoral districts (e.g., six) in an unbiased way. Let us assume that we know the population data of several small tracts. (It is 100.) These data allow us to estimate the probable number of voters in each tract. Our task is to assign tracts to districts in such a way that we create six districts with more or less the same number of voters and as spatially compact as possible. The flow diagram in Figure 17-9 indicates the basic steps. First, choose arbitrary or reasonable trial centers for each district and mark their geographic coordinates. Second, compute the matrix of distances between the centers of each tract and the trial centers. Third, assign tract populations to the nearest trial centers until each district has the same number of voters. (Split tracts are assigned to the district to which the majority of voters were previously assigned.) Fourth, compute the center of gravity for each district on the basis of all the tracts assigned to that district in step 3. Fifth, inspect the districts. If the new centers differ from the trial centers, then go back to step 2 and repeat the assignment process. The cycling process ends when the equality of the voting population and the compactness of the districts reaches an acceptable level or when there are no further shifts possible in the district boundaries. If an acceptable solution is not obtained, new trial centers should be adopted and the program begun again. See P. Haggett and R. J. Chorley, *Network Analysis in Geography* (Edward Arnold, London, 1969), Chap. 4, II.

Gerrymanders: A solution? It is one thing to identify the problem of unfair electoral boundaries but quite another to discover a solution. Just how can we arrive at a "fair" spatial arrangement? The boundaries chosen for voting districts are likely to reflect a balance between three considerations. The first is that the voting population in each district should be *equal*. The ideal concept of "one man—one vote" must be modified to offer approximate equality because uncontrollable factors such as migration and natural change are continuously modifying the voting population within a district. In practice, equality must be defined in terms of an acceptable range in the size of electorates. The second consideration is that the voting district should be *contiguous*. Districts are usually in one conterminous unit and ideally should be compact within the sense that communications among the various parts of the district is easy. The third and more debatable consideration is that of *homogeneity*, or balance. Some may think it is desirable for the parts of a district to have a common social, political, or economic characteristic. Others may argue that the district should be a balanced mix representing a wide range of communities rather than a single one. We have already seen in Figure 17-7 how these two counterstrategies may affect district boundaries.

Several computer programs have been designed for achieving equality, contiguity, and homogeneity (or balance) by a nonpartisan method. One such method is outlined by the flow chart in Figure 17-9. It takes an initial set of small tracts whose voting population is known and allocates them to electoral regions. The original allocation is successively adjusted to make the regions more equal in population while keeping them compact. (See the marginal discussion of a program for creating electoral districts.) Figure 17-10 presents the results of using such a method to redistrict a sample part of New Jersey. Six districts were to be formed from over 50 tracts [Figure 17-10(a)]. In the first allocation the largest district, with a 12,500-person electorate, was 5 percent greater than the average for all districts. In the second allocation this discrepancy was reduced to only 1 percent.

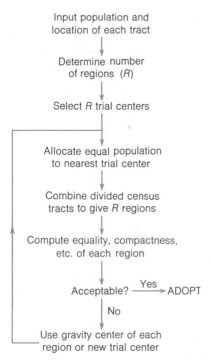

Input population and
location of each tract

↓

Determine number
of regions (*R*)

↓

Select *R* trial centers

↓

Allocate equal population
to nearest trial center

↓

Combine divided census
tracts to give *R* regions

↓

Compute equality, compactness,
etc. of each region

↓

Acceptable? ——Yes——→ ADOPT

↓ No

Use gravity center of each
region or new trial center

Figure 17-9. Electoral districting. This flow chart shows the main elements in a computer program for allocating population in census tracts to a set of compact electoral regions with a similar number of electors in each area. "Gravity center" is at the central point within a region which is most accessible for all its population.

In evaluating this approach to electoral areas, we should recall that numerous district boundaries can be drawn with approximately the same concern for equality and contiguity, and yet some will favor one party at one time rather than another. To resolve the imbalance, we need to inspect various boundary maps and adopt the one most in accord with the prevailing concept of political equity.

Regional problems: The search for efficiency

One area of government in which geographers frequently become involved is the revising of regional systems of administration. These systems include the hierarchy of local areas through which central governments keep in contact with grassroots needs. Such areas include the state, county, and township in much of the United States; the county, district, and parish in England; the *département, commune,* and *arrondissement* in France; and so on. Figure 17-11 illustrates a typical hierarchy.

Most countries, whether Western or non-Western, developed or underdeveloped, have intermediate and local authorities to which they delegate the central authority of the nation-state. The existing sequence is frequently a historical patchwork, and from time to time central governments decide to reform the hierarchy along more relevant lines. The Napoleonic reform of the ancient French system of provinces, the sequence of Soviet reforms since the early 1920s, Salazar's reforms of the Portuguese system in the 1930s, the Swedish re-

Figure 17-10. The search for improved electoral districts. These maps of Sussex County, New Jersey, in the United States, show the first stages in the allocation of census tracts to electoral regions, using the kind of program shown in Figure 17–9. The electoral districts in maps (b) and (c) are made up by combining the original tracts (a). Figures indicate the population in each electoral district in thousands. [From J. B. Weaver and S. W. Hess, *Yale Law Journal* 73 (1963), p. 306, Fig. 1.]

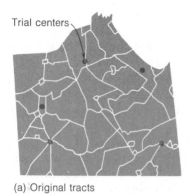

Trial centers

(a) Original tracts

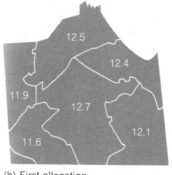

12.5
12.4
11.9
12.7
11.6
12.1

(b) First allocation

12.3
12.3
12.1
12.3
12.1
12.1
12.1

(c) Second allocation ★ ★ ★ ★

Figure 17-11. Hierarchies of administrative areas. The maps show a region-county-district-parish hierarchy in southwest England. Note how each successive area "nests" within the larger area. Similar hierarchies characterize the administrative structure of all countries; there is, for example, the state-county-township hierarchy in the United States and the département-arrondisement-commune hierarchy in France.

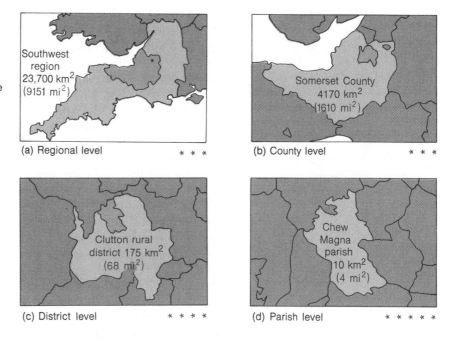

(a) Regional level   ✳ ✳ ✳

(b) County level   ✳ ✳ ✳

(c) District level   ✳ ✳ ✳ ✳

(d) Parish level   ✳ ✳ ✳ ✳ ✳

forms of the 1950s, and the British reorganization of the 1970s are each part of a recurring cycle of revision, a cycle in which geographers are increasingly being called upon to help redraw maps.

But what maps should be substituted for existing maps? On the basis of what criteria should new areas be drawn? How do we weigh a central government's demand for a few large efficient units against a local outcry for a small unit tailored to the needs of a particular community? The solution to these problems involves not just deciding on new units but arranging them into levels, or tiers. With a *single-tier system*, each local area has direct access to the central government; with a *multiple-tier system*, the local areas must work through an intervening *bureaucracy* on one or more regional levels.

The British problem We can examine the situation in a specific context by presenting a recent case of boundary reform in Britain. Until 1974 the country was split into over 1000 local districts, each with a variety of powers and each directly responsible to the central government in the performance of certain administrative functions. The system had been only slightly altered since the nineteenth century, so by the 1960s its lack of relation to the distribution of population and the needs of efficient management was becoming evident.

A Royal Commission was established to prepare recommendations. They proposed three criteria for the new system. First, they wanted the new units to be large enough to provide local government services at a low cost. "Large enough" was considered to mean a population of 250,000 to 1,000,000, depending on which government department you were talking to. The Home Office considered 500,000 the minimal population needed to support an efficient police force. The Department of Education supported this threshold for its needs but were prepared to lower it to 300,000 in sparsely populated areas. Lower figures were given by local health and welfare authorities (200,000 people) and by child-care services (250,000 people).

A second criterion for the new distribution was that the new units should be cohesive enough to reflect local community interests and to allow a local voice in political issues. In practice, cohesion and self-containment are measured by journey-to-work patterns, the range of public transport, newspaper circulation, and the organizational pattern of professional, government, and business organizations. This approach yields urban regions connected to rural areas by ties between workplaces and residences.

The third criterion for the delimitation of new units was the present pattern. Whenever possible, existing units and their boundaries, even on the county level, were used as building blocks in order to retain common interests and traditional loyalties, to preserve the skill and momentum of existing local authorities, and to minimize changeover problems.

The British solution Given the three basic criteria described above, we might assume that it would be relatively simple to find a common spatial solution. As geographers would have predicted, this is not the case (cf. Figure 10-14). The Royal Commission could not reach an agreement and proposed two alternative systems. The majority proposal was to divide the country into a few rather large local government areas (61 for England outside London itself) responsible for all local government services. In contrast, the minority proposed a two-tier system with 35 upper-level and 148 lower-level city regions. In this case, the upper-level authorities would look after planning and transportation, while the lower-level authorities would control services like education, social welfare, and housing.

We can appreciate the differences between the two proposals by looking at some maps of southwest England (Figure 17-12). The first area shown [Figure 17-12(a)] is similar in size and population to the state of Massachusetts. The existing 12 units vary considerably in size and area. If the majority proposal were adopted, a slightly enlarged

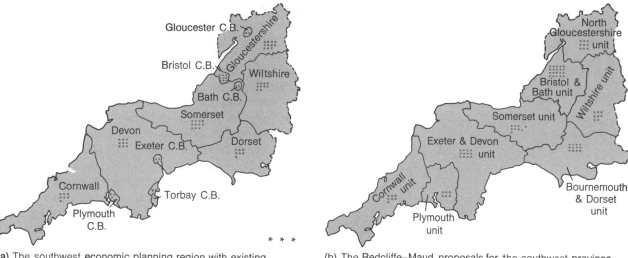

(a) The southwest economic planning region with existing county and county–borough (C.B.) boundaries

(b) The Redcliffe–Maud proposals for the southwest province, with unitary areas

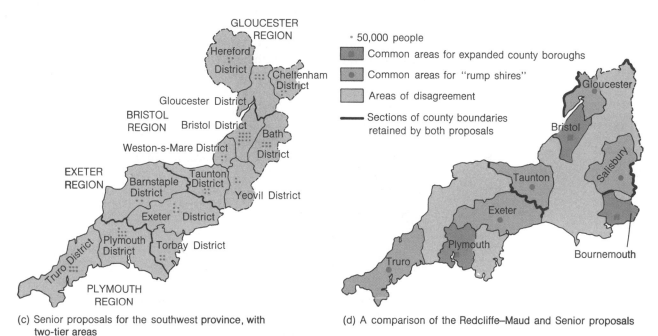

(c) Senior proposals for the southwest province, with two-tier areas

(d) A comparison of the Redcliffe–Maud and Senior proposals

Figure 17-12. Reform of administrative areas. The maps show proposed reform of boundaries in southwest England. The first three maps show alternative proposals, and the fourth, the areas of agreement and disagreement. "Rump shires" are the core areas of the old counties or shires (a) left undivided by both sets of proposals (b) and (c). [From P. Haggett, *Geographical Magazine* **52** (1969), p. 215. Reproduced by courtesy of The Geographical Magazine, London.]

southwestern province would be replaced by eight new areas [Figure 17-12(b)] larger and more uniform in size and with less extreme differences in local revenue-raising ability. If the minority proposal were adopted, a two-tier system, set within a redrawn southwestern province [Figure 17-12(c)] would result.

Despite their differences, we can see that certain recurrent geographic features dominate both the traditional map and the two proposed revisions. First, we can identify eight core areas. Five of these are largely rural and center on an existing county town; the remaining three have metropolitan centers in the leading cities of Bristol, Plymouth, and Bournemouth. Second, there are certain persistent boundaries that recur in all three maps and may represent rather fundamental discontinuities in the socioeconomic structuring of the area.

By recognizing recurrent features in proposed regional systems, geographers can isolate the broad features in any set of solutions and narrow down the areas of search. In the case of southwest England the main disputes over boundaries occur in three critical zones, as shown in Figure 17-12(d). By concentrating on these zones of disagreement and accepting that substantial areas of agreement are possible, we can reconcile a head-on clash between the proponents of apparently conflicting regional systems. The proposals now adopted for the area do retain many of the old county boundaries.

National problems: The search for stability

From the geographer's viewpoint the most important territorial unit is the modern nation-state. There are about 200 nation-states in the world today, and they range in size and importance from the Soviet Union (with one-sixth of the world's land surface) to units of a few square kilometers. Such states cover about 80 percent of the world's land surface (and the adjacent territorial waters) and divide it with a set of discrete bounded cells. (See Figure 17-13.) The areas remaining outside national sovereignty are controlled indirectly through colonial or joint-trusteeship arrangements.

In the case of an old nation-state like France, we may think of the present national boundaries as a rough approximation to the *limites naturelles* of the French nation. The idea of fitting the boundary to the ethnic group dominated the thinking of members of the Versailles Peace Conference after World War I; it played a significant part in drawing the line within the Indian subcontinent between India and Pakistan in 1948; it runs through the demands for a reunited Ireland or Germany. Yet if we consider the boundaries of most of today's nation-states, we discover that the boundaries represent arbitrary cutoff points. Many of the characteristics that give identity to a nation—

Figure 17-13. National boundaries. (a) The unmarked boundary between Canada and the United States is crossed here by the Alaska highway in the unpopulated northlands of sub-Arctic Canada. (b) The heavily protected 20 km (12 mi) Berlin Wall constructed in 1961 to regulate migration from the German Democratic Republic (East Germany) to the Federal Republic (West Germany). (c) Part of the 2400 km (1500 mi) Great Wall of China built between 200 B.C. and A.D. 600 to protect the northern boundaries of the Ch'in and Han empires against nomadic raids. [Photo (a) from Rotkin, P. F. I., (b) from United Press International, and (c) from the New York Public Library.]

language, ethnicity, a common history, and cultural traditions—do not end abruptly at its boundaries. The two German states do not include all German-speaking peoples; millions of Chinese live outside Asia (let alone a Chinese state); Israel contains only a fraction of world Jewry; and so on. Conversely, one very stable national unit (Switzerland) has several official languages.

To understand why a state like Paraguay or Iraq exists and why it has the boundaries it has, we must reconstruct its individual case history. However, we can draw a useful general distinction between boundaries on the basis of when they originated. *Antecedent boundaries* precede the close settlement and development of the region they encompass. Groups occupying the area later must acknowledge the existing boundary line. The boundary separating the relatively uninhabited land between the United States and Canada, established and modified by treaties between 1782 and 1846, is a typical example of an antecedent boundary. *Subsequent boundaries* are the converse of antecedent boundaries, in that they are established after an area has been closely settled. This type of boundary normally reflects existing social and economic patterns. The boundary between India and Pakistan, drawn at the division of British India in 1948, is an example of this type of boundary.

When boundaries are classified in this way, there is a gradation from lines drawn before there is any significant occupation of an area to those established after the main spatial pattern of settlement has been formed. We might also expect the success of boundaries, as measured by their longevity, to be related to their genesis. Subsequent boundaries may be preferable because the limits of a territorial

(a)

(b)

(c)

unit coincide better with existing sociocultural realities. It is difficult to find convincing proof of this hypothesis, however. If we examine a political map of Europe (Figure 17-14), we find a pattern of interlacing boundaries that ranges from stable boundaries like the French–Spanish one along the Pyrenees to the short-lived boundaries on France's German frontier. When we try to sort out the variations in the stability of boundaries we find two sorts of reasons. Some variations relate to the character of the boundary region itself (e.g., the fact that the Pyrenees is a mountain chain with a low population density and poor routes across it). Other variations are related to the historical relations of the two or more states whose territories border one another.

Global problems: The world's oceans

Seventy percent of the world's surface is covered by water. Up to now most conflict over territorial rights has been over land. But this situation is changing; indeed, the 1974 dispute between Greece and Turkey over the ownership of parts of the potentially oil-rich sea bed of the Aegean may typify the disputes to come.

Geographers usually divide the ocean into three zones in terms of territorial rights. First there are rights over the immediate *offshore areas* around a country. Second, there are claims for mineral rights in the *continental shelf* around a country. The continental shelf is an area of very smooth, gently sloping ocean floor which fringes all continents. Its limit is conventionally given as a depth of 100 fathoms (around 200 m or 600 ft) and its width varies from only a few kilometers to 350 km (about 200 mi). Third, there are the *ocean floors* themselves. Ocean floors make up over 90 percent of the world's ocean-covered areas and vary greatly in depth and smoothness. (Check back to Table 3-2.) We will look at these zones in turn.

Offshore areas

Offshore areas are of increasing interest to the bordering nation-states for three reasons: national defense, the enforcement of national laws, and the protection of fishing grounds and mineral resources in the shallower parts of the continental shelf.

Early claims of offshore waters were for the limited areas that could be controlled. These ad hoc "cannon shot" distances have led to a diversity of claims. Australia is bounded by territorial waters only 5.6 km (3.5 mi) in breadth, while El Salvador's claims extend to 370 km (230 mi). The Philippines claim as territorial waters all the immense sea areas between the islands of the archipelago. It was not until 1958 that the United Nations brought together 86 states for the first Law of the Sea Conference. Only slow moves toward the standardization of territorial claims have been made since that date.

Figure 17-14.   The stability of international boundaries. The maps show contrasts in the permanence of land boundaries among European countries. Some land boundaries have changed little since the fifteenth century. Sea boundaries, like that between England and France, are not shown. The four oldest land boundaries, shown in (a), are those of Spain with Portugal, of Spain with France in the western Pyrennees, of Switzerland, and of the Low Countries. Note the large number of short-lived boundaries in eastern Europe. What kind of factors make for very stable boundaries? [Based on a map by S. Columb Gilfillan. From N. J. G. Pounds, *Political Geography* (McGraw-Hill, New York, 1963), p. 29, Fig. 7. Copyright © 1963 by McGraw-Hill, Inc. Used with permission of McGraw-Hill Book Company.]

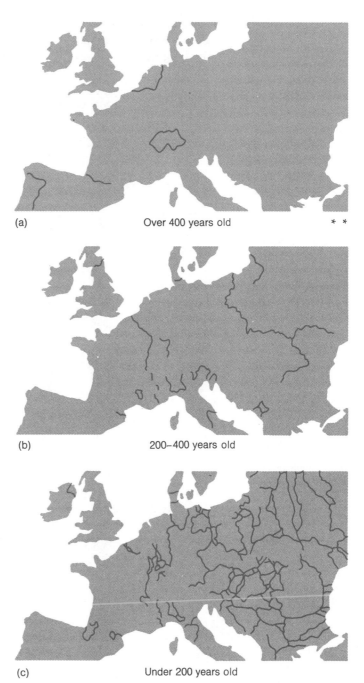

(a)   Over 400 years old   * *

(b)   200–400 years old

(c)   Under 200 years old

Continental shelves and ocean floors Although in the short term the greatest interest attaches to control of the immediate offshore waters, the long-term implications of ownership of the continental shelves and ocean floors may eventually be important. For example, if we allow the present principle of the median line [Figure 17-15(a)] to be extended to the open oceans [Figure 17-15(b)], we find some extraordinary distributions of territory. Portugal, by virtue of its ownership of the Azores and Cape Verde Islands, would stand to gain the largest

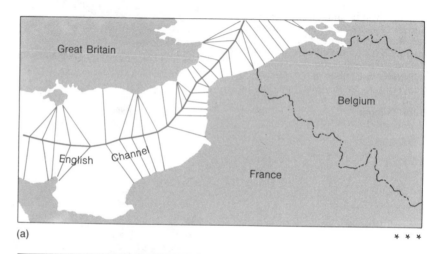

(a)

* * *

Figure 17-15. The principle of median boundaries applied to offshore areas. (a) International boundaries in the English Channel. (b) An outward extension of national sovereignty into the oceans, with each nation extending outward until it meets its adjacent and opposite neighbors. Note that while the boundaries in (a) have been ratified by the countries bordering the English Channel and are internationally recognized, those in (b) are merely hypothetical. [From (a) L. M. Alexander, *Offshore Geography of Northwestern Europe* (Rand McNally, Skokie, Ill., and Murray, London (1966), p. 58, Fig. 4; (b) T. F. Christy and H. Herfindahl, *Hypothetical Division of the Sea Floor*, a map published by the Law of the Sea Institute, Washington, D.C., 1968.]

(b)

*

share of the North Atlantic sea bed. The mineral riches of the ocean floors are only just beginning to be realized. For whereas offshore drilling for oil and gas on the continental shelf had been established from the 1920s, the present decade is seeing an increased interest in the surface deposits of the deep oceans. Minerals may accumulate on the ocean floor in the form of nodules rich in such minerals as manganese. Preliminary estimates for the Pacific suggest that such nodules may contain enough manganese to meet world demand for the next 400,000 years (at present rates of consumption) and that the nodules may be accumulating faster than they could be used. But few hard facts are known and the exploration of the ocean's mineral resources (including dissolved minerals in seawater) promises to be one of the great frontier areas of research in the remainder of this century.

If the ocean floors are going to be very important mineral resources in the future, can we allow them to be divided in the way shown in Figure 17-15? What about the problem of a landlocked country with no sea boundary? Suggestions have been made at recent United Nations conferences that the ocean floors out beyond the continental shelves should be divided up into a checkerboard of exploration areas. Then sections of the ocean would be allocated, perhaps on a random basis, to underdeveloped and nonmaritime countries. Although this is only a tentative idea, it will probably be made again as economic and technological problems in using ocean resources are solved. There is also increasing international concern by United Nations agencies in the control of pollution by oil or mineral elements like mercury, and the excessive reduction of the population of certain marine animals like whales.

17-3 | DRAWING BOUNDARY LINES

To settle a political dispute in a legal treaty is one thing, to translate that settlement into an identifiable line on the ground is another. The difficulty is most acute on the international level but is also encountered regarding interstate boundaries within the United States. Here we look at the problem of drawing boundary lines for both land and sea areas.

Dividing land areas

To clarify the problem of dividing land areas suppose we consider the internal and external boundaries of the conterminous United States. In Figure 17-16 we see that over 80 percent of the international and interstate boundaries are *geometric*. These geometric boundaries are mostly lines of latitude like 49°N, which separates the United States from Canada, and north–south meridians such as sections of the

Figure 17-16. Boundary lines. Different types of lines have been used for the international and interstate boundaries of the conterminous United States. [From N. J. Pounds, *Political Geography* (McGraw-Hill, New York, 1963), p. 89, Fig. 32. Copyright © 1963 by McGraw-Hill, Inc. Used with permission of McGraw-Hill Book Company.)

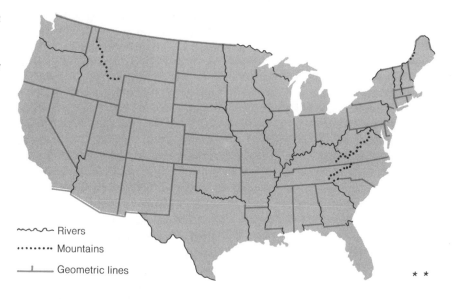

～～～ Rivers

•••••••• Mountains

⌐ Geometric lines

Oklahoma–Texas boundary at 100°W. The ease with which these lines can be drawn by standard surveying methods makes them very popular when sparsely settled or empty areas are being bounded.

Other geometric boundaries are sometimes used too. Part of the boundary between Delaware and Pennsylvania is an arc of a circle centered on the town of New Castle. Still other geometric boundaries consist of straight lines between points where locations are known. The 1853 treaty defining the boundary between Arizona and Mexico describes a straight line joining a point 31°21′N and 111°W with another point on the Colorado River. As we noted in discussing map projections in Section 2-3, tracing a straight line—simple enough on a plane—becomes complex on the earth's curved surfaces. A straight line on a treaty map may turn out to be a curve on the globe, and vice versa.

Figure 17-16 also shows some *nongeometric* boundaries; these boundaries generally follow the irregular course of natural surface features. Rivers are the most common natural feature used in drawing boundaries because they are self-evident dividing lines. The detailed demarcation of rivers causes problems for two reasons: The course of the lower portion of a river changes continually, and the river has width and may have several channels. Figure 17-17 illustrates the rapidity of changes in the course of the Rio Grande separating Texas and Mexico. Clearly, the boundaries fixed at one time may produce anomalies a decade later. Boundaries may follow the navigable channel of waterways or, in the case of wide bodies of water, some median line between the two shores. A line equidistant from both shores was constructed through Lake Erie by the International Waterways Commission for the Canada–United States border.

Mountain barriers also pose problems in deciding on the exact location of a divide or watershed. The 1782 treaty defining the northern boundary of Maine in terms of highlands divided rivers running to the Atlantic Ocean from those running north to the St. Lawrence River. But because of the complexity of the hydrology of the area, the boundary line was not finally established until the Webster-Ashburton Treaty was made in 1843.

Dividing sea areas

Lake Erie lies between the United States and Canada. Who owns it? As we saw in Figure 17-15, the principle adopted in the division of water-covered areas is for each state to extend its sovereignty at an equal rate until it meets the lake, sea, or ocean territory of adjacent and opposite states. Thus the method used for Lake Erie is an adaptation of the median-line principle used in devising Dirichlet polygons. (See Figure 17-2.)

Although simple in theory, the application of this method is complicated by ambiguous definitions of a country's coastline. Consider the map of Iceland's territorial waters in Figure 17-18, and note how the irregular shape of the coastline is approximated by a polygon. Within the straight lines of the polygon are Iceland's *internal waters*, beyond this lie the outer limits of Iceland's *territorial waters*, and beyond these outer limits lie *international waters*. The boundaries can be changed in various ways. Areas can be added by drawing the polygon in different ways (see the two areas marked x), by adopting a small offshore island or rock as one of the bases for drawing the polygon (see area y), or by extending the limit of territorial waters as was done in 1971. It now stands at 80 km (50 mi). Whether or not a bay is regarded as part of a country's internal waters, or an offshore island or sandbar is regarded as national territory, changes the baseline from which the median line with a neighboring state is determined. [See Figure 17-15(a).]

17-4	**A POSTSCRIPT: SPACE, DIVISIONS, AND CONFLICTS**

Mathematician Lewis Richardson, in a book with the intriguing title *The Statistics of Deadly Quarrels*, has considered the spatial factors that relate to conflict. Written before the intercontinental missile age, the book contends that the potential for any interregional relationship (including conflict) is a function of the number of neighbors a state has.

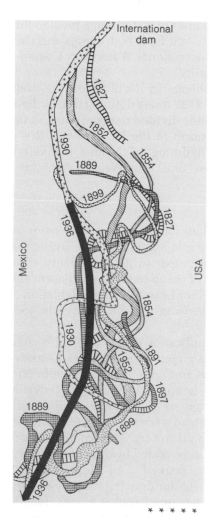

* * * * *

Figure 17-17. River boundaries. Local changes may occur in international boundaries because of geomorphic changes. The map shows shifts in the boundary between the United States and Mexico along the main channel of the Rio Grande near El Paso from 1827 to 1936. [After S. W. Boggs, *International Boundaries* (Columbia University Press, New York, 1940).]

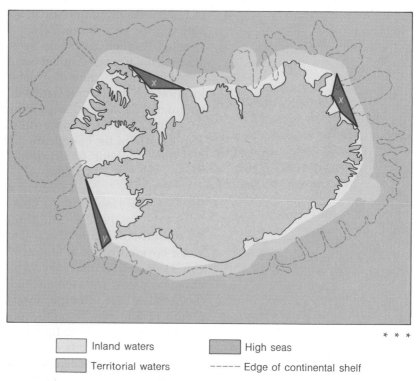

* * *

| Inland waters | | High seas | |
| Territorial waters | | ----- Edge of continental shelf | |

Figure 17-18. Offshore limits. This map of Iceland shows the effect of changes in baselines on the limits of internal waters. The 12-mile (19.3-km) limit shown has subsequently been extended to 50 miles (80 km). This unilateral extension by Iceland has caused disputes with Western European countries (notably Great Britain and West Germany) who used to fish on the continental shelf outside the 12-mile limit. [From L. M. Alexander, *Offshore Geography of Northwestern Europe* (Rand McNally, Skokie, Ill., and Murray, London, 1963), p. 109, Map 12.]

A nation-state like Germany in 1936, which had nine other nation-states abutting its own territory, has a much higher chance of conflict, according to this line of thinking, than a nation-state like Portugal with only one neighbor. If territories were closely packed in an unbounded land area, the average number of neighbors could reach as high as six. Such areas might approximate the form of Christaller's hexagonal central-place territories. If the territories were randomly assigned, then we know from statistical theory that the number of neighbors would average somewhat less than this (5.816).

Territories are established not on an unbounded plane but on a spherical globe broken into areas of land and sea. The spherical shape of the planet means that if states continue to grow larger, the number of neighbors must decrease until the world is dominated by only two states; then the number of neighbors drops to only one. Although according to Richardson's thesis this arrangement would decrease the potential for conflict, the scale of conflict in a two-state world would be immense. It might involve a global nuclear war rather than the periodic tribal skirmishing common on the six-neighbor level.

A confrontation between a land-based power dominating the world's central Eurasian landmass and a sea power controlling peninsular extremities together with the Americas, Australasia, and Africa was considered by an English geographer, Halford Mackinder, in the early years of this century. Mackinder's recognition of the strategic significance of Eastern Europe resulted in his 1904 dictum: "Who rules East Europe commands the Heartland, who rules the Heartland commands the World Island, who commands the World Island commands the World." His *heartland* included much of European Russia and Soviet Central Asia, together with what is now northern Iraq and Iran. Mackinder's thinking was taken up by a few German political geographers such as R. Haushofer; how big a role it played in Hitler's global strategy for World War II is uncertain.

Although such spatial speculations may play a part in international stress, they are overridden by far more powerful ones that lie outside the scope of this book. You may, however, care to browse through some of the writings in political geography in which these problems are explored in depth. (See "One step further . . .," p. 455.) It is particularly worthwhile to look at a book written by geographer Isaiah Bowman nearly a half-century ago. In his *New World* he reviewed, continent by continent, the stress areas of a world just recovering from the horrors of the 1914–1918 war. Current problems like those of the Middle East (see Figure 17-19) can be seen in a context of deep-seated difficulties, and some conflicts can be seen to have a spatial persistence that is both intriguing in retrospect and sobering in prospect.

If spatial factors play an integral part of war strategies, contemporary geographers would contend that they should have an equal weight in strategies of peace. Certainly Richardson's conclusions on conflict were aimed at the reduction and resolution of wars. In the early 1960s the founding of the Peace Research Society by a group of behavioral scientists, including geographers, marked one stage in restoring the balance. In this chapter we have seen the potential role that spatial design can play on levels from that of the local school or hospital district (Figure 17-20) to that of the planet as a whole. If good fences do indeed make good neighbors, then geographers need to determine from their research just where fences might best be built.

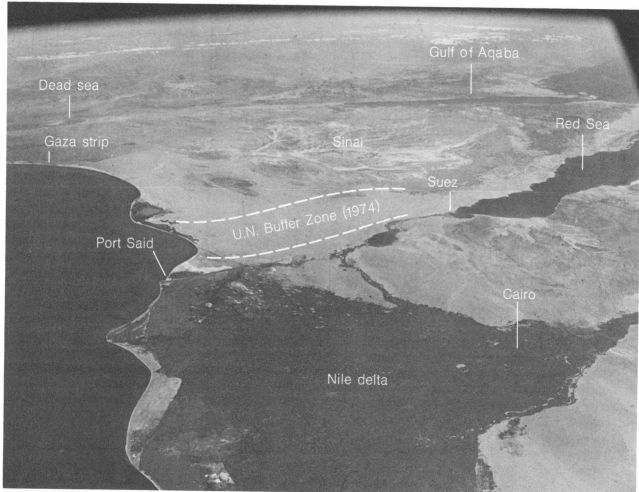

Figure 17-19. Critical areas of international stress. The opening of the Suez Canal (center) through the narrow Afro-Asian isthmus in 1869 reduced direct sea-route distances between Western Europe and India by about one-half. Arab-Israeli conflicts over the last 25 years culminated in its closure after the Six Day War in 1967. Renewed hostilities between Egypt and Israel in 1973 have led to a buffer zone being set up on the east bank of the Canal, which is to be reopened. The future of Sinai itself is likely to remain a source of conflict. This Gemini photograph looks southeast over the Nile delta and Sinai desert toward central Arabia. [Photo courtesy of NASA.]

Reflections

1. Debate the value of ethological evidence in attempts to understand human territoriality. Do you think that regarding man as a "naked ape" provides (a) insights into or (b) misleading analogies about our spatial behavior?

2. What is going on in Figure 17-4? Try introducing a third center, and plot the boundaries that might result.

3. Gather data for your local area on the boundaries of electoral districts (e.g., senatorial, parliamentary, or city council districts) and on how many votes were cast for various parties at a recent election. Was the result affected by the electoral boundaries?

4. Look at the variability in the age of the international boundaries in Figure 17-14. List factors which make for (a) stability and (b) impermanence in boundaries.

5. Select *three* areas of the world where you might expect international disputes over territory to occur in the next decade. Compare your selection with that of others in your class. Do any common patterns emerge?

6. Use the median-line principle in Figure 17-15 and a map to divide a small enclosed sea like the Mediterranean or Baltic. Would it be fair to use this principle in dividing the world's oceans?

7. Review your understanding of the following concepts:
   (a) ethology
   (b) median lines
   (c) Dirichlet polygons
   (d) gerrymandering
   (e) antecedent boundaries
   (f) subsequent boundaries
   (g) offshore limits
   (h) heartland

Figure 17-20. Hospital territories in Sweden. The boundary lines enclose areas that are closest to each of the six main medical centers for persons traveling by car. [From S. Godlund, *Lund Studies in Geography B,* No. 21 (1961), Fig. 8.]

Hours
- 0–2
- 2–4
- 4–8
- Above 8

One step further . . .

Many of the ideas presented in this chapter are part of the legacy of political geography. For general discussions of this field, see

Hartshorne, R., in James, P. E., and C. F. Jones, Eds., *American Geography: Inventory and Prospect* (Syracuse University Press, Syracuse, N.Y., 1954), Chap. 7, and

Kasperson, R. E., and Minghi, J. V., Eds., *The Structure of Political Geography* (Aldine, Chicago, 1969).

Territoriality in nonhuman animal populations is widely established. For a controversial introduction to the biological literature on this subject, read

Ardrey, R., *The Territorial Imperative: A Personal Inquiry into the Animal Origins of Property and Nations* (Atheneum, New York, 1966).

For a view of market areas and the economist's views of spatial partitions, read

Richardson, H. W., *Regional Economics* (Praeger, New York, and Weidenfeld & Nicolson, London, 1969), Chap. 2.

Problems of spatial partition within the city and the public issues to which it gives rise are being studied increasingly by geographers. See

Cox, K. R., *Conflict, Power, and Politics in the City: A Geographic View* (McGraw-Hill, New York, 1973).

Two classical studies by geographers who were intimately involved in the boundary problems that followed World War I are

Bowman, I., *The New World: Problems in Political Geography* (Harcourt Brace Jovanovich, New York, 1928) and

Boggs, S. W., *International Boundaries: A Study of Boundary Functions and Problems* (Columbia University Press, New York, 1940).

These books remain extremely relevant to the boundary problems that continue in many parts of the world today. One key problem that has come to the fore since the 1930s is that of dividing the ocean floor. For a useful introduction to this subject, see

> Alexander, L. M., *Offshore Geography of Northwestern Europe* (Rand McNally, Skokie, Ill., and Murray, London, 1963).

It is worth browsing through a historical atlas to compare the kaleidoscopic change in some parts of the earth's surface with the relative stability in others. A highly recommended atlas is

> Darby, H. C., and H. Fullard, *Cambridge Modern History Atlas* (Cambridge University Press, London, 1971).

There are no special geographic journals devoted to the topics treated in this chapter, and research is published in the general geographic serials. Journals like *International Affairs* (a quarterly) and *World Politics* (also a quarterly) often carry interesting papers. The *Journal of Peace Research* (an annual) is devoted to applying academic ideas to the resolution of conflicts.

Chapter 18

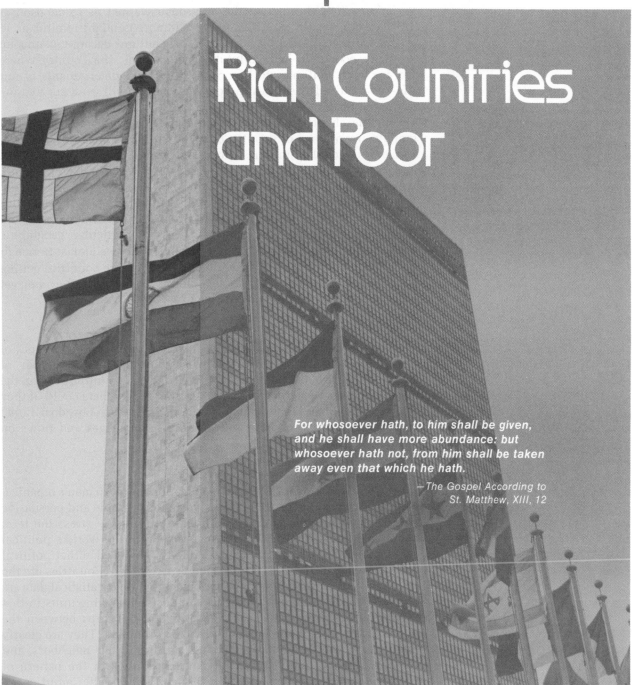

Rich Countries and Poor

For whosoever hath, to him shall be given,
and he shall have more abundance: but
whosoever hath not, from him shall be taken
away even that which he hath.

—*The Gospel According to*
St. Matthew, XIII, 12

Humorists throughout the ages have warned us to choose our parents with care. Geographic humorists might remind us to choose our birthplaces with equal caution! Because it sums up so many economic and cultural considerations, the country where we are born continues to be one of the prime determinants of our life—and indeed has a bearing on whether we will even survive the trauma of birth itself.

In this chapter, we try to answer four questions about the geographic importance of country units. First, we inquire into the relevance of national boundaries. Are countries a logical unit of analysis, and what precise role do the boundaries between them have? Second, we look at the main inequalities that exist between countries at present. What is the spatial pattern of rich and poor countries? Third, we turn to the contributions of geographers to studies of the economic-development process. Does development follow a particular geographic pattern and show a specific spatial form? Finally, we look at trends in patterns of inequality. Is the world converging? Are countries getting more like one another? Or do the rich get richer and the poor get poorer, as our opening quotation suggests?

18-1 | THE IMPACT OF NATIONAL DIVISIONS

We saw in the last chapter that the world is laced with a network of territorial divisions. Here we ask what effect the most important of these boundaries, the national divisions between countries, has on the regional mosaic. Are they so important that geographers could, if they wished, scrap their other regional divisions? If not, just how do national boundaries affect economic and cultural interchanges and flows on the planet Earth?

The country as a geographic unit

If we were searching for a single organizational unit in man's organization of the world today, there would seem to be simple and persuasive reasons for using the country as this basic unit. We stress the term "country" rather than nation because several of the world's political areas cannot be regarded as actual nations. They lack either political independence or ties of common language or blood. Countries are the principal "accounting units" for which comparative statistical data are regularly made available. Countries are decision-making units in that their central governments can affect the relationships between the population within their borders and the environment. They are clearly defined by boundaries that separate them from their neighbors, and these boundaries form noteworthy discontinuities in the pattern of human organization, sometimes in the landscape itself. Countries are increasingly being organized as integrated economic units, and they collect and publish the data we use in building up a picture of the globe. Geographers often regard countries as the individual tiles out of which the world mosaic of spatial regions is formed simply because they are the easiest units to compare. Despite its apparent attractions and considerable convenience, however, the country is not universally used as a basic regional unit by geographers. Why is this so?

If we look at a map of the world's countries (see the inside front cover), we find that there are several practical difficulties in using them as the basic spatial units in geographic studies. First there is the difference in size and population between countries. (See Table 18-1.) How can we usefully compare a country like the Soviet Union (occupying one-sixth of the world's land area) with countries like San Marino or Andorra? Second, large countries have immense internal contrasts; in Canada the population forms a thin ribbon of settlement along the southern border; most of the country is virtually unsettled. Third, we cannot be always sure that we are comparing like with like, certainly a basic principle of analysis. In France we find a country where the influence of the central government is rather uniform everywhere. In Indonesia the government has much less control over the outlying, peripheral areas. The boundaries of country units are a problem too. They are often arbitrary geometric lines wholly unrelated to either the natural environment or population characteristics. (Recall Figure 17-5 showing how the boundaries of countries in tropical Africa cut across tribal divisions.) The borders may not enclose contiguous spatial areas. (Consider West and East Pakistan before the creation of Bangladesh in 1971 or the United States and Alaska.) And they may shift violently and abruptly from time to time.

The importance of central authority within a country is related

Table 18-1 The world's largest and smallest countries[a]

Largest countries					Smallest countries				
	In area (km²)	(mi²)	In population (millions)			In area (km²)	(mi²)	In population (thousands)	
Soviet Union	22,402,000	(8,650,000)	China	751	Vatican City	0.4	(0.15)	Vatican City	1
Canada	9,976,000	(3,852,000)	India	549	Monaco	1.5	(0.58)	Nauru	7
China	9,561,000	(3,692,000)	Russia	244	Nauru	20	(8)	Andorra	16
U.S.	9,520,000	(3,676,000)	U.S.	209	San Marino	62	(24)	San Marino	19
Brazil	8,512,000	(3,287,000)	Indonesia	119	Liechtenstein	157	(61)	Liechtenstein	22
Australia	7,687,000	(2,968,000)	Pakistan	114	Barbados	430	(166)	Monaco	26
India	3,271,000	(1,263,000)	Japan	103	Andorra	453	(175)	Qatar	95
Argentina	2,777,000	(1,072,000)	Brazil	95	Singapore	580	(224)	Maldives	109

[a] Data refer to the 146 independent countries of the world, ca. 1970.

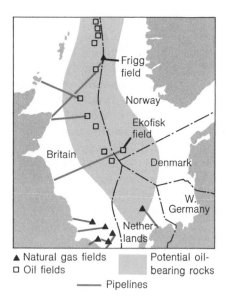

▲ Natural gas fields
□ Oil fields

Potential oil-
bearing rocks

—— Pipelines

Figure 18-1. National divisions as resource lotteries. The map shows the effect of antecedent territorial boundaries in the North Sea ownership of natural gas and oil sources. Natural gas is found mainly in the southern half of the North Sea: the earliest strike was in the Netherlands' waters, but larger sources were later found in the British area. The Frigg natural-gas field lies directly on the British-Norwegian boundary and is jointly developed by both countries. Later oil strikes have come in the northern half of the North Sea, many close to international boundaries. The Ekofisk oil field is Norwegian but is connected by pipeline to Britain because a deep trough in the seabed off the Norwegian coast would have made pipeline construction very costly. Clearly some countries have benefited more than others from the way the boundary lines have been drawn.

historically to the amount of decision-making power the central government has acquired. To an increasing extent, decisions on resource exploitation, patterns of settlement, population growth, regional development, and pollution levels are being made at the central government or "state" level. As this trend continues, boundaries between countries become ever more important to geographers in interpreting the environmental and spatial patterning of the earth's surface.

Countries as resource lotteries

We have already noted in Chapter 17 that many of the world's land boundaries and all the world's sea boundaries are antecedent; that is, they were set before the full exploitation or settlement of an area. In addition, many new nation-states in Africa are emerging within a framework of international boundaries laid down by colonial poachers from Europe in an earlier day. Nigeria, Tanzania, and Zambia each inherited land areas that make sense in terms of colonial spheres of influence but have little overlap with either environmental or cultural discontinuities.

Clearly each national territory, however defined, represents some share of the earth's stock of natural resources. In the case of agricultural resources, the pattern of environmental variation is observable and the equity or inequity of shares is well known. In the case of mineral resources, though, both the degree of geologic exploration and the changing demand for minerals make the shares more like lottery tickets. We can see this effect rather clearly in Figure 18-1. International boundaries in the North Sea were of little more than academic importance until the middle 1960s. Then natural gas and oil discoveries highlighted the advantages that countries like the United Kingdom and the Netherlands had gained from what had previously been only paper titles.

In summing up inequalities in the resource endowment of each country we should include the advantages and disadvantages conferred by the spatial configuration of its territory. For example, Figure 18-2 shows two historic tension areas where land corridors to the sea provided a source of contention: the Petsamo Corridor in northern Finland (in existence from 1920 to 1945) and the Polish Danzig Corridor to the Baltic (enshrined in Woodrow Wilson's proposals for European postwar reconstruction in 1918). To these we could add Bulgaria's short-lived Dedeagats Corridor to the Aegean Sea (held from 1912 to 1919). Each illustrates an attempt by a state in eastern Europe to secure a narrow strip of land linking its national territory to an adjacent sea. Outside Europe, the Eilat Corridor in Israel and the Antofagasta Corridoro in Bolivia serve a similar purpose.

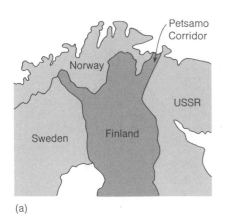

(a)

(b)

(c)   ★ ★ ★

Figure 18-2. Territory and access. Boundaries have sometimes been drawn to give particular countries access corridors to the sea or to international inland waterways. The maps show corridors held by (a) Finland from 1920 to 1945, (b) Poland from 1919 to 1939, and (c) Colombia since 1922. For more detailed maps and other historical examples, see H. C. Darby, *Cambridge Modern History Atlas* (Cambridge University Press, Cambridge, 1971).

Although most corridors have represented attempts to gain direct access to the sea, some have been aimed at indirect access by way of navigable rivers. Ecuador has a long-standing claim to a strip of northern Peru that gives access to the navigable upper reaches of the Amazon, and in 1922 Colombia obtained from Peru the narrow Leticia Corridor allowing a 120-km (75-mi) frontage on the same river [Figure 18-2(c)]. In general, most land-locked states have secured access to the sea by international conventions that allow the movement of goods across intervening territories without discriminatory tolls or taxes. Switzerland, Czechoslovakia, Austria, and Hungary are European examples of states that use such agreements in their trade.

Boundaries as filters

Another effect of boundaries is to act as a barrier or filter to spatial interaction between regions. We have already seen in Chapter 12 how absorbing and reflecting barriers were incorporated into the Hägerstrand model of diffusion to simulate the effect of boundaries on diffusion waves.

Similar indications of the distorting effect of political boundaries have been uncovered by comparing the observed and expected interactions between cities. In one study, contacts between Montreal and surrounding cities were measured in terms of long-distance telephone traffic. The expected interactions given by a population–distance formula are presented in Figure 18-3, along with the actual interactions. The traffic between Montreal and other cities in Quebec was from five to ten times greater than the traffic between Montreal and cities with comparable population–distance values in the neighboring province of Ontario. Furthermore, the strength of the provincial Quebec–

Figure 18-3. Boundaries as filters. The impact of the Quebec-Ontario boundary on spatial interaction is measured by the number of telephone calls between the city of Montreal, Canada, and other centers in the two provinces. Actual calls are plotted on the vertical axis against expected calls on the horizontal axis. (Expected interaction is given by a gravity model using the size of the other centers and the distance from Montreal.) Note how the calls from Montreal to its neighboring cities *within* the province of Quebec (a) are much higher than those to neighboring cities in the adjoining province of Ontario (b). [From J. R. Mackay, *Canadian Geographer* **11** (1958), p. 5.]

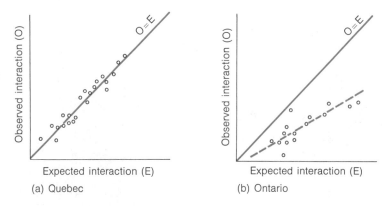

Ontario Barrier in blocking interaction was overshadowed by the blocking effect of the international boundary to the south. Montreal's traffic with comparable cities in the United States was down to only one-fiftieth that of its traffic with places in Quebec.

We can develop a general graphic model of the blocking action of both physical and political boundaries (Figure 18-4). When the barrier is a political one, marked by tariff differences, the potential field for

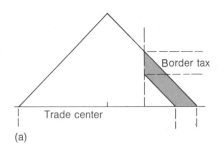

Figure 18-4. Boundaries and flows of trade. A general model of the impact of boundaries on spatial interaction. (a) A cross section of the market area around a trade center. If you follow the vertical dashed lines, you can see how this cross section is projected down to (b) to show the same situation in map form. Note how the shaded area indicates that part of the market that is lost by the effect of the border tax. Map (c) shows the still greater loss if the border can only be crossed at a single customs post. The effect of a natural boundary (hence, no border tax) with a constricted crossing point is shown in map (d).

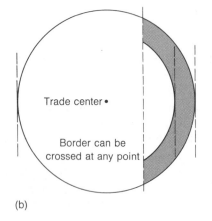

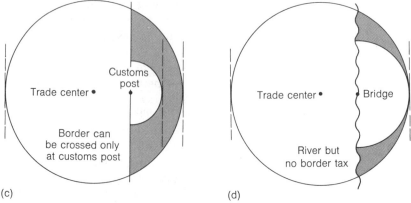

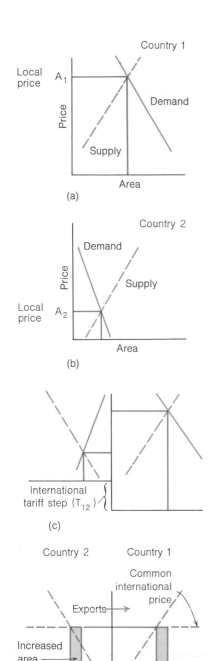

(a)

(b)

(c)

(d)

trading is restricted. It may be truncated in varying ways. Figure 18-4(b) shows the probable form if the political boundary can be crossed at all points along its length; Figure 18-4(c) gives the probable form if the boundary can be crossed only at a customs point. If the boundary is not a political one but a natural feature such as a river, with a single crossing point, the field will probably look like that in 18-4(d).

We can throw further light on the effect of tariff walls between countries by adapting some of the conventional economic models of trade. Let us assume that we have two adjacent countries, each of which produces a similar crop (e.g., wheat), separated by a tariff. In each country, the land area devoted to the crop is directly related to its price. As the price of the crop increases, the amount of land under cultivation increases, and vice versa.

Let us do as economists do and plot the relationship of the crop area to price in terms of supply and demand curves. (Readers who have already taken classes in economics will see that we are greatly simplifying the situation; those who have not need note only the conclusions we draw from the graph.) Figure 18-5 shows the area–price relationship as an upward-sloping supply curve. Price also affects the volume of demand for the crop, but in an inverse way: The demand for a high-priced good is assumed to be less than that for a low-priced good, and vice versa. We can represent this second relationship by a downward-sloping demand curve. The local price for the crop in each country is established when the demand for it and the supply of it are in balance.

If we regard the two countries as completely isolated, the demand and supply situation in each will be determined by the internal conditions in each. Let us suppose that in Country 1 farmers get a high price for the crop and a large area is devoted to it [Figure 18-5(a)]. In Country 2, conditions are different. The crop sells for a low price and

Figure 18-5. Tariffs and international trade. A simplified model of the effect of international tariff between two countries on the supply of and demand for a common agricultural product. Charts (a) and (b) show the local conditions in each country in which the local price (vertical axis) and the area of the crop under cultivation (horizontal axis) are related through the conventional economic mechanism of demand and supply. The international tariff between the two countries is shown as a "step" (T_{12}) in chart (c). In this case, the step is not high enough to keep production from being expanded in Country 1 (the low-cost producer) and the extra output exported to Country 2 (the high-cost producer). How high does the step have to be to keep out cheap imports from Country 2? This depends on the relative difference between the local price in Country 1 (A_1 in the diagram) and the local price in Country 2 (A_2). When the tariff (T_{12}) is *less* than the difference in price (i.e., $A_2 - A_1$), as it is in the diagram, then flows will occur as in (a). If, however, Country 2 raises its tariff so that T_{12} is greater than $A_2 - A_1$, no trade will take place.

the area devoted to it is smaller [Figure 18-5(b)]. What will happen if we drop our assumption of isolation? Logically, we should expect the low-priced crop from Country 2 to find its way into Country 1.

A tariff can prevent this flow of exports, however. We can see this from our model if we place our two diagrams back to back and displace them vertically [Figure 18-5(c)]. The amount of the displacement represents the size of the tariff imposed by Country 1. A flow of export occurs only if the difference in the local prices in the two countries is greater than the tariff. If the tariff is low enough for a flow to take place, a general international price will be established. The area devoted to the crop in Country 2 will expand, and the area devoted to it in Country 1 will contract [Figure 18-5(d)].

The analysis here is clearly oversimplified. The model refers to a highly simplified situation in which there is a single product and there are only two countries. Economists have developed a complex theory of international trade to explain the flow of many commodities between many nations. Nevertheless, the simple graphical example given here gives us some insight into the impact of tariffs on trade between countries. You might like to ponder the effects of tariff reductions and the establishment of common international prices (as in the European Economic Community) on the crop areas of the countries affected. Does this have implications for farm production in your own country?

18-2 | **GLOBAL PATTERNS OF INEQUALITY**

You need hardly take a geography course to learn that some parts of the earth's surface have populations more prosperous than others. *If* we make the naive assumption that richness means material wealth, even a layman will have no difficulty in separating Sweden from Senegal on one spatial level, or Beverly Hills from Watts on another. But these are extremes. If we want to distinguish between geographic areas and see how the gaps between the rich and the poor are changing, then we need a reliable ruler than can measure finer differences. What rulers have been developed, and what patterns do they reveal?

The search for a yardstick of development
Geographers would ideally like to compute and map some quantitative index that serves as an unambiguous ruler for measuring the economic or social performance of a region. For instance, we might reasonably regard a country as poor in which the low level of health services and nutrition causes many infants to die and a country as rich in which the high level of health services and nutrition prevents infant deaths. If we take available figures for infant deaths per 1000 live births in the 1950s,

we find Sweden (with 17) and the Netherlands (with 20) at one end of the scale and Tanganyika (with 170) and Burma (with 198) at the other. [See Figure 18-6(a).] But how valuable this single index is in separating poorer from richer countries is open to debate. Taiwan (with 34 infant deaths per 1000 live births) and Iraq (with 35) scores surprisingly high, while Yugoslavia (with 112) appears to be a very poor country. This difference in scores is due partly to differences in the way each country collects population data and partly to how the available wealth is distributed.

A single index of wealth, then, has evident disadvantages. The United Nations has suggested that standards of living can be properly defined only by using many indexes: indexes of health, food, education, working conditions, employment, consumption and savings, transportation, housing, clothing, recreation, social security, and human freedoms. [See, for example, Figure 18-6(b).] Although the United Nations has admitted that information is not available on many of these items, at least the principle of a multiple index has been firmly established.

If we were to measure all these indicators of wealth, they would not necessarily tell the same story. (See Figure 18-7.) Ways of combining multiple measurements into a few general indexes are now available. They involve collapsing several measurements into a smaller number of components. (See the marginal discussion of principal-components analysis.) Brian Berry at the University of Chicago compressed 43 different indexes of economic development into a single diagram (Figure 18-8). Note that this figure has two axes. The first and most important is the longer, vertical axis which measures differences in *technical development*. This single index accounts for about 84 percent of the information in the many original measures. The second and less important axis is the short, horizontal one. This index measures contrasts in the *demographic stage* a country has reached. Together, the two indexes account for 88 percent of the original contrasts between the 95 countries. Let us look at each of these indexes in more detail.

Contrasts in technical development

Figure 18-6(b) shows the world distribution of technically advanced and technically backward countries, according to Berry's first index. The term "backward" causes natural irritation to readers in those countries it is used to describe and is logically unsatisfying since the situation in such countries is a dynamic one and most of them are moving forward quickly. Let us, therefore, use the terms highly developed countries (*HDCs*), moderately developed countries (*MDCs*),

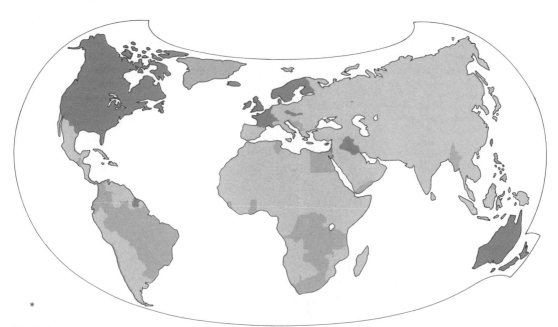

*

(a) Single index: infant deaths per 1000 live births

(b) Multiple index: Technological index based on 43 individual indexes

 Most developed 20 countries ■ Most retarded 20 countries

Figure 18-6. Development patterns. The maps show country-by-country variations in (a) infant mortality as measured by infant deaths per 1000 live births and (b) a combined index computed from 43 variables. On both maps only the countries with the 20 highest and 20 lowest scores are shown. [Data from United Nations and other sources, for late 1950s. After N. Ginsburg, *Atlas of Economic Development* (University of Chicago Press, Chicago, 1961), pp. 25, 111. Copyright © 1961 by The University of Chicago.]

and less developed countries (*LDCs*), which are more acceptable and accurate terms. Table 18-2 shows how the world's leading countries appear to fit into these three categories in the early 1970s.

Superficial theories of development A look at the map in Figure 18-6(b), the diagram in Figure 18-8, and Table 18-2 suggests that the levels of development and location on the globe are directly connected. Broadly speaking, all the LDCs have tropical locations, and all the HDCs have midlatitude locations. Such a correspondence between variables might lead us to suppose that development is a matter of natural environmental resources—and of climate in particular. Certainly the kind of conditions that we found in the climatic zones *E* and *F* in Chapter 3 are major bars to agricultural production. Natural resources do play a key role in development, but we can think of countries (Denmark, Japan, and Israel, for example) which are definitely HDCs but which have a rather limited resource base.

Two other superficial explanations of differences in development relate to race and to culture. Again, there is some apparent link between the distribution of people of Western European (Caucasoid)

Principal-components analysis

When multiple measurements are made of the same set of individuals (e.g., countries), we can usually transform the original set of variables into a new set of variables that are independent and account in turn for as much of the original variation as possible. To illustrate this concept, consider points A and B as describing the characteristics of two countries in terms of two original variables (*x* and *y*). The shaded elliptical area indicates the full set of observations, that is, the cluster of points representing all the countries examined. Using standard statistical techniques, we can identify the long axis of the ellipse (axis I, or the *principal axis*). Diagram (b) shows how this "synthetic" axis can be used as a ruler on which points A and B can be described in terms of *both* the original variables (*x* and *y*) combined. Note that this principal axis accounts for much more of the original variation within the ellipse than the secondary axis (axis II) drawn at right angles to it. This idea of "collapsing" a large set of original variables into a small number of basic dimensions or composite variables

underlies a large and expanding area of mathematics termed *principal components and factor analysis*. For a valuable nontechnical introduction to this subject, see P. R. Gould, *Transactions of the Institute for British Geography*, **42** (1967), pp. 53–86.

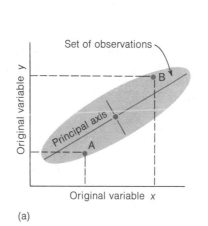

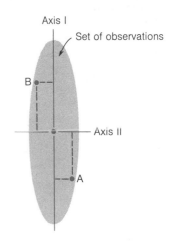

(a)   (b)

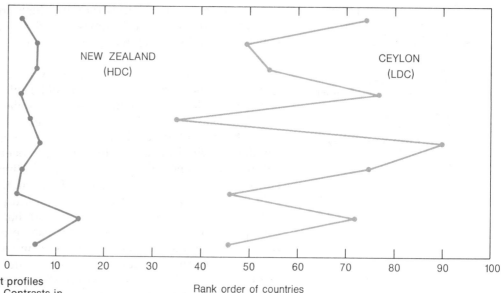

Gross national product

Infant mortality

Physicians and dentists

Food supply

Literacy

Secondary and higher education

Rail density

Motor vehicles

Energy consumption

Flow of Mail

Rank order of countries

Figure 18-7. Development profiles typical of HDCs and LDCs. Contrasts in the relative ranking of New Zealand and Ceylon on ten measures of socioeconomic development. These two countries are selected at random from Table 18–2 to illustrate the profiles of highly developed and less developed countries. Note the consistently higher performance of New Zealand on all measures compared to the more variable, but generally medium-to-low, ranking of Ceylon. All measures are calculated on a percentage or population-weighted basis to make the scores of different countries comparable.

stock and the distribution of HDCs and between people of Negroid stock and the LDCs. But again, the link is illusory, for example, Japan fails to fit into this picture. More important, people of the same racial stock have occupied quite different positions in the developmental "pecking-order" at different periods of their history and in the same locations. The economic vigor of migrant Chinese in Malaysia contrasts with the conservatism of folk of exactly the same stock in China itself.

Cultural differences also fail to provide a workable explanation of differences in development. As we saw in Chapter 10, religious beliefs do directly influence attitudes to development. A culture which is preoccupied with the hereafter or despises material prosperity is unlikely to show the same concern with development as a Rockefeller or a J. P. Morgan. But Max Weber's emphasis on the link between capitalism and the Protestant ethic looks somewhat threadbare today. The most rapidly developing countries in the last two decades (e.g., Japan, France, and the Soviet Union) have been non-Protestant.

Interlocking factors in development It is easier to destroy inadequate explanations of the contrasts in development shown in Figure 18-6(b) than it is to replace them. Climate and environment and race and culture are insufficient explanations not because they play no role in development, but because their effect is not simple. Nor is it always the same.

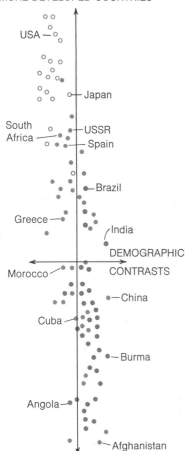

MORE DEVELOPED COUNTRIES

USA

Japan

South
Africa — USSR
Spain

Brazil

Greece
India

DEMOGRAPHIC

Morocco
CONTRASTS

China

Cuba

Burma

Angola

Afghanistan

LESS DEVELOPED COUNTRIES

● Tropical countries
○ Developed temperate countries

Figure 18-8. Generalized development yardsticks: economic and demographic scales. The vertical economic scale accounts for substantially more of the local variation than the shorter demographic scale. [After N. Ginsburg, *Atlas of Economic Development* (University of Chicago Press, Chicago, 1961), p. 113, Fig. 3. Copyright © 1961 by The University of Chicago.]

Figure 18-9 attempts to show how these factors may affect levels of development through their interaction with the four factors which economist Paul Samuelson sees as the "four fundamental factors" in understanding development—population, natural resources, capital formation (domestic or imported capital), and technology. In interpreting the diagram, it is important to note that each factor interlocks with the others. Countries showing sustained economic growth over this century (e.g., Sweden) have tended to score high on all four counts. However, countries showing slow growth may well have been held back by any one of the critical ingredients.

Contrasts in demographic stages

The second axis in Berry's development diagram (Figure 18-8) showed demographic contrasts between countries. To understand the links between the population structure of a country and its level of development, you will need to refresh your understanding of the concept of a *demographic transition,* which we first met in Chapter 6. You might like to turn back and look again at Figure 6-13.

Broadly speaking, most of the world's HDCs are in Stages III or IV of the demographic transition. They have slowly expanding or stationary populations. Conversely, most LDCs are in Stage I, and most MDCs in Stage II or III. Compare a map of the world birth and death rates (Figure 18-10) with the map of technical development in Figure 18-6(b). One of the major effects of a country's position in the demographic transition on development is related to its age distribution, which affects the size of the active labor force in relation to the total size of the population that must be supported. Consider Figure 18-11, which compares the age distribution of three countries in different demographic phases. Mexico is a country in the second phase, with a population increasing by more than 3 percent; Japan can be put in the third phase because its population increases by about 1 percent; Sweden is now increasing very slowly at a rate of less than 1 percent and is in the fourth phase. Note the contrast in the number of children (aged 0 to 14 years) in the three countries: 44 percent of the population in Mexico, 30 percent in Japan, but only 23 percent in Sweden. Likewise, the number of older folk (aged 65 years and over) is only 3 percent of the total population in Mexico, but twice this level in Japan, and three times this level in Sweden.

One of the critical questions, therefore, is whether a fall in birth rates will accompany the urbanization of the LDCs. United Nations surveys measuring the number of children 0–4 years of age in proportion to the number of women aged 15 to 44 years (i.e., potentially reproductive females) reveal a general correlation between this index

Table 18-2 Countries of the world grouped by level of economic development[a]

Less developed countries (LDCs)		Moderately developed countries (MDCs)	Highly developed countries (HDCs)
Africa		*Africa*	*Africa*
Algeria 13	Mozambique 7	Libya 2	—
Angola 5	Niger 4	South Africa 19	
Cameroon 5	Nigeria 61		*Americas*
Chad 3	Rhodesia 5	*Americas*	Canada 20
Congo 1	Rwanda 3	Argentina 23	Puerto Rico 3
Dahomey 3	Senegal 4	Chile 9	United States 209
Ethiopia 24	Sierra Leone 2	Costa Rica 2	
Ghana 8	Somalia 3	Cuba 8	*Asia*
Guinea 4	Sudan 14	Jamaica 2	Israel 3
Ivory Coast 4	Tanzania 12	Mexico 46	Japan 103
Kenya 10	Togo 2	Panama 1	Kuwait 1
Liberia 1	Tunisia 5	Uruguay 3	USSR 244
Madagascar 6	Uganda 8	Venezuela 10	
Malawi 4	United Arab Republic (Egypt) 31		*Australasia*
Mali 5	Upper Volta 5	*Asia*	Australia 12
Mauritania 1	Zaire 16	Cyprus 1	New Zealand 3
Morocco 14	Zambia 4	Hong Kong 4	
		Lebanon 3	*Europe*
Americas		Saudi Arabia 7	Belgium 10
Bolivia 4	Guyana 1	Singapore 2	Czechoslovakia 14
Brazil 95	Haiti 5		Denmark 5
Colombia 20	Honduras 2	*Australasia*	Finland 5
Dominican Republic 4	Nicaragua 2	—	France 50
Ecuador 6	Paraguay 2		Germany (East) 17
El Salvador 3	Peru 13	*Europe*	Germany (West) 60
Guatemala 5	Trinidad 1	Austria 7	Iceland 0
		Bulgaria 8	Italy 52
Asia		Greece 9	Luxembourg 0
Afghanistan 16	Laos 3	Hungary 10	Netherlands 13
Bangladesh 50	Malaysia 10	Ireland 3	Norway 4
Burma 26	Mongolia 1	Malta 0	Sweden 8
Cambodia 6	Nepal 11	Poland 32	Switzerland 6
Ceylon 12	Pakistan 65	Portugal 9	United Kingdom 55
China 751	Philippines 35	Romania 20	
India 549	Syria 6	Spain 32	
Indonesia 119	Taiwan 13	Yugoslavia 20	
Iran 26	Thailand 33		
Iraq 8	Turkey 33		
Jordan 2	Vietnam (North) 20		
Korea (North) 13	Vietnam (South) 17		
Korea (South) 30	Yemen 5		
Australasia			
Fiji 1			
Europe			
Albania 2			

[a] Latest population estimate rounded to nearest million.

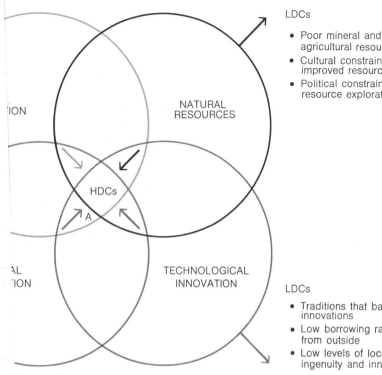

LDCs

- Poor mineral and agricultural resources
- Cultural constraints on improved resource uses
- Political constraints on resource exploration

NATURAL RESOURCES

HDCs

A

CAPITAL FORMATION

TECHNOLOGICAL INNOVATION

LDCs

- Traditions that bar innovations
- Low borrowing rates from outside
- Low levels of local ingenuity and innovation

fertility and the stage of development of a country. HDCs have generally low index levels, and LDCs generally higher levels. Some demographers have claimed that it is possible to trace the movement of countries with long demographic records, like Sweden, through the four phases. Others have suggested that birth-rate increases may follow a growing abundance of food before the death rate starts to fall. Will countries presently in the second phase move necessarily and progressively to the third and fourth phases? If so, at what rates?

We could pattern the changes in birth and death rates in Western Europe since 1700 into a general model. These changes reveal a rather consistent S-shaped fall in death rates (from about 3.3 percent to around 1.5 percent) followed by an S-shaped decline in birth rates (from approximately 3.5 percent to around 1.7 percent). The greatest rate of increase in Europe was in the mid-nineteenth century, when the lag between the two rate curves was at a maximum. But post-World-War-II fluctuations in the birth rate indicate that the demographic transition model is an oversimplification of the demographic situation in advanced countries.

| 18-3 | **SPATIAL ASPECTS OF ECONOMIC DEVELOPMENT** |

Many writers have tried to see the facts of world economic development as a linear progression through inevitable stages. For Adam Smith in the *Wealth of Nations* (1776) it was the calculus of fixed land and growing population that provided the key to a golden age in which all a country's product accrued to labor. For Karl Marx in *Das Capital* (1867) it was a one-way evolution from primitive culture, through feudalism and capitalism, to the end state of socialism and communism. For Walter Rostow in *The Stages of Economic Growth* (1960) it was again a multistage progression from a primitive society to an age of high mass consumption.

In the historical succession of economic theories, each of the major factors in Figure 18-9 has at one time or another been singled out for the dominant role. Earliest models of development stressed natural resources: Smith stressed the importance of labor, Marx of capital, and Rostow of technical innovation.

For the economic historians and development economists, the world story is a frustrating one. The facts of growth have rarely stuck

Figure 18-10. World demographic patterns. The maps show worldwide birth rates and death rates in the mid 1960s. Tints indicate different birth-rate and death-rate combinations. Consider how far the pattern of tints in any continent link with the pattern of HDCs, MDCs, and LDCs listed for that continent in Table 18-2. Countries named are analyzed in more detail in Figure 18-10. [Adapted from J. O. M. Broek and J. W. Webb, *A Geography of Mankind* (McGraw-Hill, New York, 1968), p. 442, Fig. 18-11. Copyright © 1968 by McGraw-Hill, Inc. Used with permission of McGraw-Hill Book Company.]

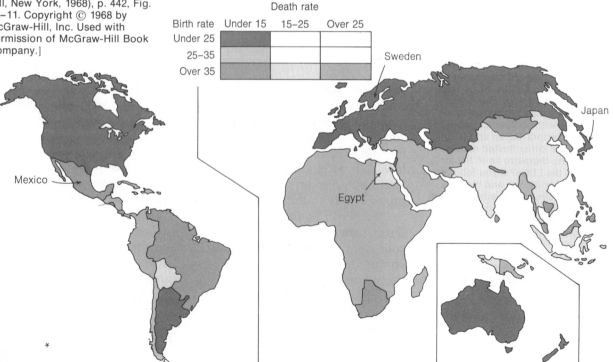

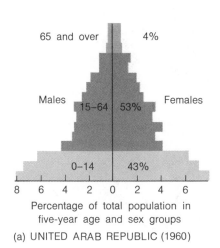

(a) UNITED ARAB REPUBLIC (1960)

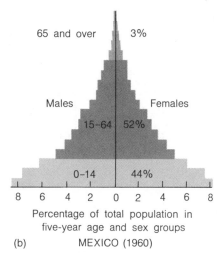

(b)   MEXICO (1960)

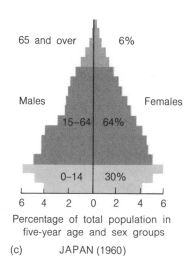

(c)   JAPAN (1960)

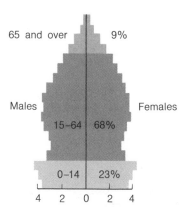

(d)   SWEDEN (1957)

Figure 18-11. Demographic contrasts. Population pyramids show the age and sex distribution of the population of four countries at different stages in the demographic transition. Note that country (a) is an LDC, country (b) is an MDC, country (c) a "recent" HDC, and country (d) a "longstanding" HDC. [Data from United Nations, Demographic Yearbooks, 1957 to 1960. Adapted from J. O. M. Broek and J. W. Webb, *A Geography of Mankind* (McGraw-Hill, New York, 1968), pp. 447–456. Copyright © 1968 by McGraw-Hill, Inc. Used with permission of McGraw-Hill Book Company.]

to the predetermined timetables of theory. But geographers have not been wholly immune to the fascination of theory building. What kind of geographic models of growth have they produced? Have they been any more successful than their colleagues in economics?

Spatial models of growth

We have already seen in this book the kind of development models a geographer builds. In Chapter 12 we reviewed the work of the Swedish geographer Hägerstrand on *spatial diffusion* models, and in Chapter 13, the parallel work on *urbanization* models. Both these sets of ideas underline the approach we shall follow, emphasizing variations in *where* development takes place and the effect it has on the changing spatial organization of the world economy.

Figure 18-12 shows a four-stage model of the spatial pattern of development of an idealized island country. It is based on the work of a group of geographers led by Edward Taaffe at Northwestern University in the early 1960s and draws heavily on Peter Gould's work on the modernization of West African countries, notably Ghana. It also

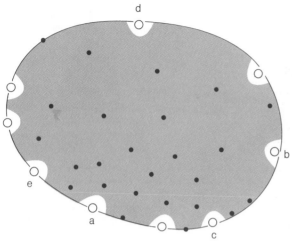

Stage I

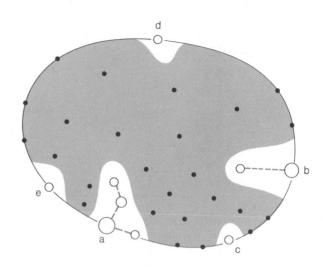

Stage II

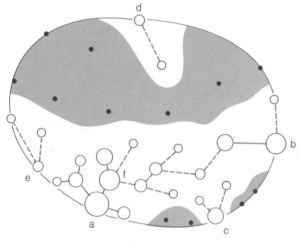

Stage III

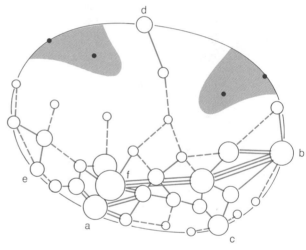

Stage IV

Outback areas not incorporated into an urban-oriented system

• Major settlements in the outback

 Westernized urban centers. Size is indicative of population at each stage

———— Transport links

Figure 18-12. The spatial structure of economic development. An idealized sequence of stages in the economic development of a hypothetical island. Note that Stage II is the critical "take-off" period in which major transport links are first driven into the interior. The four maps should be viewed as representing the processes of growth discussed in the text rather than as an exact spatial reconstruction of actual events.

has parallels with Rostow's division of development into four phases: a "traditional society," a "take-off" phase, a "drive to maturity," and a movement "toward high mass consumption."

In *Stage I* there is a scatter of small ports and trading posts on the coast. Each small port has a small inland trading field, but most of the interior villages are untouched by the coastal development. Subsistence agriculture dominates the island, apart from the few coastal pockets with trading links to the outside world. *Stage II* is the critical stage. It is roughly analogous to Rostow's "take-off" phase, which is clearly based on the analogy of an aircraft which can fly only after attaining some critical speed. Other economists have termed this stage "the spurt" or "the big push." It is marked by two geographic characteristics: first, new transport links to the interior to tap new areas of natural resource for export; and second, differential growth of the coastal centers as some expand (centers *a* and *b* in Figure 18-12), some maintain their position (*c*, *d*, and *e*), and the remainder are pinched out. Studies of African colonial areas suggest that the tapping of mineral resources and the need for political and military control are key factors in the expansion of transport links to the interior of a developing country.

Stage III is marked by the rapid growth of the transport system about each of the major ports and the emergence of important new inland centers at transport junctions (e.g., center *f*). Note also the beginnings of lateral interconnections between *a* and *b*. The northern half of the island remains isolated, but the southern half shows rapid growth and urbanization. In *Stage IV* the development of transport links continues. Note the development of high-priority shuttle lines between center *b* and center *f*. Center *f* has now taken over *a*'s role as a primate city, marking the shift from an external, export-oriented phase to one in which the island country has major internal markets of its own. The north-south transport link is now complete, and the few remaining "primitive areas" take on a new role as heavily protected wilderness areas for the overurbanized folk of the south part of the island.

Two questions are raised by this sequence. First, what processes are shaping the pattern? Second, does the sequence described match the events we actually observe?

An attempt to answer the first question is presented in Figure 18-13. This summarizes the four basic processes that have been built into our four-stage model. None of them are new to you. Each has been described in earlier parts in this book (see especially Sections 13-1 and 14-1), and you may wish to use this as an opportunity to review them. The four processes are (1) an S-shaped increase in population,

Figure 18-13. Spatial development processes. Summary diagrams of the four basic growth processes underlying Figure 18–12. Numbers I through IV relate to the four phases of growth shown in that figure.

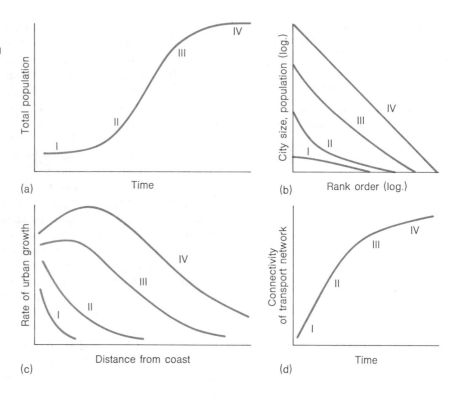

reflecting the demographic transition; (2) the emergence of a family of rank-size rules with an early primate pattern in Stage II, and a more regular form in Stage IV; (3) the launching of a series of diffusion waves for rates of urban growth; and (4) an increase in the connectivity of the transport network as development proceeds. In Figure 18-13(c), the peak of the urbanization wave moves inland in sympathy with the faster growth of the inland center at location *f*.

How far does the historical pattern of development support our idealized model? We shall mention here two pieces of evidence from the many studies geographers have done on this topic. Figure 18-14 shows the ways in which the importance of New Zealand ports has changed over a 100-year period. Note the pinching out of small ports, particularly on the west coast of the South Island, and the increasing dominance of the Christchurch area.

The second piece of evidence is more general. You will recall from the discussion of transport networks in Chapter 16 (Section 16-2) the ways in which geographers use graph theory. One simple measure of increasing connectivity is the ratio between the number of links in a system and the number of nodes. This is termed the *Beta index*. Thus

Figure 18-14. The spatial structure of economic development. A century of development in New Zealand's South Island shows some of the integration of transport links and the concentration of port facilities predicted by the model in Figure 18–13. [From P. Rimmer, *Annals of the Association of American Geographers* **57** (1967), pp. 21–27.]

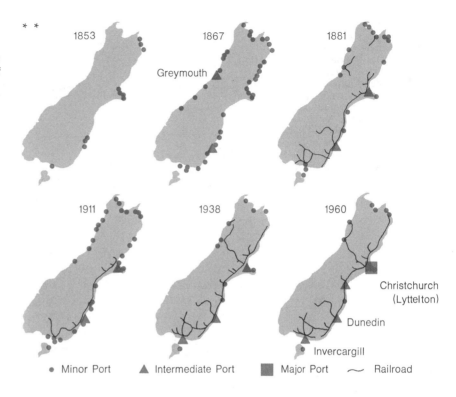

● Minor Port   ▲ Intermediate Port   ■ Major Port   ⌁ Railroad

if we have a railroad system with 12 links and 8 nodes, its Beta index is $^{12}/_8$, or 1.50. Figure 18-15 presents Beta indexes for the railway systems of several countries. The values range from around 1.33 to 0.50; when the index is less than 1.00, it indicates that the network is split into several separate subsections as in Stage II in Figure 18-15. Highly developed countries like France have high Beta indexes, whereas poorly developed countries like Ghana have low Beta indexes. The relation between economic development and the connectivity of transport networks is also shown by the changing position of a single country (French Indochina) over time. In the case of developed countries in the last few decades, we should expect the relationship of the railway network to growth to be weaker. The closure of some rail links (and therefore a decreased Beta index) would be compensated for, in such cases, by the increased connectivity of other transport links (highways, air routes, etc.).

Center-periphery models of development

An alternative approach to modeling the spatial pattern of economic development has been proposed by planner John Friedmann of UCLA.

Figure 18-15. Connectivity and economic development. Connectivity values (as measured by the Beta index) are shown for the railway systems of countries with different levels of economic development. The maps on the left indicate the evolution of a simple transport system and the resulting connectivity values. N = the number of nodes, L = the number of transport links, and β = the ratio of links to nodes (i.e., the Beta index). [After K. J. Kansky, *Structure of Transportation* (Department of Geography, Research Paper 84, University of Chicago, Chicago, Ill., 1963), p. 99, Fig. 25.]

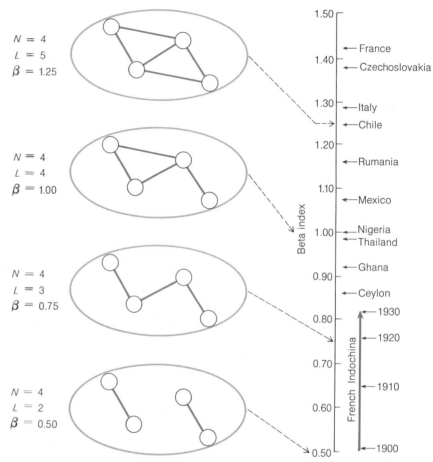

He maintains that we can divide the global economy into a dynamic, rapidly growing central region and a slower-growing or stagnating periphery. There are four main regions in Friedmann's scheme.

First, Friedmann describes *core regions*, which are concentrated metropolitan economies with a high potential for innovation and growth. They exist as part of a city hierarchy and can be distinguished on several levels: the national metropolis, the regional core, the subregional center, and the local service center. On the international level, the North Atlantic community comprising the metropolitan clusters of both eastern North America and western Europe may be regarded as a core region for development in the Western world.

Friedmann's second and third regional elements are growth regions. *Upward-transition regions* are peripheral areas whose location is relative to core areas, or whose natural resources lead to a greatly intensified use of resources. They are typically areas of immigration, but this is spread over numerous smaller centers rather than being concentrated at the core itself. *Development corridors* are a special case of upward-transition regions that lie between two core cities. A typical example of an expanding corridor region is the Rio de Janiero–São Paulo corridor in Brazil.

Resource-frontier regions are peripheral zones of new settlement where virgin territory is occupied and made productive. In the nineteenth century, the midcontinental grasslands of the world provided such a frontier region for grain and livestock production. In the present century, agricultural colonization on this massive scale is not occurring. New agricultural zones are being opened up but through much effort (e.g., the Soviet occupation of the virgin lands of Siberia or the colonization of the Oriente, the trans-Andes lowland areas of Colombia, Ecuador, and Peru). Currently, resource-frontier areas are commonly associated with mineral exploitation (the North Slope of Alaska is a good example) and commercial forestry. The continental shelves are likely to be the important frontiers of exploitation by A.D. 2000. In a similar fashion, more intensive development of unused mountain, desert, and island areas for recreational resources is rapidly increasing their status and bringing them into this category.

The fourth element in Friedmann's model is the *downward-transition region*. These regions are peripheral areas of old, established settlements characterized by stagnant or declining rural economies with low agricultural productivity, by the loss of a primary resource base as minerals are depleted, or by aging industrial complexes. The common problems of such regions are low rates of innovation, low productivity, and an inability to adapt to new circumstances and to improve their own economies.

Outside the four main types stand a few zones with special characteristics. Regions along national political borders or watershed regions fall into this group. Friedmann suggests that the four main regional types exist on various spatial scales. For instance, downward-transition areas exist on the global level (the rural part of much of the "underdeveloped world" of Latin America and Afro-Asia) and within the cities themselves (blighted and ghetto areas), as well as on the national level (the Italian South, the Mezzogiorno). They may also vary with respect to the status of the general spatial economy of which they are a part. Thus, the problems of Appalachia, a depressed area within a core region, must be distinguished from those of depressed areas *within* downward-transition regions.

Friedmann's center-periphery model is linked directly to Thünen's zoning. (See Section 15-1.) Thünen himself thought of old Western European cities and the growing centers of eastern North America as proving a North Atlantic "world city" about which global land-use zones developed. By the 1860s the fall in transport costs—both ocean rates and rates for overland travel by rail—had made the sheep and wheat lands of central North America, the Pampas, and Australasia equivalent to the outer rings in the Thünen's model. In this historical context, Friedmann's scheme fits firmly within the evolving sequence of ideas about the impact of changing accessibility of places on patterns of world development.

18-4 | CONVERGENCE OR DIVERGENCE?

Most countries are richer now than they were at the beginning of the century. The real pattern of development is thus a dynamic one, and the question of the direction of change is important. Are the rich countries growing richer—and the poor, poorer? We can use two lines of evidence in trying to decide whether international contrasts in wealth between countries are deepening or lessening: the historical evidence of statistical trends and the arguments of theoretical growth models.

Evidence from historical trends

One major block to using historical evidence is its highly variable quality. Estimates for current levels of income and the gross national product for most of the less-developed countries are crude, and reconstructions for earlier time periods vary with the assumptions made. Although the present differences between advanced Western countries and the Third World are clear, the historical trends in those differences are obscure. Even when information is available on income or production, we lack the data on comparative costs needed to translate income or production figures into meaningful comparisons of regional welfare. There is simply not enough quantitative evidence to confirm the impression of an increasing difference between the "haves" and "have nots."

In the case of continental blocks and individual countries, the situation is more hopeful. In the United States regional incomes for the nine main census divisions have converged since 1880. The trend was not, however, a steady one: In the 1920s, the regional contrasts in income actually increased rather than decreased. This was, however, an isolated phase linked to the different ways in which parts of the USA withstood the Great Depression. In the case of a smaller spatial economy, Great Britain, during the last 25 years there were rather weak tendencies toward convergence despite a very active regional equalization policy. The gap between the poorest and richest regions narrowed a bit, but the actual magnitude of this gap in Britain, as in most Western European countries and particularly Scandinavia, was already narrow. In developing Afro-Asian and Latin American countries, where regional differences in income are greater, the data available are not sufficiently accurate to make firm estimates. The impact of equalization policies on long-term trends within the Soviet Union is not really known either.

Historical and empirical studies provide no strong convergent or divergent trends in regional income. What evidence there is points to rather weak and unsteady changes rather than strong and headlong processes. Regional changes have a complex spatial pattern with different trends operating in various ways on various scales; thus it is probable that divergence and convergence are occurring simultaneously on different spatial levels. Which appears to be taking place may be a function of the levels for which data are available.

Evidence from theoretical models

Theoretical models of regional growth have been largely produced as byproducts of general economic theory. They cover only certain parts of the regional growth and have little geographic detail. Here we look at the conditions of a few of the economic models.

Swedish economist Gunnar Myrdal has stressed that economic market forces tend to increase, rather than decrease, regional differentiation. The buildup of activities in prosperous, growing regions influences the less prosperous, lagging regions through two types of induced effects: spread effects and backwash effects.

Spread effects The positive impacts on all other regions of growth in a thriving region are called, by Myrdal, *spread effects*. This impact comes from the stimulation of increased demand for raw materials and agricultural products and the diffusion of advanced technology. Thus, to give a simple example of a spread effect, the medical services in a poor country may gain from the advances in drug therapy conducted in an advanced country without itself having to meet the high costs of the initial research.

Backwash effects *Backwash effects* of agglomerated growth are net movements of population, capital, and goods that favor the development of the growing area. One classic example of the backwash effect is the "brain drain," typified by movements of medical doctors to the United States from poorer countries. In this and similar selective migration flows the poorer region loses its most highly skilled workers. In a more extreme form, it may also lose its most active population (say, aged 20 to 40 years), leaving only the young and the old behind.

These two opposing forces do not imply the existence of an equilibrium situation. Indeed, Myrdal maintains that the two effects balance each other only rarely. What is more likely is a cumulative upward or downward movement over a considerable time that leads to long periods of increasing regional contrasts (Figure 18-16).

Although Myrdal's model of economic growth has been criticized for its qualitative nature and lack of econometric substance, the more formal models of regional economic growth *also* fail to demonstrate conclusively the direction of movement. One modern model of the regional system (the Harrod–Domar model) maintains that interregional growth leads to divergence. Rapidly growing regions have high levels of income and net inward movements of labor and capital. In contrast, more traditional models of regional growth indicate that, though fast-growing regions have a net inward movement of capital, income levels in these regions are low and the net movement of labor is outward. Table 18-3 summarizes the findings of three models of regional growth.

Figure 18-16. A model of economic development. Myrdal's model of cumulative upward causation. Note that the word "former" in a box refers to the industries described in the preceding box, that is, the box from which the impulse (shown by the arrow) is being received. By reversing the values in the boxes, a cumulative downward spiral can be substituted for the upward spiral shown here. [From D. E. Keeble, in R. J. Chorley and P. Haggett, Eds., *Models in Geography* (Methuen, London, 1967), p. 258, Fig. 8–4.]

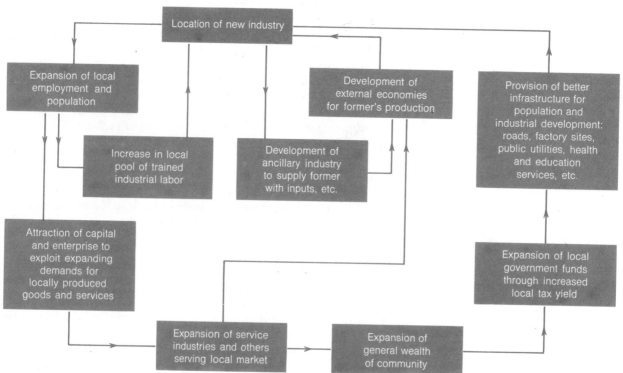

Table 18-3 Alternative economic models of regional growth*[a]*

	Characteristics of fast-growing regions			
Economic model*[b]*	Income levels	Direction of labor flows	Direction of capital flows	Direction of interregional growth
Model 1 (Neoclassical)	Low	Outward	Inward	Convergence
Model 2 (Harrod–Domar)	High	Inward	Inward	Divergence
Model 3 (Export base)	Unspecified	Inward	Outward	Unspecified

[a] Data from H. W. Richardson, *Elements of Regional Economics* (Penguin, London, 1969), p. 58, Table 2. Copyright © Harry W. Richardson.
[b] For a description of the three models and their theoretical assumptions, see Richardson, pp 47–55.

The conflicting conclusions of these economic models is disturbing, yet the results follow logically from the assumptions in each case. What is significant is that, on the basis of theoretical analyses, we can argue that either regional convergence or regional divergence will result from interregional growth. It all depends on which model we believe best describes the present pattern of the world economy. Like the historical evidence on trends, theoretical evidence on where economic growth is heading is less conclusive than we would expect from our intuitive sense of an ever-widening gap between rich and poor countries.

Future trends
Despite the scarcity of evidence and the lack of agreement between economists, there is still an abundance of projections of future trends. Most of these start with trends in population. By A.D. 2000 it seems likely that the world population will have nearly doubled and that there will be 6.4 billion of us. The average annual rate of growth assumed in this estimate is a little below 2 percent. According to current United Nations estimates, Africa and Latin America will be the two fastest-growing continents (both with a 2.7-percent annual increase). Yet the bulk of world population will continue to be in Asia, which has 58 percent of the world total. If we use the definitions in Table 18-2, then LDCs have about two-thirds of world population today. By A.D. 2000 this share will have risen to three-quarters.

What makes these figures disturbing is that they are not matched by equivalent projected changes in the gross national products of the countries in the two groups. (*Gross national product* [GNP] is the economist's term for the total value of goods and services produced by a country during a given time period, usually a year.) The present ratio between the GNP of the LDCs and the developed world (HDCs and MDCs) is 15 percent to 85 percent, and this seems unlikely to change much by the end of the century. For both worlds the per capita GNP is likely to rise—by about 3 percent a year—so standards of living will more than double by the end of the century. The rise will probably be faster in developed countries. The present gap in living standards between the two worlds, now standing at 12:1, will widen in favor of the developed world to around 18:1.

Individual world regions display enormous variations within this range. Figure 18-17 presents the population and per capita GNP of 13 major regions in 1965 and the projected populations and GNPs for the year 2000. In studying the graph, note that both axes are logarithmic. The five development classes (pre- to post-industrial) are arbitrary. If these projections are correct, there will be strong growth process accompanied by divergence not only between the richer and poorer countries of the world but also among the current leading countries. In other words, the rich areas will get richer and the poor areas will become poorer. As we have noted, however, these projections are based on incomplete evidence and probably exaggerate the degree of divergence.

Zero economic growth?

In this chapter we have looked at some of the current patterns of rich and poor countries in economic terms. Most of the world's countries are very poor compared to the United States. Less than ten of the world's countries have gross national products greater than that of the state of California. Given their high standard of living and the concern over population pressures, resources, and pollution that we described in Part Two of this book, it is wholly understandable that a zero economic growth (ZEG) movement should have arisen in the Western world. The idea that we must return to a self-sustaining economic system to prevent an ecological disaster has been strongly argued by figures like Rachel Carson, Paul Ehrlich, and Barry Commoner. We shall look at some computer models of the future growth dilemma in Chapter 21.

Whatever the appeals of ecological arguments for ZEG, a geographer must note that they are not equally convincing all over the globe. Readers of this book from LDCs (and such readers are likely to be very few) will be unimpressed by antipollution arguments when they still want to develop the industry that will pollute and use resources. Concern about pollution is understandably low in countries where famine and disease are the front-line problems. And do the readers from the fortunate lands of North America, Western Europe, and Australasia

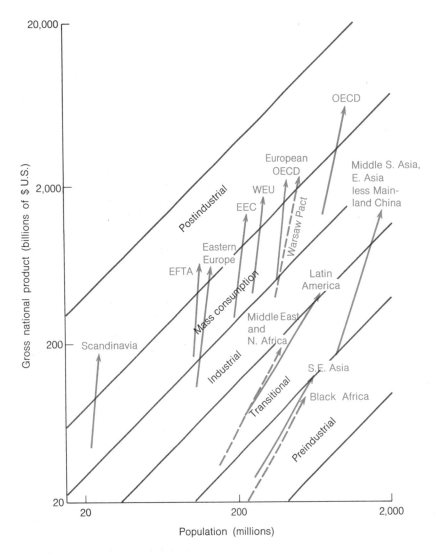

Figure 18-17. Projected changes in economic development for major world regions. The arrows indicate changes in population and gross national product between 1965 and 2000. (See also Figure 21–15.) EEC = the original six common market countries of the European Economic Community. EFTA = the European Free Trade Association. WEU = the Western Economic Union. OECD = the Organization for Economic Cooperation and Development (18 European countries plus the United States and Canada). [From H. Kahn and A. J. Weiner, *The Year 2000* (Macmillan, New York, 1967), p. 155, Fig. 13. Copyright © 1967 by the Hudson Institute, Inc.]

really want their incomes to be frozen for the next few decades — or want half their own country's GNP diverted to international aid?

In the longer run, many changes are possible. Once we start measuring growth in terms of net social welfare (NSW) rather than GNP, the developed lands may look considerably less affluent than we now believe them to be. The possibility of regional revenue sharing and equalization appears, currently, to be greater at the more restricted spatial scale of the region than on the global scale. It is to this within-country level of spatial organization that we turn again in the next chapter.

Reflections

1. Why are so many of the world's less developed countries located in the tropics? Do you rate the climate of tropical countries (a) a major factor or (b) an irrelevant point in your explanation?

2. What is happening in Figure 18-5? Redraw the diagram, making the tariff high enough to prevent a flow of exports from Country 2 to Country 1.

3. What effect does a country's population pyramid have on its level of economic development and vice versa? Draw population pyramids to illustrate your points.

4. Consider the changing connectivity of the networks in Figure 18-15. Increase the number of towns (nodes) from 4 to 6, and draw links to illustrate a range of Beta index values from 0.5 to 1.5.

5. Write down your own views on zero economic growth (ZEG). Debate within the class whether this is (a) feasible and (b) desirable. Do you think you would hold the same views if you lived in a less developed country?

6. How do you expect your own country's share of the world's wealth to change between now and the end of the century? Why? Compare your views with those of your classmates.

7. Select any one LDC from Table 18-2 that interests you. Use encyclopedias and reference books to look into its background. Which factors appear to be the most critical in accounting for its low level of development?

8. Review your understanding of the following concepts:
   (a) tariff barriers
   (b) technical development indexes
   (c) demographic indexes
   (d) population pyramids
   (e) the Beta index
   (f) core-periphery models
   (g) spread effects
   (h) backwash effects
   (i) gross national product (GNP)
   (j) zero economic growth (ZEG)

One step further . . .

A world view of regional inequalities and their spatial distribution is given in
   Ginsburg, N., Ed., *Atlas of Economic Development* (University of Chicago Press, Chicago, 1961) and
   Stamp, L. D., *Our Developing World* (Faber, London, 1963).

Conditions that lead to the emergence of regional problem areas are summarized in
   Hoover, E. M., *Location of Economic Activity* (McGraw-Hill, New York, 1948), Chaps. 9–11, and
   Perloff, H. S., *et al.*, *Regions, Resources and Economic Growth* (Johns Hopkins Press, Baltimore, Md., 1960).

A variety of possible approaches to regional planning problems is provided in

Friedmann, J., and W. Alonso, Eds., *Regional Development and Planning: A Reader* (M.I.T. Press, Cambridge, Mass., 1964).

Economists' views of the economic-development process are given in
Hirschman, A. O., *The Strategy of Economic Development* (Yale University Press, New Haven, Connecticut, 1958) and
Myrdal, G., *Rich Lands and Poor* (Harper & Row, New York, 1957).

The case against "growth" is presented in
Mishan, E. J., *Costs of Economic Growth* (Penguin Books, Harmondsworth, 1967).

Turn to the regional readings listed in Appendix C for guidance on the geographic literature for individual countries.

Chapter 19

Inequalities Within Countries

WELFARE ISSUES IN REGIONAL PLANNING

"I have a dream. . . . I've been to the mountain top . . . I've seen the promised land"
—MARTIN LUTHER KING, JR.
Speech from the Lincoln Memorial, Washington, D.C., August 28, 1963.

At the time of his death, Martin Luther King was preparing for a "poor people's march" on Washington, D.C., in early 1968. This march was to focus public attention on the deeply complex problems of poverty in America. Readers of the previous chapter, with its graphs and charts showing the United States heading the table of highly developed nations, must find it paradoxical that we begin a chapter on regional inequalities by focusing on the richest country in the world. When we talk about a country as being an HDC, MDC, or LDC, we are referring to the *average* conditions of the country taken as a whole. This ignores the strong regional variation within the country. To find pockets of poverty within the United States (say, parts of rural Appalachia) is no more noteworthy than to find pockets of affluence within an LDC (say, the rich Copacabana Beach area within Brazil). Variation in development occurs at all spatial levels, both within and between countries.

In this chapter we concentrate on differences *within* countries, focusing especially on the spatial inequalities in welfare within a country. But what is an inequality—and what is welfare? And once we define these things, how do we define a just distribution of welfare? These issues dominate the first part of this chapter and form a backdrop to the study of regional-planning practices in the second part. There we look at how Western European countries tackle the problem of poor regions. Finally, we illustrate the kind of issues raised at the local level within the broader framework of regional planning.

19-1 SPATIAL INEQUALITIES AND WELFARE ISSUES

Before we look at the details of regional-planning practices, we must investigate three more general issues. What is spatial inequality? What is social welfare? And what is a geographically just distribution? None of these questions can be answered precisely, since the answer to each depends on the reader's view of society itself. This first section may perhaps serve as a framework for a debate which runs far outside the bounds of our particular focus of inquiry.

Lorenz curves: Measuring inequalities in welfare

Our first question is a technical one, and produces much less emotional reactions than the others. Given that differences between regions exist, how should we measure them?

One of the most useful measures of inequality is the *Lorenz curve* (Figure 19-1). This is a graphic representation of the distribution of any measure of welfare (e.g., income). If it is perfectly straight, the distribution is perfect. The more bowed a Lorenz curve is, the more unequally the proxy for welfare is distributed. The difference between an actual

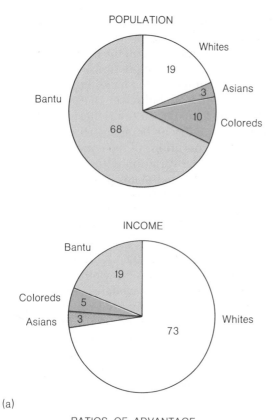

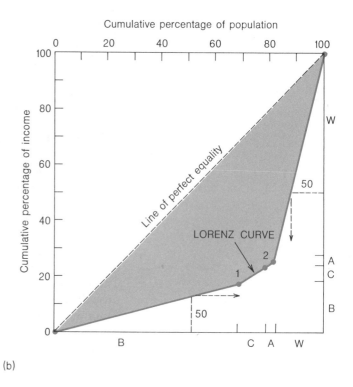

RATIOS OF ADVANTAGE

White (W) 73/19 = 3.84   Coloreds (C) 5/10 = 0.50
Asians (A) 3/3 = 1.00   Bantu (B) 19/68 = 0.28

Figure 19-1. Measures of inequality.
(a) The distribution of population and income among the four main ethnic groups in South Africa. (b) A Lorenz curve showing the inequalities between the four groups. The shaded area represents an "inequality gap." [South African data for 1967 from D. M. Smith, *An Introduction to Welfare Geography*, University of Witwatersrand, Johannesburg, Department of Geography, Occasional Paper No. 11, p. 95, Table 5.]

Lorenz curve and a straight line is called an *inequality gap.*

The construction of Lorenz curves is illustrated in Figure 19-1, using the distribution of the income in South Africa as a measure of the distribution of welfare there. If we divide up the South African population (some 21.5 million in 1970) into major ethnic groups, there are strong contrasts in income per capita. The Bantu peoples make up about two-thirds of the population and receive about one-fifth of the income. Conversely, the 3.7 million white South Africans make up about one-fifth of the country's population but control nearly three-quarters of its national income.

By dividing their share of the country's income by the percentage of the population they include, we can compute a *ratio of advantage* for each ethnic group. Thus for the Bantu, this ratio is 19 (their share of income) divided by 68 (their share of the population), or 0.28. Ratios above 1 indicate that a group is better off than the average group in the nation, and ratios below 1 that it is worse off. To draw a Lorenz curve, we take the group with the lowest ratio, in this case, the Bantu, and plot its position on a graph of population and income (point 1 in Figure 19-1). We then take the group with the next-lowest ratio, the black population (about 2 million strong) and add its shares of population and income to those of the Bantu. We then plot the position of the two groups together (their cumulative position). This is point 2. The Bantu and the blacks together make up 78 (68 + 10) percent of the population and share 24 (19 + 5) percent of the country's income. The shares of remaining groups are added in the same way, and further cumulating percentages of population and income are plotted.

In addition to being simple to construct, Lorenz curves have a number of properties that are useful in studying inequalities. If we look at the two 50-percent marks in Figure 19-1, we can see that the lower half of the population receives only around 13 percent of the country's income; conversely, half the income goes to only around 15 percent of the population. South Africa is used here only as an example. Lorenz curves for all the world's countries have this characteristic convexity. Inequality exists everywhere, though the degree of inequality varies. This variation can be shown by comparing countries with very different inequality gaps. An LDC like Thailand has a much greater gap than the HDCs. Even within the advanced countries, there are strong differences in the size of the gap. Sweden, with its progressive taxation system, has a Lorenz curve much closer to the line of perfect equality than the United States.

Alternative indicators of welfare

So far we have used income to measure welfare. It is, however, only one index, and a very crude one, of the social welfare of an area. A *social welfare indicator* is simply a term for a statistic which shows the change from "worse-off" to "better-off" areas. As you might guess, it has proved extremely difficult to get people to agree on such measures, since they must first agree on what social welfare is.

From the geographic viewpoint, there are two aspects of social indicators we should note. First, no two indicators tell exactly the same spatial story. Consider Figure 19-2, which shows contrasts in the ten provinces of Canada in the 1960s. These contrasts have been measured by four different indicators of social welfare: (a) per-capita income; (b) higher education, as measured by the number of college enrollments; (c) average unemployment rates; and (d) average job-participation rates for males. Each of these indicators tells us some-

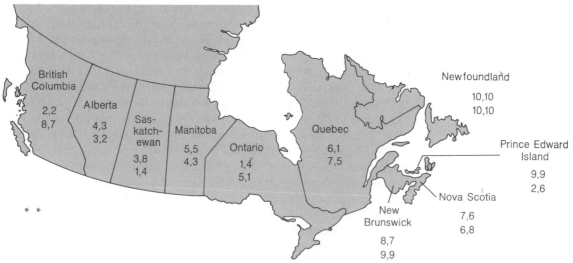

British
Columbia

2,2
8,7

Alberta

4,3
3,2

Saskatch-
ewan

3,8
1,4

Manitoba

5,5
4,3

Ontario

1,4
5,1

Quebec

6,1
7,5

Newfoundland

10,10
10,10

Prince Edward
Island

9,9
2,6

Nova Scotia

7,6
6,8

New
Brunswick

8,7
9,9

* *

Figure 19-2. Spatial variations in socioeconomic health. Each of the ten Canadian provinces is ranked using four alternative measures of income and job opportunities. "Healthy" provinces have low rank scores, and "unhealthy" provinces have high ones. Only Newfoundland, which ranks 10th every time, has wholly consistent scores. [Data from P. E. Lloyd and P. Dicken, *Location in Space* (Harper & Row, New York, 1972), p. 207, Table 10–5.]

thing about the "health" of the provinces in terms of, say, jobs and the potential skills of the population. In Figure 19-2 we have superimposed on each province its position in the marked list of Canadian provinces with respect to each indicator: A 1 indicates a position at the top of the table (the "best" score), and a 10 indicates a position at the bottom of the table (the "worst" score).

Of the ten provinces, only Newfoundland has a consistent set of scores. The problems of the Maritime provinces show up, as does the relative prosperity of Ontario and Alberta. British Columbia is strikingly well off in some ways but badly off in others.

Ranking is of course a rough measure of contrasts, since small differences may force a province into a low position. "Roughness" is also a problem in Figure 19-3, which moves south of the border and looks at interstate contrasts in the United States. Here the nine measures are much more complex. They were produced by Kansas economist J. O. Wilson from the "domestic goals" proposed by President Eisenhower's 1960 Commission on National Goals. Each index was created by collapsing a large number of raw measures of welfare in much the same way measures of wealth were collapsed in Chapter 18.

The profile of each state shows its rank position on each index. Profiles are plotted for three states with different average performances. Minnesota is in the upper half of the ranked list of all states, with particularly high rankings on indicators II and IX. North Carolina is in the lower half and Missouri is consistently in the middle of the table. No state has an entirely consistent pattern of rankings. For example,

SOCIAL WELFARE INDICATORS

I   Status of the individual

II   Racial equality

III   Democratic process

IV   Education

V   Economic growth

VI   Technological change

VII   Agriculture

VIII   Living conditions

IX   Health and welfare

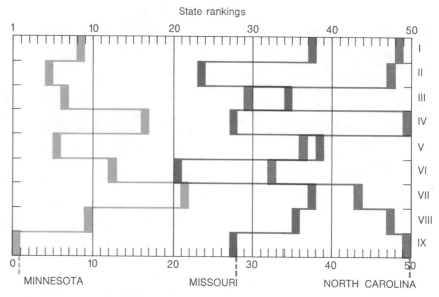

Figure 19-3. Interstate differences in the quality of life. The graph shows the relative ranking of three states (out of the 50) in terms of nine social welfare indicators. Low rankings indicate a high quality of life; high rankings indicate poor conditions. The nine indicators were derived by statistical procedures from 85 different variables. [Data from J. O. Wilson, *Quality of Life in the United States,* Midwest Research Institute, Kansas City, Missouri, 1969, p. 13.]

California, whose position on the first eight indicators ranges from first to fourth, plunges to 14th position on the health and welfare index. Hawaii ranks first (ties with Utah) with respect to racial equality but 40th with respect to technological change.

The second general question raised by social welfare indicators is one of stability. Do we wish to measure inequality in terms of an area's position in any one year or decade? Or should we be more interested in rates of change? Figure 19-4 illustrates the contrast between the answers to the two questions in a study of the quality of life in 18 metropolitan areas of the United States. Urbanists Jones and Flax, on whose data the figure is based, wanted to measure how Washington, D.C., stood in relation to other major cities and to see whether living conditions there were getting relatively better or worse. Using seven measures of the quality of life (ranging from housing costs through robbery rates to pollution levels), they came up with the rankings shown on the longer, horizontal axis of the chart. Minneapolis headed the list of high-ranking cities, and Los Angeles trailed the list of low-ranking ones. Washington, D.C., did not do too badly, but San Francisco—a shock to Bay Area fans (including the author)—was next to the last.

How were conditions in these centers changing? The results of a second study on changes during the 1960s are shown on the shorter, vertical axis of the chart. While conditions in a number of low-scoring

Figure 19-4. Intercity comparisons of social welfare. An attempt to see how Washington, D.C., ranked in relation to other metropolitan areas with respect to living conditions produced the results shown here. Each city is ranked on both the quality of life there (on the horizontal axis) and on changes in that quality in the 1960s (on the vertical axis). The high scores of Minneapolis and Boston were no surprise, but the poor showing of San Francisco was wholly at variance with its image. Note that all urban data are sensitive to the exact city boundaries used and how much of the suburban area is included. [Data from M. V. Jones and M. J. Flax, *The Quality of Life in Metropolitan Washington, D.C.: Some Statistical Benchmarks,* The Urban Institute, Washington, D.C., 1970.]

cities like New York and Chicago were getting better, the quality of life in Washington, D.C., was getting relatively worse. In San Francisco, according to the study, conditions were deteriorating more rapidly than anywhere else. Of course, measuring changes over such a short time period may be misleading, since the results may reflect short-term influences such as a particular mayoral program.

Spatial aspects of social justice

Let us assume that we have agreed on a proper measure of welfare and that we have found marked inequalities between provinces, states, metropolitan areas, or other regions. What then? A concern with spatial inequalities in welfare must raise, for geographers, ethical issues which have troubled philosophers from Aristotle to Marcuse. As Johns Hopkins geographer David Harvey puts it, how *do* we achieve "a just spatial distribution, justly arrived at."

What kind of claims can the people of any disadvantaged region make on the larger, national community? From the welter of conflicting suggestions, three major ideas stand out. First, they can make claims based on *need.* We may argue that all parts of a country should have a basic right to a certain standard of education or medical care, regardless of spatial differences in the cost of providing those services. Postal costs in remote rural areas in most countries are well above the national norm, but postal charges are usually common all across the country. Second, they may make claims based on their *contribution to common good.* Areas which contribute greatly to the good of the whole nation might be expected to receive a greater-than-average payment. A special subsidy to a city faced with the problem of preserving costly old buildings of historical value (e.g., as Amsterdam is in Holland or Venice is in Italy) would be merited because of the city's contribution to the heritage of the whole country. Third, the people of a region may make claims on the larger community on the basis of *merit*. The environmental challenge to human life varies from region to region. The allocation of extra resources to a region might be justified to protect areas of great potential stress from natural hazards (e.g., to protect marginal agriculture in semiarid lands) or from social hazards (e.g., to reduce crime in high-risk areas in inner cities and ghettos). Of these three bases for the distribution of social resources, need is arguably the primary one, with contributions to the common good and merit running second and third, respectively. Readers interested in following up these lines of thinking should definitely browse through David Harvey's *Social Justice and the City*. (See "One step further . . ." at the end of the chapter for bibliographic details.)

Let us stay with the simple concept of need and look at its spatial implications. Figure 19-5 shows how different philosophies might af-

Figure 19-5. Enigmas in spatial justice. The charts reveal the spatial implications of four alternative approaches to revenue-sharing in a simplified five-region country. The arrows indicate the directions of transfers to the subsidized areas (shaded). Rearranging the diagrams (putting the poorer areas at the center and the richer ones on the periphery) would give us a picture of problems in big cities today.

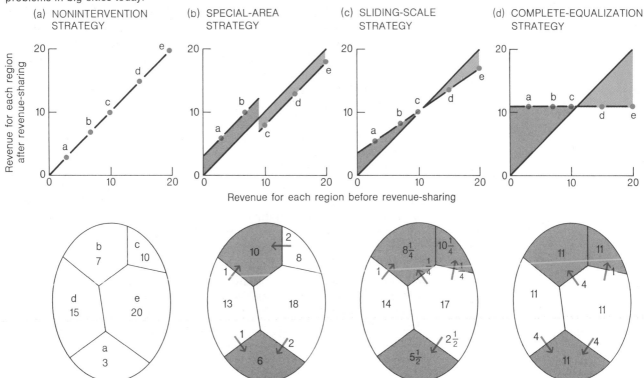

fect a spatial redistribution of wealth in an idealized five-region country. Four situations are described. In the first, the government's policy is a "laissez-faire" one of nonintervention, with no revenue sharing [Figure 19-5(a)]. In the second, problem areas (the two poorest areas) are designated as needing special help [Figure 19-5(b)]. This kind of approach—having a special strategy for special areas—poses problems we shall consider later in the chapter. In the third situation, there is a sliding-scale approach to regional needs [Figure 19-5(c)]. This is rather like the negative-income-tax approach, in which rich areas subsidize poor areas on the basis of need. In the fourth, a complete-equalization approach is taken and taxes are adjusted so that all areas are brought to the same income-level [Figure 19-5(d)]. In each case, a different interpretation of social justice leads to a different spatial pattern of levels of welfare. These cases are, however, highly simplified, and deal with a hypothetical country. We turn now to realistic examples of current regional-intervention policies in Western Europe and North America.

19-2 | INTERVENTION AT THE REGIONAL LEVEL

Should central governments intervene to adjust regional inequities—or should they allow the normal equilibrium-seeking forces like migration to operate? If the locational theory stressed in Part Four of this book tells us anything, it is this: First, spatial specialization occurs because it is an efficient way to use immobile resources; and second, the basis of specialization is continually changing. That is, the spatial economy of a country appears to be both specialized and dynamic. We should expect that the emergence of new centers would be accompanied by the obsolescence of earlier, and now less efficient, centers of regional production.

What appears to be at issue is not the necessity of spatial change but who should pay the costs involved. On the local scale the blight and decay in our city centers are part of the price paid for the benefits the automobile has brought to suburban areas. On the regional scale the limited opportunities of some old coalfield areas are a legacy of their specialized role in preceding decades. Few people would support a complete equalization policy like that in Figure 19-5(d) which, by moving resources to the population, effectively froze the present spatial pattern; but more observers are asking whether the costs of spatial changes should fall solely on a local part of a city any more than they should fall on a local part of a country. The logic of the argument does not stop at national boundaries; it has implications for future revenue-sharing on the international level.

Tools of regional policy

If central governments wish to intervene in the regional growth process, what tools can they use? Three main strategies are currently in use.

The first is investment in the public sector. Regional investment of this type spans the construction of entire new cities in underdeveloped regions, like Brasilia in west-central Brazil [Figure 19-6(a)], to the building of new schools in a city ghetto. It is commonly aimed at improving the basic infrastructure of a region. Transport and power facilities are generally the prime targets for improvement. The dam-building by the Tennessee Valley Authority [Figure 19-6(b)] and the road-building in the current program for developing Appalachia exemplify the emphasis in regional-support legislation.

Inducements to business in the private sector to invest in a region are a second strategy. These inducements may be positive, such as capital grants or tax concessions to industries already operating or willing to operate in undesirable areas, or negative, such as controls and penalties for companies in fast-growing areas. Companies operating in rapidly developing areas may have proportionately higher taxes and face legal restrictions on their expansion. Positive and negative inducements can be combined, as in Great Britain, where factories have been built in stagnating areas to encourage companies to locate there.

Inducements to individuals and households to locate in or leave a region are a third strategy. Migration from a declining area may be hindered by the inability of a would-be migrant to sell his house or land. Compensation to those farmers willing to move (plus aid to enlarge the farms of those who stay) is offered by the governments of both Ireland and Sweden to make migration from agricultural areas with low and declining prospects easier. When the need is to attract population, similar monetary and other inducements are used. An example is the Ceylonese government's scheme for agricultural colonization of the island's dry zone. This kind of regional strategy is usually linked with a strategy of investment in the public sector so that an improved infrastructure, such as land clearance and irrigation in Ceylon, is created before the colonists arrive.

The choice of regional policy tools depends largely on the resources available in the country as a whole and, more importantly, on its sociopolitical system. For instance, Britain has tried to resolve the problem of unemployment in its peripheral coalfield areas by loans for new industry (initiated in 1934), building factories and industrial estates (in 1936), providing tax incentives for new industry, including depreciation allowances (in 1937), controlling industrial buildings outside the problem areas (in 1945), and awarding standard grants for new

(a)

(b)

(c)

Figure 19-6. Tools of regional policy. (a) A large-scale national reorientation of resources underlay the building of a new federal capital at Brasilia by the Brazilian government. The new capital lies on the central plateau of Goias, about 900 km (560 mi) inland from the old coastal capital of Rio de Janeiro. Begun in the late 1950s, the new capital city has a population now approaching 500,000. (b) Public power and transportation program sectors are typified by the Tennessee Valley Authority program in the 1930s. (c) New agricultural settlements in Israel's southern arid zones are financed by the government. [Photographs (a) by Erwitt, Magnum, (b) courtesy of the United Nations, and (c) courtesy Israel Information Services.]

industrial buildings and new machinery (in 1960). The measure of central government control of industry's location in Britain is representative of that in a politically middle-of-the-road state with a mixed economy and a strong commitment to equalizing job opportunities between regions. States following a less socialist policy tend to allow industry to decide its location on commercial grounds; those following a more socialist policy favor a greater degree of direct control by central government.

Defining areas of need

To allow regional policies to be introduced, we must draw a clear distinction between problem and nonproblem areas. The boundary chosen should clearly reflect the character of the inequality. To take a simple example, in the 1930s the Brazilian government defined the boundaries of the Nordeste (a problem region in the northeast "bulge" of that country) by rainfall figures. There the key problem was recurrent drought, leading to crop failure and famine, so it made sense to draw a boundary around the aided area in rainfall terms. Regions within the drought boundary received special government help, while regions outside did not. Since planning regions involve drawing lines through areas which may or may not show sharp breaks on the map, they raise just the same questions that we considered in our study of environmental and cultural regions earlier in this book. (See Sections 5-2 and 10-5.) Table 19-1 shows two examples of planning regions. Example 1 is distinguished by a single feature (a watershed), is based on a nodal unit, and is built by "splitting" (i.e., the watershed is split away from the rest of the area round about it). By contrast, Example 2 is distinguished by several measures of distress, is based on uniform criteria, and is built by grouping (local districts are grouped together to construct the planning region).

We encounter much greater difficulties when a problem region is defined by population characteristics and levels of distress. In Great Britain the basis for preferential regional treatment is largely unemployment. Recent legislation has scheduled special aid for those localities where, in the opinion of the government, a high rate of unemployment exists or is imminent. In practice, an unemployment rate of 4.5 percent has been adopted as the critical cutoff point.

We can think of a number of drawbacks to this index. Unemployment rates are very unstable over time. They tend to underestimate job opportunities; that is, areas of high unemployment also have out-migration and a low proportion of folk in jobs (e.g., there may be no jobs for women in the labor market). This means that the official unemployment figures are too optimistic in problem areas. Thus, before

Table 19-1 Types of regions

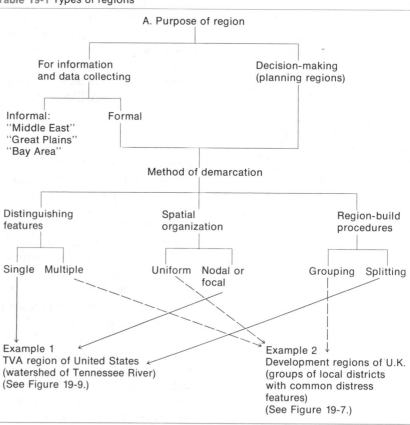

we can establish any realistic index of relative welfare, we need to modify our definition of unemployment to include other variables related to migration or wage levels. We noted this problem earlier in relation to defining rich and poor countries. We can also develop general indexes of economic health or distress on a smaller spatial scale.

No matter what index of regional need we adopt, we are left with the question of where to draw the line. If we adopt a 4.5 percent unemployment rate to define the regions which will receive aid, then we invite protests from regions just below the threshold. An extreme if simplified illustration is shown in Figure 19-5(b) where unaided region c is made worse-off than aided region b. The limit may conceal sharp contrasts within the distressed region itself. The difficulty is made more acute by the fact that the same index may yield different values if different regional subdivisions are chosen. Generally, the smaller the system of subdivisions, the greater is the spread of index values, and vice versa. This gives great scope for gerrymandering the boundaries of a distressed region to make it qualify for government aid.

One reaction to this problem has been the replacement of a twofold division (between regions that receive aid and those that do not) by three or more levels of aid. Such a system has the theoretical advantage of matching the amount of aid to the degree of distress, but is objected to on the grounds of increasing administrative costs. In England and Wales there is now virtually a four-level system of regions: (1) areas of rapid growth and high prosperity subject to negative controls on further expansion (e.g., Greater London), (2) "normal" areas which require neither positive nor negative intervention (e.g., most of southern England), (3) areas with moderate economic difficulties (e.g., Plymouth), and (4) depressed areas (e.g., South Wales) receiving the full range of government assistance. (See Figure 19-7.) France has a similar five-stage series of zones, with metropolitan Paris and rural Brittany at opposite ends of the scale.

The difficulty of having a fully differentiated scheme of regional assistance is that, at one extreme, it heavily penalizes the very productive regions, while at the other it grants massive help to the least productive ones. If it was successful, it would freeze the existing geographic pattern of production. Such a freezing is, of course, wholly at variance with the change and diffusion already discussed. Indeed, it would be disturbing to think what effect a fully successful regional aid policy would have had on the United States if it had been in operation in 1920, or in 1820! There are clearly some regions (e.g., worked-out mining areas) where the levels of unemployment are so high and the prospects so poor that it is hard to justify a policy other than strategic withdrawal. However, *within* problem areas discriminatory policies should favor those parts of the problem regions that have the greatest growth potential. It is to this idea that we now turn.

Growth poles

The notion of productive points within, or as near as possible to, distressed areas as centers of new investment is an attractive one. It has been strongly urged by French regional economist Francis Perroux in his concept of the growth pole (*pole de croissance*). Basically, a *growth pole* consists of a cluster of expanding industries which are spatially concentrated (usually within a major city) and which set off a chain reaction of minor expansions throughout a hinterland.

A growth-pole policy in regional development means the deliberate selection of one or a few potential poles in a problem area. New investment is concentrated in these areas rather than being spread thinly

Figure 19-7. Zones of government aid. The United Kingdom is divided into four types of areas which receive varying degrees of incentives to industrial development. Northern Ireland has its own system of incentives, broadly similar to those of the British developmental areas.

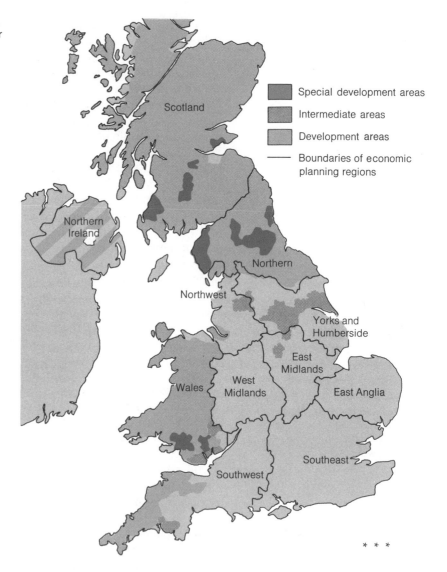

"in penny packets" over the whole area. The arguments in favor of this policy are that public expenditures are more effective when they are concentrated in a few clearly defined areas and that new industries there will stand a better chance of building up enough agglomeration economies to achieve some degree of self-generating growth. Agglomeration economies are the benefits that come from the sharing of common infrastructure (roads, power supplies, water, etc.), the increased size of the labor market, and reduced distribution costs. Although the evidence is conflicting, the necessary population of a growth pole may lie between 150,000 to 250,000 people. Only in cities of this size will there be the basic ingredients of large-scale diversified growth. Whether spatial concentration is an *essential* part of growth pole theory is open to doubt. Some plants—petrochemical or metallurgical smelting plants, for example—may need specific locations away from urban concentrations of population. They bring benefits to the poorer region in terms of increased income (particularly through contributions to local taxation) rather than through more jobs.

In practice, a growth-pole policy is likely to run into two kinds of difficulties: the technical ones of selecting the best potential pole, and the political ones of convincing the unsuccessful poles of the wisdom of the policy. The situation is typified by the Mezzogiorno, the southern half of Italy, in which the income level is barely half the Italian average. Here are the classic symptoms of many regional distress areas: high agricultural employment (up to 55 percent of the rural population), low productivity, and high emigration rates. Although development of the area requires a growth-pole approach, in reality the principal industrial projects—the steelworks at Taranto, the automobile plant in Calabria, and the petrochemical complex in Sicily—are geographically separated. This situation is explained more by economic reasons than by political ones. For whatever its long-term benefits, the growth-pole concept implies denying investment, certainly in the short run, to other parts of the regional problem area. In 1970 the announcement of the selection of one city, Catanzaro, as a regional center was followed by rioting in another city, Reggio di Calabria, which had been passed over.

Strategic withdrawal

An inverse growth-pole policy may be used when the general economy of an area is experiencing a long-term structural decline. For example, Figure 19-8 presents the British government policy toward the mining villages in the Durham coalfield in the northeastern part of England. This area had its heyday during the last century when Durham coal was in demand in a country that was a world leader in the Industrial Revolution. Today most of the mines are closed or closing, made uneconomic by the competition from more efficient fuels and from better coalfields in other parts of England. Unemployment rates are high, job opportunities low, and there is a steady out-migration from the mining villages. An extreme interventionist might argue that the economic and social waste of unused manpower and an underused infrastructure (roads, railways, schools) should be prevented by subsidizing new

Figure 19-8. A local strategy of spatial concentration and withdrawal. This map of Durham County in northeast England is based on a development plan adopted in the 1950s. Since then, some settlements have been reclassified. The fourth group of villages, termed "write-offs" by the plan's critics, are mainly small mining villages in which the colliery has closed. In these settlements new capital spending was to be limited to the social and other facilities needed for the life of the existing property. [Data courtesy of County Planning Department, Durham County Council, England.]

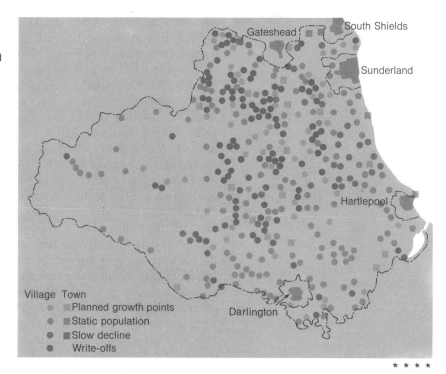

★ ★ ★ ★

employment opportunities and by fixing high prices for coal. Conversely, we could argue that migration is an effective solution to the problems of the area, that subsidies can only be granted at the expense of the more productive sections of the country, and that the new generation (and particularly their children) will lead happier and fuller lives once they have established themselves in the more prosperous and job-rich parts of the country.

Government policy, as Figure 19-8 shows, has reflected a mixture of both viewpoints. On the regional level the response has been interventionist. A new town has been built, new industries have been attracted to it, new roads constructed, and so on. On the local level, however, the situation is different. The smaller and more remote villages are being progressively abandoned, and people are being encouraged to move to the larger and better endowed areas. What we are seeing is, in effect, a policy of strategic withdrawal that balances the social costs of nonintervention (borne locally) against the economic costs of intervention (paid largely by the rest of the country).

Regional planning in the USA

The notion of regional planning in the context of the five-year plans of communist states like Soviet Russia or the social-democratic traditions of Sweden or Britain is a familiar one. The political institutions of a country like the United States make it harder for the central government to intervene at the regional level. What kind of regional planning occurs within a more capitalistic state like the United States?

In the 1930s regional planning in the United States was rather narrowly confined to the development of water resources, with special attention to river-basin planning. President Franklin Roosevelt's establishment of the Tennessee Valley Authority was an outstanding example of this kind of intervention, which was widely copied by other countries. Federal intervention in the economic life of the nation was, however, seen as the thin end of the wedge of a socialist-style intervention in some quarters. Thus it was not until 1961 and the Kennedy Administration that the Area Redevelopment Act was passed. Under this Act, areas with unemployment rates greater than 6 percent or running at specified levels above the national average were eligible for assistance. Although over 1000 counties were designated to receive aid (see Figure 19-9), the total impact of the Act was rather small. Some industrial parks were established, but the major share of the Act's resources went for recreation and tourist projects.

The major contribution of the Johnson Administration was the

Figure 19-9. U.S. regional planning. The map shows two examples of federal programs in the United States. The shaded areas indicate counties designated to receive aid under the Kennedy Administration's Area Redevelopment Act. Heavy boundaries enclose the six interstate planning regions set up by the Johnson Administration's Economic Development Act. The main federal effort has been in the depressed Appalachian area, where it overlaps with part of the TVA area set up in the early 1930s.

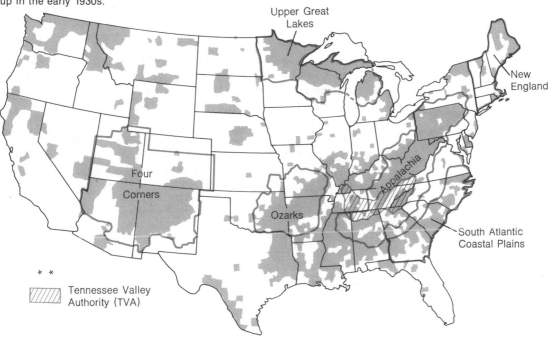

passage of the Appalachian Regional Development Act in 1965, which involved federal aid being coordinated on an interstate basis. The main areas of public investment were transportation (particularly through extensions of the interstate highway program), the development of natural resources, water-control schemes, and social and educational programs. By 1968, five further interstate areas had been set up under the Economic Development Act.

Aid appears to have been spread somewhat thinly, and the total funds available were, by Western European standards, small in relation to the United States' huge GNP. By the late 1960s, public concern had shifted somewhat from the plight of the less prosperous rural regions to the more concentrated problems of the inner cities. Currently it is the plight of the cities and the problems of metropolitan revenue-sharing that dominate the scene. As the suburbs have sprawled outward and industry has decentralized, more American cities have been left to cope with a severe housing blight, high crime rates and other innercity ills, and a dwindling tax base.

19-3 | LOCAL PLANNING PROBLEMS

Intervention in regional problems may range all the way from international aid programs affecting half a continent down to decision-making on specific issues of local significance. Here we look at three examples of intervention on a limited spatial level.

Single versus multiple land uses

As population density on the fixed area of the earth has increased, so has the pressure to use land in intensive ways. In Chapter 7, we saw something of the effect of this increasingly intensive land use on the level of environmental change. Nevertheless, the idea of using the same piece of land for a variety of purposes is very attractive.

Such ideas are not new. Indeed, many so-called "primitive" ways of using land, such as the shifting cultivation systems in the humid tropics, represent an integration of crop production, grazing, and timber use within a limited area. In the context of contemporary Western society, however, the term multiple land use refers to the integration of major uses like forestry and recreation. Often, but not always, it implies that the lands being used are publicly owned. The subject is a hotbed of controversy. What kind of land uses can be mixed? Which mixtures are long-term and stable ways of using the land, and which are dangerous and short-sighted? And who decides what is an optimal land-use strategy?

At the outset, we should separate two different types of multiple land uses. First, there is the *common use* of the same tract of land for two or more purposes. For example, the same river basin might be used for recreation, forestry, and grazing. The physical possibilities and limitations of such common uses depend on their *compatibility*. Table 19-2 indicates the physical compatibility of nine land uses with one another. For example, there is high compatibility between forestry as a primary use and wildlife refuges as a secondary use for land primarily devoted to agriculture. About some combinations no generalizations can be made at all. For instance, some urban areas (city parks) very definitely have secondary recreational uses, while others have not. The compatibility of reservoirs and water management with other secondary uses of land may vary considerably from place to place.

What the table makes clear is that certain land uses are highly intolerant of others. When land is used primarily for urban and transport purposes, usually secondary use for the other seven types is ruled out. In contrast, grazing and forestry are tolerant both of each other and of numerous secondary uses. Of course conditions vary within each category; for example, agriculture may be very intensive (and intolerant of other uses) or very extensive (and tolerant). Note also that the table re-

Table 19-2 Compatibility of major land uses

Primary land use	Physical compatibility with secondary use for								
	Urban purposes	Recreation	Agriculture	Forestry	Grazing	Transport	Reservoirs and water management	Wildlife	Mineral production
Urban purposes	Complete	High for city parks; zero for others	None	None	None	Very poor, except city streets	None	Very poor	Very poor
Recreation	None	Complete	None	Poor to moderate	Very poor to none	Very poor	Poor	Fairly high	Very poor
Agriculture	None	Very poor	Complete	Zero	Zero	Zero	Very poor	Poor to moderate	Poor
Forestry	None	High	None	Complete	Variable— none to fairly high	Zero	Zero	High	Poor to moderate
Grazing	None	High	None	Usually very poor	Complete	Zero	Poor to fairly high	High	Poor to high
Transport	None	None directly; incidental on rights of way	None	None	None	Complete	None	None	None
Reservoirs and water management	None	Poor to high	Very poor	Very poor	Poor to moderate	None	Complete	Poor to high	Very poor
Wildlife sanctuaries	None	High	Very poor	Moderate	Moderate	None	Poor	Complete	Very poor
Mineral production	None	Poor	Poor	Fair	Fair to moderate	Fair	Poor	Poor to fair	Complete

SOURCE: M. Clawson *et al.*, *Land for the Future* (Johns Hopkins University Press, Baltimore, Md., 1960).

fers only to physical compatibility of different land uses. Some of the economic, social, and legal problems of common use are formidable.

A second main type of multiple land use is *parallel use* of the same tract of land for two or more primary purposes. In this case, individual land uses are kept spatially separate but are intermingled within a given tract and administered as a single unit. National forests, for example, contain limited recreational strips (e.g., along highway margins or lakeshores), and commercial forestry is confined to the back country. In practice, the parallel use of intermingled tracts has often involved publicly owned land, usually in areas with relatively low land values. Privately owned land in areas with higher land values is usually divided into small parcels; parallel land use in these areas poses more difficult economic and legal issues.

How do we decide whether a particular strategy of multiple land use is efficient? Ideally, we should survey the benefits and costs that would arise from a given land-use combination. However, the process of defining, let alone measuring, the costs and benefits is complicated by several practical difficulties. Some land uses produce their benefits in spatially distant areas but bring few local benefits. Thus, watersheds create benefits like water not only for local residents but also for remote cities. In a similar fashion, recreation and wildlife benefits can accrue to the whole population and not only to local residents. In the same way, some types of land use may bring penalties for the local population. A watershed-control program that would bring great benefits to the water supply of a distant city might sterilize land for local agriculture. Wildlife reserve areas in East Africa restrict the herding and hunting possibilities for the local population; urban freeways bringing suburban commuters to a CBD increase the noise and pollution levels and decrease the visual amenities for the population in the inner areas of a city. Finally, some benefits from multiple land uses may be so intangible or long term that they cannot be enjoyed by the local population.

Locating international airports

Analogous problems of balancing costs and benefits are posed by the location of unwanted facilities. By "unwanted facilities" we mean facilities like new freeways, prisons, or sewage disposal works which a large community may want, but which may impose unfair burdens (real or imagined) on those who live near them. Among the largest of these unwanted facilities are international airports like Kennedy Airport in New York (Figure 19-10). We can use the conflict surrounding these facilities as an example of more general locational conflicts.

Consider the frantic efforts to find a suitable location for a third

Figure 19-10. Locating international airports. This view of the central terminal area of New York City's John F. Kennedy International Airport underlines the massive land-use demands of this type of facility. This is only the visible tip of an iceberg of related problems—road links, noise levels, pollution problems, and associate industrial growth—that makes the location of new airports an intractable planning problem. [Photography courtesy of the Port of New York Authority.]

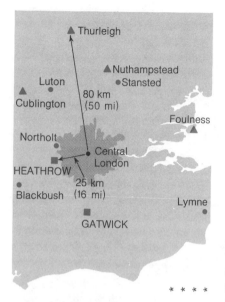

■ Main London airports
● Small airports
▲ Potential sites for new airport

Figure 19-11. The search for airport cities. The map shows the short list of four alternative sites for a third London airport put forward by the Roskill Commission. The Commission's choice, on the ground of costs, was an inland site, Cublington; the final government choice, on environmental grounds, was a coastal site, Foulness.

London airport within the crowded southeastern part of England (Figure 19-11). An initial decision to locate a new airport at Stansted, north of London, was abandoned in 1967 after loud local protests. In 1968, a government commission was set up to advise on the site. The commission examined 78 potential sites and rejected most on grounds that ranged from meteorological conditions to accessibility to London to local construction costs. No ideal site was found, but four sites—Cublington, Thurleigh, Nuthampstead, and Foulness—were considered reasonable possibilities. Estimates of the total costs and benefits associated with each of these sites were made. As Table 19-3 shows, two kinds of cost were dominant: the costs to the airlines of aircraft movements and the cost to the passenger of movements from his home to the new airport. Note that the first type of cost showed much less variation from site to site than did the second. Costs falling on the local community are relatively small in terms of the total (i.e., unless you happen to live in that community!), but vary greatly from one site to another. After many calculations of this kind, the commission recommended in 1971 that the third London airport be located at Cublington.

The selection of the inland site at Cublington raised a storm of protest. In a minority report, one member of the commission described selecting an inland site in a rural area with a high level of amenities as "an environmental disaster." Citizens questioned the utility of applying cost–benefit analysis, an economic tool, to an essentially political situation, and they challenged the detailed calculations of the costs falling on the local areas. (For example, the value of a Norman church that would have to be bulldozed despite its considerable historical and architectural interest was costed at its fire-insurance value.) In response to this protest, the government decided to adopt the coastal Foulness site, despite higher construction costs there. Even this location has met a storm of protests and it may be the late 1970s before construction anywhere can begin.

The sheer scale of an international airport lies at the heart of the locational problem. In this case, although the area within the airport perimeter would be only a few square kilometers (about the size of central London itself), the direct and indirect effects of the facility would be felt over a far larger area. Some 65,000 people would be directly employed at the airport, and a new city of around 275,000 (or an equivalent expansion of nearby urban centers) would be needed to cope with the population directly or indirectly generated by the airport. The new highways and railways linking the airport with London would take up thousands of acres of land well outside the immediate neighborhood of the airport. Above all, a noise umbrella would extend

Table 19-3 Costs of alternative airport sites in southeast England[a]

Type of cost	Size of cost (in millions of pounds)							
	Cublington		Thurleigh		Nuthampsted		Foulness	
Movement of aircraft from new airport to their destinations	960	[$2304]	972	[$2333]	987	[$2369]	973	[$2335]
Travel of passengers from their homes to new airport	887	[$2129]	889	[$2134]	868	[$2083]	1041	[$2498]
Disruption of the local community to be displaced by the new airport	55	[$132]	55	[$132]	66	[$158]	45	[$108]
Noise costs inflicted on the residents of the immediate airport vicinity	14	[$34]	14	[$34]	24	[$58]	11	[$26]
Capital costs of building new airport	289	[$694]	288	[$691]	285	[$684]	252	[$605]
Total net costs	2265	[$5436]	2266	[$5438]	2274	[$5458]	2385	[$5724]

[a] All costs are given in millions discounted to a common base year (1975). Bracketed numbers indicate millions of dollars.
SOURCE: Roskill Commission, *Report on the Third London Airport* (H.M.S.O., London, 1970).

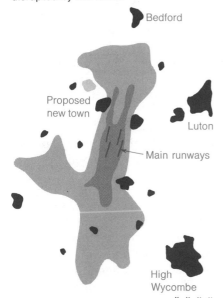

Figure 19-12. Environmental hazards around major airports. The map shows the noise umbrella (lightly shaded) that would surround the projected London airport site at Cublington. Normal residential life in the inner zone (heavily shaded) would be severely disrupted by the noise.

Bedford

Proposed new town

Luton

Main runways

High Wycombe

* * * *

over an area of 750 km² (290 mi²), and an inner zone of 125 km² (48 mi²) of land would be virtually uninhabitable as a normal residential area. (See Figure 19-12.)

The conflict over all unwanted facilities—be they on the massive scale of an airport or the local scale of a penitentiary—is that while the benefits arising from them would be *regional* in scale and affect a large population, the costs in disruption and a loss of amenities would be *local* in scale and would fall on a rather small proportion of the population. Thus, in England there is general agreement that the London region needs expanded airport facilities, but no particular part of the region wants the airport. In this situation no amount of detailed locational analysis will resolve the problem. Rather, the whole community must reevaluate its system of calculating costs and benefits so that those who suffer from the creation of the facility will be appropriately and realistically compensated. (See the marginal discussion of alternative ways of measuring costs.)

Interlinked ecological problems

Although in textbooks we can separate one planning issue, like multiple land uses, from another, like airport sites, in practice we find that problems of location and land use are closely intertwined. Let us consider a small-scale English example, one that could be repeated many hundreds of times in other countries. Figure 19-13 shows a small manmade lake, the Chew Valley Reservoir. Despite its pastoral

Alternative ways of measuring costs

In calculating the cost of ecological or spatial alternatives, geographers should be aware of the different types of costs accountants measure. These include:

1. *Opportunity costs.* These measure the loss or sacrifice involved in using a resource or location for one purpose rather than another. Opportunity costs are the best measure of the cost of using a resource or space for one purpose when there are competing alternatives and the resource or space is being *fully used.*

2. *Marginal costs.* These measure the extra costs incurred in using spare resources or space for a particular purpose. Marginal costs are important when an area is not fully used and the issue is whether or not to utilize *spare capacity.*

3. *Full costs.* This is an accounting concept for comparing the full costs (both fixed and variable) of using a particular resource or space with the full benefits (both long term and short term). Full costs have the advantage of being relatively easy to calculate in a standard way, thereby allowing the interregional comparison of alternatives. The concept is common in *cost-benefit* analysis.

and apparently peaceful quality, this 12-km² stretch of water has been the center of three conflicts, each of which involved ecological and spatial issues so critical that it is worth examining them as a microcosmic example of conflicts on all spatial levels.

The first conflict broke out in the late 1950s, when the lake was being planned. The city of Bristol, about 15 km (9 mi) distant, was demanding ever more water, and a new storage reservoir was urgently needed. But where should it be located? The Chew Valley was ideal because of its hydrology, its unpolluted catchment, and its nearby location. However, the area to be flooded was officially designated as first-class agricultural land. The conflict of interests between the city (containing around 400,000 people) and the area to be flooded (containing only a few farms) raised significant questions about the location of unwanted facilities, the permanent elimination of food-producing land, and individual-versus-state rights. The final decision to build the reservoir in the Chew Valley hinged on opportunity costs; similar reservoirs in other sites would have caused still more difficulties.

The second conflict, in the late 1960s, concerned the ecology of the newly formed lake. To increase their agricultural yield, farmers in the watersheds around the lake were using larger amounts of artificial fertilizers. Creeks draining the watersheds carried downstream to the lake water that had absorbed small quantities of fertilizers. The newly formed lake acted as a trap in which the nutrient levels in the lake built up. (Nutrients are the chemical elements essential for the growth of plants.) Water lost by evaporation, overflow, or abstracted as a water supply was replaced by inflowing enriched waters from the catchment. In the enriched waters new organic food chains (see Section 5-1) were created. These chains produced spectacular color changes in the lake as large colonies of algae turned the

Figure 19-13. Problems in land-use planning. The photo shows the manmade Chew Valley Lake created in 1952 to meet increasing demands for water from cities in southwest England. Its location, its pollution, and its multiple uses have made it a source of considerable conflict among different land and water users. [Ordnance Survey. Crown copyright reserved.]

water green. Oxygen levels in the deeper lake waters decreased, and the fish population fell. This process in which lake water is enriched by nutrients which cause excess algae growth is called *eutrophic.* Although the long-term process of eutrophication is difficult to stop or reverse, cooperation between the water authorities and local landowners has reduced the rate at which it is occurring.

The third conflict in the early 1970s centered on the use of the lake environment. Its original and primary purpose was simply to provide an urban water supply. Fishing was permitted and the lake

Eutrophication and "Lake Death"

A *eutrophic* lake like that of the Chew Valley is simply a nutrient-rich lake. As the nutrient supply (particularly nitrogen and phosphorus) of a lake builds up from the inflow of fertilizer-enriched water from the surrounding fields, some major changes in the lake's biology and chemistry take place. Water weeds become very abundant. Later, thick mats of blue-green algae (called blooms) also become evident. As the algae die, they sink to the lower layers of water, decaying and consuming the oxygen dissolved in the waters. Eutrophic lakes are undesirable for man in several ways. The surface mats of algae interfere with recreational use, choke out game fish (trout are particularly affected in the Chew Lake), and affect the water's taste. Decayed algae clog water treatment filters and slowly make the lake more shallow. Thus, in the very long run, the lake literally dies.

Eutrophication is not a new phenomenon. Blue-green algal blooms were reported from Switzerland's Lake Zurich in 1896. There is no doubt that eutrophic processes have been greatly speeded-up in the last few decades, due to increasing use of fertilizers in agriculture and the introduction of synthetic detergents (containing phosphorus) into domestic use from the 1950s. Phosphorus levels in Lake Erie have increased threefold since the 1940s. Eutrophication may also occur in the estuaries (mouths) of major rivers. Algal blooms were first observed on the Potomac in 1925 and have been massive and persistent since 1962.

Attempts to control eutrophication and stop "lake death" are difficult. Nitrogen has so many natural and artificial sources and is so critical in agricultural production that its control is not generally feasible. Attempts to reduce phosphorus, and particularly the tight control of synthetic detergents, is more promising. For a good nontechnical account of eutrophication, see Donald E. Carr, *Death of the Sweet Waters* (Norton, New York, 1966).

stocked, and ornithologists were allowed to observe the wildfowl population. Pressure to use the lake for other purposes has been successful. Sailing was permitted in 1968, and other recreational uses will be allowed in the near future. Not all the uses proposed for the lake are compatible with each other, or with the primary use of the area for a domestic water supply. Figure 19-14 presents some zones of activity that now allow parallel use of the lake for different purposes.

The three conflicts over this small English lake raise a number of questions about conflicts among interest groups. The location of the reservoir created a controversy over the importance of the city's need for water compared to the economic and social costs that would be inflicted on a small farming community by the creation of the reservoir. The eutrophication problem stirred up the conservationists against farmers who wanted to improve the productivity of their land. The lake's potential uses created cross conflicts among groups wishing to use the lake primarily for their own purposes. Each conflict involved spatial considerations at different geographic scales and pointed up the need for a long-term ecological viewpoint. And each conflict was exacerbated by the crowding caused by a high population in a limited and intensively used land area.

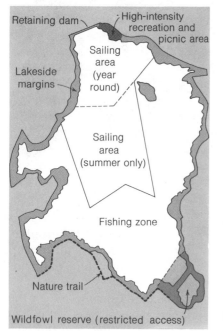

Figure 19-14. Zoning solutions for multiple land uses. The map shows the proposed zoning of Chew Valley Lake for competing uses. [Data from Bristol Waterworks Company. From C. Hartley, *Countryside Community Recreation News Supplement,* No. 3 (1971), p. 9, Fig. 1.]

* * * * *

In this chapter we have discussed how central governments intervene to adjust regional differences within a country. But the issues which intervention raises have as much to do with ethics and politics as with economics. Insofar as these matters of capital and conscience have a spatial framework, they are of concern to geographers because their resolution affects the world distributions he studies. Communities are showing increasing concern with the consequences, both ecological and socioeconomic, of locational decisions and spatial inequalities. Geographers view this concern with mixed reactions, since they are keenly aware of the spatial nature of the inequalities, yet equally skeptical of any naive argument for blanket uniformity. Regional policies which aim either at achieving greater uniformity or at avoiding unpalatable locational decisions (be they the choice of growth centers at a regional level or of sites for reservoirs at a local level) may well be justified on social grounds in the short-term; but they may prove very expensive for society as a whole in the long run. A balanced assessment of the costs and benefits involved demands that we try and see the future spatial and ecological consequences of present decisions. In the Epilogue, we shall take a look at some first faltering steps in this direction.

Reflections

1. How do geographers measure spatial inequality? Assume that a country is divided into five regions, each with 20 percent of the country's population but with unequal shares of its total income (say, 50, 25, 15, 8, and 4 percent, respectively). Plot this variation in regional incomes as a Lorenz curve. (Go on to plot actual curves for your own province or state, if you have the time.)

2. Gather data on the different levels of welfare in counties in your home state. Plot the values on a map and study the resulting distributions. Identify high and low areas, and attempt to explain their distribution.

3. Examine Figure 19-5. What kind of regional strategy would you favor for your own country? On what basis would you give some areas special aid? Which areas would receive this aid? Compare your list of needy areas with the areas suggested by others in your class.

4. List the arguments you would use to (a) support and (b) oppose a policy of strategic withdrawal of population from an area.

5. Examine the factors that determine the location of a major international airport. Should the local population of an area be able to veto the construction of an airport there?

6. What do we mean by "a just spatial distribution"? Can the geographer contribute to this debate?

7. Review your understanding of the following concepts:
   (a) Lorenz curves   (f) growth poles
   (b) inequality gaps   (g) strategic withdrawal
   (c) ratio of advantage   (h) compatible land uses
   (d) social welfare indicators (i) benefit-cost analyses
   (e) revenue-sharing programs   (j) eutrophication

One step further . . .

The general philosophical issues in regional-help programs are discussed in
   Harvey, D. W., *Social Justice and the City* (Edward Arnold, London, 1973), esp. Chap. 3.

Problems and issues in the United States are discussed in
   Morrill, R. L., and E. H. Wohlenberg, *The Geography of Poverty in the United States* (McGraw-Hill, New York, 1971) and
   Smith, D. M., *The Geography of Social Well-being in the United States* (McGraw-Hill, New York, 1973).

Many texts review national styles of regional economic planning. For a discussion of French and British styles, see
   Boudeville, J. R., *Problems of Regional Economic Planning* (Aldine, Chicago, and Edinburgh University Press, Edinburgh, 1966) and
   McCrone, G., *Regional Policy in Britain* (Verry, Lawrence, Mystic, Conn., and Allen & Unwin, London, 1969), Chap. 3.

Ways of measuring economic stress and diagnosing the limits of regional problems are given in
   Wood, W. S., and R. S. Thomas, Eds., *Areas of Economic Stress in Canada* (Queen's University, Kingston, Ont., 1965), Chap. 3.

For a discussion of the issues raised by planning with respect to land and water resources and biological resources in America, look at
   White, G. F., *Strategies of American Water Management* (University of Michigan Press, Ann Arbor, Mich., 1969),
   Tunnard, C., and B. Pushkarev, *Man-Made America: Chaos or Control?* (Yale University Press, New Haven, Conn., 1963), and
   Ehrenfeld, D. W., *Biological Conservation* (Holt, Rinehart & Winston, New York, 1970).

Regional problems are well covered in the regular geographic periodicals, but *Antipode* (a quarterly) is the forum where the issues of welfare geography and spatial justice are argued out in the most challenging manner. You might also like to look at a few of the wide variety of specialized journals: *Regional Studies* (a quarterly), the *Journal of Regional Economics* (another quarterly), and *Papers of the Regional Science Association* (a biannual publication) are all interesting.

Epilogue

The Future Task

The Epilogue is concerned with the future in three senses. *Outer Space, Inner Space* (Chapter 20) describes important and rapidly developing areas of geographic research. The discussion ranges from the use of remote-sensing devices in earth-orbiting satellites to, at the other end of the spectrum, research into the "inner world" of human spatial and ecological behavior. *Ways of Looking Forward* (Chapter 21) describes the increasing concern of geographers with the multiple worlds of the future. We look at the techniques geographers can use to project the future spatial patterns of man on the planet on all scales from the global to the local, and we draw together some of the speculations on man's future tenure and organization that have come up in earlier chapters. *On Going Further in Geography* (Chapter 22) is concerned with the future in a more personal sense. It has been written specifically for the student considering whether to pursue further courses in geography and tries to outline the main structure of the field, its development, its internal controversies, its job possibilities, and so on. It returns, after one complete whorl of the helix, to some of the issues raised in *On the Beach* (Chapter 1) and serves as a beginning to the more advanced courses that lie on the next and higher circuit.

Chapter 20

Outer Space, Inner Space

Huck Finn to Tom Sawyer in their flying boat:
"We're right over Illinois yet. And you can see
for yourself that Indiana ain't in sight. . . .
Illinois is green, Indiana is pink. You show me
any pink down there, if you can. No, sir; it's
green."
"Indiana pink? Why, what a lie!" "It ain't no
lie; I've seen it on the map, and it's pink."

—MARK TWAIN
Tom Sawyer Abroad (1896)

n the last four sections of the book we have been looking at the environmental challenge the earth poses for man. We have asked four questions. What is man's ecological response? What is his cultural response, and how has it affected the mosaic of world regions? How have these world regions been shaped into a hierarchy of city regions? What conflicts and stresses have been set up between these regions? All four questions underline the critical need for more and more geographically relevant information.

Here we are faced with a paradox: growing information, yet growing uncertainty. Let us take the first point. In the 1970s the level of existing information is, of course, higher by many orders of magnitude than it was a century ago. From wide-ranging surveys of the growth of scientific information, we know that information grows in an exponential fashion; that is, the greater the amount of information that exists, the faster it grows. Depending on what we measure, it is possible to estimate, roughly, that the amount of environmental information tends to double within a period of 10 to 15 years or so. If one accepts the general form of this growth curve, the amount of information available to geographers such as Alexander von Humboldt and Carl Ritter in the early part of the nineteenth century was about 1000 times smaller than the amount available to the current generation of geographers. To create the images produced in a few seconds in our opening photograph would have required decades of survey and calculation by an earlier generation of mapmakers.

Despite this dramatic increase in data, the demand for geographic information is likely to be far higher in the last quarter of the twentieth century than at any previous time. For, despite its impressiveness, the increase in total geographic information has scarcely kept pace with the growing realization of our ignorance about the earth. Shifts in the geographic focus of man's activities—to the jungles of Cambodia, the subarctic steppes of the Alaskan North Slope, or the inner blighted areas of American cities—may show how sparse is the stock of information available. Also, shifts in emphasis on different resources have swung the spotlight to some areas where there is a critical lack of information, areas such as the changing levels of air pollution or the exhaustion of resources. More important still is the growing evidence that man filters and bends evidence to fit preconceived mental images of the world.

In this chapter we focus on two views of the information problem. First, we look at the world from outer space and examine the contributions of remote sensing to our understanding of the world. Second, we see the world from the inner space of men's minds and discuss the contributions of perception studies to geographic understanding. Finally, we look at the problem of mapping information and ask how we can formulate a valid picture of the whole world from the limited evidence of a few sample investigations.

20-1 | REMOTE SENSING: THE VIEW FROM OUTER SPACE

The first man to try to photograph the earth's surface from a balloon was probably a Parisian photographer, Gaspard Félix Tournachon. His attempts to capture "nothing less than the tracings of nature herself, reflected on the plate" date from 1858. The mania for balloon photography caught on (Figure 20-1), and by July 1863 even Oliver Wendell Holmes was conceding that Boston "as the eagle and wild goose see it" was a very different place from that seen by its solid citizens on the ground.

As we saw in Chapter 2, it was World War I, with the advent of aircraft and the widespread use of military intelligence, that converted aerial photography from a pastime into a program. The war led to a growing interest in aerial photography for resource surveys from the 1920s onward. And over the last 15 years the development of satellites has provided a new dimension of airborne surveillance. Artificial satellites circle the earth along orbital paths and stay in space for lengths of time dependent on their size and distance from the earth. The useful life of satellites is usually determined by the running down of their batteries or by electronic failures rather than by the gradual slowing down or decay of their orbits. Small satellites in high orbits have longer lives than large satellites in low orbits.

Figure 20-1. The beginnings of remote sensing. Jacques Ducorn using a dry-plate camera from a balloon in 1885. [From Gaston Tissandier, *La Photographie en Ballon* (Gauthier-Billars, Paris, 1885).]

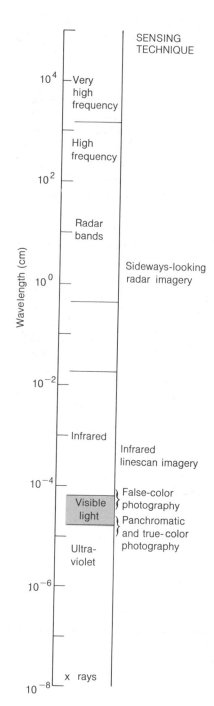

Improvements in sensors

The impact of satellites on the provision of geographic information has come from two factors: first, the rapidly extending range of sensors, and second, improvements in the spacecraft themselves.

Photography Sensors are instruments used to detect the electromagnetic energy associated with a particular object on the earth's surface. Black and white (panchromatic) aerial photography has for long been the principal sensing technique for the geographic study of the earth's surface. This uses the energy reflected by visible light falling on that surface. However, during the last 25 years there have been significant extensions in both the range and the capability of sensors. The array of remote-sensing systems now available has been expanded by the exploitation of different parts of the electromagnetic spectrum. (See Figure 20-2.) The term *spectrum* refers to the wide range of electromagnetic waves that vary in length from less than one billionth of a centimeter (cosmic rays) to hundreds of kilometers (alternating electric current). That narrow part of the spectrum which our eyes can see is visible radiant energy (or light). This section of the spectrum was the first part to be used for airborne records, which were made by conventional photographic techniques.

Developments in photographic chemistry have vastly extended the amount of information that can be captured on photographs. For example, look at the series of photographs in Figure 20-3. The subject is the same in all three photos, but the film picks up and accentuates different aspects of the scene. Black-and-white panchromatic film (used in the second photo) is sensitive to all colors, whereas black-and-white orthochromatic film (used in the third photo) is sensitive to all colors except red. Thus, the panchromatic film represents all colors as shades of gray, and the orthochromatic film records red objects as black. New types of film are sensitive to electromagnetic waves on the borders of visible light.

Geographers' interest in variations in land use makes them especially concerned with films that can differentiate between vegetation

Figure 20-2. Sensing techniques and the electromagnetic spectrum. Remote sensing was at first confined to emissions of visible light and conventional photographic techniques were used. Equipment in use today can also detect infrared and radar waves. As more sophisticated sensors become available, an ever-wider range of electromagnetic emissions from the earth's surface can be recorded and mapped. *False color* photographs are used to emphasize and separate important features by giving them distinctive and contrasting colors. In such photos, healthy vegetation, for example, may show up as bright red rather than green.

(a)

(b)

Figure 20-3. Film types and landscape images. Photographs of mixed woodland using (a) infrared, (b) panchromatic, and (c) orthochromatic film. Note the differing capability of the films to reduce haze and differentiate broadleaf from needleleaf species. Compare these photos with the chart in Figure 20-4. [Photos by Tony Philpott.]

(c)

colors, notably shades of green. Infrared photography uses film that picks up varieties of tone and color that are not evident in either black-and-white or true-color photographs. Figure 20-4 shows how the reflective properties of forest-covered terrain vary for emissions of different wavelengths. Differentiation between the two types of forest is clearly much easier than in the visible sections, where the properties of the two types are similar and partly overlap. Most landscape features appear to have similar distinctive spectral signatures. These *signatures* are the unique pattern of wavebands emitted by a particular environmental object.

Other sensors Newer, nonphotographic sensing systems have advantages and liabilities when it comes to the analysis of terrestrial

Figure 20-4. Image separation at different wavelengths. The light reflected by the foliage of broadleaf and needleleaf trees is shown by the two broad curves. Note that the curves for the two types of foliage slightly overlap at the shorter wavelengths of visible light, making foliage difficult to separate on ordinary panchromatic or orthochromatic film. However, the two curves are widely separated in the longer wavelengths of near infrared light. It is therefore easier to map the different species using infrared film. [After R. N. Colwell, in R. U. Cooke and D. R. Harris, *Transactions of the Institute of British Geographers*, No. 50 (1970), p. 4, Fig. 3.]

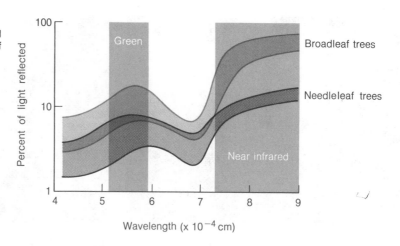

phenomena. *Radar sensors* direct energy at an object and record the rebounded energy as radio waves. There are two advantages of radar images over conventional photographic images for mapping purposes. First, radar imagery is independent of solar illumination and is unaffected by darkness, cloud cover, or rain. This means that any part of the globe can be scanned on demand, including the zones especially difficult or impossible to photograph by conventional methods — such as the cloud-covered humid tropics and regions of polar night. The time available for radar scanning in the humid midlatitudes is five or ten times greater than that available for taking aerial photographs of acceptable quality. Second, radar imagery gives greater detail of the terrain. For example, radar images on a scale of about 1:200,000 provide information on drainage patterns which is roughly equivalent to that derivable from a 1:62,500 topographic map. High-resolution radar can pick up very fine irregularities in the ground surface, and even show differences in subsurface conditions to a depth of a few meters. Figure 20-5 gives an example of a radar image of farmland in Kansas. The various crops appear in different shades of gray; the lightest shades indicate sugar beets.

Improvements in satellites

Early satellites The first American satellite (Explorer I) was launched in January 1958. The first generation of observation satellites was formed by the eight members of the Tiros (Television and Infrared Observation Satellites) family launched between 1960 and 1963. As their name implied, the satellites carried two types of sensing devices. First, there were television cameras that transmitted pictures of the visible part of the spectrum back to earth. The first detailed weather pictures

Figure 20-5. Radar images of landscapes. Farmland in Kansas as seen by radar. Different crops are distinguished by various tones. The lightest areas are planted with sugar beets. [Photo by Westinghouse Electric Corporation.]

Terms used in remote-sensing studies

Bands are sections of the electro-magnetic spectrum with a common characteristic, such as the visible band.

Enhancement refers to processes which increase or decrease contrasts on received images (e.g., photos) so as to make them easier to interpret.

Ground truth is information about the actual state of any environment at the time of a remote-sensing flight overhead.

Imagery is the visual representation of energy received by remote-sensing instruments.

Line scanning produces an image by viewing and recording a picture one line at a time, as on a cathode-ray tube (or TV set).

Multispectral sensing is the recording of different portions of the electro-magnetic spectrum by one or more sensors.

Platforms are objects on which a remote sensor is mounted, usually an aircraft or satellite.

Radar is a sensor which directs energy at an object and records the re-bounded energy as radio waves.

Resolution is the ability of a remote sensing system to distinguish signals which are close to each other, in time, space, or wavelength.

Sensors are instruments used to detect the electromagnetic energy associated with a particular object on the earth's surface.

Signatures are the unique pattern of wavebands emitted by a particular environmental object.

Synoptic images are those giving a general view of a part of the earth's surface, usually from high-altitude satellites.

Thermal infrared records the thermal energy *emitted* by objects of different temperatures on the earth's surface.

of cloud patterns were received as early as 1960, and Tiros III discovered its first hurricane (hurricane Esther) in July 1961. Second, Tiros carried infrared detectors that measured the nonvisible part of the spectrum and provided information on local and regional temperatures on the earth's surface.

The drawback of the earliest type of satellite (see Figure 20-6) was that its camera pointed through the baseplate, and so it could make undistorted vertical observations of the earth's surface only once on each of its orbits. The improved cartwheel Tiros photographed two continuous strips of the earth's surface as it rolled along its orbital path. The problem was completely overcome with the introduction of a new type of observational satellite in 1964. Nimbus satellites were stabilized without spinning about their own axis and were earth-oriented, so the three television cameras they carried could always view the earth from vertical or near-vertical angles. Following a north–south orbit timed to coincide with the earth's rotation, these satellites could photograph the entire earth's surface once every day.

The earth resources technology satellite (ERTS) From a geographer's viewpoint, the most important development in global satellites was

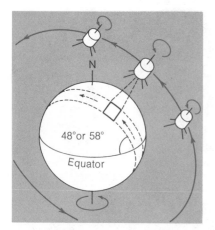

(a) Space-oriented Tiros

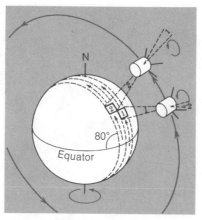

(b) Cartwheel Tiros

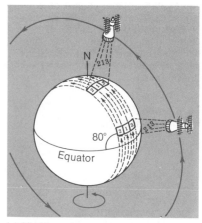

(c) Nimbus

Figure 20-6. Earth-orbiting satellites. The charts show the increasing capacity of the first three generations of satellites to monitor the earth's surface. The original Tiros satellite could make undistorted vertical observations of the earth's surface only once during each orbit. [From E. C. Barrett, *Science Journal* **3** (1967), p. 75.]

the ERTS. This satellite was launched from California in July 1972. It makes 14 revolutions a day around the earth, its sensors covering a series of 160-km (100-mi) wide strips. The strips overlap, so that the whole surface of the earth is covered once in every 18 days. Thus in the first year of the satellite's life each part of the planet came within range of its sensors twenty times. Of course, many areas were cloud covered, but it is estimated that in the first year the ERTS provided cloud-free coverage of about three quarters of the world's land masses.

Pictures reaching the satellite are converted into electronic signals, stored on tape, and then broadcast back to three ground stations, at Fairbanks, Alaska, Goldstone, California, and Greenbelt near Washington, D.C. Unlike the images recorded by earlier satellites, the images received by ERTS are passed through a *multispectral scanner* which is sensitive to different parts of the electromagnetic waveband shown in Figure 20-2. This means that the signals transmitted to Earth can be recombined in many ways to bring out unsuspected features of the planet's surface. For example, when the signals are decoded and exposed on photographic film, they may produce monochrome, natural color, or false color images. False color photos, as we have noted, are used to emphasize and separate important terrain features. For example, vigorously growing vegetation may show up as bright red and diseased crops as pale yellow. Clear water may show up as black, while contaminated water carrying silt and sewage may appear bright blue.

Potential research from spacecraft

From the geographic viewpoint, satellites are immensely useful because of their very wide coverage and the fact that difficult or inaccessible terrain presents no bar to data collection. Three areas of research applications seem exceptionally promising at this time.

The first is studies that take advantage of the potential *worldwide coverage* of satellite systems. Typical are proposals for a worldwide study of surface temperatures using infrared scanning systems (Figure 20-7). For despite a growing network of information on atmospheric temperatures, the data on temperatures of the ground surface itself are rather sparse and largely available only for developed areas. Records of daily and seasonal changes in temperature would tell us more about the heat and water balances between the earth and the atmosphere. Such data bear directly on human needs by helping to identify areas where the temperature is suitable for certain crops.

There is a similar lack of worldwide information on precipitation. Rainfall is normally recorded daily by meteorological bureaus' gauges scattered over the earth's surface. Satellites promise the rather exact location of rain and snow as they are actually occurring. Through links with conventional ground stations and surface radar stations, the type and intensity of precipitation can be measured, and improved forecasts can be given. Other proposals for taking advantage of the global coverage of the satellite system relate to the completion of the World Land Use Survey and to cloud interpretation studies. The World Land Use

Figure 20-7. Satellite images. (a) The eastern seaboard of the United States from Cape Cod to Chesapeake Bay, recorded by Tiros VII. (b) Infrared data on Hurricane Camille (1969) with differences in temperature showing up as different shades. Note the characteristic vortex shape with violent winds moving around a calm central area. [Photographs courtesy of NASA.]

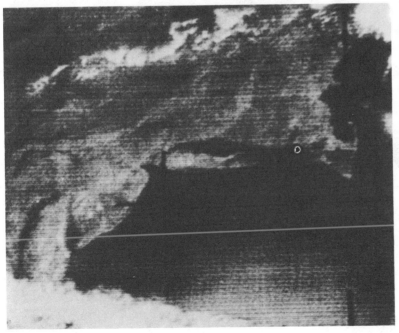

(a)

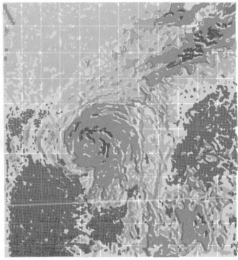

(b)

Survey was initiated by British geographer Sir Dudley Stamp to provide a series of maps, on a scale of 1:1,000,000, of the human use of the earth's land surface. Examples of cloud interpretation studies are presented in Figure 20-8.

The second promising area of research is the study of phenomena that occur mainly in *inaccessible* and therefore *sparsely monitored* parts of the earth's surface. Most of the ice masses of the world, for instance, are located either in polar or high mountain areas where both ground and airborne surveys pose logistical problems. With satellite surveillance we can regularly record changes in the extent of ice masses, their surface characteristics, or the presence of new snow deposits. Such studies are of value because glaciers contain three-quarters

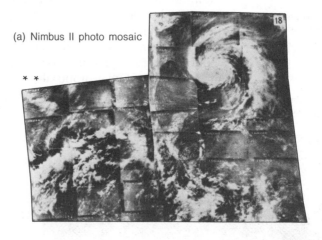

(a) Nimbus II photo mosaic

Figure 20-8. Cloud analysis using satellite photographs. (a) Nimbus II photographs of the central American region, June 11, 1966, compiled into a photo mosaic. (b) The identification of different cloud types, each characteristic of different meteorological conditions. This is termed nephline analysis. The line of crosses (+++++) shows the position of the intertropical convergence on this day. (See Figure 3–6(c) for the average position of this convergence zone.) (c) The smoothing of nephlines to indicate the main patterns of air circulation. [From E. C. Barrett, *Progress in Geography* **2** (1970), pp. 192–193, Figs. 19, 20.]

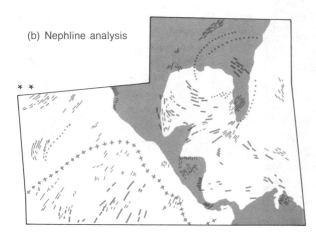

(b) Nephline analysis

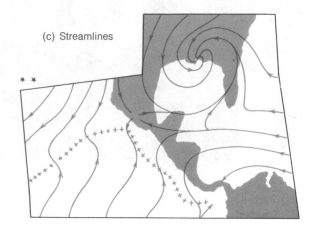

(c) Streamlines

of the world's fresh water and are critically connected to short-term changes in the earth's hydrology and climatology, as well as to long-term changes in sea levels. Studies of glacial budgets (that is, gains or losses in the volume of ice in a glacier or ice sheet), and rates of iceberg measurement at the seaward margins, are all feasible with existing satellite and sensor technology. Work is going ahead in this field.

The third promising area of satellite research is the study of *ephemeral* or *highly mobile* phenomena that cannot be recorded with present survey techniques. We can include in this area proposals to monitor the distribution of bush fires and their effect on natural vegetation. Although the role of fire in the formation of systems of vegetation such as those in the savannahs has been fiercely argued, little quantitative data exist on the timing of such fires, their location and extent, and their relationship to later changes in vegetation. High-resolution images with sharp detail also can record the occasional, but ecologically critical, human use of wildlife and recreational areas. Photographs from aircraft are already helping to determine the location and intensity of vehicle parking in national forests and wildlife areas. Satellite photographs also can check on the increasing pressures on these areas and give us early warning of their overuse.

More modest projects use spacecraft for recording changes in the distribution of airborne sediments (e.g., dust clouds and pollution), movements of coastal sediment, movements of traffic in metropolitan areas, and the distribution of seagoing craft. One particularly intriguing suggestion in the realm of historical geography is that sensing devices could help us reconstruct caravan networks in the western Sahara and Takla Makan deserts. Trails frequented by animals have a surface composition and chemical content unlike the untrodden and unfertilized terrain around them, and it seems possible that such differences are detectable by appropriate sensors.

20-2 | PERCEPTION: THE VIEW FROM INNER SPACE

Electronics have shrunk world space in the 1970s to the size of a TV screen. Satellites and television allow each of us an unprecedented opportunity to peer into our neighbor's backyard. On an average day's TV program we find items which range from hotel fires in São Paulo, to floods in the Australian outback, to a nature film shot in the Canadian tundra. Never before have man's senses been stretched over the whole globe in this way.

Yet how do we view the world around us? The evidence at hand from geographers and psychologists suggests that we retain primitive, twisted, and biased pictures of our local "world" that may be far removed from the "real" world shown by the satellite sensor. In this section we look at some of the early results in an important and rapidly growing area of geographic research.

The image of the city

Consider the way we look at the city in which we live. How much of it do we know well? How much do we not know at all? Is the geographic view of the city from the central areas the same as that from the suburbs? If not, then how does it differ? Figure 20-9 presents three maps of Los Angeles that show the perception of the city by people residing in three district areas within it. A sample of respondents in each area was asked to sketch maps of the city based on their own travels within the city and the contacts they had made. Their combined views are summarized in the maps. Those living in a poor inner section of the city near downtown Los Angeles, Boyle Heights, have a quite limited view of the city dominated by the nearby city hall, railroad station, and bus terminal [Figure 20-9(a)]. Suburban residents of Northridge in the San Fernando Valley had a limited but spatially more extensive view than the Boyle Heights residents [Figure 20-9(b)]. They had an extensive perception of their own valley and its facilities, but little real familiarity with the main city, beyond the Santa Monica Mountains. The respondents around the University of California Campus at Westwood, an upper-class sector just west of Beverly Hills, had a wide-ranging and detailed image of almost the whole Los Angeles metropolitan area [Figure 20-9(c)].

This kind of research indicates something about the warped mental maps we have of cities. Groups from higher-income areas have a wider

Figure 20-9. Images of a city. The maps show Los Angeles as seen by a cross section of its residents from three contrasting districts (marked by a star). Boyle Heights (a) is a largely black neighborhood near the downtown; Northridge (b) is a suburban residential community in the San Fernando valley; and Westwood (c) is a high-class housing area near the UCLA campus. Each cell indicates a different section of the city; tints indicate the proportion of those interviewed who were familiar with that section of the city. Unshaded cells were unfamiliar and lie outside the residents' area of perception. [Data by R. Dannenbrink, Los Angeles City Planning Commission. From *National Academy of Sciences Publication No. 1498* (1967), pp. 107–112, Figs. 2–4.]

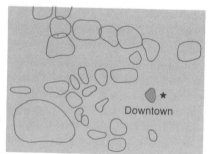

(a) Boyle Heights

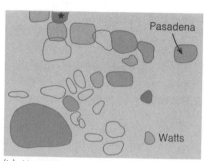

(b) Northridge

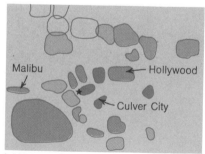

(c) Westwood

Familiarity (percent)

< 25	50-74
25-49	>75

view of the city, and more educated groups have a more accurate view of the city. On the other hand, this research fails to explain why certain parts of the city environment are commonly known, or why some cities produce more positive and memorable mental images than others. One day in Cincinnati may leave a person with a clearer mental map of the city than one day in Kansas City, for example. Why?

Some clues to the answer to this question have been provided in work carried on by Kevin Lynch at the Massachusetts Institute of Technology. Residents of three contrasting North American cities (Boston, Los Angeles, and Jersey City) were interviewed and asked to sketch a map of their city, to provide descriptions of several trips through the city, and to list and comment upon the parts of the city they felt were the most distinctive. With these documents Lynch pieced together the public image of each city held by its inhabitants.

There were considerable, and interesting, variations in the responses of individuals linked to their age, sex, length of residence, area of residence, and so on. But enough common ground was found to allow some citywide generalizations. Lynch organized the common elements of the mental maps into five types of spatial phenomena. The results for Boston (Figure 20-10) are shown in Figure 20-11.

The five types of elements can be defined as follows. *Paths* [Figure 20-11(a)] are the channels along which we customarily, occasionally, or potentially move within the city. They range from streets to canals and are the reference lines we use to arrange other elements. *Edges* [Figure 20-11(b)] are linear breaks in the continuity of the city. They may be shorelines, railroad tracks, or barriers to movement. The Charles River in Boston and the lakefront in Chicago have all the abrupt barrier qualities of an edge.

Nodes [Figure 20-11(c)] are focal points within the city. They are commonly road junctions or meeting places. Louisburg Square in Boston or Times Square in Manhattan are typical nodes. *Districts* [Figure 20-11(d)] are medium-to-large sections of the city that we can mentally enter "inside of" and that have some common identifying character. Beacon Hill or South End in Boston are typical districts. Finally, *landmarks* [Figure 20-11(e)] are also reference points but much smaller in size than nodes. A landmark is usually a simple physical object: a building, a store, a mountain. It may be memorable for its beauty or its ugliness. The gold dome of Boston's state house and the old Hall of Records in Los Angeles illustrate the role of landmarks within the city.

The relative strength and richness of these five spatial elements give coherence and character to a city. Cities with strong elements may be interesting environments to live in, despite dilapidation or deterio-

Waterfront   CBD   State House   Beacon Hill

Charles River

Figure 20-10. Images of Boston. An aerial view of the city of Boston looking across the Charles River. But how do the residents of the city see it? What landmarks, markers, and districts do they recognize and use to orient themselves? Figure 20–11 suggests some of the answers from research by Kevin Lynch. [Photograph from Rotkin, P. F. I.]

ration. San Francisco, Cincinnati, New Orleans, Montreal are North American cities that come into this first category. Cities with weak elements may be formless, monotonous, and lacking in character. You may like to provide your own candidates for this type of city.

Lynch's approach has been repeated in many cities around the world on a variety of scales. Dutch geographers have emphasized that although regular street structures help us to find our way easily, difficulties in orienting ourselves arise when the overall structure of a city is clear but the individual elements are too uniform to be individually distinguished. It appears that we construct mental maps more readily in an older European city (a baffling jumble of streets, but with distinctive elements to serve as locational clues to where we are) than in a newer North American city (a regular street pattern, but too many roads looking like each other).

Figure 20-11. Elements in mental maps of Boston. Five main elements in the image of Boston as seen by its residents. [From K. Lynch, *The Image of the City* (MIT Press, Cambridge, Mass., 1960), p. 147, Fig. 37. Copyright © 1960 by The Massachusetts Institute of Technology.]

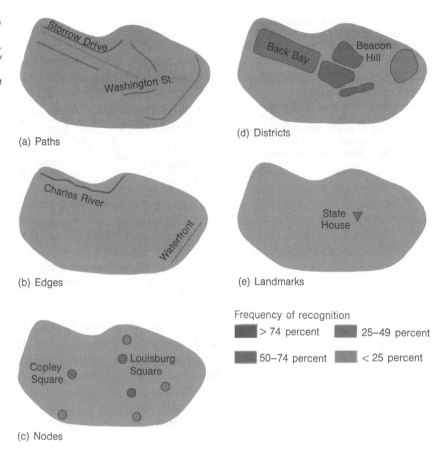

(a) Paths

(b) Edges

(c) Nodes

(d) Districts

(e) Landmarks

Frequency of recognition

> 74 percent   25–49 percent

50–74 percent   < 25 percent

Perception and wilderness areas

At the opposite pole to research on urban perception stands the work on wilderness areas. Just how do we shape our images of "the great outdoors"? Let us begin with a simple case. Figure 20-12 shows an empty forest and lake area on the borders between Canada and the United States made up of the Quetico Provincial Park in Ontario and the Boundary Waters Area in Minnesota. This area has been officially designated as a wilderness by the two state governments, so its boundaries represent the official image of a wilderness area. But how do these boundaries match the views of the people who actually use the area? What areas do the campers, canoeists, and fishermen consider to be wilderness, and what essential qualities should a wilderness environment have?

The main research problem is one of converting anecdotal obser-

Figure 20-12. Wilderness images. The map shows canoeists' perceptions of "wilderness" areas within the Quetico-Superior Park on the boundaries of Minnesota and Ontario. Shading indicates the percentage of canoeists who recognized the area as a "wilderness." This map differs from that constructed by other campers using automobiles or boats. [From R. C. Lucas, *Natural Resources Journal* **3** (1964), p. 394, Fig. 3. Reprinted with permission.]

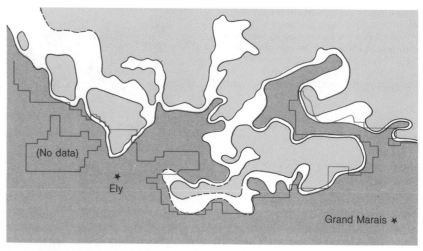

Proportion of canoeists regarding area as wilderness ✳ ✳ ✳

Over 90%   Under 50%   Park boundary

vations and subjective hearsay into some form of quantitative yardstick that allows different assessments of environments to be metered and mapped. The solution adopted was to interview a representative sample of the groups using the area. Each respondent was invited to say whether he thought he was in a wilderness area and to indicate on a map where the boundaries of the area lay. The term "wilderness" was always left to the respondent himself to define, implicitly or explicitly.

By superimposing the maps of the respondents on one another we can construct contour maps of the area showing the percentage of parties visiting the area that described it as being "in the wilderness." The canoeists' contour maps have an intricate pattern (Figure 20-12) closely related to the waterways and sensitive to distance from the canoe trails. The boundaries of the wilderness for other users (campers, motorists, resort guests) were less sharply defined and were related more to roads, crowding, and noise levels in the less remote areas.

From a planning viewpoint, it is worth recording that all the boundaries drawn by users of the area differed from those drawn by the resource managers of the wilderness area. Such users' responses, when translated into isarithmic maps, are clearly of considerable potential value in the future zoning of areas for recreational use. Also, these responses imply that some measurement of natural resource potential is possible even for such elusive things as the quality of a landscape.

Regional images

If we can map the way in which we perceive wilderness areas and the built-up environments of cities, then it should be possible to map larger spatial areas too. How do we view the different parts of our own countries? What is our mental map of northern New England or the bush country of Australia?

Geographers already have attempted to establish the mental maps of wider spatial areas than the city. Cross sections of young people (aged 16 to 18 years) from 20 schools in Britain were used to determine views of the country. All the individuals were asked to rank the counties of England, Wales, and Scotland in terms of their desirability as places to live and work.

Statistical analysis of the results for each school yielded surface maps of the areas' perceived desirability (Figure 20-13). Although the maps for each school differed, they had several common elements. First, each map gave a high rating to areas immediately adjacent to the school. This attachment to the home country was present in all the groups studied. Second, the south of Britain was perceived as being generally more desirable than Scotland and the North, especially by students at southern schools like Falmouth. Third, London was seen as either very attractive or very unattractive. Other maps indicated a strong preference for well-known vacation areas such as the southwest peninsula.

This kind of research has been repeated in areas such as the United States, Europe, and regions of Nigeria. The results confirm the British findings, but there are obviously variations in the strength of attachment to the home area. There are some doubts about the use of the ranking as a way of distinguishing between the desirability of many alternatives, however, and geographers are having to develop more refined methods based on psychological tests.

Figure 20-13. Regional images. The maps show contrasts in the images of British regions held by senior pupils at four schools in different locations. The "most desirable" areas of the country according to pupils at each school are shaded. The shading reflects both an affection for the local area surrounding each school and a generally higher ranking of the southern part of the island. [After R. R. White, Pennsylvania State University, unpublished master's thesis, 1967.]

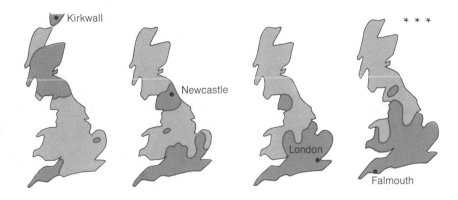

The results achieved to date raise a number of questions. How persistent are our mental maps? Do they correctly evaluate the spatial area within which our locational decisions to migrate are made? Can similar maps be made of spatial concepts of particular groups, say American senators, Wall Street investors, or Iowa farmers? In all these areas the mental maps are largely unknown and the potential for further research is enormous.

20-3 | MAPPING GEOGRAPHIC SPACE

Since the Greek geographers first conceived of the planet earth as a sphere and set about measuring its dimensions, one geographic goal has been to fill in the world map as accurately as possible. As we have seen in this chapter, the advent of remote sensing has allowed that goal to be achieved. One might imagine that, as the last blank areas in the world map are filled in, all other geographic puzzles will be solved too. No longer can we debate, as did our forebears, the location of the source of the Nile or whether a northwest passage between the Atlantic and Pacific Oceans really existed.

But even though one set of spatial puzzles has been solved, new ones have been uncovered. As we have just seen, geographers are becoming increasingly aware of mental maps. We have also noted, in earlier chapters, the ways in which travel times and transport costs can crumple and distort the familiar world map. The conventional world map describes a space that is continuous, isotropic (i.e., movement is equally possible in all directions), and three-dimensional. Nevertheless, the real space in which man moves is discontinuous, anisotropic (i.e., the costs of movement vary markedly over the map), and change rapidly over time—which means it has four dimensions rather than three. Mapping this real space poses fundamental questions for mapmakers and requires a reassessment of conventional Euclidean geometry. (See the marginal discussion of non-Euclidean space.)

Spatial transformations

Some analysts view this break from traditional mapping as paralleling the changes in theoretical physics when concepts of *absolute space* (in which the spatial coordinates have a fixed structure) were replaced by *relative space* (in which the coordinates reflect the structure of what is being described). The distinction between the two is shown by the maps of Sweden in Figure 20-14. Here we see a transformation from a rectangular coordinate system in absolute space into one in which locations are plotted relative to their direction and distance from a given center. Swedish geographer Torsten Hägerstrand used this transformation to describe the migration field of a small parish in the south

Non-Euclidean space

The geometric concepts synthesized in Euclid's *Elements* (written about 300 B.C.) form the basis of geographical measurement of the globe. Consider, for example, the distance between locations A and B on a plane.

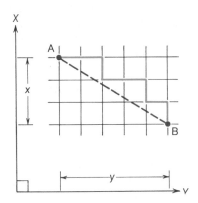

In Euclidean space the distance between these two points (d_{AB}) is given by the Pythagorean theorem as

$$d_{AB} = \sqrt[2]{x^2 + y^2} = \sqrt[2]{3^2 + 5^2} = 5.8 \text{ units.}$$

The variables x and y measure the differences between the two locations. If we superimpose on our continuous plane a Manhattanlike grid of streets, the distance from A to B becomes

$$d_{AB} = \sqrt[1]{x^1 + y^1} \quad \text{or} \quad x + y = 8 \text{ units,}$$

as we can no longer walk directly from A to B.

When we compare the formulas for estimating the distance between the same two points in Euclidean space and "Manhattan" space, we see that the difference lies in the exponents; these have a value of 2 in the first case but 1 in the second. Formal geometries have been developed to handle spaces where distances are both greater and less than those given by the Pythagorean theorem, but so far these non-Euclidean geometries have been little explored by geographers. See D. W. Harvey, *Explanation in Geography* (Edward Arnold, London, 1969), Chap. 14.

central part of his country. Most of the migration movements were over short distances, but a few migrants traveled large distances, such as 5000 km (3000 mi) to the United States. Local movements could be clearly distinguished only if they were plotted on a large-scale map, but movements to the main Swedish cities and to overseas countries needed to be plotted on small-scale maps. The problem of matching the different scales was overcome by using a special projection centered on the village from which migration was occurring. In this projection all radial distances from the center are transformed to logarithms, while all directions remain true. The result is a map on which local and international movements can be plotted on a single sheet. Although the transformed map of Sweden is unfamiliar and distorts its familiar outline, it reflects a limited field of space rather accurately.

This simple illustration shows just one of the ways maps can be transformed for different geographic uses. Considerable geographic research effort is going into ways of plotting far more complex rela-

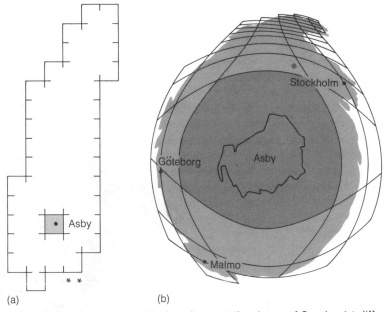

(a) (b)

Figure 20-14. Sweden: a migrant's view. A conventional map of Sweden (a) differs greatly from the same space as viewed by migrants from a single Swedish parish. The map of Sweden centered on the parish of Asby (b) can show population migration over both short and long distances. Note the correspondence between the central cell with curved boundaries and the corresponding square cell on the conventional map. [From T. Hägerstrand, *Lund Studies in Geography B*, No. 13 (1957), p. 73, Fig. 38.]

tions (e.g., the "time maps" of New Zealand and the South Pacific shown in Figure 13-8). Since this research area is rather mathematical, we must leave its consideration to more advanced courses.

Computer mapping

The growing need for environmental information in new map forms has thrown special emphasis on the role of the electronic computer in map production. Although digital computers were available around 1945, only in the last decade or so have geographers exploited their enormous capability for the storage, transformation, and retrieval of vast amounts of environmental data.

Figure 20-15 presents a typical computer map. Many are produced according to a system pioneered by a group at the Harvard Laboratory for Computer Graphics. The laboratory's work has been built largely on the Synagraphic Mapping Technique (SYMAP), a set of programs for creating a diversity of maps by combining alphabetical and numerical keys in fast-line printers. By superimposing combinations of numbers and letters on the printer, one can develop a series of "tones"

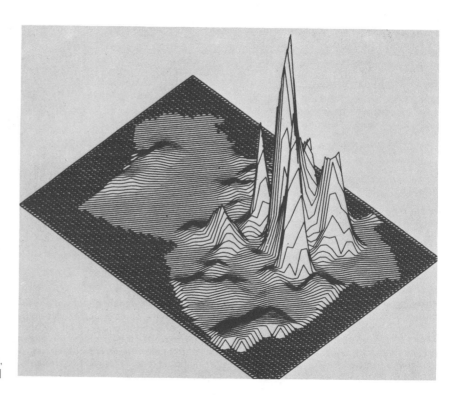

Figure 20-15. Computer cartography. This three-dimensional map, produced on the high-speed graph plot of a computer, shows the percentage of blacks in the population of different areas of New Haven, Connecticut, in 1967. [From D. L. Birch, *Journal of the American Institute of Planners* **37** (1971), p. 85, Fig. 5. Reprinted with permission.]

that simulate the gray scales of conventional isarithmic and choropleth maps. (For example, if you print a single character, say, an "O," this gives a very light tone. If you overprint "X" onto "Z" and then onto "H," you get something which gives a much darker tone.) Maps can be produced quite rapidly. In fact, geographers call for maps of environmental conditions from a stored set of data (a data bank) and have them within a few seconds.

The fastest and inherently most flexible output is provided by cathode-ray tube (CRT) displays. Maps are displayed on a television tube and can be photographed to provide permanent copies. As well as the line printer and the CRT display are numerous incremental graph plotters that can quickly produce maps and graphs from data stored in the computer.

In addition to being used for direct mapping, digital computers are increasingly being relied upon to prepare spatial data in map form. For example, the transformation of map coordinates from a system of Cartesian coordinates to one of polar coordinates is a trivial task even for mechanical calculators. The digital computer allows much more difficult transformations of geographic data from one map projection system to another. This means special projections can be tailored to particular jobs; we can draw a map for New Yorkers on a projection system centered on New York, a map for Muscovites centered on Moscow, and so on.

Yet another research byway that the computer has opened is in historical geography. By reading in the coordinates of places given on old maps and charts and comparing them with the true coordinates, we can recreate the original projection system used on the old maps. This allows us to estimate the probable location of settlements long since abandoned and provides some clues for archeological research. The use of this method for locating the position of old shipwrecks, however, has yet to be reported!

Mapping from samples

In their drive to make valid worldwide statements about the earth as the home of man, geographers often assume the need to inspect all corners of the world and to make a complete environmental inventory. When we compare our situation with that of other investigators, we see that this assumption can be relaxed. The geologist makes inferences about rock strata from the records of a few bore holes; the Gallup pollster makes predictions about our voting behavior by interviewing a few thousand citizens, not millions.

In a way, the earth's surface is analogous to a population. To be sure, the population is an unusual one because it does not consist of

Sampling designs

Geographers use different sampling designs to investigate particular spatial distributions. Each involves randomization procedures.

Stratified sampling divides the study area into separate strata, and individual sampling points are drawn randomly from each strata. The number of sampling points is made proportional to the area of each strata. Stratification can be used when major divisions within the area are already known to the investigator.

Systematic sampling employs a grid of equally spaced locations to define the sampling points. The origin of the grid is decided by the random location of the first sampling point. The grid also can be randomly oriented. Systematic location gives an even spatial coverage for mapping purposes.

Nested, hierarchic, or *multistage* sampling divides the study area into a hierarchy of sampling units that nest within one another. Random processes select the large first-stage units, the smaller second-order units within these, and so on. The final location of points within each small unit can be randomly determined. Nested sampling is useful in reducing field costs and in investigating variations at different spatial levels within the same area but it entails higher sampling errors. For a discussion of the applications and limitations of these and other designs see B. J. L. Berry and D. F. Marble, Eds., *Spatial Analysis* (Prentice-Hall, Englewood Cliffs, N.J., 1968), Chap. 3.

a finite number of individuals (for instance, the 200 million individuals who make up the U.S. population) but is continuous. Therefore, when we want to designate an individual within this spatial population, we have to do so in an arbitrary way, by specifying a point location like 68°30′N 27°07′E, an aerial unit like 1 km × 1 km, or a census division like a tract or county. These individuals, however defined, form *sampling units*, from which we can construct a picture of the population as a whole.

Making inferences about a population from a small part of it is a dangerous procedure. We may select samples that are biased in some way or the sample may be too small for a reliable estimate. But how do we decide what is "too small"? Fortunately, many of the rules governing the relationships between a sample and a population were worked out by mathematicians like Karl Pearson and R. A. Fisher in the first half of this century. Today there is a large body of well-substantiated sampling theory to which geographers can turn in setting up their investigations. Sampling theory helps us to decide how much error is likely to occur with sample designs of different kinds. For example, in simple *random sampling* (Figure 20-16), the accuracy or sampling error is proportional to the square root of the number of observations. This means that if we were to increase the number of sampling points in Figure 20-16 from 25 to 100 we could expect to improve our accuracy not by four times as much, but by only $\sqrt{4}$, or 2.

Geographers design their sample surveys of environmental characteristics in close cooperation with their colleagues in statistics, and most of the problems they encounter are general statistical ones we need not be concerned with here. Geographic research has been confined mainly to working out the efficiency of different kinds of spatial arrays of sampling points. Each spatial sampling design has particular disadvantages and advantages and is used for specific types of en-

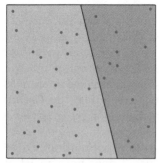

(a) Stratified sampling

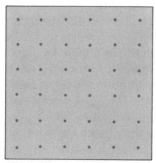

(b) Systematic sampling

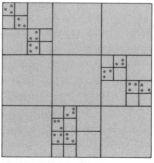

(c) Nested sampling

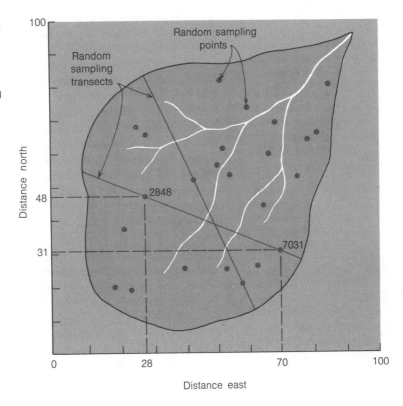

Figure 20-16. Simple random sampling. It is frequently too costly to inspect every part of a region before making generalizations about it, so geographers rely increasingly on a series of sample observations. But how should such observation points be selected? To avoid bias, the location of each sampling point within the watershed shown here is given a pair of random numbers on the two axes of the coordinate system.

vironmental investigations. (See the marginal discussion of sampling designs.)

Space-time sampling

Conventional sampling designs are useful if we are trying to measure some aspect of environmental quality that varies over *space*. How do geographers cope with qualities that are also varying over *time*? Let us take as an example the problem of measuring the pollution levels in the atmosphere over a forested watershed adjoining an urban area. Any of the watersheds in the Catskill Mountains near New York City or the Coast Ranges near Los Angeles will serve as an example. The chemical composition of the air may vary over the different parts of the basin (over space), and it may also vary from day to day and from hour to hour. A complete survey of the variations would require a rather dense pattern of recording stations across the area, each continuously monitoring atmospheric changes. Yet if we assume a finite budget, there is a limit to the number of such stations that we can afford. Sampling theory indicates where a small but representative

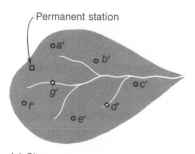

Permanent station

(a) Sites

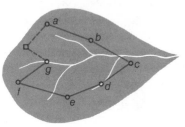

(b) Monday *(a″)*

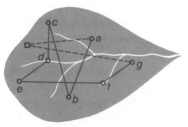

(c) Tuesday *(b″)*

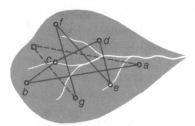

(d) Wednesday *(c″)*

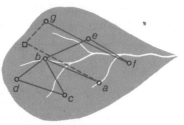

(e) Thursday *(d″)*

number of stations can be located. If we find that even this reduced number of stations is too costly, then we make further reductions by using a mixed strategy of fixed and mobile recording stations.

One of the ways in which this can be done is shown in Figure 20-17. Here the atmosphere throughout the target area is sampled by a combination of one permanent recording station with a series of seven occasional recording stations (*a′* through *g′*). The number of occasional recording stations (seven in this case) is determined by the convenience of measuring environmental levels on each day of the week. Table 20-1 shows how the time available for taking measurements can be divided into seven equally spaced days (*a″* through *g″*) and equally spaced hours within each day (*a* through *g*). By visiting the seven out-stations in a different but predetermined sequence each day, the observer can build an unbiased picture of environmental fluctuations. Note that arranging the recording sequence in a symmetric square—a *Latin-square* design in the language of statistics—ensures that each day's pollution is the average of a balanced mix of locations and hours, each hour's pollution is the average of a balanced mix of locations and days, while each location's pollution is the average of a balanced mix of hours and days. As a further bonus, the total recording effort is reduced to one-seventh of what it would be plus of course the time taken by an observer to move from one temporary station to the next.

By arranging mixtures of recording stations in this way, we can construct a much finer mesh of recordings of variations over both time and space, and move further toward the goal of accurately mapping environmental patterns on the earth's surface.

Conclusion: The unfinished map

Cartography was one of the earliest areas of research by geographers and played a key part in the geographic training of the Greeks. In recent years its study has declined and it has become something of

Figure 20-17. Sampling in space and time. A Latin-square sampling design is used here to reduce the time needed to monitor pollution at seven sites within a watershed. (See also Table 20–1.)

Table 20-1 A Latin-square design for space-time sampling

	Location of sampling sites						
	a′	b′	c′	d′	e′	f′	g′
Days of the week	*Times of the day*						
Monday (a″)	a	b	c	d	e	f	g
Tuesday (b″)	b	e	a	g	f	d	c
Wednesday (c″)	c	f	g	b	d	a	e
Thursday (d″)	d	g	e	f	b	c	a
Friday (e″)	e	d	b	c	a	g	f
Saturday (f″)	f	c	d	a	g	e	b
Sunday (g″)	g	a	f	e	c	b	d

a Cinderella subject in geography departments. The world map appeared to be nearly complete, and there were few discoveries to be made: *Terra Australis Incognita* (the "unknown southern land") had been added to the map in the eighteenth century, darkest Africa had been opened up in the late nineteenth century, and the polar areas had been explored in our own century. Today the situation has changed. Just as one mapping task appears to have been successfully completed, another and more difficult one of mapping in new kinds of economic and social space has appeared. The research frontier of inner space has succeeded the closed frontier of outer space. Geographers find that they are back with the Greeks, pondering just what kind of spatial world they are really living in.

Reflections

1. Use the addresses in Appendix B at the back of the book to obtain some satellite photos of your own area on different scales. What kinds of information do these photos show that cannot be obtained from conventional maps?

2. Radiation-sensitive sensors now enable satellites to capture differences in *heat* on the earth's surface on film. List ways in which this capacity provides valuable information for the geographer.

3. Select ten areas (these may be as small as parts of a city or as large as countries), and have several of your friends place them in order in terms of their desirability as places to live. Comment on the results.

4. Consider the maps of Los Angeles in Figure 20-9. Assume you are teaching a group of high school children in this city. How would you go about widening their spatial image of the city?

5. List the problems encountered by geographers trying to map an individual's view of space.

6. Review your understanding of the following concepts:
   (a) remote sensing
   (b) infrared sensors
   (c) radar images
   (d) false color photographs
   (e) mental maps
   (f) isotropic and anisotropic space
   (g) Euclidean space
   (h) random sampling
   (i) Latin squares

One step further . . .

An introduction to remote sensing, stressing its geographic applications, is given in

Barrett, E. C., and L. F. Curtis, Eds., *Environmental Remote Sensing: Applications and Achievements* (Edward Arnold, London, 1974).

Excellent examples of earth photographs from spacecraft are available in a number of NASA publications—for example,

National Aeronautics and Space Administration, *Earth Photographs from Gemini III, IV and V* (Government Printing Office, Washington, D.C., 1967).

An introduction to the study of "inner space" and the overlap of geographic and psychological research in this area is provided by

Downs, R. M., and D. Stea, Eds., *Image and Environment: Cognitive Mapping and Spatial Behavior* (Edward Arnold, London, 1974).

For a discussion of mental maps and spatial preferences at the level of the individual and the city, see

Gould, P. R., and R. R. White, *Mental Maps* (Penguin, 1974), and Baltimore, Md.,

Lynch, K., *The Image of the City* (MIT Press, Cambridge, Mass., 1960).

Some of the problems involved in handling the more complex kinds of geographic space are reviewed in

Harvey, D. W., *Explanation in Geography* (St. Martin's Press, New York, 1970), Chap. 14.

Although the results of current research are given in the regular geographic journals, you will need to look at specialized technical journals such as *Photogrammetric Engineering* (a quarterly) and *Remote Sensing of Environment* (also a quarterly) in order to keep abreast of the most recent developments in remote sensing. Look at *Environment and Behavior* (another quarterly) for an example of the growing range of journals concerned with the inner-space interface between geographic and psychological research.

Chapter 21

Worlds Present, Worlds Future

WAYS OF LOOKING FORWARD

After performing the most exact calculation
possible in this sort of matter, I have found
that there is scarcely one tenth as many
people on the earth as in ancient times. What
is surprising is that the population of the earth
decreases every day, and if this continues, in
another ten centuries the earth will be
nothing but a desert.

— MONTESQUIEU
Lettres persanes (1721)

About a decade ago leading scientists throughout the world were quizzed by the Rand Corporation. They were asked to name the major scientific advances that they expected in their own field during the next few decades. Their replies ranged from organ transplants to ocean farming. Panels of specialists were also asked to give their views of the timing of a short list of expected breakthroughs — and there was a fair scatter of opinion between the optimists and those who included "never" in their estimates.

Some possible developments are of profound interest to geographers because they are likely to change the present balances of population and environment; some examples are large-scale weather control, effective and simple contraceptives, economic ocean farming, or synthetic protein substitutes for food. We should certainly like to know when we can expect breakthroughs to begin to affect the regional and global situations we study. But does Rand's "Delphi technique" give us the answer? (Delphi is the site in ancient Greece, famed for its oracle able to foretell the future.) Let us take the first predicted development, large-scale weather control. Only one-quarter of those consulted thought that this would be feasible before 1986, and another quarter felt that it could not come until after 2000. The general view was that weather control on an extensive basis should be possible by the last decade of this century.

The lack of a single, clear answer underlines the nice linguistic distinction that the Romans drew between past events and future events. Past events (in Latin, *facta*) are accomplished and unchangeable, whereas future events (*futura*) are still fluid. Only the *facta* can be known; the *futura* can merely be estimated, projected, or divined. Thus, the absence of a cross-channel bridge between Denmark and southern Sweden in 1972 was a fact, unchallengeable and 100-percent known. That a bridge may be built by 1981 is one of a set of possibilities. But this future possibility is important enough to the pattern of accessibility in this part of Scandinavia to be worth studying and mapping. (See Figure 21-1.) Geographers are increasingly concerned with the multiple worlds of the future rather than the single world of the past.

In this chapter we look at four questions. First, what effect does uncertainty about the future have on human decision-making — and, in the longer run, on the human geography of the globe? Second, how can we reduce that uncertainty by making forecasts for the short term? Third, what chance have we of looking still further ahead toward the end of the century? Fourth, what effect does the growing interest in forecasting have on the way geographers go about their work? We have already met some aspects of forecasting — for example, in our study of population growth in Chapter 6 and resource uses in Chapter 8. Here we try and draw some of these threads together in looking at the way geographers are responding to the challenge of the future task.

21-1 UNCERTAINTY AND HUMAN DECISION-MAKING

We have already noted in this book the role of our concept of the future in determining attitudes to space and resources. (See Chapter 10-3.) Here we expand this concept and illustrate its effect over time and over space.

Uncertainty over the timing of natural events

Many human groups occupy environments that are dangerously unpredictable. Figure 21-2 illustrates two such environments: a coastal area subject to periodic hurricanes and flooding, and a city built in an earthquake zone. There are many more. The floodplains of major rivers, the semiarid margins of the continental grasslands, the slopes of volcanoes, all present risks to the populations that settle them.

Floods How do these groups react to the uncertain environmental threat that hangs over them? Geographers Gilbert White and Ian Burton have investigated the ways in which flood dangers are viewed. They began by looking at the likelihood of flooding in different areas and constructed a *risk curve* [Figure 21-3(a)] for the 498 urban communities in the United States that are built on river floodplains (and have

Figure 21-1. Conditional forecasts. Most geographic statements about the future are conditional, that is, they relate to what is expected to happen *if* certain actions are or are not taken, or certain conditions are fulfilled or not fulfilled. These typical projective maps are from a Swedish study of the impact of proposed links with Denmark. The contours indicate the hours needed to travel by car from any location in eastern Denmark or southern Sweden to a market of 3 million people given (a) a water barrier between the two countries and (b) a bridge and ferry service. [From T. Hägerstrand, *Transactions of the Institute of British Geographers*, No. 42 (1967), p. 15, Fig. 14.]

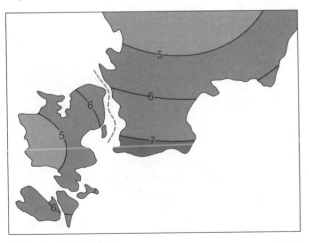

(a) Barrier assumption

* * *

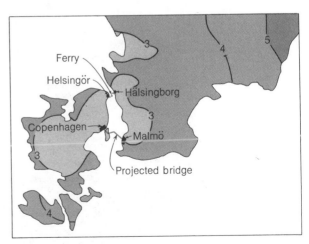

(b) Bridge assumption

Figure 21-2. Uncertainties in the natural environment. Extreme and irregular natural events such as (a) hurricanes and (b) earthquakes provoke a wide variety of reactions related to the perception of the groups affected and the magnitude, frequency, duration, and spacing of the events. (See Table 22–1.) [Photographs (a) from United Press International, (b) by J. Eyerman, Black Star.]

(a)

(b)

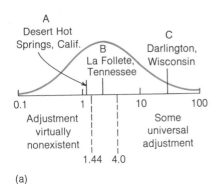

(a)

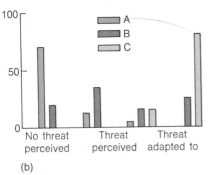

(b)

Figure 21-3. Flood frequencies and the perception of future flood hazards. (a) The curve shows the frequency of floods in 496 urban places in the United States. Most places for which flood-frequency data were available have two or three floods each year. (b) The degree of adjustment to the hazard in three places with three different experiences of flooding is illustrated. The height of each column reflects the number of respondents in each place who fail to perceive a threat, who perceive a weak or strong threat (two levels of "perceived"), or who adjust to the hazard. The letters and shading identify the three locations named in (a). [From R. W. Kates, *University of Chicago Department of Geography Research Paper 78* (1962), Fig. 9.]

flood-frequency data). Some of these urban places were liable to have floods only once in ten years, while in others waters rose to danger levels dozens of times in a single year. For most cities in the sample, the likely number of floods each year was two or three.

As Figure 21-3 indicates, human responses and adjustments to the known danger of flooding do not increase consistently as the risk becomes greater. Until the environmental stress builds up to the point where the likelihood of damage is regular and recurrent, little or no adjustment to the possibility of flooding takes place. Figure 21-3(b) shows the degree to which inhabitants of three United States communities where the likelihood of flooding is different "perceive" a flood threat. Darlington, Wisconsin, can expect 20 floods in any 10-year period, whereas Desert Springs, California, can expect only one flood in the same period. When the probability of the recurrence of a hazard is high, as it is in Darlington, the danger is widely perceived but evaluated in different ways. Table 21-1 lists four of the ways in which individuals respond to the possibility of recurring natural disasters. Each response represents an optimistic rationalization for continuing to live in a hazard area. It is interesting that the range of responses is great when the probabilities of a recurrence are moderate; responses tend to be more uniform in high-risk and low-risk areas.

Droughts Studies of drought hazard areas in the Great Plains support the results of the floodplain research. A person's perception of the risk of a drought is directly related to the degree he is likely to be affected by it; wheat farmers are far more aware of the likely incidence of droughts than cattle ranchers. But why do persons, in spite of previous exposure to a hazard, return to hazard-prone sites? It seems likely that risks act as filters, attracting settlers who have the inclination or capability (because of their personality, financial resources, etc.) to cope with a threatened hazard but repelling the more timid souls.

Low-risk environments Uncertainty over the future is not, however, confined to "risky" environments. All human behavior is based on assumptions about unknown future events; we select a college course, or a husband, on the basis of our intuitive forecast of whether we will be pleased with our choice! Figure 21-4 illustrates this fact, using data from a study of the Mellansverige region of central Sweden, a region which would be regarded as having a rather stable low-risk environment. Geographer Julian Wolpert compared the actual farming pattern in the area with that which would have occurred had farmers made the theoretically "best," or *optimum*, use of resources. He did this by a detailed study of the productivity of farm labor for a sample of 17

Table 21-1 Human reactions to irregular natural hazards

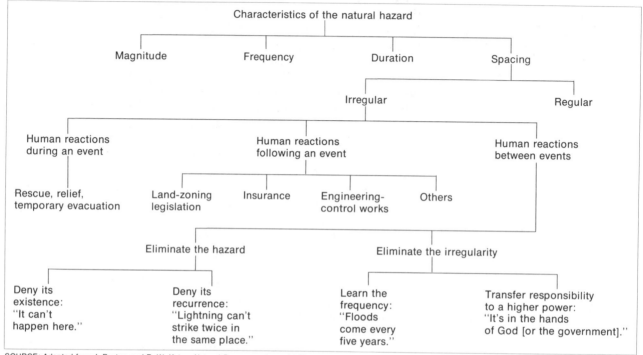

SOURCE: Adapted from I. Burton and R. W. Kates, *Natural Resources Journal* **3** (1964), p. 435.

representative farms. Given the mix of resources available to each farmer, Wolpert was able to use a mathematical *optimization technique* to compute the theoretical highest yield from the resources. The results for the sample farms were then extrapolated for an additional 500 farms and the findings mapped.

The maps in Figure 21-4 reveal that half the area was producing only 70 percent of what it could produce, and some local pockets were producing only 40 percent. The method of estimating how much could be produced may cause doubts about the exact values obtained, but the width of the gap separating the two maps is too great and too consistent to be explained by errors from this source.

Then what is causing the gap between the two maps? Wolpert's analysis showed that there were also regional variations in the farmers' "knowledge situation" that corresponded to time lags in the diffusion of information, such as data on recommended farm practices, from centers like Stockholm and Uppsala, and there was uncertainty as to which crop and livestock mixes would be profitable. There was un-

Figure 21-4. The impact of uncertainty on resource use. The maps show the actual and potential productivity (indicated by numbers and tints) of farms in the Mellansverige area of Sweden. The actual productivity, shown in (a), is well below the estimated maximal productivity, shown in (b). [From J. Wolpert, *Annals of the Association of American Geographers* **54** (1964), pp. 540, 541, Figs. 2, 3.]

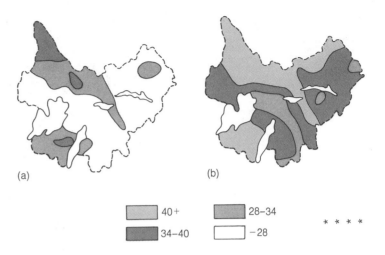

(a)   (b)

40+   28–34

34–40   −28

* * * *

certainty also not only about external factors like fluctuations in weather or market prices, but about personal things like future health or finances. As a result of all these factors it is clear that Swedish farmers in this part of Sweden (and perhaps, farmers everywhere) were not aiming at peak productivity but were satisfied to be able to maintain production at an acceptable but suboptimal level. This kind of behavior is termed "satisficing" behavior.

Uncertainty over the location of natural resources

So far we have restricted our concern with uncertainty to uncertainty about the future. But we know that uncertainty has a spatial dimension too. Information is a scarce commodity. When we choose a university or try to find a local apartment, we do so on the basis of limited information. There comes a point in our search when we decide, rightly or wrongly, that we have read enough brochures or knocked on enough doors, and we make our choice at that point.

Johns Hopkins University geographer David Harvey has attempted to build this uncertainty over location into a general model of search behavior. He suggests that in any environment a group will search for the best way to use natural resources to meet its needs. It conducts this search either by making its own experiments or by copying other groups coping with the same environment.

Consider the fate of the first group of settlers to arrive at an area (group 1) in Figure 21-5. They chance on a relatively good resource-use pattern in the first time period, but later experiments bring them below the subsistence line and they fail to survive. From the experi-

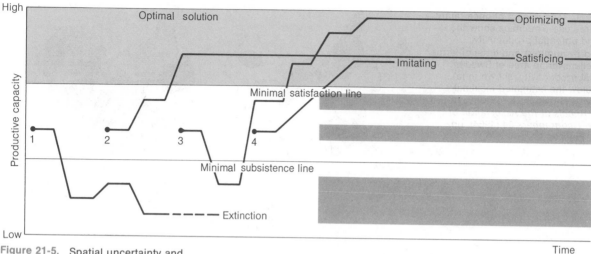

Figure 21-5. Spatial uncertainty and search behavior. Each line indicates the migration of a human group (1 through 4) over time. Note that the productive capacity of the environment varies from low (at the bottom) to high (at the top). The experiences of unsuccessful groups (e.g., those on track 1) teach later groups to avoid unproductive areas. Note the contrasts in the vigor of the search by group 3, which is looking for the best possible environment, and that by group 4, which is happy to imitate group 2 and settle for a possible (but suboptimal) location. The shading below the minimal satisfaction line indicates areas eliminated from the search by the experience of the groups. [From D. Harvey, in R. J. Chorley and P. Haggett, Eds., *Models in Geography* (Barnes & Noble, New York, and Methuen, London, 1967), p. 594, Fig. 14–11.]

ence of the first settlers, group 2 learns which resource-use combinations to avoid (the shaded areas of the diagram). Experiments bring them above the minimal satisfaction line, and they settle down happily. Group 3 is an active group that explores several resource-use combinations and comes very close to achieving the theoretically optimum resource-use pattern. Group 4 is an imitative group with low goals that moves rapidly to follow the pattern set by the successful, though unambitious, group 2.

If you look at historian W. P. Webb's book, *The Great Plains*, you will find an excellent example of how different groups responded to the resource mix provided by this grassland area. The pre-Columbian Plains Indian, the sixteenth-century Spaniard, the cattleman in the 1840s, and the wheat farmer in the 1880s all experimented with the Great Plains environment. In the withered hedgerows (a pre-barbed-wire experiment at fencing, British style), in the bankruptcies that followed long droughts, and in the blowing topsoil, we see the aftermath of the experiments that failed. Each cultural group saw in the Great Plains environment different possibilities, linked to the current technology and the group's cultural background.

21-2 | SHORT-TERM FORECASTING

How can geographers reduce uncertainty by making statements about the future? Here we look at ways of making forecasts for both the short and the long term. Contemporary methods of forecasting represent the modest survivors of a long history of research—much of which has

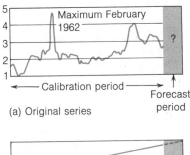

(a) Original series

(b) Linear trend

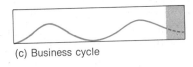

(c) Business cycle

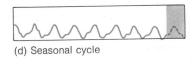

(d) Seasonal cycle

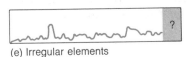

(e) Irregular elements

Figure 21-6. Projected time series. Graphs (b) through (e) show how the time series in (a) can be broken down into regular and irregular components. The regular components may be either linear trends or recurrent cycles (e.g., business cycles or seasonal cycles). Regular components can be projected forward as elements in a forecast. [Data are monthly unemployment figures for Bridgewater, England (Figure 21–10), for 1960–1969 supplied by the Department of Employment.]

proved disappointing. Here we illustrate a few of the methods that have proved most useful.

Projecting past trends

The simplest forecasting models are those in which we already have a long set of records from the past and we wish to look only a short way into the future. Such *forecasting models* are simplified representations of how the real world works, used in attempts to predict the future. Such models may be expressed by mathematical equations, graphs, stated in words, or drawn on a map. We have already seen how geographers use past climatic records (in Chapter 4) and past population censuses (in Chapter 6) in this way. Now we shall look at this kind of trend-projection method more closely.

Consider Figure 21-6, which shows the record of unemployment from 1960 to 1968 for a small English town (Bridgewater) of around 25,000 people. (See also Figure 21-10, page 559.) Data are available for the percentage of the local labor force out of work for each of the 108 months in this period, and they form an erratic picture [Figure 21-6(a)]. The forecaster's problem is this: Given this information can we make any useful predictions about the likely trends in jobs in 1969?

The language of forecasting

Bellwethers are areas within a country that show trends earlier than the rest of the country. Literally the term means a sheep (wether) wearing a bell that walks ahead of the flock.

Business cycles are recurrent fluctuations in general economic activity. They may be observed in fluctuations in the production or employment within a country or region.

Delphi forecasts are qualitative statements about the future based on the averaged views of experts in a particular field. They are usually confined to forecasting the timing of technical "breakthroughs."

Forecasting models are simplified representations of how the real world works, used in attempts to predict the future. Such models may be expressed by mathematical equations, graphs, stated in words, or drawn on a map.

Naive models are forecasting models that do not involve any theoretical basis. For example, we may use a straight line projection where one looks at how a region's economy has grown in the past and projects that it

will grow at the same rate in the future.

Postdictions are the opposite of predictions. They forecast "in reverse" by using present and past records to project a situation at a still earlier period.

Projections are the forward extension of trends recognized from past and present records.

Scenarios are informal forecasts of future situations based on a sequence of arguments rather than on mathematical equations.

Simulation models are models which reproduce complex, real-world situations in terms of a computer program. Literally, simulation is the art or science of "pretending."

Time series are data gathered over a period of time, usually at regular time intervals.

Turning points are points in time when economic activity in a region or country changes direction. When activity is falling during a recession and then swings upward again, the precise time at which it begins to swing is the turning point.

One way is to look for any apparent patterns or trends in the past record. We find, for example, that unemployment in this town has been getting worse, and we can generalize this rise as a long-term trend. This trend is represented in Figure 21-6(b) by a straight line. The line was fitted to the past values for unemployment by the method of least squares (see the marginal discussion of choosing trend curves) and is the best estimate we can make on the linear element in the rise in joblessness. By continuing this line forward into 1969, we can see how long-term trends will affect the job situation in the town if, and only if, the 1969 trend is generally the same as in the rest of the decade.

Although the trend line gives a useful general indication of change, it ignores shorter term *cyclic* variations. For example, there are also important seasonal variations. More people are unemployed in the winter, which is a slack period for the construction industry and the tourist industry. February tends to be the worst month of unemployment, and July the best. With other statistical devices we can pick out these seasonal swings and, again, project them into 1969 as in Figure 21-6(d). In addition to this seasonal cycle of work, there are other recurrent changes that have a longer duration. In Bridgewater, business-cycle variations had a duration of about 5 years, with peaks in the summer of 1961 and the winter of 1964. Again, this fact can be recognized by statistical methods, and its effect projected forward as in Figure 21-6(c). If we now combine the long-term trend, the business cycle, and the seasonal cycle, we have a better chance of projecting the 1969 values accurately. However, we are still left with irregular elements, like the closure of a single plant, whose recurrence cannot be predicted from past records [Figure 21-6(e)].

The question of accuracy

How can we be sure that our projections are accurate ones? The only reliable way is to wait to see what happens and compare the actual values with the projected ones. If we do this for Bridgewater, as in Figure 21-7, we find that although the projections are reasonably good for the whole of 1969, they become progressively worse in the following two years. You can think of this increasing error as a "cone" of increasing uncertainty that widens the farther ahead we try to forecast.

In this particular instance a comparison of actual values with forecast values was possible. But this would not be so for long-range forecasts. If we are forecasting the population of New York State in 1990, we want to judge its probable accuracy now, not in 1991 or whenever the relevant evidence is in. Comparing real values with forecast values is therefore a poor way of evaluating the utility of a forecasting model. We can arrive at a verdict only after the results have happened. Also,

Figure 21-7. Projected and actual trends. (a) Here, the shaded "forecast period" of Figure 21-6 is enlarged to show the three regular trends being projected forward over one year. (b) These three trends are summed to give a projection that can be compared with actual values for unemployment.

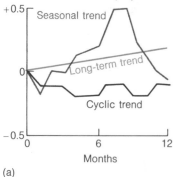

(a)

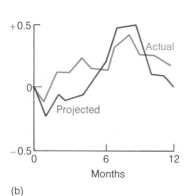

(b)

Choosing trend curves

The trend curve in Figure 21-6 is only one of a series of possible trends that can be fitted to the monthly observations. In addition to the straight line with the form

$$Y = a + bt,$$

we might also fit a simple exponential curve,

$$\log Y = a + bt,$$

or a parabola,

$$Y = a + bt + ct^2,$$

as well as more complex logistic or Gompertz curves to the data. In the equations Y is the variable we are examining, t is time, and a, b, and c are constants. There are methods for checking the relative suitability of the different curves, but a wide area for personal judgment of likely trends remains. In using the trend curves for projecting future values, we may find errors because (1) the trend in the future does not continue to follow the kind of curve chosen, (2) the exact values of the constants (a, b, and c) are uncertain because they were computed from a limited amount of data, and (3) individual points vary about the fitted curve. For an introductory discussion with worked examples, see J. V. Gregg *et al.*, *Mathematical Trend Curves* (Oliver and Boyd, Edinburgh, 1964), Chap. 1.

all trend projections assume that conditions in the forecast period are substantially the same as those in the period over which the trend model was fitted.

For these reasons, we need tests for projections that are applicable *before* and not *after* a forecast is made. Such tests depend partly on common sense and partly on statistics. We look first at the internal logic of a forecasting model, that is, whether it seems to make reasonable sense in terms of what we know about a situation. Alternatively, we can use the past record itself as a test by arbitrarily separating it into two halves and using one half to project the other. Thus, in the case of Figure 21-6(a) we can use the first 54 months and make a model that fits it as closely as possible. When we use this model to project the second half of the record, its success will provide some clues to its probable performance in a real forecasting situation. Tests with split series are named *Janus tests*, after the Roman god of doorways with two faces, one looking backward and one forward.

Forecasting contagious diffusion

By combining independent trend projections for a set of small areas and mapping the results, we can produce a map of future conditions over a larger region. Geographers are not content to work in this way, however. We've noted throughout this book that events which are near in space tend to have closely related values (that is, they are autocorrelated). Therefore, we wish to introduce this additional spatial factor into our forecasts.

Let us return for a bit to the models of spatial diffusion discussed in Chapter 12. We noted there that computer simulation plays a vital part in the geographer's study of diffusion processes. For example, in the United States, Richard Morrill used computers to simulate the block-by-block northward expansion of the black ghetto in the city of Seattle. The basic Hägerstrand model of diffusion fitted to the actual pattern of expansion from 1940 to 1960 by using ten 2-year cycles. Rules were created to model the in-migration of a black population from outside Seattle, to calculate the critical pressures within the ghetto, and to trace the direction and distance of spread of families from the ghetto into surrounding city blocks. The rules were modified to allow rapid absorption into the ghetto of middle-class, single-family houses to the north, but only slow inroads by blacks to the west and along the lake to the east (where there was rather expensive housing and apartment blocks). Figure 21-8 presents a typical 2-year cycle and shows the direction of family movements.

Having built the model to fit past processes accurately, Morrill could run the simulation for 2-year cycles into the future, tracking the

Figure 21-8. Simulation of ghetto expansion. The map shows one two-year period from a computer simulation of the expansion of the northern part of Seattle's black ghetto. The arrows show simulated movements of families within the ghetto. Shading indicates city blocks around the edge of the ghetto where housing is desired by black families. A distinction is made between an unsuccessful attempt to move (light shading) and a successful attempt (dark shading, with an arrow entering the block). [From R. L. Morrill, *Geographical Review* **55** (1965), p. 357, Fig. 13. Reprinted with permission.]

▮ Blocks newly entered   ▯ Contacts only   ✶ ✶ ✶ ✶ ✶

expansion of the ghetto. Computer generated maps will, like trend projections, contain increasing errors over time. Yet despite the dangers and difficulties of using such simulations to project regional futures, they open the door to exciting possibilities. If we can identify the factors that allow an innovation to diffuse rapidly from one center but only slowly from another, we can use that information in two ways. When we wish to speed up the spread, we can concentrate research funds on the promising centers. For instance, geographers are currently trying to distinguish the best location for new family planning clinics in the western region of Nigeria.

In some cases we may be interested in slowing down a diffusion process and limiting its spatial spread. For example, Canadian geographer Roland R. Tinline devised optional strategies for the control of foot-and-mouth disease in England. In 1967 England suffered an epidemic that forced the slaughter of about 443,000 animals and cost millions of dollars. The control strategy involved slaughtering infected animals and banning the movement of all animals in a cordon around the outbreak. Tinline devised a mathematical model of the spread of the actual outbreak. This model can be used to create control strategies for further outbreaks by simulating a mathematical outbreak and by testing its spread over different parts of England given various kinds of control strategies.

Bellwethers and "early warning" regions

Stock-market analysts keep a watchful eye on certain shares that tend to move a little ahead of the rest of the market. Similarly, electoral pollsters tend to watch closely the returns from particular primary elections in the United States (e.g., the slogan "As Vermont votes, so votes the nation"). Geographers use similar bellwethers, looking for regions that are consistently ahead of others in terms of spatial diffusion. So *bellwethers* are areas within a country that show trends earlier than the rest of the country. Hopefully the changes in such regions will, like dips provide some early warning of storms ahead. We can illustrate this idea with a map of southwest England (Figure 21-9), which shows the timing of upturns in two business cycles for each of 70 local areas. A ring of small towns 20–50 km (12–31 mi) from the regional capital (Bristol) tends to respond several months earlier than the capital city itself, and nearly half a year ahead of some of its more remote westerly brethren.

For geographers, as for stock-market analysts, the key issue is to judge how consistent these early warning signs are. In the case of regional business cycles, attempts to isolate lead areas have produced conflicting results. Over a 26-year period (1919–1945), the cyclic pattern of several American cities with different industrial structures reveals wide differences in turning points for relatively minor cycles; but it is difficult to discern the leaders and laggards from one minor cycle to another. (Note that *turning points* are times when economic activity in a region or country changes direction.) During major cycles turning points tend to be coincidental. However, there are some weak indications that cities whose economy is linked to certain industries (e.g., Cleveland and Detroit, with their steel and automobile industries) did lead the other cities for most of the period.

How optimistic are geographers about research in this area? Work in the 1950s suggests that spatial differences in turning points among regions within the United States have not been proved.

However, more recent investigations lead to somewhat different conclusions. A study of unemployment in Midwestern cities in the early 1960s shows that a group of cities around Pittsburgh regularly led the Detroit and Indianapolis areas by 3 to 5 months. Regional unemployment data for 10 British regions shows the Midland region leading most other regions by 3 months and Scotland and the North leading by 6 months. A lot more work needs to be done before we can be sure, however.

★ ★ ★

Forecasting hierarchic diffusion

Similar lead–lag relationships are being explored in other fields. Geog-

Figure 21-9. Lead and lag areas. These maps of southwest England indicate areas that reached low points on the business cycle three or more months after Bristol (lag areas). Bristol, the regional capital and major employment center, reached its own trough in the months indicated. [From K. Bassett and P. Haggett, in M. D. I. Chisholm, A. E. Frey, and P. Haggett, Eds., *Regional Forecasting* (Butterworth, London, 1971), p. 398, Fig. 6.)

Lead areas
Lag areas

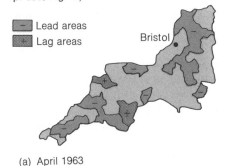

(a) April 1963

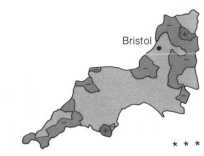

(b) July 1967

raphers working with epidemiologists on the spread of infectious diseases have mapped reservoir areas from which epidemics may erupt periodically to invade communities where the disease is not endemic. In spreading outward, an epidemic may move along consistent pathways. For example, research on the cholera outbreaks in the eastern United States in 1832, 1849, and 1866 showed consistent elements. All three epidemics started in the East, with early cases in New York City, and all tended to hit large and accessible cities a few months before smaller or more remote ones. In 1832 the links in the chain of outbreaks were clear: The disease followed the main arteries of travel, the waterways, and moved from New York by way of the Hudson–Erie Canal to the Greak Lakes, and down from the headwaters of the Ohio to the lower Mississippi. By 1868 the urban areas were more tightly bound together by the railway system, and the spread of the disease reflected the urban hierarchy more closely. As each town acquired the disease, so it was passed on to nearby smaller settlements, until by the end of the year it had blanketed the eastern half of the country.

In these spectacular cases of hierarchic diffusion, the centers of diffusion, the mechanisms of transfer, and the potential paths of spatial spread are clear. Not all diffusions are as straightforward. If we consider the changing job opportunities in Bridgewater, the small English town whose unemployment pattern we studied earlier (Figure 21-10), we can see that its local fortunes are bound up with those of its local region, and the national and international economy of which it forms a small part. Decisions on British aircraft engines by Lockheed in southern California cause significant changes in the fortunes of the Bristol engine plants in southwest England. These changes are passed on in muted form to engine part suppliers in Bridgewater itself, to be reflected as flickers on its employment curve [Figure 21-6(a)].

How can geographers handle this multilevel, multinational, multi-industry diffusion? Figure 21-11 shows, in a simplified form, one of the possible ways to do it. We assume a three-level urban hierarchy in which changes in the upper level are passed on to the two lower levels. The changes are represented by a simple wave indicating varying levels of economic activity in the city. As each wave passes down to the next city it may be *lagged* (occur later in time) and *modulated* (be damped down or amplified). We know that small swings on the national level may be transformed into large swings in the local economy. Areas with large numbers employed in a heavy industry like steel may have economic cycles more pronounced than that of the nation as a whole; that is, the rises may be higher and the slumps lower. Conversely, areas with service industries like universities may experience milder than average swings in economic activity.

Figure 21-10. Interdependence of local centers. Employment trends in the small town of Bridgewater, England (population 25,000), reflect in a subdued and lagged fashion trends in both the national economy and the neighboring cities. See Figure 21–6 for Bridgewater's unemployment history in the 1960s. [Aerofilms photo.]

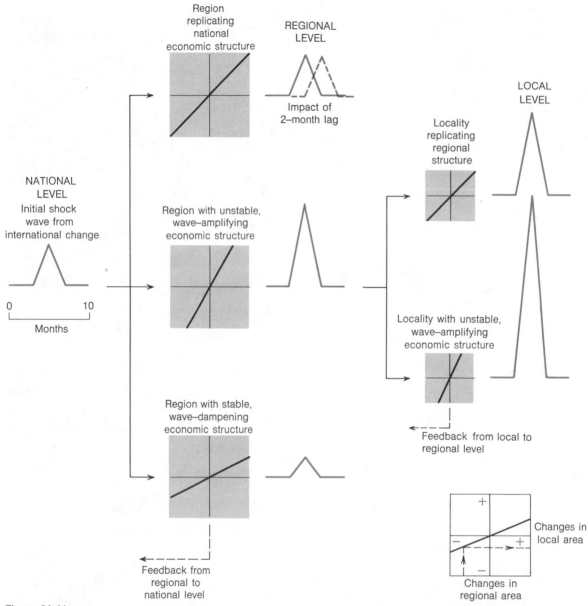

Figure 21-11. Hierarchic transmission of shocks. This schematic model shows how changes in economic activity on one spatial level may be transmitted to others. The *transformation boxes* (shaded) describe the degree of amplification or dampening caused by the economic structure at the regional or local level. The diagram is, of course, greatly simplified. In practice, the waves are highly irregular and may be lagged in time, the shocks may start at levels other than the national one, and the effects of several shocks may intermingle and overlap. The box at the lower right shows how a major fall in economic activity in a regional area is reflected in a dampened form at the local level.

Examples of regions that leave economic structures that dampen or amplify natural waves are illustrated in Figure 21-11. Note how the economy of a region modifies the initial impulse transmitted from places higher in the hierarchy. The degree of amplification or dampening can be described simply enough by a *transformation box*, and we can think of the urban hierarchy as a set of stations in a radio network—passing on a signal but in a distorted form. To this simple image we must add the complication that each city is not only receiving signals but emitting signals of its own. Teasing out some spatial order from this more complex situation takes geographers into mathematical areas that lie outside the scope of this book.

We have developed the idea of hierarchical links between cities as one example of the way in which geographers work to understand how shocks are passed from one area to another. Similar ideas are used in physical geography by climatologists studying heat and moisture transfers between one region and another, and by hydrologists examining the relations between precipitation and streamflows in sets of river basins.

21-3 | **LONG-TERM SCENARIOS AND SPECULATIONS**

Thus far we have been considering short-term methods of forecasting that are essentially projections firmly based on past records. They are largely dependent on the numerical analysis of historical records of rainfall, riverflows, regional employment levels, and so on. But what if these records are inadequate indicators of the future, or what if we wish to forecast changes that are difficult to quantify? Then we must proceed in a more speculative manner. We look here at three cases in which different methods are used to predict changes on various regional scales. You may wish to compare these three cases with the long-term forecasts of physical geographic phenomena we encountered in Chapter 4.

Urban futures: A life-cycle approach

We begin by looking at a model of the economic life and death of a city built by an MIT team led by systems analyst Jay Forrester. As Figure 21-12 shows, it projects changes in three aspects of a city—business, jobs, and homes—over a 250-year period. The term "year" refers to the time periods in the model and may not precisely fit any exact historical time span. What does this approach tell us about the evolution of the city?

The cycle begins with an *urban growth phase* (years 0–100), marking the foundation and growth of the city. All the graphed indexes increase, although each reaches its maximum at a different time; for

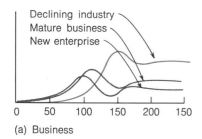

(a) Business

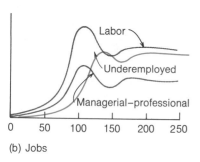

(b) Jobs

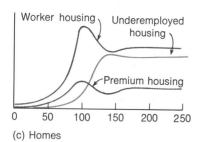

(c) Homes

Figure 21-12. An urban life cycle. The graphs cover the projected life cycle of a Western industrial city, simulating trends of jobs, business, and homes over a 250-year period. Note the period of stagnation setting in after about 200 years. [From J. W. Forrester, *Urban Dynamics* (MIT Press, Cambridge, Mass., 1969), p. 4, Fig. 1–1. Copyright © 1969 by The Massachusetts Institute of Technology.]

example, the number of new enterprises peaks in year 100, the total labor force in year 115, and so on. In geographic terms, this phase is characterized by the establishment of businesses, an influx of workers, and the construction of housing. The growth phase is succeeded by an *urban decline phase* (years 100–180), characterized by a trough in all types of indexes. The trough lasts generally around 60 years, with variations of as much as 50 years; for instance, the trough for new enterprises is reached in year 160 and that for declining industry in year 180. In spatial terms, this phase is marked by the deterioration of the original housing around the commercial core and the construction of successive rings of new housing on the periphery of the original city. Old housing near the center is occupied by new in-migrants with lower skills and a lower income. The presence of large numbers of these in-migrants prevents the demolition of the older residential buildings and restricts the expansion of the commercial core. New industry locates on the fringe of the city, and activity in the core begins to be dominated by office and professional activities drawing skilled labor from the suburban areas. The original housing near the commercial core is eventually replaced by expensive apartments and office buildings. The zone of deteriorating housing shifts to the inner suburbs, where a new ring of slum housing reaches outward.

A final *urban stagnation phase* (from year 180 onward) is one of slower change. Underused housing and underemployment both reach a maximum in year 250; the working population is at about 80 percent of what it was in peak years and the number of new enterprises is around 50 percent. Geographically the stagnation phase is marked by a slow peripheral expansion, a decreasing central population, increasing average population densities, and escalating journey-to-work problems.

How were these results obtained? Forrester used numerous observed relationships among different levels of housing, employment, and industry in contemporary American cities. These scores of individual relationships were combined in a simulation model so that changes in one set of values triggered reactions in others. The whole chain of changes yielded the trajectories in Figure 21-12.

What is the significance of Forrester's results? Opinions differ sharply. The model appears to take a gloomy view of the future of the city. However, it was not built to project a future that would occur, but to project the future implication of present relationships. Geographers, too, are unhappy with its assumption of fixed rather than variable city limits; by allowing the city to grow spatially, we can modify the trajectories in Figure 21-12. The model's most important role is as a vehicle for testing policies. With simulated planning, alternative housing policies or tax systems can be tried and evaluated without the cost, inconvenience, or time delays involved in experimenting with a real city. By changing values in the model, we can also study regional differences in types of cities.

World futures: A pessimistic model

Although the work of Forrester's MIT team on urban systems was well known, it was the publication of *World Dynamics* in 1971—followed by an extended and popularized version, *The Limits to Growth*—that caught the public imagination. This book extended the same systems approach used in Forrester's city model to the global level. Special attention was paid to the interrelationship of five fundamental factors: global population, agriculture, resource use, industry, and pollution. Figure 21-13 shows the links between these elements as a series of positive and negative feedback loops. (We have already met and discussed these terms in our study of ecological systems in Chapter 5.) Even in this highly simplified diagram it is possible to see how population, agriculture, and industry tend to be linked via positive loops, so that increases in one tend to drive up the level of the others. An exception is the link between industry and population. Industrialization is associated with lower rates of population increase. The addition of the resource and pollution factor brings major "governors" into the system, slowing down and even reversing growth through powerful negative feedback loops.

Figure 21-13 is a very simplified picture of the real MIT model, which linked the five basic elements through a complex network of over 130 rates, multipliers, and equations. These were linked in a

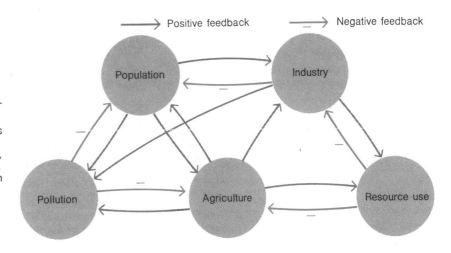

Figure 21-13. Main elements in the MIT model. This chart gives a highly simplified view of the five main elements in Forrester's global model. Most of the links connecting population, agriculture, and industry are positive feedbacks (i.e., self-reinforcing over time), whereas both pollution and resource use create negative feedbacks (self-canceling over time). The actual computer model had around 140 elements, rather than the few shown here.

computer program to project trends over a past 70-year period (1900 to 1970) over the next 130 years (i.e., 1970 to 2100).

Findings of the MIT model What findings did the MIT team come up with? Figure 21-14 attempts to summarize their results by reproducing some of the many computer runs of the model. In looking at the charts, remember that the vertical axis has deliberately been left unlabeled to underline the uncertainty attaching to the values.

The team studied first a "standard" world model. This model assumed "no major change in the physical, economic, or social relationships that have historically governed the development of the world system." As Figure 21-14(a) shows, population peaks out about the year 2050 as the rapidly dwindling resource base forces a slowdown of industrial growth. Industrial output (measured in per capita terms) peaks out around 2000, but the momentum of both population and pollution continues to drive them upward for another generation.

Given this first set of gloomy results, the MIT team then looked at responses to the crisis the model was predicting. Four strategies for overcoming the crisis were examined: increasing capital investment, reducing birth rates, cutting pollution, and stepping up agricultural production. Figure 21-14(b) shows the impact of the first strategy. It assumes that "unlimited" nuclear power will double the world's resource reserves and make extensive recycling possible. But population again tops out around the year 2050, its growth halted this time by accelerating pollution and dwindling per-capita food production. The other three strategies were found to be equally unrewarding. But suppose that, instead of trying one strategy at a time, we impose all four together. What then? The computer output in Figure 21-14(c) tells the story. The situation is much improved, in that population stays at a plateau for most of the next century. After that, however, pollution rates again begin to soar, industrial and agricultural production nosedives, and population again begins to fall.

This last result is disappointing since a stable population was held for only 80 years. What do we need to do to create long-term stability? Figure 21-14(d) shows an answer. If we raise food production by 20 percent and cut everything else (pollution by 50 percent, natural resource use by 75 percent, capital investment by 40 percent, and birth rates by 30 percent), the global population drops to slightly below the

Figure 21-14. Global alternatives. The charts show simulated growth paths for world population and resources projected by the MIT model. The vertical scale in each diagram is left purposely blank to indicate the lack of numerical precision in the forecasts. For a discussion of each alternative and the basis of the projections, see the text. [From J. Forrester, *World Dynamics* (Wright-Allen Press, Cambridge, Mass., 1971).]

(a) WORLD MODEL, STANDARD RUN

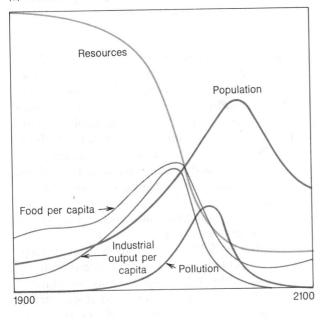

(b) "UNLIMITED RESOURCES" ASSUMPTION

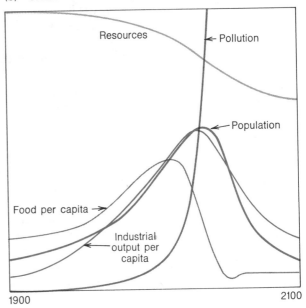

(c) FOUR-POLICY CONTROL PROGRAM

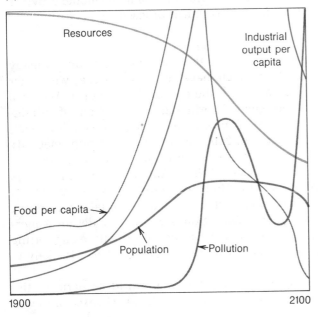

(d) STABILIZED WORLD MODEL

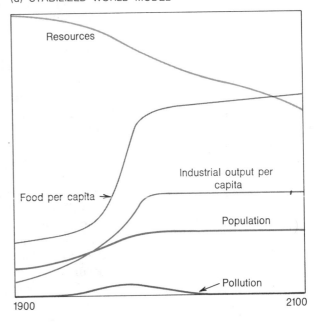

1970s level. While resources are still declining, the effects of this decline are largely offset by recycling and the substitution of one resource for another.

Evaluation of the MIT model How much importance can we attach to these projections? There are two kinds of evaluations we can make. The first is a technical one. If we look carefully at the model, we can find a number of debatable assumptions. For example, we have already seen in this book that natural resources are *not* a stockpile wasting away (Section 8-1) and that world population will probably follow a logistic rather than an exponential path (Section 6-3). Also, the structure of the computer model has been widely criticized as being both too simplified and structured so that small errors can be compounded into major trends. In addition, economists have been particularly concerned by the absence of any recognition of the fact that scarcity of a resource brings a chain of counterbalancing reactions.

Second, we can evaluate the model in terms of its general message. It is successful insofar as it has drawn massive public attention to the long-term problems of sustaining life on this planet. More particularly, it has shown that attempts to intervene on one front (e.g., through population control) may be counterproductive and self-defeating. Stabilization as a goal is shown to be achievable only through the coordination of very many measures in a careful and integrated way. Even if we reject the specific projections of the MIT team, the model provides a benchmark from which further work can proceed.

World futures: An optimistic scenario?
The MIT team gives us one view of the future. But it is one of many views. Scholars in this last decade have been obsessed, as were their nineteenth-century forebears, with the end of their century. We have seen a succession of experiments in bringing together teams of individuals to think about the form of society, technology, the environment, or the world community in A.D. 2000. Sometimes the groups met only for a short time for a specific purpose, for example, in the 1960s conference on the future environment of North America. In other instances more permanent groups like the Hudson Institute in New York or the Futuribles group in Paris were established to undertake continuing studies. One of the methods of such groups has been to design *scenarios*, that is, alternative pictures of the future, each with a supporting document stating how it was constructed—and whether it is likely to occur.

As an example, Figure 21-15 presents a world scenario of the probable economic ranking of nations in A.D. 2000. This scenario was developed by Herman Kahn, director of the Hudson Institute, and assumes a world population of slightly less than 6.4 billion. The nations are divided into six categories, based on their gross national products per capita: postindustrial, early postindustrial, mass consumption, mature industrial, transitional, and preindustrial economies. Most of the world's population will live in countries in the fifth, or transitional, group, which will include some of the world's largest countries— China, with an estimated 1.3 billion people; India, with 0.95; Pakistan, with 0.25; and Indonesia, with 0.24.

Kahn stresses that the scenario has three main elements. First, there are *fixed elements*, which include the fixed locations of large regional agglomerations of population. Second, there are what he terms *surprise-free projections* like decreasing interregional transport costs. Third, there are *variable-choice elements* such as a growth-oriented society versus a stability-oriented one. The map contains significant long-term and surprise-free elements, and we would not expect this pattern to be totally reversed. But there is room for substantial changes in the individual ratings. The increase in China's industrial capacity may well be greater than Kahn suggests; whereas the current Japanese boom may be partly offset by stronger competition from other East Asian countries. This world picture is accompanied by other scenarios

Figure 21-15. World scenario, A.D. 2000. This projection of the state of the world in A.D. 2000 by the Hudson Institute is more optimistic than that of Forrester's MIT team. The map shows the economic ranking of the nations of the world at the end of the century, based on projected per capita earnings. Six types of economy are shown. (Compare with Figure 18–17.) [From H. Kahn, *Science Journal* **3** (10) (1967), p. 120, Fig. 1.]

Visibly postindustrial

Early postindustrial

Mass consumption

Mature industrial

Transitional

Preindustrial

for individual resource elements that show sharper differences. For example, the scenario for the world's fuel resources is affected by whether the forecasters assume a primary role for nuclear power or whether they believe that we will continue to be heavily dependent on traditional fuels.

21-4 | **FORECASTING AND GEOGRAPHY**

What effect does this growing interest in forecasting have on the way geographers go about their work? We shall look, in this section, at its effect on the study of the way geographers look at man's past as well as his future on the globe. We shall note also the emergence of new and important traditions in our ways of working.

Forecasting in reverse

One of the curious fallouts of research in forecasting has been to stimulate projections of past events. When data are sparse the past may be almost as bare of facts as the future, and historical geographers are interested in filling in blank areas on their maps too. How can they do this?

One example of this "postdictive" (as opposed to predictive) research is geographer William Black's study of the spatial extension of a railway network into southern Maine during the decade 1840–1850 (Figure 21-16). He found that an 1840 settlement with 2000 inhabitants or more could be regarded as a potential node for a railway system. With 72 nodes defined in this manner, the maximal number of possible city-to-city links (2556) was very large from a purely geometric viewpoint. Most of the potential links were unrealistic in terms of practical railway building, however, and their number was reduced to 196. Of these potential links, only 27 had actually been constructed by 1850.

Black addressed himself to two questions: Why were some links between some centers built? Why were others not built? He argued that the probability of a connection being built increased with: (1) nearness to the point where the network began (i.e., the city of Portsmouth in nearby New Hampshire); (2) the shortness of the link; (3) the importance of the centers connected (defined as a product of the two populations, using 1840 census data); (4) the potential local interaction (given from a gravity model); (5) the absence of intervening opportunities in the form of alternative and nearer nodes; (6) the potential regional interaction; and (7) the closeness of a link's orientation to the orientation of all the Maine cities.

A statistical sorting procedure was used to separate the characteristics of those links which were built from those which remained unbuilt. This showed that six of the seven hypotheses contributed significantly to the forecast pattern of the railroad construction. Figure 21-16

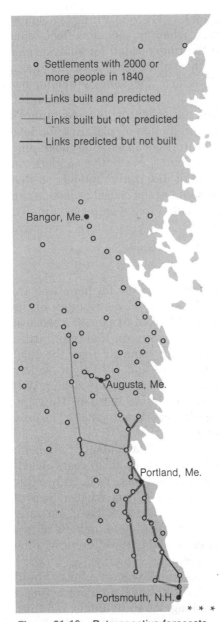

Figure 21-16. Retrospective forecasts. The map shows the actual railroad network in the state of Maine, in the northeastern United States, in the 1840s, and the network that "postdicted" on the basis of a forecasting model. [After W. R. Black, *Iowa Discussion Papers*, No. 5 (1967), Fig. 1.]

shows how accurate this "forecast in reverse" turned out to be. Most of the links that were actually built in the 1840s were correctly predicted by the model; only four of the links actually built were left out.

A number of debatable assumptions and simplifications were made in this study. For instance, all distances were measured in terms of direct overland distance, so the variable landscape of southern Maine was reduced to a uniform and flat plane. The value of Black's method lies in its comparability. Results for the 1840s in one corner of the United States can be compared with results for other regions and other time periods. The model also illustrates how methods designed to help predict future decisions can be reversed to help interpret decisions already made.

Geographers are not concerned only with the recent past. Archeologists and prehistorians have shown increasing interest in the locational models developed by geographers to describe the spatial distribution of the human population. For example, if we take archeological data on the size of Roman cities in Gaul as measured by the area enclosed by city walls (see Figure 21-17), we find that it shows the same kind of rank-size regularities we described in Chapter 14 in the discussion of modern city distributions. Similar work on Roman settlements in England and Mycenaean settlements in the Aegean shows some evidence that they conform to a Christaller-like pattern. What is the significance of findings of this kind? Broadly, they hold out the promise that we can predict which sites—from a series of archeological possibilities—are most likely to be worthwhile places to dig. New frontiers in joint research by geographers, and archaeologists and other students of the past are being opened up.

Changes in research traditions

Concern with the geography of the future is a relatively recent phenomenon among geographers. A review of the geographic publications in any single year would reveal an overwhelming commitment on the part of geographers to the study of the recent past. The books, journals, and maps published in 1975 are largely concerned with the world of the late 1960s—its spatial patterns, its ecological relationships, its regional systems. A proportion of the research is concerned with describing and interpreting earlier decades, and with the geography of still earlier centuries. Very few indeed are concerned with the world beyond the 1970s. Thus, we can represent the general traditional pattern of research by the kind of skewed distribution shown in Figure 21-18.

We do not have to look hard to find the reasons for such a pattern. Geographers have been heavily dependent, in their research, on em-

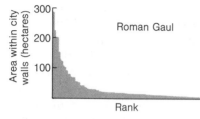

Figure 21-17. Structural regularities in past human settlements. The diagram shows rank-size regularities in early European cities. The area enclosed by the walls of cities in Roman Gaul is illustrated by shading. [From N. J. G. Pounds, *Annals of the Association of American Geographers* **59** (1969), p. 153, Fig. 8.]

Figure 21-18. The characteristic time distribution of geographic research. Most published geographic research deals with the recent past; very little is concerned with the future. The horizontal scale is in years.

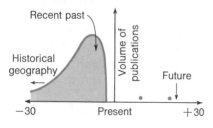

pirical data published by official agencies. There is a time lag, sometimes several years in length, between when an agency collects data and when it publishes it. If we add the time needed for analysis and writing, this alone is almost enough to explain the pattern in Figure 21-18. But not quite. Econometricians face the same problem, but spend more of their time in forecasting. Part of the reason geographic research seems to favor analysis of the past might be the geographer's concern with descriptive rather than predictive or planning models. *Descriptive models* replicate selected features of an existing geographic system and attempt to show how it operates. *Predictive models*, on the other hand, rearrange the structure of descriptive models so that the value of variables of interest at the end of the causal sequence can be predicted from the value of variables earlier in the sequence. Finally, *planning models* incorporate alternative decisions into a predictive model so that we can evaluate their effect.

To forecast, or not to forecast?

Forecasting is a risky business—and gets riskier the further ahead we try to look. Let us take the example of the spatial distribution of the United States population. We know what it looked like in 1970 because of the census taken that year. Given information on trends in births, deaths, and migration, we could predict with some accuracy the 1980 map. But the 1990 map poses more problems and is likely to be less accurate. Beyond that date our forecasts get progressively riskier. Will many more of us be living in Colorado or Florida in 2030? How important will New York City be as a population center in 2080? The only honest answer is that nobody now knows.

Just how far ahead we can, or should, look is a matter of judgment. Certainly, over the short and medium term (say, up to 20 years ahead) some projection is possible. Note that forecasts made even within this period are *conditional*. Geographers are not saying what a map of future population *will* look like, but what it *would* look like *if*, and only *if*, the trends on which it were drawn continue to hold. In view of this critical limitation we may legitimately ask whether the effort now going into geographic forecasting is worthwhile.

But surely there are persuasive reasons why geographers should continue, and indeed intensify, their efforts in this field. Forecasting is an integral part of human decision-making. As individuals, families, societies, and nations, we order our lives in terms of what we think will happen in the next day, month, year, or decade. We constantly make forecasts, so that the question we must really ask is not "Should geographers make forecasts?" but "How can geographers make their forecasts more accurate?"

Reflections

1. Most of you reading this book will still be alive in A.D. 2025. What changes do you expect to see in the landscape of your own area by that date?

2. Gather data on the economic development of your own region or country over the last decade. (Figures for unemployment rates are useful, and they are often available on a monthly or quarterly basis.) Plot the data on graph paper and look for a pattern. Is a long-term trend clear? Can you spot any seasonal or other cyclic influences?

3. Suppose there was a severe flood in your locality. Would you, because of it, consider a similar flood next year (a) more likely or (b) less likely? Why? Would past records help you decide the likelihood of another flood?

4. Review the somewhat pessimistic graphs of the MIT model in Figure 21-14, and compare them with Figure 21-15. Are you (a) very worried or (b) complacent about these doomsday philosophies? Debate your views with the rest of the class.

5. Has geographers' concern for the future increased their interest in the past? Why?

6. Review your understanding of the following concepts:
   (a) "satisficing" behavior
   (b) search behavior
   (c) long-term trend components
   (d) cyclic variation
   (e) bellwether regions
   (f) simulation models
   (g) scenarios
   (h) postdiction

One step further . . .

Philosophical problems in forecasting and distinctions between long- and short-term projections are discussed in

Boulding, K. E., *The Meaning of the Twentieth Century* (Harper & Row, New York, 1964) and

De Jouvenal, B., *The Art of Conjecture* (Basic Books, New York, 1967).

Uncertainty and environmental risks and their impact on human attitudes are delightfully exposed in

Tuan, Yi-Fu, *Topophilia: A Study of Environmental Perception, Attitudes, and Values* (Prentice-Hall, Englewood Cliffs, N.J., 1974).

Examples of how forecasting may actually be achieved can be found in the flexibly constructed, long-term scenarios of

Kahn, H., and A. J. Wiener, *The Year 2000: A Framework for Speculation on the Next Thirty Years* (Macmillan, New York, 1967) and

Hall, P., *London 2000* (Praeger, New York and Faber, London, 1970), 2nd ed.,

For more formal computer-dependent models, look in particular at the MIT team's work, reported in

Forrester, Jay W., *World Dynamics* (Wright-Allen Press, Cambridge, Mass., 1971) and

Meadows, D. H. *et al.*, *The Limits to Growth* (Universe Books, New York, 1972), Chap. V.

A mixture of the two approaches, with a variety of geographic applications to regional planning problems, is given in

Chisholm, M. D. I., A. E. Frey, and P. Haggett, Eds., *Regional Forecasting* (Butterworth, London, 1971).

The growing volume of research in this area is beginning to appear in the regular geographic journals. There are also special journals devoted to forecasting, such as the *Journal of Long Range Planning* (a quarterly), which has occasional papers of geographic interest.

Chapter 22

On Going Further in Geography

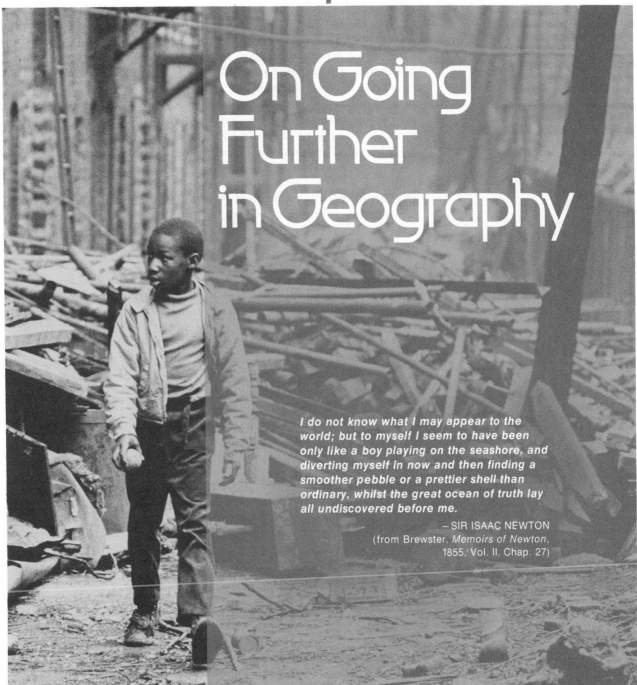

I do not know what I may appear to the world; but to myself I seem to have been only like a boy playing on the seashore, and diverting myself in now and then finding a smoother pebble or a prettier shell than ordinary, whilst the great ocean of truth lay all undiscovered before me.

—SIR ISAAC NEWTON
(from Brewster, *Memoirs of Newton*, 1855, Vol. II, Chap. 27)

On the bachelor's level not even one graduate in every 150 in United States universities majors in geography. Even within the social sciences division, in which it is often placed, the numbers of geography majors are dwarfed by those in psychology or economics. On the graduate level, figures for the 1960s indicate that for every doctorate granted in geography there were 10 in physics, nearly 20 in chemistry and engineering, and 30 in education. Within the social sciences division, geographers were only one-seventh as numerous as their colleagues in economics or history.

This small population of newly-hatched geographers should be seen in perspective, however. We need to see the situation in the United States' position in an international context. Geography as a modern university subject had its origins in the institutes of Germany and France during the early nineteenth century. The German contributions to the field, certainly until World War II, were dominant ones, and a majority of the classic works in geographic literature appeared first in the German language. In the universities of Western Europe, and those of Britain and the Commonwealth (to which geography spread, largely, in the first half of this century) the number of geography students is relatively large. Canada and Australia are strong research centers. The situation in the Soviet Union is also much more advanced; Soviet geographers outnumber their American colleagues by about three to one.

Also, the situation is changing. The number of geography students enrolled in U.S. colleges and universities has more than doubled over the last decade, and the membership of the leading professional society, the Association of American Geographers, has increased by more than three times.

University geography is a curious mixture. In size it is perhaps comparable with subjects with small enrollments like anthropology or archeology rather than major fields like mathematics or history. In its recent spurt of growth, however, it is more like a modest version of biochemistry or computer science.

This final chapter is addressed to students who have found enough in the preceding 21 chapters to interest them in taking further courses in geography. Its philosophy is shown in Figure 22-1. We shall look first at the past growth of the field to try to understand the path followed in getting to the present position. From the present we shall look forward to the future; we shall project the present drift and contrast this projected future with other targets for the field. Then we shall ask four basic questions: How did geography evolve as a separate field of study? What is its present structure? What is its likely future—if present trends continue? What should its future role be?

Figure 22-1. Trends in geography. In this chapter, we begin by studying past processes of growth in geography as a way of understanding the present position. Once this is achieved, we look forward to the future growth of the field. Here we are faced with a choice. Will the future be simply a forward projection of existing trends flaring out into a "cone of uncertainty" (shaded) as we get further away from the present? Or, will the current generation of geographers—particularly the younger ones—set up alternative goals toward which the subject will be steered? Such "programmed" growth could reverse some of the ongoing trends (A, B, and C) or reinforce them (D).

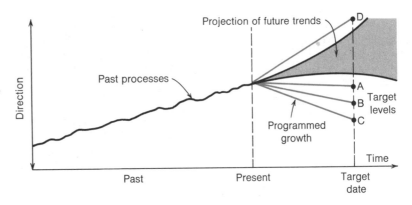

22-1 | THE LEGACY OF THE PAST

We can understand the character of geography as an academic field in the 1970s only if we see it as one scene in a lengthy play. It is useful to divide the play into three acts. The first act is dominated by isolated research by individual scholars, the second by organized research by groups and societies, the third by the incorporation of research into national and international organizations. It is clear that the stages cannot be precisely fixed in time; each is continuing in different subdivisions of the field or in different countries in various stages of growth.

Act I: The individual scholar

The first growth period, from the beginning of formal geographic study in ancient Greece to the mid-nineteenth century, can be characterized by geographic studies sporadically distributed in time and space. The number of scholars who would have counted themselves as geographers was always small, and it was only occasionally that clusters of workers formed—in Alexandria in the second century B.C., in Portugal in the fifteenth century, or in the Low Countries in the sixteenth century. The patronage that encouraged these groups to join together usually came from an interest in practical problems: methods of surveying the earth, instruments for marine navigation, mapmaking, and the printing of atlases. In this early period men found the answer to many questions about the general shape of the earth and ways of putting spatial information on maps. The maps of the period include some of the most majestic products of Renaissance Europe. (See Figure 22-2.) Most of the geographic schools were short-lived, however, and had fluctuating fortunes.

Figure 22-3 shows some of the leading scholars in geography since 1770. Remember, in interpreting this diagram, that the number of names

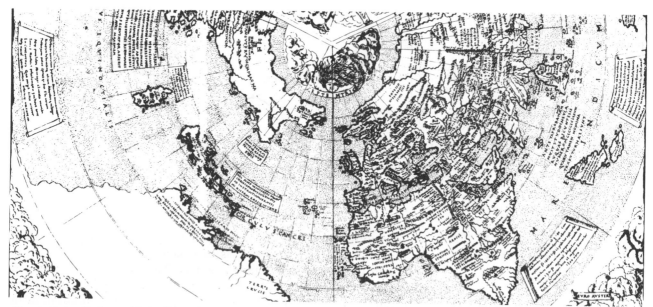

Figure 22-2. Cartographic traditions in geography. Mapmaking and exploration of the world played a key role in the early development of geography. This late-Renaissance world map was the first printed map to show America. It was made by an Italian, Contarini, in 1506 and illustrates the compromise between accuracy and adornment typical in maps of this period. For a description of modern development in cartography, see Chapter 2. [Map courtesy of the Trustees of The British Museum.]

that might have been included was less in 1800 than it was in 1900. Indeed, as we shall see later, the general growth of geography over the two centuries spanned by this diagram has been logarithmic. Most of the geographers who have *ever* lived are alive today! Note too that boundaries between disciplines were loosely drawn. It is not surprising that individuals like Immanuel Kant, Alexander von Humboldt, and T. R. Malthus played notable roles in the growth of various sciences. The diagram also underlines the significant part played by Germany in the early part of the period (until 1929 nearly half of all geographic publications were in the German language).

The exact significance of individuals in an overall growth process is difficult to assess. Science often involves a snowball effect, because of which more than due emphasis is placed on the contributions of a few individuals, for example, on the work of the few physicists or chemists who win Nobel Prizes. This so-called "Matthew effect" (after the Gospel's remark that to him that hath more shall be given) also occurs in geography, and Figure 22-3 inevitably reflects it. Yet it would be impossible to think of German geography in the mid-nineteenth century without Carl Ritter and Friedrich Ratzel, France without Vidal de la Blache, the United States without W. M. Davis, or Britain without Halford J. Mackinder (Figure 22-4). Our perspective on the present century is too short to allow the key figures to be discerned with certainty.

Figure 22-3. Geography, 1770–1970. The chart shows some leading scholars in the most recent period of geographic research. It indicates the emergence of some major schools and the changing national balance of research. (Compare the German names in the nineteenth century with the American ones in the twentieth.) The geographers' names have been taken from a list in Sir Dudley Stamp's *Longman Dictionary of Geography* (Longmans, London, 1966), largely concerned with Western scholarship. No comparative information was available on the growth of, say, Chinese geography. Some names have been included to illustrate the major contributions to geographic thought by scholars from other disciplines, particularly earlier in the period when the boundaries of fields were loosely drawn. Names have been chosen with a view to illustrating the emergence of significant research themes. Inevitably, other geographers would choose other names. Yet, while there might be only a modest overlap, the same general pattern of evolution of research themes over time would probably emerge. Very recent trends (within the last decade or two) are discussed in Section 22-3.

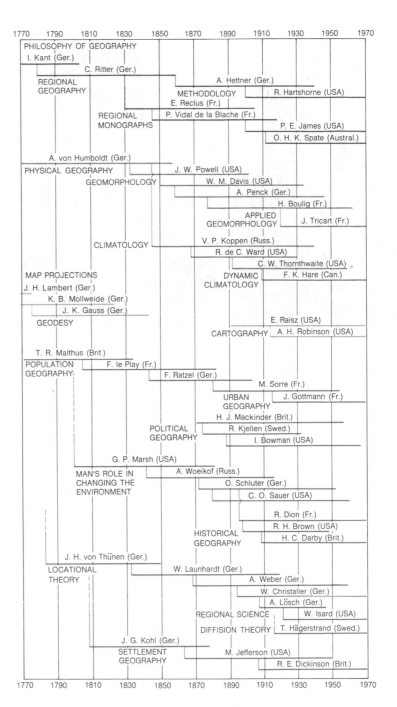

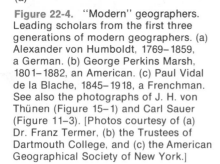

(a)

(b)

(c)

Figure 22-4. "Modern" geographers. Leading scholars from the first three generations of modern geographers. (a) Alexander von Humboldt, 1769–1859, a German. (b) George Perkins Marsh, 1801–1882, an American. (c) Paul Vidal de la Blache, 1845–1918, a Frenchman. See also the photographs of J. H. von Thünen (Figure 15–1) and Carl Sauer (Figure 11–3). [Photos courtesy of (a) Dr. Franz Termer, (b) the Trustees of Dartmouth College, and (c) the American Geographical Society of New York.]

Act II: Groups and societies

The second period of growth, beginning in the early 1800s, was marked by the organized interlinking of research. One of the earliest methods of linkage was the foundation of societies to foster common interests in geographic research. Such geographic societies generally fall into four groups. The first group, the national societies, emerged in the early half or middle years of the nineteenth century. These societies have a strong interest in global exploration. For instance, the Royal Geographical Society, in London (Figure 22-5), dates from 1830 and marks the merger of several early exploring clubs like the Association for Promoting the Discovery of the Interior Parts of Africa, founded in 1788. The American Geographical Society of New York was founded in 1852 by a group of businessmen to provide a center for accurate information on every part of the globe.

The second main group of societies is national professional groups, largely dominated by university and research geographers. These groups are usually later in date, smaller in membership, and less catholic in scope than the national societies. The Association of American Geographers (1903), the Institute of British Geographers (1933), and the Regional Science Association (1954) are typical of this group.

Figure 22-5. Geographical societies. The Royal Geographical Society typifies the groups formed in the middle of the nineteenth century to foster the research and exchange of geographic information. Its parkside headquarters in west London house a major library, map collection, and archives. [Aerofilms photo.]

The third group consists of societies devoted primarily to promoting geographic education in schools; the Geographical Association is an example. The fourth and most rapidly expanding set of groups originated during the 1950s. These organizations are subgroups within national professional organizations concerned with a particular aspect of geography (e.g., cartography, geomorphology, or quantitative methods). Geography appears to be following the pattern of other sciences in the rapid growth of this fourth group.

The prime function of the societies was to foster common research interests through the reading of papers and the publication of journals. The establishment of journals like the *Annals of the Association of American Geographers* represented key breakthroughs in the circulation of research findings. Other journals were published by interested

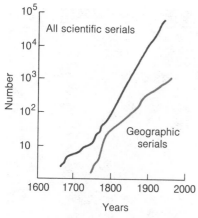

Figure 22-6. The growth of geographic research. The chart shows the cumulative total of scientific and geographic periodicals founded since the middle of the seventeenth century. The vertical scale of the graph is logarithmic: the number of both scientific and geographic periodicals has been growing exponentially. [From D. R. Stoddart, *Transactions of the Institute for British Geographers*, No. 41 (1967), p. 3, Fig. 1–3.]

individuals, as was *Petermann's Geographische Mitteilungen* in 1855, or by small groups, as was Ohio State University's *Geographical Analysis* in 1969. The growth of journals provides a useful indicator of the increasing volume of geographic research. As Figure 22-6 indicates, the field has been rapidly expanding since the seventeenth century. The number of all scientific periodicals doubles about every 15 years, and the number of geographic periodicals increases at about half that rate. The slower rate of increase in geography is typical of all the older, established sciences like geology, botany, and astronomy because the total increase in scientific publications reflects the high birth rate of new scientific fields like endocrinology and computer science.

Act III: National and international organizations

A critical role of geographic societies was to convey the importance of the problems they studied to the rest of the community. Their partial success was marked by the beginning of a third phase of geographic study, overlapping the second, in which geography departments were formally established in major universities and in which some countries set up government-sponsored research centers.

In the university sector, Germany again took the lead, with a considerable number of departments established by 1880. Developments in France were only slightly less rapid, but developments in the United States, Britain, and the Commonwealth lagged considerably. New geography departments often had an irregular spatial diffusion pattern, marked by curious regional concentrations and sparse areas; Figure 22-7 shows the distribution of degree-giving departments in the United States, where there is presently a strong Midwestern emphasis. Meanwhile, the need for national geographic research centers has led to the emergence of institutes like Brazil's *Instituto Brasileiro de Geografia e Estatistica* or the Soviet Union's *Akademiya Nauk SSSR*, each charged with the investigation and publication of regional data within their vast national territories. The latter has a staff of over 300 geographers working in ten divisions and a massive publishing program, including the bimonthly *Izvestiya, Seriya Geograficheskaya*. Even when separate geographic institutes have not been set up, geographic research has been increasingly incorporated into such organizations as Britain's Ministry of Town and Country Planning (now part of the Department of the Environment) and Australia's Commonwealth Scientific and Industrial and Research Organization (CSIRO).

Since 1923 the initiation and coordination of geographic research requiring international cooperation has been handled through the International Geographical Union (IGU). This organization holds meetings at intervals of four years. In between congresses, it appoints

Figure 22-7. Geography in American universities. The map shows the location of all U.S. universities with masters or doctoral programs in the mid-1960s. For further information, check the current *Guide to Graduate Departments in the United States and Canada*. (See "One step further . . ." at the end of this chapter.)

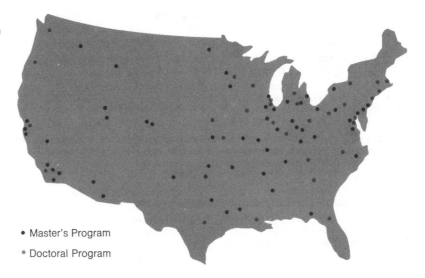

• Master's Program

• Doctoral Program

commissions to study special subjects like arid zones, quantitative methods, or economic regionalization. The number of member countries has now swollen to over 60. Different member countries have a different interest in various areas of research; for example, problems in applied geography and regional planning dominate much Eastern Europe research. If we compare member countries, we can see that the size of their geographic research effort is broadly related to their overall scientific budget. However, some smaller countries play a role in research out of proportion to their size. One outstanding example is Sweden, which, with few universities and a small group of geographers, has led research in several important areas. The volume and quality of research from New Zealand is also remarkably high in relation to its small number of research centers.

22-2 | **THE PRESENT STRUCTURE**

Geographers have repeatedly tried to define their field at each stage of its growth. For those who like formal definitions, Table 22-1 gives a variety of often-quoted ones. None of them will satisfy all geographers, but all geographers will recognize some common identifiable elements.

Unity and diversity in geography

Let us try to summarize what these common elements in definitions of geography are.

Table 22-1 Some contemporary definitions of geography

Definition	Source
"Geography is concerned to provide an accurate, orderly, and rational description of the variable character of the earth's surface."	R. Hartshorne, *Perspectives on the Nature of Geography* (Murray, London, 1959), p. 21.
"Its goal is nothing less than an understanding of the vast, interacting system comprising all humanity and its natural environment on the surface of the earth."	E. A. Ackerman, *Annals of the Association of American Geographers* **53** (1963), p. 435.
"Geography seeks to explain how the subsystems of the physical environment are organized on the earth's surface, and how man distributes himself over the earth in relation to physical features and to other men."	Ad Hoc Committee on Geography, *The Science of Geography* (Academy of Sciences, Washington, D.C., 1965), p. 1.
"Geography is concerned with giving man an orderly description of his world . . . [however] the contemporary stress is on geography as the study of spatial organization expressed as patterns and processes."	E. J. Taaffe, Ed., *Geography* (Prentice-Hall, Englewood Cliffs, N.J., 1970), p. 1.
"Geography . . . a science concerned with the rational development, and testing, of theories that explain and predict the spatial distribution and location of various characteristics on the surface of the earth."	M. Yeates, *Introduction to Quantitative Analysis in Economic Geography* (Prentice-Hall, Englewood Cliffs, N.J., 1968), p. 1.

First, we have seen that geographers share with other members of the earth sciences a concern with a common arena, the earth's surface, rather than abstract space, but that they look at that arena from the viewpoint of the social sciences. They are concerned with the earth as the environment of man, an environment that influences how he lives and organizes himself, and at the same time an environment that man himself has helped to modify and build.

Second, geographers focus on man's spatial organization and his ecological relationship to his environment. They seek ways of improving how space and resources are used, and emphasize the role of appropriate regional organization in reaching this end. Their work provides a perspective of man's tenure on the earth and various forecasts—both optimistic and pessimistic—of his future on the planet.

Third, we have seen that geographers are sensitive to the richness and variety of the earth. They do not believe in blanket solutions to development problems; instead, they feel that policy should be carefully tuned to the spatial variety concealed by terms like "tropics," "Appalachia," and "ghetto." On each geographic scale, they seek always to disaggregate and dissect the uniform space of the legislator within the complex space of the real world.

Within this broadly defined area of agreement, different branches of geography have proliferated, each concerned with a limited research topic. Table 22-2(a) gives a summary of the orthodox division of geography into the study of regions (regional geography) and an

TABLE 22-2 The internal structure of geography

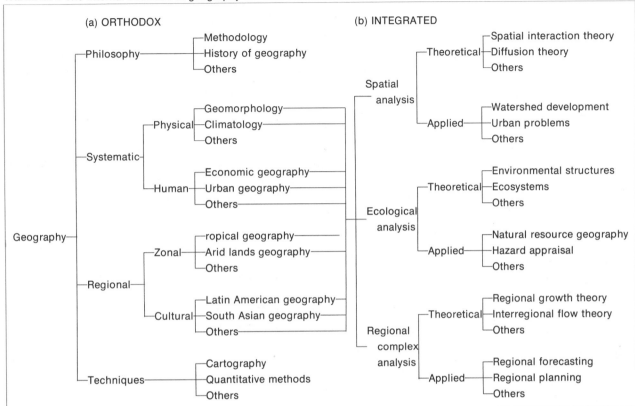

analysis of their systematic characteristics (systematic geography). Each may be further subdivided into more specific fields like the regional geography of Latin America or urban geography. Some geographers recognize as a separate branch the study of the regional or systematic geography of past periods (historical geography), but others argue that time is an essential component in all geographic studies.

These conventional divisions are important, not least because most university catalogues describe courses in these terms. But perhaps a more helpful way of structuring geography is by how it approaches its problems [Table 22-2(b)]. In this book we have distinguished three different approaches.

Spatial analysis The first approach, termed *spatial analysis,* studies the locational variation of a significant property or series of properties. We have already encountered such variations in interpreting the distribution of population density or of rural poverty. Geographers ask what factors control the patterns of distribution and how these patterns can be modified to make distributions more efficient or more equitable.

Ecological analysis A second approach to geography is through *ecological analysis,* which interrelates human and environmental variables and interprets their links. We have already studied such linkages in the hydrologic cycle and land-use cycles. In this type of analysis geographers shift their emphasis from spatial variation between areas to the relationships within a single, bounded, geographic area.

Regional complex analysis The third approach to geography is by way of *regional complex analysis,* in which the results of spatial and ecological analysis are combined. Appropriate regional units are identified through areal differentiation, and then the flows and links between pairs of regions are established. We discussed some of the difficulties in regional complex analysis in (Chapter 10) and looked at a few of its possible applications in regional planning (Chapter 19).

The advantage of looking at geographic problems in terms of these three approaches rather than the orthodox divisions is that they stress the unity of physical and nonphysical elements rather than their diversity. Geographers concerned with water resources or human settlement may find a common ground in the ways in which systems are studied or in their parallel search for efficient regional units.

The work of most geographers falls within the "triangle" formed by these three approaches to the field. Some may specialize, or, if you like to think of it this way, move toward one of three corners of the triangle. Indeed, the whole subject appears to have zigged and zagged over the decades, sometimes staying in the regional corner (as in the 1930s), sometimes lurching toward spatial analysis (as in the 1950s and 1960s). In the present decade it seems on the move again, now headed for the ecological corner.

Geography and supporting fields

Geography is particularly dependent on the flow of concepts and techniques from more specialized sciences. For example, in regional climatology we adapt models originally developed by meteorologists, who in turn draw their concepts from basic physics. Likewise, our models of regional growth borrow from the econometrician. But because the physicist and the econometrician both use systems of equations developed by mathematical research, we can trace both back to a common origin. As Figure 22-8 makes clear, the same concepts come together on very different levels: Mathematics and geography represent two such levels. In mathematics the links come from the common logical structure used to derive equations; in geography the links come from the common regional context to which equations are applied. Each region introduces a set of disturbance terms into the general equations, so that it stands out as a residual or anomaly. If we can define what these anomalies are and give them a rational structure, we are well on the way to defining the elusive quality of individual regions.

This dependence underlines the fact that sovereign subjects are about as irrelevant as sovereign states. Good botanists need to be reasonable biochemists, good engineers need to be fair mathematicians, and so on. Familiarity with these supporting fields is normally obtained by taking parallel courses in other departments. Table 22-3 lists some of the subjects in supporting fields taken by geography majors in American universities. Of these, special importance attaches

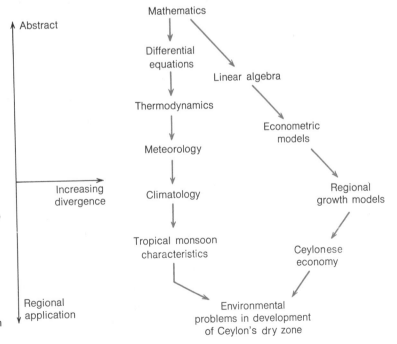

Figure 22-8. Levels of integration. Academic subjects may come together and be integrated at quite different levels of abstraction. A geographer may integrate by focusing the results of a number of systematic sciences on a specific regional problem, that is, how to overcome the environmental problem of low and unreliable rainfall in developing Ceylon's dry zone. The diagram shows two of the strands in such an analysis, using the findings of two different disciplines (meteorology on the left and econometrics on the right). Both disciplines come together again at a very abstract level through their common dependence on mathematics.

Table 22-3 Supporting fields in U.S. undergraduate majors programs

Percentage frequency of occurrence in survey of department programs			
Below 20	20–40	40–75	Above 75

Biology
Botany
Physics
Planning
Population
Zoology
	Anthropology
	History
	Political
	  science
	Sociology
		Economics
		Geology/
		  Geomorphology
			Mathematics/
			  Statistics

SOURCE: The Association of American Geographers panel on Program Inventory and Development Questionnaire refers to practice in the period 1960–1965.

to mathematics because it provides a common language in which geographers can express spatial, ecological, and regional concepts in a concise and comparable way. Outside mathematics, the choice is much wider and depends more on your personal interests and requirements.

Let us assume that your particular interest is in the humanities and that you are especially attracted to the study of one of the major cultural areas like China. Then the appropriate courses to support regional courses in East Asian geography might include the history and economic structure of China; clearly, courses in Cantonese or Mandarin would be needed for more serious research. By contrast, if you wanted to specialize in environmental problems, you would need earth science courses like hydrology or oceanography, perhaps supported by a course in resource economics (Figure 22-9). Which courses you elect to take will depend on a combination of factors: your interest, your ability and previous training, and your long-term expectations. It is to the third of these we now turn.

Jobs in geography

Although it is now less fashionable to earmark university courses as "career oriented" or "general education," the distinction still lingers. Law, medicine, and geology typify fields where most graduates have a rather well-defined occupational outlet ahead of them. By contrast, fields like philosophy or political science have a high educational content but less obvious career prospects. Where does a degree in geography take us?

Jobs in geography can be broadly divided into two categories. First, one can find a job that is "field sustaining" in the sense that it is concerned directly with supporting the continuing study of the field itself. Thus, many geographers find jobs in geographic teaching or research.

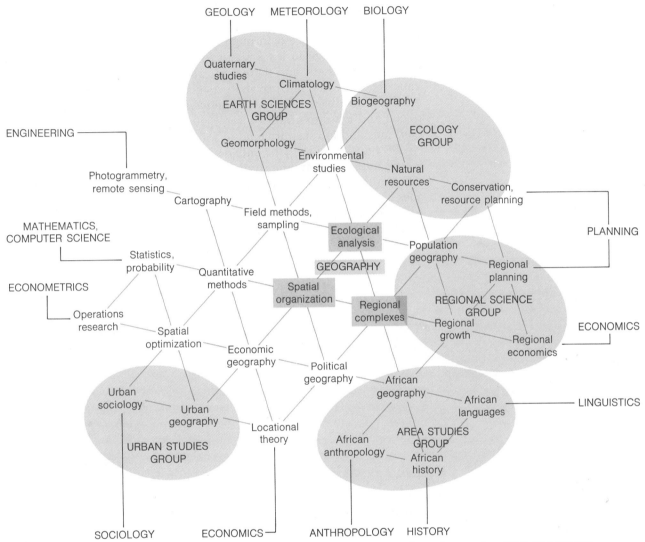

Figure 22-9. Links between geography and supporting fields. The courses indicated are intended to be representative rather than exhaustive. Departmental boundaries are drawn differently in some universities and colleges. For example, geomorphology tends to be taught within geography departments in British universities but within geology departments in America. African studies are used to illustrate area studies; a similar diagram could be constructed, for example, for Latin American or South Asian studies.

Teaching opportunities exist on all levels from the elementary school to the postdoctoral level. In the countries of Western Europe, geography is an important high school subject, and many graduates go back to teaching on this level. The position of geography in United States schools is less strong, though an important new curriculum for this level has been initiated through the High School Geography Project. The 2-year college forms a significant and expanding job market for geography teachers, and there was a recurrent shortage of PhDs for advanced teaching on the university level in the 1960s. There is, however, a delicate feedback relationship between the number of graduates produced and job opportunities (see Figure 22-10), and periods of abundant jobs and job scarcity tend to follow a cyclic pattern.

Second, there are jobs outside geography itself. Geographers have a long tradition of public service on all levels, from global agencies to city planning commissions. On the *international* level, the tradition goes back at least to World War I, when geographer Isiah Bowman (later president of Johns Hopkins University) was prominent among the geographers advising the American delegation at the Versailles Peace Con-

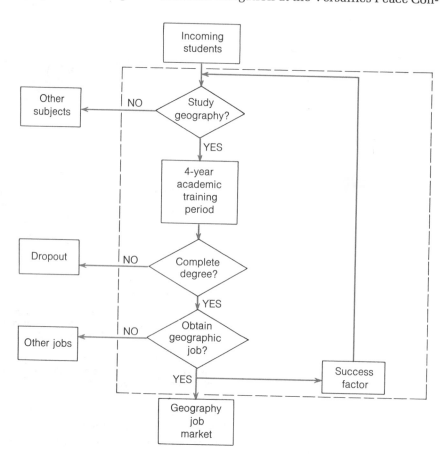

Figure 22-10. The geographic job market. The chart shows the feedback between job opportunities and the number of graduate students. The negative feedback tends to introduce a short-term cyclic element into the general long-term expansion in numbers. [After J. W. Harbaugh and G. Bonham-Carter, *Computer Simulation in Geology* (Wiley-Interscience, New York, 1970), p. 15, Fig. 1–11.]

ference. This tradition continues today, with Russian geographers like Yuri Medvedkov working in the World Health Organization at Geneva and the British geographic team advising on the settlement of the Argentine–Chilean boundary dispute.

On the *national* level, geographers are well represented in federal departments of the United States; indeed, nearly 10 percent of the American Association of Geographers membership works in federal establishments. This sort of work began in the 1920s, when geographers played an important part in the Soil Conservation Service. In the succeeding years, their involvement has widened to include work for several agencies, such as the Geography Division of the Census Bureau and the Water Resources Board. Despite their growing role, however, American geographers are not as well represented in national planning affairs as they are in countries like the Soviet Union, Sweden, and Britain. In Britain many geographers are working in the newly formed Department of the Environment, where they are engaged in research that ranges from regional planning to local land use. Below the national level extends a hierarchy of regional commissions and local planning agencies, each with geographers represented on them.

Besides working in the public sector, geographers work in the *private* sector as independent consultants or employees of large corporations. The optimal location of a model city, new airstrips for an expanded flying doctor service, or new census units for urban population all involve the kinds of problems discussed in this book and analyzed more deeply in more advanced geography courses. More geographers are establishing their own consulting agencies and so multiplying the opportunities for applied research.

But these broad categories hardly do justice to the range of opportunities that individual geographers—like individuals in any other field—create and develop. All university geographers' mail regularly includes letters (more often, postcards) from former students using their geographic training in careers as unlike as the administration of a regional hospital system and the planning of highways in New Guinea. Geography graduates are also giving more of their spare time to community counseling and are becoming increasingly interested in local planning issues involving ecological or locational decisions at the community level.

22-3	THE FUTURE PROSPECT

As we noted in Figure 22-1, there are two elements in the future prospects of geography: first, the trends that can be foreseen from a continuation of movements already occurring; second, the goals toward which we wish to steer. Here we look at each element in turn.

Projections of existing trends

Perhaps the only sure forecast we can make about geography is that it will persist. The questions geographers ask are so basic that it is impossible to imagine a world without them, a world where regional differentiation is not significant or where general theories fit local circumstances with sheathlike precision. We certainly expect the causes of differentiation to change, for as we gain uniformity in one sphere we lose it in another. However specialized science becomes, some scholars will still want to integrate and synthesize in the traditional geographic manner.

Beyond this, four more precise trends seem likely. The first projection concerns quantity. As we saw in Figure 22-6, the quantity of geographic research as measured by the number of specialized geographic serials has been doubling every 30 years since around 1780. For the increase in more recent years, a variety of other measures is available, including memberships in professional societies, enrollments in classes, degrees granted, research published, and so on. These indicators all confirm the general exponential rate of past growth. Even though individual decades are likely to have spurts or slowdowns above or below a long-term rate of increase in geographic research, it would be unreasonable not to expect a substantial increase over the next decade.

A second projection relates to the fission of geography into specialized subdisciplines. Each individual scholar finds increasing economies of scale by specializing in a limited range of problems; his limited time and resources (books, equipment, maps, computing time, or whatever) can be concentrated on in-depth study of a limited topic or region. As a result, an increasing number of geographers tend to think of themselves as South Asian geographers, or diffusion specialists, or arid-zone geomorphologists. This trend is not confined to geography. The general botanist or zoologist, not to mention the still more general biologist, has long since been displaced in most universities by endocrinologists, conchologists, and what have you.

A third projection relates to quantification. One of the most striking differences between the research papers in geographic journals of the 1970s and those of the 1950s is the greatly increased proportion of research using mathematical techniques. In the earlier period, the main mathematical applications were of spherical geometry in cartography and surveying, and of probability and statistics in climatology. Today the range of mathematical models has significantly expanded, and the applications now affect most branches of the field. We find historical geographers fitting polynomial surfaces to the spread of early settlement, or industrial location analysts modeling

Philosophical schools in human geography

Man's relationship to geographic space and to the natural environment has been viewed differently in various periods of the growth of geography. Some of the "schools" of thought you may encounter in reading further are listed below.

Environmentalism is the view that natural environment plays the major role in determining the behavior patterns of man on the earth's surface.

Normative describes approaches to geography that establish a norm or standard. Thus, they are largely concerned with establishing some optimum condition (e.g., the "best" location or the "best" settlement pattern).

Phenomenology is an existential philosophical school which admits that introspective or intuitive attempts to gain geographic knowledge are valid.

Positivism is a philosophical school which holds that man's sensory experiences are the exclusive source of valid geographic information about the world.

Possibilism, in contrast to environmentalism, stresses the freedom of man to choose alternative patterns of behavior despite geographic location. It is linked especially to the ideas of the French geographer Paul Vidal de la Blache (see page 578).

Probabilism is a compromise position between environmentalism and possibilism that assigns different probabilities to alternative patterns of geographic behavior in a particular location or environment.

Methodology and *epistemology* are terms used to describe the study of these different schools of thought and their contribution to the philosophy of geography.

siting decisions as Markov chains. During the 1960s there was a skirmish between those geographers anxious to innovate with mathematical methods and those skeptical of their usefulness in solving orthodox problems. (See Figure 22-11.) Today the general acceptance of such techniques, the more complete mathematical training of a new generation, and the widespread availability of standard computers on campuses, make the conflict of a decade ago seem unreal. The first few years of overenthusiastic pressing of quantitative methods on a reluctant profession have given way to the present phase, in which mathematical methods are just one of many tools for approaching geographic problems.

The fourth trend is of more direct concern to human geography, rather than to the whole field. Human geographers are getting increasingly uneasy about the "positivist" nature of much geographic research. *Positivism* is a philosophical approach which holds that our sensory experiences are the exclusive source of valid information about the world. This attitude developed in the natural sciences (like physics) but has been borrowed by geographers working in social-science areas. A positivistic approach leads to the discussion of human behavior in terms of analogies drawn from the natural sciences. Thus in Chapter 16 we discussed human migration in terms of Newton's laws of gravity. Much geographic effort of the 1960s went into trying to explain patterns of human behavior with neat, lawlike statements. Ultimate causes and the essential nature of phenomena like migration were put aside as being unknowable or inscrutable.

Given its historical pattern of evolution, it is understandable that positivism should have played a major part in geographic explanations of phenomena. Whatever its virtues in the more physical parts of the field, its effect on human geography was to lead to a stylized, sometimes overacademic kind of research. What then, should we replace it with? The current decade favors renewed interest in a phenomenological approach. *Phenomenology* is an existential philosophical approach which admits that introspective or intuitive attempts to gain knowledge are valid. It accepts subjective categories as appear to be in the experience of the person behaving. The Great Plains are a "desert" if the person or group settling them believes them to be one and acts accordingly! Phenomenologists look with scepticism at the attempts at lawlike statements so characteristic of the previous decade.

This shift in philosophical position has also occurred in other fields like social anthropology and social psychology and has reinforced the links between geography and other social sciences. Of course no geographer's work is ever "purely" positivistic or "purely"

Figure 22-11. Quantification. A cartoon of the quantitative revolution tearing geography away from its conventional qualitative traditions. In reality, "Geographia" is somewhat older and more able to take care of herself than the cartoonist suggests. [From L. Curry, *Canadian Geographer* **11** (1967), p. 265, Fig. 1.]

phenomenological. Most geographers adopt a position between these two extremes, with systematic approaches to the field (like that in this book) tending to stress the positivistic side and works with a regional emphasis adopting a more phenomenological viewpoint.

New goals for geography?

Each new generation of geographers builds on earlier work but reinterprets its goals to match the prevailing scientific and social mores. Judging by the interests of the current generation of graduate students or current issues of radical journals such as *Antipode*, geography is likely to become more strongly oriented toward applied fields and practical problems. The questions debated over coffee cups are outward looking, socially relevant, and action oriented.

Can we help society to chart a middle way between the mindless exploitation and pollution of the natural world implicit in a short-run materialist economy and the unrealistic pipe dreams of a pastoral, protected, but unproductive world? Can we predict the welfare implications of different types of locational situations or spatial arrangements? Can we help to redraw political boundaries in order to equalize resource opportunities and also minimize the likelihood of future conflicts? What will the world geography or that of the United States look like by A.D. 1990 or A.D. 2030? These and many other questions are being asked today.

If these questions are typical of the broad, long-term issues young geographers wish to tackle, there will have to be some significant departures from the presently projected trends in the field. For example, instead of increasing specialization, there will have to be a greater emphasis on extending the ecosystems approach (see Chapter 5, Section 5-1), still largely confined to the physical and biological world, to include man's own environment-modifying activities. Physical and human geographers will have to spend more time at each other's seminars to exploit their unique opportunities for cooperative research. A rapprochement will be needed between those interested in building quantitative models and those interested in the realities of individual regional complexes. Geographers will also need to stop acting as "terrographers" and devote more time to at least the continental margins of the ocean-covered 70 percent of the earth. Finally, more geographers will be needed outside the classroom, concerning themselves with the consequences of their learning.

That there are problems to be overcome is undeniable if geographers are to make their fullest contribution — improving, as far as they are capable, the relationship between man and his environment. Geographers have no special philosopher's stone that gives them instant insight or a divine right to be heard or consulted. The full credentials of painstaking research and tested theory are still being established. Their special curiosity is at once a strength and a weakness, and geographers are conscious that their own studies often lack the rigor of an econometrician's or the temporal perspective of a historian.

We have tried to show areas where these difficulties are being faced and overcome — and others where significant progress remains to be made. The book will have served its purpose if it attracts students from other disciplines into those areas that have, for too long, remained a narrow geographer's monopoly. These students may help to provide new insights and solutions to some of the puzzles that have, for too long, baffled the small group of geographic scholars. We hope that you will accept the invitation — and the implied challenge offered here — and go on from this brief and prefatory book to the rewarding geography courses that lie ahead.

Reflections

1. Look back at the notes you made in response to Question 1 at the end of the first chapter. How accurate were your predictions? Compare your own findings with those of the rest of the class.

2. "Adopt" any one of the individuals named in Figure 22-3. Browse through at least two of his or her works, and, if the person is now deceased, read through his or her obituary notice in one of the leading geographic journals. (The *Annals of the Association of American Geographers* is a particularly useful source of information on North American geographers.) List three major influences on the person's work.

3. Read through the definitions of geography in Table 22-1. Which is closest to your own definition of geography? Why?

4. Which subjects in other fields do you consider most useful for geography students? Justify your choice.

5. How should the research agenda of geography (or any academic field) be decided? Evaluate the relative importance of (a) experience (i.e., the past history of research endeavors, successful and unsuccessful), (b) the views of the present generation of geographers, and (c) the needs of society in determining this agenda.

6. Review your understanding of the following concepts:
   (a) The Matthew effect
   (b) spatial analysis
   (c) ecological analysis

(d) regional complex analysis
(e) quantification
(f) positivism
(g) phenomenology

One step further . . .

By the time you reach this point in the book, many of you will have had your fill of "further readings" in geography. But those few of you who are thinking of going further in the field might like to look at a publication that provides useful information on geography departments in North America:

> Association of American Geographers, *Guide to Graduate Departments in the United States and Canada* (AAG Washington, D.C., annual).

For information on departments in other countries turn to Appendix C. The best general history of the early growth of geography is still probably

> Dickinson, R. E., and O. J. R. Howarth, *The Making of Geography* (Clarendon, Oxford, 1933).

Yet, this should be supplemented by more recent works on the history of the modern period. One excellent study is

> James, Preston E., *All Possible Worlds: A History of Geographical Ideas* (Odyssey, New York, 1972).

A brief but clear introduction to the current philosophy of the field is given in
> Ackerman, E. A., *Geography as a Fundamental Research Discipline* (University of Chicago, Department of Geography, Research Paper 53, Chicago, 1958).

Serious scholars will want to delve into the two classic statements on the field by a leading geographic philosopher, Richard Hartshorne:
> *The Nature of Geography: A Survey of Current Thought in the Light of the Past* (Association of American Geographers, Lancaster, Pa., 1946) and
> *Perspectives on the Nature of Geography* (Rand McNally, Skokie, Ill., 1959).

Geographers' ideas about their subject are constantly changing and evolving. *The Annals of the Association of American Geographers* (a quarterly) and the *Publications of the Institute of British Geographers* (a semiannual) carry reviews of current changes in the AAG or IBG presidents' "State of the Union" addresses to the annual conference. You might also look at *Progress in Geography* (an occasional publication), an international journal devoted exclusively to papers reviewing current developments in specialized fields within geography.

Appendix A

Conversion Constants

Geographers are by definition internationally minded people. Most of the world's population measures its environment in terms of the metric system, and an increasing proportion of scientific research is now reported in metric terms. Measurements in this book are therefore given in the metric form. Nonmetric equivalents are given, in most cases, in parenthesis. In a few cases where the historical contexts demand it—as in the discussion of the township and range system—the original nonmetric forms are retained in the discussion. Temperatures are given in degrees centigrade, with fahrenheit equivalents in parenthesis.

MEASURES OF DISTANCE

Kilometers to miles (1 km = 0.62137 mi)

1	2	3	4	5	6	7	8	9	10	km
0.62	1.24	1.86	2.49	3.11	3.73	4.35	4.97	5.59	6.21	mi

Meters to feet (1 m = 3.28084 ft)

1	2	3	4	5	6	7	8	9	10	m
3.28	6.56	9.84	13.12	16.40	19.69	22.97	26.25	29.53	32.81	ft

Centimeters to inches (1 cm = 0.393701 in.)

1	2	3	4	5	6	7	8	9	10	cm
0.39	0.79	1.18	1.57	1.97	2.36	2.76	3.15	3.54	3.94	in.

Square kilometers to square miles (1 km^2 = 0.386102 mi^2)

1	2	3	4	5	6	7	8	9	10	km^2
0.39	0.77	1.16	1.54	1.93	2.32	2.70	3.09	3.47	3.86	mi^2

Hectares to acres (1 hectare = 2.471054 acres)

1	2	3	4	5	6	7	8	9	10
2.47	4.94	7.41	9.88	12.36	14.83	17.30	19.77	22.24	24.71

MEASURES OF TEMPERATURE

Degrees centigrade to degrees fahrenheit [$°C = \frac{5}{9} (°F - 32)$]

−80	−70	−60	−50	−40	−30	−20	°C
−112	−94	−76	−58	−40	−22	−4	°F
−10	0	10	20	30	40	50	60	°C
14	32	50	68	86	104	122	140	°F

EXPONENTIAL SCALES

Large numbers are frequently easier to describe in an exponential form. Thus scientists will often say 10^6 instead of 1,000,000, 10^7 instead of 10,000,000, 10^8 instead of 100,000,000, and so on. The number 10 is the most convenient base for an exponential scale, since conversions can be accomplished merely by moving decimals. Thus, typical conversions are

$$10{,}800{,}000 = 1.08 \times 10^7$$
$$2{,}370 = 2.37 \times 10^3$$
$$-481{,}000 = -4.81 \times 10^5$$
$$0.00171 = 1.71 \times 10^{-3}$$

Appendix B

Sources of Geographic Information

The following notes give a brief guide to some of the more frequently required sources of information for students and instructors.

1. COLLEGE AND UNIVERSITY DIRECTORIES

Specific information on geography departments in both the United States and Canada is given in the Association of American Geographers' *Guide to Graduate Departments in the United States and Canada* (Washington, D.C., annual). More general guides to the institutions of which these departments are a part are given in *American Universities and Colleges* (American Council on Education, revised every two or three years) and in J. Cass and M. Birnbaum, *Comparative Guide to American Colleges* (Harper & Row, New York, 1973). Information on universities with geography departments in British Commonwealth countries is available in *Commonwealth Universities Yearbook* (Association of Commonwealth Universities, London, annual).

The most comprehensive list of geography departments and geographic institutions in the rest of the world is given in E. Meynen, Ed., *Orbis Geographicus* (Franz Steiner, Wiesbaden, 1968), which is now somewhat dated. More up-to-date information can be obtained from *World of Learning* (Europa Publications, London, annual). Data is arranged by countries, and the listing of universities and colleges, their senior staff, and their subject areas gives an indirect guide to geography departments.

2. PROFESSIONAL SOCIETIES

The major professional body for geographers in the United States is the Association of American Geographers. Information may be obtained by writing to the Executive Secretary, 1710 Sixteenth St., Washington, D.C., 20009. In Canada the parallel society is the Association of Canadian Geographers. Correspondence should be addressed to the group's secretary, Burnside Hall, McGill University, P.O. Box 6070, Montreal 101. These bodies play a key role in the organization of geographic activity through their annual conferences, regional meetings, journals, and listings of job opportunities. Other geographic societies in Britain (Institute of British Geographers, c/o Royal Geographical Society, 1 Kensington Gore, London, SW7 2AR) and Australia (Institute of Australian Geographers, Secretary, Department of Geography, University of Sydney, Sydney, N.S.W. 2006) play a similar role. Most societies have reduced membership rates for student members.

Alongside the professional societies stand the associations founded for members with general geographic interests and those concerned with specialized subsectors of the field. Information on these groups is available in *Orbis Geographicus*. Students wishing to explore the situation in their own countries are advised to consult bulletins produced by the major international body coordinating geographic work, the International Geographical Union. Information may be obtained by writing to the Secretary-General, Prof. Channoy D. Harris, Department of Geography, University of Chicago, Chicago, Illinois 60637, USA.

3. THE LITERATURE OF GEOGRAPHY

Those who have read the section "One step further . . ." at the end of each chapter in this book will have noticed the diversity as well as the richness of literature on geography. More than most fields, geography overlaps with and interpenetrates neighboring disciplines. As a result, locating literature in geography is likely to be more taxing than in other fields. Indeed, students consulting the "geography" section in their college library may be greatly perplexed. Most library classification systems — for example, the Dewey or the Library of Congress system — scatter much geographic work through nongeographic categories and encumber the so-called geography category ("G" in the Library of Congress system) with books only loosely connected with modern conceptions of the field. A useful guide through the jungle is provided by J. G. Brewer, *The Literature of Geography: A Guide to Its Organization and Use* (Bingley, London, 1973), which gives a very full introduction to source material on all sections of the field. For a guide to sources of information on regional work and suggestions for specific books, consult Appendix C of this book.

Information on the ever-increasing flow of geographic books and papers is provided in two major bibliographies: *Current Geographical Publications* (American Geographical Society, New York, ten issues each year) and *Geographical Abstracts* (GeoAbstracts Ltd., University of East Anglia, Norwich, six issues a year for each of four separate series).

4. GEOGRAPHIC DATA

One of the primary sources of geographic data is maps. Information on the current map coverage of a country is obtainable from its major map-producing agency. Index maps will normally be issued without charge by agencies such as the *U.S. Geological Survey* (Map Information Office, U.S.G.S., Washington, D.C. 20242) or Britain's *Ordnance Survey* (Romsey Road, Maybush, Southampton SO9 4DH). A useful guide to the current map coverage of all countries is given in *International Maps and Atlases in Print* (Bowker, London, 1974). New maps and atlases are regularly listed in *Bibliographie Cartographique Internationale* (Paris, annual) and *Bibliotheca Cartographica* (Deutsche Gessellschaft für Kartographie, published twice yearly).

Information about conventional aerial photographs is normally available from the same sources as maps. Thus students in the United States should first consult the Map Information Office in Washington, D.C., and ask for a publication entitled *Current Status of Aerial Photography*, which is issued free of charge. Aerial photographs may also be purchased from commercial air-survey organizations in various countries. For example, an unusually wide range of photographs is available from the Aerofilms Library (Aerofilms Ltd., 4 Albermarle Street, London, W1X 4HR, England).

Satellite photographs of all parts of the earth have been taken as part of the NASA program. The U.S. Department of the Interior established the EROS (Earth Resources Observation Systems) program to gather data on natural and manmade features on the earth's surface by remote sensing in 1972. Information about the coverage of satellite photographs is now available from the EROS Data Center (Sioux Falls, South Dakota 57198, USA).

Statistical sources are playing an increasingly important role in geographic research. Some idea of the richness and range of published statistical material is given in J. B. Mason, *Official Publications: U.S. Government, United Nations, International Organizations and Statistical Sources* (American Bibliographical Center, Clio Press, 1971). However, comparison of figures for different countries' boundaries and different periods can be dangerous. Students using such materials are advised to consult the statistical librarian in their College library for advice on particular sources. A valuable general guide is given in J. M. Harvey, *Sources of Statistics* (Bingley, London, 1971).

Appendix C

Supplementary Regional Reading

Although an attempt has been made in this book to draw regional examples from a wide variety of locations, scales, and time periods, some regions have had, inevitably, to be underrepresented. Thus, about one-third of the cases cited in this book deal with North America, another third deal with Europe, and the remaining third deal with the remaining world regions. Instructors will doubtless wish their classes to follow up the discussion of some of the topics raised in this book by investigating cases where the environmental and spatial issues are most relevant, particularly in supplementary regional courses. The material listed below has been selected from a wide range of possible texts and supporting material. No more than three items for each region have been included under the four main headings. The regional division used follows broadly that of *A Geographical Bibliography for American College Libraries* (Association of American Geographers, Commission on College Geography, Publication No. 9, Washington, D.C., 1970), and instructors are referred to this excellent volume for further details.

Atlases[a]	Serials	Texts	Subregional and systematic studies
ANGLO-AMERICA			
Atlas of the Historical Geography of the United States (Carnegie Institution of Washington, D.C., and the American Geographical Society, Washington, D.C., and New York, 1932)	*Association of American Geographers, Annals* (Association of American Geographers, Washington, D.C., 1911–), quarterly	Paterson, J. H., *North America: A Geography of Canada and the United States* (Oxford University Press, London, 1966)	Gottmann, J., *Megalopolis: The Urbanized Northeastern Seaboard of the United States* (Twentieth Century Fund, New York, 1961)
* *National Atlas of the United States* (Washington, D.C., 1955–)	*Geographical Review* (American Geographical Society, New York, 1916–), quarterly	Watson, J. W., *North America: Its Countries and Regions* (Praeger, New York, 1967)	*Canada: A Geographical Interpretation*, J. Warkentin, Ed. (Methuen, London, 1968)
* *Atlas of Canada* (Queen's Printer, Ottawa, 1957	*Canadian Geographer* (University of Toronto Press, Toronto, 1951–), quarterly	White, C. L., E. J. Foscue, and T. L. McKnight, *Regional Geography of Anglo-America* (Prentice-Hall, Englewood Cliffs, N.J., 1964)	Brown, R. H., *Historical Geography of the United States* (Harcourt Brace Jovanovich, New York, 1948)
LATIN AMERICA			
* *Atlas Historico Geografico y de Paisajes Peruanos* (Presidencia de la Republica, Lima, 1970)	*Ibero Americana* (University of California Press, Berkeley, 1932–), occasional publication	West, R. C., and J. P. Augelli, *Middle America: Its Lands and Peoples* (Prentice-Hall, Englewood Cliffs, N.J., 1966)	Parsons, J. J., *Antioqueño Colonization in Western Colombia* (University of California Press, Berkeley, 1969)
* *Atlas Nacional do Brasil* (Instituto Brasileiro de Geografía e Estatística, Rio de Janeiro, 1966)	*Revista Geográfica* (Instituto Pan-americano de Geografía e História, Rio de Janeiro, 1941–), biannual	James, P. E., *Latin America* (Odyssey Press, New York, 1969)	Friedmann, J., *Regional Development Policy: A Case Study of Venezuela* (MIT Press, Cambridge, Mass., 1966)
	Revista Brasileira de Geograf a (Conselho Nacional de Geograf a, Rio de Janeiro, 1939–), quarterly	Cole, J. P., *Latin America* (Butterworth, London, 1970)	Robock, S. H., *Brazil's Developing Northeast* (Brookings Institution, Washington, D.C., 1963)
WESTERN EUROPE			
Oxford Regional Economic Atlas of Western Europe (Oxford University Press, London, 1971)	*Geografiska Annaler* (Generalstabens Litografiska Anstalt, Stockholm, 1919–), quarterly	Gottmann, J., *A Geography of Europe* (Holt, Rinehart & Winston, New York, 1969)	Mead, W. R., *An Economic Geography of the Scandinavian States and Finland* (University of London Press, London, 1958)
Atlas of Britain and North Ireland (Clarendon Press, London, 1963)	*Institute of British Geographers, Transactions* (Institute of British Geographers, London, 1935–), biannual	Church, R. J. H. et al., *An Advanced Geography of Northern and Western Europe* (Longmans, Harlow, Essex, England, 1953)	Houston, J. M., *The Western Mediterranean World: An Introduction to Its Regional Landscapes* (Longmans, Harlow, Essex, England, 1964)

[a] Representative national atlases are marked with an asterisk.

Atlases[a]	Serials	Texts	Subregional and systematic studies
* *Atlas över Sverige* (Generalstabens Litografiska Anstalts Förlag, Stockholm, 1953–)	*Petermanns Geographische Mitteilungen* (Petermanns Geographische Mitteilungen, Gotha, German Democratic Republic, 1855–), quarterly	Dickinson, R. E., *The West European City* (Routledge & Kegan Paul, London, 1961), 2nd ed.	Smith, C. T., *An Historical Geography of Western Europe Before 1800* (Praeger, New York, 1967)
EASTERN ERUOPE AND U.S.S.R.			
Atlas SSR (SSSR Glavnoe Upravlenie Geodezii i Kartografii, Moscow, 1962)	*Soviet Geography, Review and Translation* (American Geographical Society, New York, 1960–), monthly	Hooson, D. J. M., *The Soviet Union: People and Regions* (Wadsworth, Belmont, Calif., 1966)	Hamilton, F. E. I., *Yugoslavia: Patterns of Economic Activity* (Praeger, New York, 1968)
Oxford Regional Economic U.S.S.R. and Eastern Europe (Clarendon Press, London, 1956)	*Akademiia Nauk SSSR, Izvestiia, Seriia Geograficheskaia* (Akademiia Nauk, SSSR, Moscow, 1951–), bimonthly	Berg, L. S., *Natural Regions of the U.S.S.R.* (Macmillan, New York, 1950)	Pounds, N. J. G., *Eastern Europe* (Aldine, Chicago, 1969)
* *Atlas Ceskoslovenských Dejin* (Geodetic and Cartographic Institute, Prague, 1965)	*Geographia Polonica* (Polish Academy of Sciences, Warsaw, 1964–), occasional publication	Lydolph, P. E., *Geography of the U.S.S.R.* (Wiley, New York, 1970)	Harris, C. D., *Cities of the Soviet Union* (Rand McNally, Skokie, Ill., 1970).
SOUTH AND EAST ASIA			
Oxford Regional Economic Atlas for India and Ceylon (Oxford University Press, London, 1953)	*Economic Survey of Asia and the Far East* (United Nations Economic Commission for Asia and the Far East, Bangkok, 1947–) annual	Ginsburg, N. S., et al., *The Pattern of Asia* (Prentice-Hall, Englewood Cliffs, N.J., 1958)	Spate, O. H. K., and A. T. A. Learmonth, *India and Pakistan* (Barnes & Noble, New York, 1967)
An Historical Atlas of China (Aldine, Chicago, 1966)	*Journal of Asian Studies* (Russel H. Fifield, Ann Arbor, Mich., 1941–), 5 issues a year	Spencer, J. E., *Asia, East by South: A Cultural Geography* (Wiley, New York, 1954)	McGee, T. G., *The Southeast Asian City* (Praeger, New York, 1967)
* *Regional Structure of Japanese Archipelago* (Japan Center for Area Development Research, Tokyo, 1965)	*Association of Japanese Geographers, Special Publications* (Department of Geography, University of Tokyo, 1966–), occasional publication	Fisher, C. A., *Southeast Asia: A Social, Economic, and Political Geography* (Barnes & Noble, New York, 1966)	Buchanan, K., *Transformation of the Chinese Earth* (Bell, London, 1970)

[a] Representative national atlases are marked with an asterisk.

Atlases[a]	Serials	Texts	Subregional and systematic studies
MIDDLE EAST AND NORTH AFRICA			
Oxford Regional Economic Atlas of the Middle East and North Africa (Oxford University Press, London, 1960)	*Middle East Journal* (Middle East Institute, Washington, D.C., 1947–), quarterly	Fisher, W. B., *The Middle East: A Physical, Social, and Regional Geography* (Dutton, New York, 1971)	Mikesell, M. W., *Northern Morocco: A Cultural Geography* (University of California, Berkeley, 1961)
		Longrigg, S. H., *The Middle East: A Social Geography* (Duckworth, London, 1963)	Fisher, W. B., ed., *The Land of Iran,* (Cambridge University Press, London, 1968)
		Brice, W. C., *Southwest Asia* (University of London Press, London, 1967)	Orni, E., and E. Efrat, *Geography of Israel* (Israel Program for Scientific Translations, Hartford, Conn., 1966)
AFRICA SOUTH OF THE SAHARA			
Oxford Regional Economic Atlas of Africa (Oxford University Press, London, 1960)	*Journal of Modern African Studies* (Cambridge University Press, London, 1966–), quarterly	Hance, W. A., *African Economic Development* (Praeger, New York, 1967)	Wellington, J. H., *Southern Africa: A Geographical Study* (Cambridge University Press, London, 1955)
* *Atlas of Uganda* (Uganda Department of Lands and Surveys, 1962)	*South African Geographical Journal* (South African Geographical Society, Bloemfontein, 1917–) annual	Grove, A. T., *Africa South of the Sahara* (Oxford University Press, London, 1967)	Mabogunje, A. L., *Urbanization in Nigeria* (University of London Press, London, 1968)
* *Development Atlas of South Africa* (Department of Planning, Pretoria, 1966–)	*Nigerian Geographical Journal* (Nigerian Geographical Association, Ibadan, 1957), biannual	Kimble, G. H. T., *Tropical Africa* (Twentieth Century Fund, New York, 1960)	Soja, E. W., *The Geography of Modernization in Kenya* (Syracuse University Press, Syracuse, N.Y., 1968)

[a] Representative national atlases are marked with an asterisk.

Atlases[a]	Serials	Texts	Subregional and systematic studies
AUSTRALASIA			
* *Atlas of Australian Resources* (Department of National Development, Regional Development Division, Canberra, 1962–)	*Australian Geographer* (Geographical Society of New South Wales, Sydney, 1928–), biannual	Dury, G. H., and M. I. Logan, eds., *Studies in Australian Geography* (Heinemann Educational Australia, Melbourne, 1968)	Clark, A. H., *The Invasion of New Zealand by People, Plants, and Animals: The South Island* (Rutgers University Press, New Brunswick, N. J., 1949)
	New Zealand Geographer (New Zealand Geographical Society, Dunedin, 1945–), biannual	Cumberland, K. B., and J. W. Fox, *New Zealand: A Regional View* (Tri-Ocean, San Francisco, 1965)	Fosberg, F. R., ed., *Man's Place in the Island Ecosystem: A Symposium* (Bishop Museum Press, Honolulu, 1963)
	Pacific Viewpoint (Department of Geography, Victoria University of Wellington, 1960–), biannual	Brookfield, H. C., with D. Hart, *Melanesia* (Methuen, London, 1971)	Ward, R. G., *Land Use and Population in Fiji: A Geographical Study* (Her Majesty's Stationery Office, London, 1965)
TROPIC, ARID, AND POLAR ZONES			
Antarctica Map Folio Series (American Geographical Society, New York, 1964–)	*Journal of Tropical Geography* (Departments of Geography, University of Singapore and University of Malaya, Singapore, 1953–) occasional publication	Gourou, P., *The Tropical World: Its Social and Economic Conditions and Its Future Status* (Wiley, New York, 1966)	American Association for the Advancement of Science, *The Future of Arid Lands* (AAAS, Washington, D.C., 1956)
	Polar Record (Scott Polar Research Institute, Cambridge, England, 1931–), quarterly	Hills, E. S., ed., *Arid Lands: A Geographical Appraisal* (Methuen, London, 1966)	Lee, D. H. K., *Climate and Economic Development in the Tropics* (Harper & Row, New York, 1957)
	Arctic (Arctic Institute of North America, 1948–), quarterly	Baird, P. D., *The Polar World* (Wiley, New York, 1964)	Kimble, G. H. T., and D. Good, eds., *Geography of the Northlands* (Wiley, New York, 1955)

[a] Representative national atlases are marked with an asterisk.

Appendix D

Using the Book in Introductory Courses

Introductory courses in geography vary from country to country, between universities and colleges, and from department to department at the same level. The following charts outline ways in which this book may be modified from its original purpose—to serve as the basis of a full one-semester introduction to geography—in order to meet the needs of shorter and more specialized introductory courses. Syllabuses and teaching strategies for such courses are discussed at length in the excellent series of publications of the Association of American Geographers Commission on College Geography. See especially *New Approaches in Introductory Courses* (#4, 1967), *Introductory Geography: Viewpoints and Themes* (#5, 1967), and *Geography in the Two-Year Colleges* (#10, 1970).

In using the course outlines below, instructors should note the following points:

1. The sequences charted here are *only* suggestions. In any book, chapters have to be arranged in a linear (and hopefully logical) se-

quence. The material in this book has, however, a much more complex structure, and the sequence of chapters represents only one of many possible compromises. Indeed, the twenty-two chapters could be arranged in no less than 1,124,000,000,000,000,000,000 different ways! The text has been redesigned into more modules than the first edition so that instructors can more easily rearrange and structure material to meet individual needs.

2. Each sequence assumes that one week of each teaching period (either a semester or a quarter) will be used for reviews, tests, and the like.

3. The symbols used in the sequences are interpreted as follows:

Two-week period in which the text is supplemented by materials listed in the section "One step further..." or suggested by the instructor

One-week period in which students are assigned limited supplementary work

Half-week period

△ Chapter assigned as collateral reading

☐ Chapter skipped or assigned as optional reading

The numbers within the circles refer to the *chapters*.

Course G–1. INTRODUCTION TO GEOGRAPHY

The book is designed to provide material for a full 20-week, one-semester course introducing beginning students at the college level to the richness and range of geography. (The basic philosophy behind this approach is discussed in the *Preface*. The chapter sequences below suggest how the book can be adapted for use in shorter introductory courses.

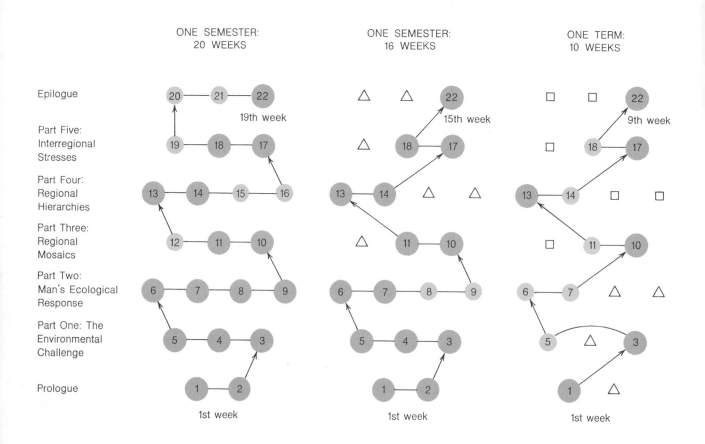

<table>
<tr><td></td><td>ONE SEMESTER:
20 WEEKS</td><td>ONE SEMESTER:
16 WEEKS</td><td>ONE TERM:
10 WEEKS</td></tr>
</table>

Epilogue

Part Five:
Interregional
Stresses

Part Four:
Regional
Hierarchies

Part Three:
Regional
Mosaics

Part Two:
Man's Ecological
Response

Part One: The
Environmental
Challenge

Prologue

Course G–2. INTRODUCTION TO CULTURAL/HUMAN GEOGRAPHY

In many colleges the introductory geography courses are split into two halves—an environmental sciences course (an "Introduction to Physical Geography") and a social sciences course (an "Introduction to Human Geography" or an "Introduction to Cultural Geography"). The following two plans suggest ways in which this book might be used in such courses. How many of the chapters not directly included should be assigned for collateral reading or skipped entirely is a matter for the instructor to judge in the light of time constraints and the background of members of the class.

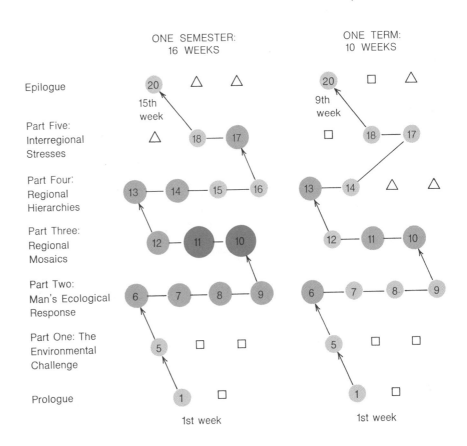

ONE SEMESTER:
16 WEEKS

ONE TERM:
10 WEEKS

Epilogue

Part Five:
Interregional
Stresses

Part Four:
Regional
Hierarchies

Part Three:
Regional
Mosaics

Part Two:
Man's Ecological
Response

Part One: The
Environmental
Challenge

Prologue

Course G–3. INTRODUCTION TO SPATIAL ANALYSIS/ ECONOMIC GEOGRAPHY

One common variant on Course G–2 (above) is to introduce beginning students to geography through the medium of spatial analysis. Two schemes for using this book in courses that take this approach— usually listed in the catalog as "Introductions to Spatial Analysis" or "Introductions to Economic Geography"—are outlined below. Again, the emphasis is on introducing students to the most elementary principles of the discipline, and the course should be regarded as a broadly based forerunner of more detailed specialized courses.

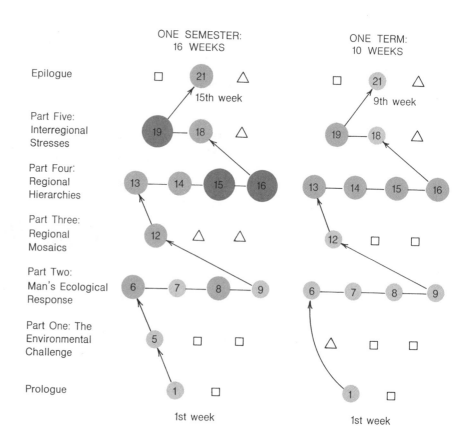

OTHER SPECIALIZED ONE-QUARTER (10-WEEK) INTRODUCTORY COURSES

A fourth possible use of this book is in brief introductory courses in particular areas. Three examples of such use usage are shown below. Only limited aspects of regional structure and organization are treated in the book, and supplementary material on specific areas for the "Introduction to World Regional Geography" course is suggested in Appendix C. More advanced topics in each chapter can be pursued further in the "Senior Seminar/Colloquium."

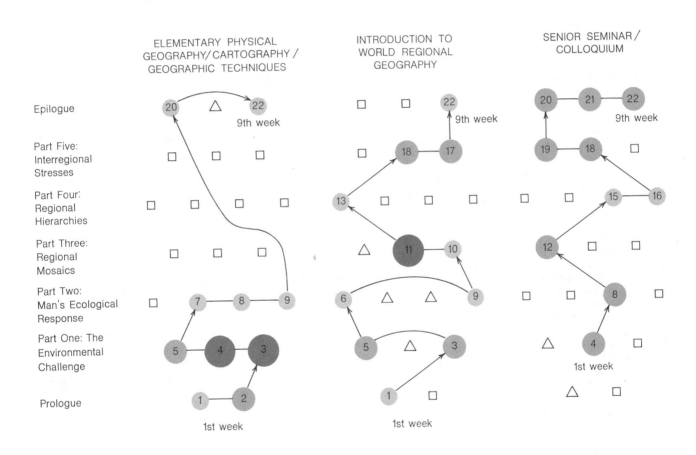

Index

75 76 77 7 6 5 4 3 2 1